PREDATION

Direct and Indirect Impacts
on Aquatic Communities

*Edited by W. Charles Kerfoot
and Andrew Sih*

University Press of New England

HANOVER AND LONDON, 1987

University Press of New England
Brandeis University
Brown University
Clark University
University of Connecticut
Dartmouth College
University of New Hampshire
University of Rhode Island
Tufts University
University of Vermont

Printed in the United States of America

Library of Congress Cataloging-in-Publication Data

Predation: direct and indirect impacts on aquatic communities.

   Includes bibliographies and index.
   1. Aquatic ecology. 2. Predation (Biology)
3. Biotic communities. I. Kerfoot, W. Charles,
1944– . II. Sih, Andrew.
QH541.5.W3P74   1986   574.5'3   86-40113
ISBN 0-87451-376-6

5 4 3 2 1

Figures 5.4, 5.6, 23.1, and 23.2 copyright Ecological Society of America, reprinted with permission

Contents

Preface

The central theme of this volume, that the mere presence of a predator can alter the relationship between two or more competitors in a variety of important ways, comes from the symposium "Competition, Predator Avoidance, and the Traits and Distributions of Aquatic Organisms," sponsored by the Ecological Society Of America during its 1984 meetings in Fort Collins, Colorado. At the time, the senior editor (WCK) had just returned from the Ecosystems Research Center, Cornell University, where he had investigated the subject of indirect food web interactions, whereas the junior editor (AS), chiefly responsible for organizing the symposium, was in the midst of preparing a series of review articles on indirect effects beyond food web interactions. At the conference and during subsequent exchanges, it became clear that indirect effects were very common in experimental studies but that the term "indirect effects" clearly meant different things to different people.

There are at least four to five types of interactions currently described under "indirect effects." Some of these are important phenomena. For example, the notion of indirect effects in "top-down" trophic impacts embraces an entire school of aquatic research. The classical "keystone predator" effect in intertidal studies describes a situation in which a predator has an indirect, beneficial effect on a suite of inferior competitors by depressing the abundance of a superior competitor (e.g., Paine 1966; Lubchenco 1978; Dungan, this volume). This "enemy of my enemy is a friend of mine" scenario certainly has intuitive appeal, aside from being an important organizing concept in intertidal studies. A second related, but less prominent, indirect interaction develops in cascading effects when a top consumer benefits prey two trophic links lower in a food chain because it reduces the abundance of an intermediate consumer (Abrams 1984).

A third and even more basic indirect conflict is found in pure exploitative competition. If resources are explicitly incorporated in feedback dynamics, then the interaction between competitors is transmitted through quantitative and qualitative changes in that resource base, making the interaction an indirect one (see Abrams, Levitan, Vanni, this volume).

A fourth and entirely different kind of indirect effect results if the mere presence of a predator alters the distribution of prey in time or space (Sih 1980). These behavioral responses constitute a qualitatively different type of indirect effect (see Miller and Kerfoot, Sih, this volume). Here the predator acts more like a catalyst, altering interactions without necessarily directly influencing species. A fifth kind of indirect action involves cases of chemical induction (see Stemberger and Gilbert, Havel, this volume). In the latter type, certain species of prey alter their morphology and physiology in the presence of predators, thus modifying encounter or removal rates.

Two goals emerged from the symposium: (1) a desire to clarify the kinds of indirect "effects" and (2) a need to determine the importance of indirect interactions in community dynamics. Consequently, we expanded the original 6 contributions to the present 24 by soliciting manuscripts in three main subject areas: (1) direct and indirect influences of top predators on food web dynamics, (2) chemically mediated interactions between predators and prey, and (3) research on the variety of ways that predators modify prey behavior, life-styles, and morphology. We tried to restrict subject areas to the traditional definition of predation, i.e., to effects of organisms that seize and ingest food items, although many treatments included primary producers and nutrient cycling dynamics. Although most contributions were easy to pigeonhole, others combined one or more types of indirect effect. Some centered on stability analysis (Murdoch and Bence, Abrams, Levitan), whereas others marveled at the natural fluctuations in systems (Mills et al., Power). In terms of response time, contributions spanned the range from minutes (Levitan), days (Kerfoot), months (Hairston), to millions of years (Aronson and Sues).

Collectively, the contributions emphasize that predation *and* competition represent strong vertical interactions within aquatic food webs and that the term "predation effects" has been used in an overly restricted fashion. In the past, predation was demonstrated to have major effects on organismal traits (e.g., Edmunds 1974; Jeffries and Lawton 1984), population dynamics (Murdoch and Oaten 1975; Hassell 1978; Taylor 1984), and community structure (e.g., Paine 1966; Connell 1975; Zaret 1980; Sih et al. 1985). To date, most studies of predation have emphasized direct lethal effects. Recently, however, there has been a growing awareness that indirect effects also might play an important role (e.g., Holt 1977; Abrams 1984).There are even those who maintain that indirect effects are some of the central elements in ecosystems (Patten 1983) and in community interactions (Boucher 1985). Thus, we see this volume as a kind of introduction to a developing field, one that shows several promising lines of investigation.

We would like to extend thanks to several individuals and institutions that aided our efforts. The senior editor (WCK) acknowledges a debt to both Al Beeton (director, GLMWC) and George Gamota (past director, IST), who provided encouragement and, at times, financial support. Richard Holmes (chairman, Department of Biological Sciences, Dartmouth College) and the ecology group of that department helped greatly by ensuring a cooperative and stimulating environment during the initial phases of contract negotiations. Simon Levin (director, ERC) and his staff and postdoctorates provided some of the original inspiration for this effort. The junior editor (AS) thanks the Ecological Society of America for its original sponsorship of the symposium that led to this volume. Last of all, we both thank our wives, Lucille Zelazny and Marie-Sylvie Baltus, for their patience through this professional gestation.

REFERENCES

Abrams, P. A. 1984. Foraging time optimization and interactions in food webs. *Amer. Naturalist* 124 : 80–96.

Boucher, D. H. (ed.). 1985. *The biology of mutualism.* London: Croomhelm, Ltd.

Connell, J. H. 1975. Some mechanisms producing structure in natural communities. In M. L. Cody and J. M. Diamond (eds.), *Ecology and evolution of communities,* pp. 460–90. Cambridge, Mass.: Belknap Press.

Edmunds, M. 1974. *Defence in animals.* New York: Longman.

Hassell, M. P. 1978. *The dynamics of arthropod predator-prey systems.* Princeton, N.J.: Princeton University Press.

Holt, R. D. 1977. Predation, apparent competition, and the structure of prey communities. *Theor. Pop. Biol.* 12 : 197–229.

Jeffries, M. J., and J. H. Lawton. 1984. Enemy free space and the structure of ecological communities. *Biol. J. Linn. Soc.* 23 : 269–86.

Lubchenco, J. 1978. Plant species diversity in a marine rocky intertidal community: Importance of herbivore food preference and algal competitive abilities. *Amer. Naturalist* 112 : 23–39.

Murdoch, W. W., and A. Oaten. 1975. Predation and population stability. *Adv. Ecol. Res.* 9 : 1–131.

Paine, R. T. 1966. Food web complexity and species diversity.

Patten, B. C. 1983. On the quantitative dominance of indirect effects in Ecosystems. In W. K. Lauenroth, G. V. Skogerboe, and M. Elug (eds.), *Analysis of ecological systems: State-of-the-art in ecological modelling,* pp. 27–37. Amsterdam: Elsevier Scientific Publishing Company.

Sih, A. 1980. Optimal behavior: Can foragers balance two conflicting demands? *Science* 210 : 1041–43.

Sih, A., P. Crowley, M. McPeek, J. Petranka, and K. Strohmeier. 1985. Predation, competition, and prey communities: A review of field experiments. *Annu. Rev. Ecol. Syst.* 16 : 269–305.

Taylor, R. J. 1984. *Predation.* New York: Chapman and Hall.

Zaret, T. M. 1980. *Predation and freshwater communities.* New Haven, Conn.: Yale University Press.

Ann Arbor, Michigan   W. C. K.
Lexington, Kentucky   A. S.
Spring 1986

I. Direct Interactions: Components and Collective Properties

1. Planktivory by Freshwater Fish: Thrust and Parry in the Pelagia

W. John O'Brien

Planktivorous fish have evolved a variety of means to search out and attack their zooplankton prey. But zooplankton species also have evolved structures and behaviors that, at least partially, foil this predation. Determining the predation cycle between planktivorous fish and their zooplankton prey is an effective means of understanding this evolutionary thrust and parry. For planktivorous fish, the predation cycle consists of location, pursuit, attack, and retention of prey. Different species of fish vary in their capabilities at each step, and different species of zooplankton may greatly lessen predation by foiling their predators at one or more steps in the cycle.

Resource limitation and predation are two major forces that act to limit animal and plant populations and hence structure biotic communities. Resource acquisition is a constant requirement of all organisms and, in a sense, provides an absolute limit to population density. Because there is little a resource-limited population can do to create more resources, evolutionary responses to resource limitation tend to be straightforward: take as fast as possible and store as much as feasible. Evolutionary responses to predation, however, are far from straightforward. Predators have evolved a variety of means to search out and stalk their prey, and as many means have evolved to foil, at least partially, each predatory scheme. As Forbes (1887) wrote long ago in *The Lake as a Microcosm*, "Every animal has its enemies, and Nature seems to have taxed her skill and ingenuity to the utmost to furnish these enemies with contrivances for the destruction of their prey in myriads. For every defensive device with which she has armed an animal, she has invented a still more effective apparatus of destruction and bestowed it upon some foe, thus striving with unending pertinacity to outwit herself; yet life does not perish in the lake, nor even oscillate to any considerable degree." In this chapter I will discuss this evolutionary move–countermove, using freshwater zooplankton and planktivorous fish as examples.

Zooplankton are food for many freshwater predators. These commonly include invertebrate predators such as copepods (Kerfoot 1977, Luecke and O'Brien 1983a), insects such as *Chaoborus* and notonectids, and occasionally invertebrates such as hydra (Cuker and Mozley 1981), jellyfish (Dodson and Cooper 1983), flatworms (Maly, Schoenholtz and Arts 1980), rhabdocoels (Schwartz and Hebert 1982), amphipods (Anderson and Raasveldt 1974), mites (Riessen 1982), and odonates (Johnson and Crowley 1980). Vertebrates include nonvisual filter-feeding fish (Drenner, Strickler, and O'Brien 1978; Janssen 1978), amphibians (Attar and Maly 1980), and birds (Dodson and Egger 1980). But the most studied, if not always the most important, planktonic predators are visually feeding planktivorous fish (Hrbacek 1962; Brooks and Dodson 1965; O'Brien 1979; Zaret 1980b).

Many studies of predation on zooplankton, including those of planktivorous fish, have used comparisons between water bodies in which a predator was present or absent to emphasize that predator's potential impact on its prey (Brooks and Dodson 1965; Hall, Cooper and Werner 1970; O'Brien 1975). Likewise, the effect of a predator on zooplankton has been shown by losses of prey. Neither of these approaches, however, describes the mechanism by which the skills of a particular predator are matched to the vulnerability of a zooplankton prey species. Holling (1959) demonstrated an

3

approach to the study of predation that can aid in pulling apart the specific interaction between predator and prey when he formulated the predation cycle, the series of sequential behaviors a predator must go through to locate, pursue, attack, and ingest its food. For a planktivorous fish, the predation cycle includes location of prey, pursuit, attack, and retention (O'Brien 1979). This is a useful means of analysis because different predators may have varying skills in one of the components of the predation cycle, and particular prey species, or certain size prey, may vary in their vulnerability to specific stages of the cycle. In this chapter I will analyze the ability of planktivorous fish to complete the predation cycle successfully, the abilities of freshwater zooplankton species to foil their predators at one or more stages in the predation cycle, and the effect of various environmental factors on the abilities of predator and prey.

PREY LOCATION

Visual stimulus is the major determinant of vulnerability for zooplankton exposed to visually feeding fish. There are several aspects of planktivorous fish vision, physiology, and behavior, and aspects of zooplankton morphology and behavior, that contribute to ease or difficulty of location.

Capabilities of Planktivorous Fish

There are few generalities about factors that affect the visual abilities of planktivorous fish. It is known that bluegill sunfish have the longest reported reaction distance to a given size of zooplankton prey, almost 30 cm for a 2.0-mm *Daphnia pulex* at a light intensity a little over 100 lux, while arctic grayling have the shortest distance for these conditions, around 8 cm. White crappie and several trout species have intermediate reaction distances under similar conditions (fig. 1.1). Yet there is almost no physiological information that explains why these fish species differ in ability to locate zooplankton prey visually.

FISH SIZE. The size of the fish has been shown to influence its ability to locate zooplankton prey. Hairston, Li, and Easter (1982) reported that the visual angle of prey that small bluegill sunfish react to decreased from 30 min of arc

for a 3.5-cm fish to 15 min of arc for a 6.0-cm fish. However, Wright and O'Brien (1984) could demonstrate no statistically significant difference in reaction distance to similar-size prey of white crappie 7 to 8 cm long versus white crappie 15.5 to 16 cm long. If these two studies are comparable, it would seem that fish size influences visual ability up to a size of approximately 6 to 8 cm and then has no further effect as the fish grow larger.

LOCATION VOLUME. The volume within which various planktivorous fish can locate zooplankton prey has received considerable attention recently. Luecke and O'Brien (1981b) found that bluegill sunfish that feed on solitary 2.0-mm *D. pulex* had a search volume shaped much like a forward-directed hemisphere. Confer et al. (1978) and Dunbrack and Dill (1984) found search volumes for lake trout *(Salvelinus namaycush)* and coho salmon *(Oncorhynchus kisutch)* shaped somewhat like a forward-directed hemisphere. In the case of the lake trout, however, the visual field projected to the rear of the fish, and in the case of the salmon it projected further above than below the fish. Further complicating matters, Evans and O'Brien (1986) found the shape and size of the search volume of white crappie to vary with different prey species, prey densities, and light intensity. For example, the search volume of white crappie had a reduced vertical component. With larger prey the shape was more semicircular, similar to bluegill sunfish but with a more reduced vertical dimension. But when white crappie fed on small prey, the search volume was more constricted resembling a pie shape in the horizontal plane (Evans and O'Brien in press).

Environmental Effects on Location

LIGHT LEVEL. Because fish locate their zooplankton prey visually, any environmental factor that alters the quantity or quality of light in lakes may affect the ability of visually feeding planktivores to locate prey. Several studies have demonstrated the importance to fish of light intensity in locating zooplankton prey. Vinyard and O'Brien (1976) showed that bluegill sunfish had a threshold for location of *D. pulex* at 10 lux and that their abilities to detect prey rapidly declined as light intensities were reduced below 10 lux. This general pattern of a

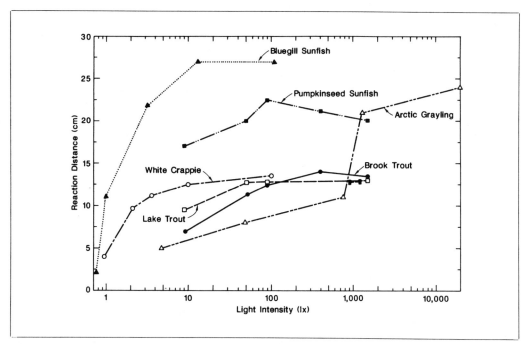

Figure 1.1. *Reaction distance of six species of fish to 2.0-mm* Daphnia pulex *at varying light intensities. The fish species are shown: bluegill* (Lepomis macrochirus) *from Vinyard and O'Brien (1976), arctic grayling* (Thymallus arcticus) *from Schmidt and O'Brien* (1982), *pumpkinseed* (Lepomis gibbosus), *lake trout* (Salvelinus namaycush), *and brook trout* (Salvelinus fontinalis) *from Confer et al. (1978), and white crappie* (Pomoxis annularis) *from Wright and O'Brien (1984).*

threshold of light intensity above which there is little further increase in location ability has been confirmed in other species of fish (fig. 1.1), although the actual threshold intensity varies: 10 lux for white crappie (Wright and O'Brien 1984), but 100 lux for rainbow trout and pumpkinseed sunfish (Confer et al. 1978) and 1,000 lux for arctic grayling (Schmidt and O'Brien 1982). In terms of ingestion, Zaret and Suffern (1976) demonstrated that golden shiner feeding rates increased with increasing light intensity up to a range where feeding rate remained nearly constant, between 95 to 800 ergs • cm^{-2} • sec^{-1}.

WATER CLARITY. Any factor such as water color or turbidity that reduces light transmission within a lake also should affect visual predation on zooplankton. Vinyard and O'Brien (1976) demonstrated that bluegill reactive distance to *D. pulex* declined dramatically as turbidity levels rose from clear to quite murky. McCabe and O'Brien (1983) suggested that *D. pulex* could exist only in very turbid Midwest

reservoirs only because of reduced vulnerability to visual feeding fish in such murky, low-light environments. An alternative explanation for the presence of *D. pulex* in turbid reservoirs was set forth by Arruda, Marzolf, and Faulk (1983), who suggested that the organic molecules adsorbed onto silt and clay particles provide a nutrient supplement to filter-feeding zooplankton. However, there seems little doubt that *Daphnia* and other zooplankton are less easily located by planktivorous fish when suspended silt and clay concentrations are high.

Few experimental studies have considered the effects of water color and stain on predation of zooplankton, although there is mention of this subject in the literature. For example, Kitchell and Kitchell (1980) drew some elegant inferences about how water clarity influenced *Daphnia* abundance, drawing from remains preserved in the sediments of two basins experimentally produced by division of one lake. Both lakes were initially brown-stained dystrophic lakes with very shallow euphotic zones. The original fish fauna was removed, and both lakes were

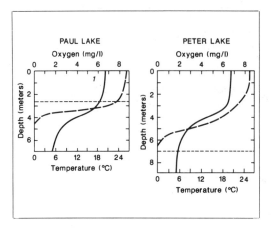

Figure 1.2. *A midsummer oxygen and temperature profile for Peter and Paul Lakes. The relative position of the thermocline with respect to the euphotic zone (straight dotted line) is shown. From Kitchell and Kitchell (1980).*

stocked with rainbow trout. One, Peter Lake, was treated for 27 years with hydrated lime to precipitate the humic colloids, and that treatment did indeed deepen the euphotic zone to 7 m, compared with 2.5 m for Paul Lake, the other (fig. 1.2). *D. pulex*, which was initially very abundant in both lakes, remained so in the stained lake, but was almost completely eliminated from the treated lake. Kitchell and Kitchell (1980) suggested that the extension of the euphotic zone well into the anaerobic hypolimnion eliminated a dark, daytime refuge for *D. pulex* and that visual feeding predation by rainbow trout caused the elimination of the large daphnid in the lime-treated lake.

Zooplankton Adaptations to Reduce Location

SIZE. Larger zooplankton can be located at greater distances than can smaller zooplankton (Werner and Hall 1974, Confer and Blades 1975, Vinyard and O'Brien 1976, Eggers 1977, Schmidt and O'Brien 1982, Wright and O'Brien 1984). Thus, in the presence of intense planktivorous fish predation, it is only small-bodied zooplankton species that remain (Brooks and Dodson 1965). Thus, the visual stimulus provided by larger zooplankton is a major reason that they are fed upon preferentially by fish. The obvious strategem for zooplankton faced with intense planktivorous fish

predation is small size. There is considerable evidence that smaller species (Brooks and Dodson 1965) and smaller individuals within a species (Threlkeld 1979) result in the face of planktivorous fish predation. These community and population changes seem to be the result of predation, not an *a priori* defensive response to predation. However, M. Vanni and S. Dodson (personal communications) have evidence that in the presence of bluegill sunfish *D. pulex* will produce smaller offspring and grow to a smaller size (for other examples of induced responses, see Havel, this volume).

Despite benefits, small body size may confer several disadvantages on zooplankton. Brooks and Dodson (1965) suggested that large zooplankton were superior competitors and that they predominated in the absence of visually feeding fish predators because of competitive exclusion (for a review, see Hall et al. 1976). Dodson (1974) suggested that large-size zooplankton were much less vulnerable to predation from tactile-locating invertebrate planktivores.

CORE BODY SIZE. Although prey size is an important component that determines zooplankton vulnerability to visually feeding vertebrate planktivores, it is by no means the only factor. The actual object located by planktivorous fish may not be the entire body but the more visually obvious portion of the body (Zaret 1972; Zaret and Kerfoot 1975; Kettle and O'Brien 1978; O'Brien, Buchanan, and Haney 1979; and Kerfoot 1980).

Zaret (1972) first pointed out the importance of pigmentation in that differential eye pigmentation of two *Ceriodaphnia* morphs in Lake Gatune was responsible for the higher predation rate on the nonhelmeted but more pigmented morph. This same mechanism was used to explain differential mortality on a morph of *Bosmina longirostris*, which also showed variable eye pigmentation (Zaret and Kerfoot 1975). However, Confer, Applegate, and Evanik (1980) suggested that such observations might be explained by daily variation in eye pigmentation within individuals (but see Konecny et al. 1982).

O'Brien et al. (1979) demonstrated that the distance at which a small lake trout could locate *D. longiremis* with exuberant but relatively clear head structures was less than predicted by total body length (fig. 1.3). They termed the visible portion of the body "core body" and further demonstrated its applicability to *Holopedium*, which have large but clear gelatinous sheaths surrounding the core body.

PIGMENTATION. Most lake-dwelling zooplankton are not pigmented but are quite clear and transparent, presumably because of the greater vulnerability of pigmented individuals to visually feeding planktivores. There are, however, cases of pigmented zooplankton. These include pigmented ephippial eggs, hemoglobin development, dark carapace pigmentation, and pigmentation due to ingested food or large eyespot.

Some *Daphnia* may be pigmented due to coloration of the carapace (Dodson 1984). Dodson and others have noted that the carapace of *D. middendorffiana* occurring in Arctic ponds is darkly pigmented; however, when this same species occurs in deep lakes that contain planktivorous fish, its carapace is quite clear (Luecke and O'Brien 1983), probably an adaptation to reduce location by planktivorous fish. The predatory copepod *Heterocope septentrionalis* shows a similar pattern; it is bright red in Arctic ponds but pale green in deep Arctic lakes. Luecke and O'Brien (1981a) demonstrated that the red color morph of *Heterocope septentrionalis* was more vulnerable to fish predation than was the pale green color morph.

Several *Daphnia* species have been shown to develop hemoglobin at oxygen concentrations below 3.0 mg/L (Fox and Phear 1953; Green 1956; Kring and O'Brien 1976; Landon and Stasiak 1983). Hemoglobin can deeply pigment the entire body, making the pigmented individuals more easily located by fish (O'Brien 1979). However, planktivorous fish rarely tolerate oxygen concentrations as low as 3.0 O_2/L. Thus, it would be rare for such hemoglobin-pigmented daphnids to suffer increased predation (Kring and O'Brien 1976).

The eggs of many zooplankton species are pigmented, especially ephippial eggs of *Daphnia* species toward the end of egg development. Mellors (1975) showed in laboratory experiments that pumpkinseed sunfish and red-spotted newts preyed selectively on ephippial-carrying *D. pulex* versus those without ephippia, and Tucker and Woolpy (1984) demonstrated that bluegill could locate *D. magna* with parthenogenic eggs at greater distances than non-gravid *D. magna* of a similar size. Further, Mel-

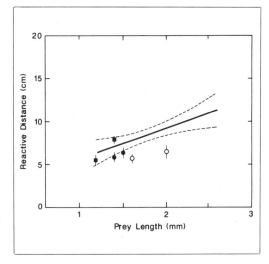

Figure 1.3. *Reactive distance of small lake trout to different sizes of* Daphnia middendorffiana *and different sizes and morphs of* D. longiremis. *Solid line, linear regression of reaction distance to a clear, lake-dwelling type of* D. middendorffiana; *dotted line 95% confidence interval around this line; solid square, mean values for the nonhelmeted morph of* D. longiremis; *open circles, mean values for two sizes of the helmeted morph of* D. longiremis; *solid circles, mean values of the core body size of the helmeted morph of* D. longiremis. *From O'Brien, Kettle, and Riessen (1979).*

lors (1975) found that planktivorous sunfish and perch collected from a small lake had ephippia in their digestive tracts. Although the presence of the eggs may increase predation on female *Daphnia*, Mellors also showed that *Daphnia* ephippial eggs could hatch after fish gut passage.

Some zooplankton, such as mites, are darkly pigmented but are not eaten by fish (Riessen 1984). Kerfoot (1982) has shown that these mites are distasteful to fish, and thus the pigmentation seems to be aposomatic, that is, a warning coloration. In general, however, mites are not a common component of open-water lake zooplankton communities but can be common along the littoral margins.

Thus, it seems certain that zooplankton should reduce pigmentation as much as possible in the face of fish predation. However, this cannot be taken to an extreme because pigmentation may confer certain benefits. For example, certain zooplankton are susceptible to photo damage from either ultraviolet or far blue light (Hairston 1979, 1980). But various types of pigments in zooplankton either screen this light (such as the melanin pigment of *D. middendorffiana*; Luecke and O'Brien 1983b) or serve as oxygen radical scavengers (such as carotenoid pigment of *Heterocope*; Luecke and O'Brien 1981a).

Another situation in which pigmentation is beneficial is eye pigmentation. It is adaptive for zooplankton to be quite clear, but a clear structure cannot detect light. To detect and assess light intensity it is necessary to capture some of the photons, which requires pigmentation. Zooplankton vertically migrate, primarily in response to visual predation from fish (Wright et al. 1980). They time the migration by assessing the level and rate of change of light intensity; hence, eye pigmentation reduces predation by mediating migration but may increase predation by making the animal more visible.

PREY MOTION. Ware (1973) first demonstrated that rainbow trout could locate moving prey at a greater distance than stationary prey. Wright and O'Brien (1982, 1984) showed that white crappie could locate moving *Chaoborus* and diaptomid copepods at considerably greater distances than when they were motionless. Zaret (1980b) invoked differential motion of zooplankton in Lake Gatun, Panama, to account for selective predation on one of the species, which, though smaller, moved considerably more. Thus, there is considerable evidence that zooplankton motion does increase vulnerability to fish predation through increasing the likelihood of being located. *Daphnia*, which move continuously, can be located at greater distances than copepods of a similar size that are not in motion (Wright and O'Brien 1984). Why *Daphnia* continually move and thus greatly increase their vulnerability to fish predation is still unknown, although food-gathering dynamics are suspect.

VERTICAL MIGRATION. Many aspects of zooplankton vulnerability to visual predators are not under control of the prey. However, zooplankton individuals and populations can exert behavioral control of light intensity by migrating into the lower light found deeper in the lake. There is considerable field evidence that zooplankton species do this. Larger zooplankton species and individuals are most at risk from visual predation; thus, larger zooplankton should migrate deeper into lakes than

small ones, and this is found. Hutchinson (1967) reported that when several *Daphnia* congeners occur in the same lake, the largest species is found deepest and the smallest migrates the least. Furthermore, larger individuals within a species are found deepest (Zaret and Suffern 1976; Wright, O'Brien, and Vinyard 1980). Wright et al. (1980) found *D. parvula* to migrate to depths where light intensities were less than 10 lux, the light intensity where sunfish planktivores begin to have difficulty locating zooplankton prey (fig. 1.1). Zaret and Suffern (1976) found *D. galeata mendotae* to be located well below that light intensity (2,200 ergs $\cdot$ cm^{-2} $\cdot$ sec^{-1}) at which golden shiner feeding rates were highest.

PURSUIT

Pursuit is the phase of the planktivorous fish–zooplankton predation cycle that begins after the prey has been located and continues until the fish swims to within 1 cm or less of the prey. Pursuit distances can be as high as 30 to 40 cm for large prey at high light; however, under typical lake conditions the distances are generally no greater than 5 to 10 cm (Wright and O'Brien 1984).

Fish Pursuit Capabilities

CHOICE OF PREY. For the fish, pursuit is the point in the predation cycle where choice is made. Two types of choice are possible: choice not to pursue a located prey or choice of which among several located prey to pursue. Choice not to pursue a located prey, i.e., deletion of a particular species or size class of prey from the diet, has been suggested as a possible mechanism involved in the optimal foraging of planktivorous fish (Werner and Hall 1974). However, O'Brien et al. (in press) have presented evidence that there is very minimal energy cost to pursuit and none to handling, suggesting that all solitary located prey should be pursued.

There are a variety of suggestions as to factors governing the choice among several located prey. Eggers (1982), in formulating the reactive field volume model as a hypothesis to account for large-size-selective predation on zooplankton, suggested that fish can locate large prey in a larger volume than small prey. Thus, given comparable zooplankton densities, planktivorous fish would locate more large prey than small prey. However, Eggers claimed the choice of which located prey to pursue was random. Wetterer and Bishop (1985) pointed out the inefficiency of random choice and suggested that evolutionary selection probably improved on such a mechanism.

However, Vinyard and O'Brien (1975) demonstrated that bluegill sunfish actually expressed preference for large prey. They exploited the variable nature of dorsal tilt of fish when exposed to a light from the side (Von Holst 1950) to evaluate bluegills' preference for different prey items. The change in tilt of the fish was found to increase in direct proportion to prey size (Vinyard and O'Brien 1975) (fig. 1.4), but not in response to more darkly pigmented prey. The authors contended that size was the sole determinant of this index of interest.

Vinyard and O'Brien (1975) also demonstrated that bluegill sunfish chose the larger of two offered *D. magna* prey, and O'Brien, Slade, and Vinyard (1976) presented evidence that it was not absolute size but relative, or apparent, size that determined bluegill sunfish choice of prey (fig. 1.5). Wright (1981) with white crappie (fig. 1.5) and Gibson (1980) with sticklebacks observed similar choice of the apparently largest prey. The apparent size choice mechanism has been widely debated (Gardner 1981; Gibson 1980; Bartell 1982; Mittelbach 1981; Werner, Mittelbach, and Hall 1981), and a recent study (O'Brien et al. 1984) showed that bluegill sunfish, when presented two daphnid prey each within 10 cm of the fish and where the absolutely bigger was the apparently smaller, chose the absolutely larger. This effect was more pronounced as the difference in absolute size increased.

Motion also seems to influence choice of prey by planktivorous fish. O'Brien et al. (1984) presented white crappie a binary choice between a diaptomid copepod and a daphnid. They found that if the copepod did not move at the time of choice, the daphnid was chosen 80% of the time. If the copepod moved, however, the choice was arbitrary, with both chosen about 50% of the time. This same kind of experiment was repeated using heat-killed *D. pulex* that were artificially moved like a daphnid or copepod; the "copepod mimic" was sometimes kept motionless. The results were very similar to the study that used live prey (table 1.1).

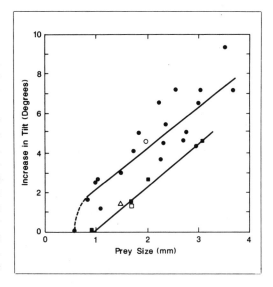

Figure 1.4. *The change in dorsal tilt of two sizes of bluegill presented various sizes of* Daphnia pulex *of varying pigmentation. Solid circles, average response of small fish; solid squares, average response of large fish; open squares and circle, average response to D. pulex with ink in their gut by large and small fish, respectively; open triangle, average response by large fish to hemoglobin-containing D. pulex. From O'Brien and Vinyard (1975).*

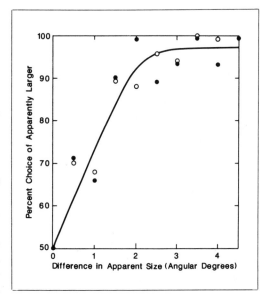

Figure 1.5. *Apparent size choices of white crappie and bluegill presented two D. magna. Each point represents the mean of observations within a 0.05 difference in apparent size. Solid circles, observations on white crappie from Wright (1981); open circles, observations on bluegill from O'Brien, Slade, and Vinyard (1976). The line is fitted by eye.*

Table 1.1. *Importance of motion (see text for details)*

Copepod	% Choice of *Daphnia*	% Choice simulated *Daphnia*
Moving	46	38
Stationary	82	86

Zooplankton Defense Capability

Because no small lake zooplankton have eyes capable of forming an image, none can detect the approach of a planktivorous fish. Thus, once located, there is little the prey can do to swim away from a pursuing fish; hence, evasion must await the attack phase.

Zooplankton can, of course, minimize the likelihood of being chosen out of several located individuals. Most of the factors that increase the likelihood of location by planktivorous fish also increase the chance that a given zooplankton is chosen from among several located. This is definitely true of large body size. Regardless of whether planktivorous fish choose on the basis of apparent or absolute size, an increase in body size will increase the chance of being chosen. Thus, reduced size should minimize predation at the pursuit stage as well as the location stage of the predation cycle (O'Brien et al. 1984). Likewise, increased motion by zooplankton will increase the likelihood of choice from among several located prey (O'Brien et al. 1984). Whereas daphnids move continuously, many zooplankton species are motionless much of the time (Wright and O'Brien 1984). Calanoid copepods and *Diaphanosoma* are notable in moving only infrequently. However, it is not clear whether increased pigmentation increases the likelihood of choice or whether increased pigmentation just increases the likelihood of location. For example, Vinyard and O'Brien (1975) exploited bluegill dorsal tilt as an index of fish preference and found no increase in dorsal tilt when the fish was exposed to pigmented versus nonpigmented prey.

ATTACK

The attack phase of the predation cycle occurs once the fish has moved to close quarters with the prey and inhales the water surrounding the prey into the buccal cavity. This is typically the first sign the zooplankton prey have that things may be amiss.

Fish Attack Capability

Planktivorous fish vary in their ability successfully to attack evasive zooplankton prey. If a zooplankton prey can sense the attack and swim out of the parcel of water being inhaled, that prey item will be unavailable to planktivorous fish. There are three obvious aspects involved in evasion that planktivorous fish may be able to regulate: the timing and strength of the signal given off to the prey, the amount of water inhaled, and the speed with which the water is inhaled.

Of those three, the volume of water inhaled seems the most fixed. The amount of water inhaled is a function of the volume of the buccal cavity, which varies dramatically with fish size. Wright and O'Brien (1984) found white crappie buccal volumes increased from 0.5 ml for 7-cm fish to 10 ml for 16-cm fish. Thus, larger fish can inhale greater volumes of water; and indeed, Schmidt and O'Brien (1982) found larger arctic grayling better able to capture a large predaceous copepod. However, it is commonly observed that the opercals flare after planktivorous fish attack an evasive prey (personal observation) leading to the impression that the fish may be using a two-step pump to inhale more water than can be accommodated solely in the buccal cavity. The first step of such a two-step pump would be filling the buccal cavity, with the second step being the pulling some of that water out of the buccal cavity over the gills by flaring the opercals. But Lauder (1980) reports that the volume of water inhaled during an attack is no greater than the buccal volume.

However, fish can use different attack strategies to fill the buccal cavity more quickly and with less turbulent signals to the zooplankton prey. Vinyard (1982) has nicely shown that not only do planktivorous fish use different attack strategies toward evasive versus nonevasive prey, but they can learn to switch from one strategy to the other within 20 encounters of an evasive prey type. He observed that the Sacramento perch (*Archoplites interruptus*) would greatly increase the curvature of its body so as to leap through the space occupied by the prey during an attack. The duration and speed of this lunge were shorter and faster when feeding on *Diaphanosoma* (0.047 sec and 987 cm/sec), an evasive prey, than when feeding on *Daphnia* (0.132 sec and 172 cm/sec), a nonevasive prey.

This lunge by planktivorous fish through the water occupied by the zooplankton prey is commonly observed. Wright and O'Brien (1984) report it for white crappie and Evans (1986) has observed it with bluegill sunfish. The lunge may increase the speed of intake of the volume of water, but such a lunge should be most effective in reducing the signal given to the zooplankton (Strickler, personal communication). This is so because the fish is moving over what is being inhaled; thus, the water surrounding the prey is minimally displaced, giving the prey no sheer forces or turbulent motion to cue to the attack phase.

Vinyard (1980) has shown fish learn which prey evade and alter their choice of prey on this basis. He vital-stained *D. pulex* red and blue and gave the fish a binary choice. He removed one color prey as it was chosen 50% of the time, simulating an evasion. The fish soon learned to attack the nonevading color. In a second series of experiments he caused the other color to evade with the identical result.

INCIDENTAL INGESTION. One generally thinks of the attack of a visually feeding planktivorous fish as directed toward a visually located prey. However, any individual zooplankton within the volume of water to be inhaled is also under attack. Wright, O'Brien, and Luecke (1983) termed this *incidental ingestion* and felt it could be quite common when small prey, not likely to be targets of direct attack, were dense.

Zooplankton Defense Capability

There is little zooplankton prey can do to reduce the impact of attack except to swim out of the volume of water being inhaled by the fish. To do this successfully the zooplankter must first detect the water currents indicating the suction attack of a fish and then swim out of the volume of water about to be inhaled.

In a series of simple but clever experiments, Szlauer (1965) demonstrated that copepods swimming in a beaker could avoid a rapidly lowered glass tube, whereas *Daphnia* could not. Drenner et al. (1978) devised a simulated suction device and reported the percentage of times different species could evade it. Again

they found calanoid copepods had the best evasion success, whereas *Daphnia* species were captured nearly 100% of the time, i.e., at the same capture efficiency as gas bubbles or heat-killed zooplankton. Drenner and McComas (1980), using the same device, found *Diaphanosoma* to have evasion skills intermediate to calanoid copepods and cladocerans (fig. 1.6). The value of this approach is that the volume and duration of the suction attack is known, whereas with empirical observations of fish attacks the volume is inferred from buccal volume estimates.

There have been several empirical studies of zooplankton evasion from the suction attack of visually feeding planktivorous fish. As with other studies, Wright and O'Brien (1984) found calanoid copepods, but not daphnids, could evade the attack of white crappie. Schmidt and O'Brien (1982), in studying evasion of a large copepod (*Heterocope septentrionalis*) from the attack of arctic grayling, observed a strong effect of temperature. At 15°C the copepods could successfully evade the attack of a 8.5-cm grayling 45% of the time. However, this declined to 12% at 5°C. The authors suggest that this may be due to more successful temperature acclimation by grayling. More intriguing is the possibility that the decline in water temperature and resulting increased water viscosity may have shifted the copepod swimming into a domain of low Reynolds number while the suction attack of the larger, faster fish remained in a domain of high Reynolds number.

RETENTION

The final stage of the predation cycle for planktivorous fish is the retention or straining phase. Here planktivorous fish must strain the prey from the water inhaled to eat the prey. A considerable volume of water, compared with the volume of the prey, may be inhaled during attack. A 15-cm white crappie, for example, has a 10-ml buccal volume, and when feeding on a 1-mm-long prey weighing 0.05 mg, it takes in 200 times more water than prey.

In most planktivorous fish it is the gill rakers that perform this straining function. Gill rakers are bony or cartilaginous structures shaped much like a crescent comb, with the teeth or rakers lying across the path of the water as it flows out through the gill arches.

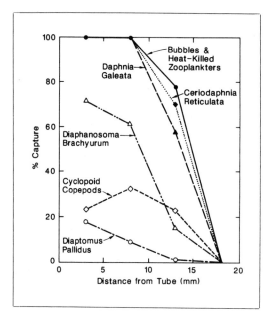

Figure 1.6. *Percentage of organism or particle capture by the siphon-system as distance from the opening of the siphon varied. Points plotted at distances 3, 8, 13, and 18 represent averages of capture success at 5-mm intervals. From Drenner and McComas (1980).*

Thus, the minimum spacing between the teeth of the rakers sets the minimum size of prey that can be eaten.

Fish Retention Capability

Fish vary considerably in their straining or retention capability because the spaces between the rakers vary greatly among species. There is a great deal known about the number of gill rakers for a wide variety of fish species because gill raker number is an important taxonomic character for a number of fish groups. However, without knowing both the thickness of the rakers and the overall length of the raker arch, gillraker number gives only a crude idea as to interraker spacing.

Galbraith (1967) suggested that selective gill raker retention of large prey might account for large-size selective predation observed for rainbow trout. However, he noted that there were few prey in the diet near the minimum size that could be retained by the rakers; yet many zooplankton of this size occur in the lake. He concluded that some other mechanism must be causing the size-selective predation.

Schmidt and O'Brien (1982) measured the distance between the gill rakers of arctic grayling 3 to 20 cm in length. They found the median interraker width increased from 0.17 mm for 3-cm-long fish to 0.48 mm for 13-cm-long fish and then remained unchanged for larger fish. Thus, grayling larger than 10 to 14 cm would have difficulty in retaining small zooplankton prey. Evans (personal communication) found guts of larger (15–20 cm) arctic grayling from Toolik Lake, Alaska, to be full of *Heterocope septentrionalis*, a larger (2–3 mm) copepod. These fish had ingested no small zooplankton, and yet a small daphnid and *Bosmina* are more common in this lake than *Heterocope septentrionalis*.

Wright and O'Brien (1984) measured the interraker spacing of white crappie and found the median spacing of the gill rakers from the first gill arch of white crappie was 0.05 mm for an 8-cm fish and increased to 0.2 mm for 16-cm fish. Thus, white crappie would seem capable of retaining much smaller zooplankton than comparably sized arctic grayling.

Such measurements of gill raker spacing from preserved fish can give only a static picture of gill raker retention. The spacing could vary or fluctuate as water passes through the rakers. Wright, O'Brien, and Luecke (1983) obtained a dynamic estimate of white crappie retention by observing the size of small prey in the diet taken in by incidental ingestion. The minimum size ingested through this mechanism was much larger than would be predicted from a measure of interraker size distribution (fig. 1.7). The reason for this considerable discrepancy is unclear.

Zooplankton Defense Capability

SMALL SIZE. Once again small size is advantageous for zooplankton because it may allow passage through the gill rakers. As mentioned above, small size reduces vulnerability to planktivorous fish through reduced location or choice, and it is not surprising that we find primarily small-size zooplankton in the presence of intense planktivory.

However, unlike development of small size to reduce location and choice, small size for reduced retention need be only in one dimension. That is, a laterally compressed body shape could slide through the gill rakers while still presenting a large visual image. A number of

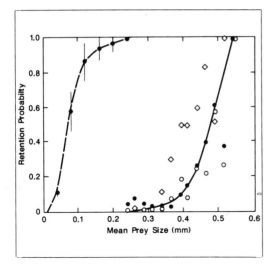

Figure 1.7. *Retention probability versus mean prey size estimated by gill raker measurement and incidental ingestion. Both curves were fitted by eye. From Wright, O'Brien, and Luecke (1983).*

Daphnia species are quite laterally compressed, and reducing vulnerability to gill raker retention may be part of the selection force involved.

SUMMARY

For planktivorous fish, the predation cycle consists of location, pursuit, attack, and retention of prey. Different species of fish vary in their capabilitites at each step, and different species of zooplankton may greatly lessen predation by foiling their predators at one or more steps in the cycle.

Visual stimulus determines the likelihood of location and may be estimated by measuring the reaction distance of fish to a given prey. Reaction distance increases with fish size when fish are small and with increasing prey size. However, reaction distance decreases with decreasing light intensity below a light-intensity threshold. The threshold varies with different species of fish.

Zooplankton have evolved several structural adaptations that reduce the likelihood of location, such as small body size and minimal pigmentation. Some have evolved behavioral adaptations to reduce location, such as minimal body motion and vertical migration to depths of low light intensity.

The pursuit phase of the predation cycle is

where active choice of several located prey may occur. Several species of fish have been shown to choose the apparently largest prey sighted and to choose a moving versus non-moving prey. Planktivorous fish vary in their success at attacking evasive prey, with larger fish generally better at this stage of the predation cycle. Likewise, zooplankton species differ in their evasion skills, with diatomid copepods often able to evade attack, whereas daphnids have no evasion skills.

The retentive abilities of fish rest primarily on the closeness of the spacing between the gillrakers. The ability of zooplankton to reduce retention is a function of overall body size and perhaps of lateral compression. Thus, the predator–prey evolutionary interplay between planktivorous fish and zooplankton is complex and full of reciprocal adjustments.

REFERENCES

Anderson, R. S., and L. G. Raasveldt. 1974. *Gammarus* predation and the possible effects of *Gammarus* and *Chaoborus* feeding on the zooplankton composition in some small lakes and ponds in western Canada. *Canadian Wildlife Service Occasional Paper* 18 : 1–23.

Arruda, J. A., G. R. Marzolf, and R. T. Faulk. 1983. The role of suspended sediments in the nutrition of zooplankton in turbid reservoirs. *Ecology* 64 : 1225–35.

Attar, E. N., and E. J. Maly. 1980. A laboratory study of preferential predation by the newt *Notophthalmus v. viridescens*. *Can. J. Zool.* 58 : 1712–17.

Bartell, S. M. 1982. Influence of prey abundance on size-selective predation by bluegills. *Trans. Amer. Fish. Soc.* 111 : 453–61.

Brooks, J. L., and S. I. Dodson. 1965. Predation, body size, and the composition of the plankton. *Science* 150 : 28–35.

Confer, J. L., G. Applegate, and C. A. Evanik. 1980. Selective predation by zooplankton and the response of Cladoceran eyes to light. In W. C. Kerfoot (ed.), *Evolution and ecology of zooplankton communities*, pp. 604–608. Hanover, N.H.: University Press of New England.

Confer, J. L., and P. I. Blades. 1975. Omnivorous zooplankton and planktivorous fish. *Limnol. Oceanogr.* 20 : 571–79.

Confer, J. L., G. L. Howick, M. H. Corzette, S. L. Kramer, S. Fitzgibbon, and R. Landesberg. 1978. Visual predation by planktivores. *Oikos* 31 : 27–37.

Cuker, B. E., and S. C. Mozley. 1981. Summer population fluctuations, feeding, and growth of *Hydra* in an arctic lake. *Limnol. Oceanogr.* 26 : 697–708.

Dodson, S. I. 1974. Zooplankton competition and predation: An experimental test of the size-efficiency hypothesis. *Ecology* 55 : 605–13.

Dodson, S. I. 1984. Predation of *Heterocope septentrionalis* on two species of *Daphnia*: Morphological defenses and their cost. *Ecology* 65 : 1249–57.

Dodson, S. I., and S. D. Cooper. 1983. Trophic relationships of the freshwater jellyfish *Craspedacusta sowerbyi* Lankester 1880. *Limnol. Oceanogr.* 28 : 345–51.

Dodson, S. I., and D. L. Egger. 1980. Selective feeding of red phalaropes on zooplankton of arctic ponds. *Ecology* 61 : 755–63.

Drenner, R. W., and S. R. McComas. 1980. The role of zooplankter escape ability and fish size selectivity in the selective feeding and impact of planktivorous fish. In W. C. Kerfoot (ed.), *Evolution and ecology of zooplankton communities*, pp. 587–93. Hanover, N.H.: University Press of New England.

Drenner, R. W., J. R. Strickler, and W. J. O'Brien. 1978. Capture probability: The role of zooplankter escape in the selective feeding of planktivorous fish. *J. Fish. Res. Bd. Can.* 35 : 1370–73.

Dunbrack, R. L., and L. M. Dill. 1984. Three-dimensional prey reaction field of the juvenile coho salmon (*Oncorhynchus kisutch*). *Can. J. Fisheries Aquat. Sci.* 41 : 1176–82.

Eggers, D. M. 1977. The nature of prey selection by planktivorous fish. *Ecology* 58 : 46–69.

Eggers, D. M. 1982. Planktivore preference by prey size. *Ecology* 63 : 381–90.

Evans, B. I. 1986. Strategies and tactics of search behavior in salmonid and centrarchid planktivorous fish. Ph.D. thesis, University of Kansas, Lawrence, Kan.

Evans, B. I. and W. J. O'Brien. (in press) An analysis of the feeding rate of white crappie. *Environ. Biol. Fish.* submitted.

Forbes, S. A. 1887. The lake as a microcosm. *Bulletin of Science Association of Peoria* 1887 : 77–87.

Fox, B. M., and E. A. Phear. 1953. Factors influencing haemoglobin synthesis by *Daphnia*. *Proc. R. Soc. Lond. [Biol.]* 141 : 179–89.

Galbraith, M. G. 1967. Size-selective predation on *Daphnia* by rainbow trout and yellow perch. *Trans. Amer. Fish. Soc.* 96 : 1–10.

Gardner, M. B. 1981. Mechanisms of size selectivity by planktivorous fish: A test of hypotheses. *Ecology* 62 : 571–78.

Gibson, R. M. 1980. Optimal prey-size selection by three-spined sticklebacks (*Gasterosteus aculeatus*): A test of the apparent-size hypothesis. *Z. Tierpsychol.* 52 : 291–307.

Green, J. 1956. Variation in the haemoglobin content of *Daphnia*. *Proc. R. Soc. Lond. [Biol.]*, 145 : 214–33.

Hairston, N. G., Jr. 1979. The adaptive significance of color polymorphism in two species of *Diaptomus. Limnol. Oceanogr.* 24 : 38–44.

Hairston, N. G., Jr. 1980. The vertical distribution of diaptomid copepods in relation to body pigmentation. In W. C. Kerfoot (ed.), *Evolution and ecology of zooplankton communities*, pp. 98–110. American Society of Limnology and Oceanography Special Symposium 3. Hanover, N.H.: University Press of New England.

Hairston, N. G., Jr., K. T. Li, and S. S. Easter, Jr. 1982. Fish vision and the detection of planktonic prey. *Science* 218 : 1240–42.

Hall, D. J., W. E. Cooper, and E. E. Werner. 1970. An experimental approach to the production dynamics and structure of freshwater animal communities. *Limnol. Oceanogr.* 15 : 839–928.

Hall, D. J., S. T. Threlkeld, C. W. Burns, and P. H. Crowley. 1976. The size-efficiency hypothesis and the size structure of zooplankton communities. *Annu. Rev. Ecol. Syst.* 7 : 177–208.

Holling, C. S. 1959. The components of predation as revealed by a study of small-mammal predation of the European pine sawfly. *Can. Entomol.* 91 : 293–320.

Hrbacek, J. 1962. Species composition and the amount of zooplankton in relation to the fish stock. *Rozpr. Cesk. Akad. Ved.* 72 : 1–116.

Hutchinson, G. E. 1967. *A treatise on limnology. Vol. 2: Introduction to lake biology and the limnoplankton.* New York: John Wiley and Sons.

Janssen, J. 1978. Feeding behavior repertoire of the alewife, *Alosa pseudoharengus*, and the ciscoes *Coregonus hovi* and *C. artedii*. *J. Fish. Res. Board Can.* 35 : 249–53.

Johnson, D. M., and P. H. Crowley. 1980. Odonate "hide and seek": Habitat-specific rules? In W. C. Kerfoot (ed.), *Ecology and evolution of zooplankton communities*, pp. 569–79. Hanover, N.H.: University Press of New England.

Kerfoot, W. C. 1977. Implications of copepod predation. *Limnol. Oceanogr.* 22 : 316–25.

Kerfoot, W. C. 1980. Commentary: Transparency, body size, and prey conspicuousness. In W. C. Kerfoot (ed.), *Ecology and evolution of zooplankton communities*, pp. 609–17. Hanover, N.H.: University of New England Press.

Kerfoot, W. C. 1982. A question of taste: Crypsis and warning coloration in freshwater zooplankton communities. *Ecology* 63 : 538–54.

Kettle, D., and W. J. O'Brien. 1978. Vulnerability of arctic zooplankton species to predation by small lake trout (*Salvelinus namaycush*). *J. Fish. Res. Board Can.* 35(11) : 1495–1500.

Kitchell, J. A., and J. F. Kitchell. 1980. Size-selective predation, light transmission, and oxygen stratification: Evidence from the recent sediments of manipulated lakes. *Limnol. Oceanogr.* 25 : 389–402.

Konecny, M. J., C. A. Lanciani, and C. P. White. 1982. Habitat-specific variation in eyespot size of a benthic cladoceran. *Oecologia* 55 : 279–280.

Kring, R. L., and W. J. O'Brien. 1976. Effect of varying oxygen concentrations on the filtering rate of *Daphnia pulex. Ecology* 57 : 808–14.

Landon, M. S., and R. H. Stasiak. 1983. *Daphnia* haemoglobin concentration as a function of depth and oxygen availability in Arco Lake, Minnesota. *Limnol. Oceanogr.* 28 : 731–37.

Lauder, G. V. 1980. The suction feeding mechanism in sunfishes *(Lepomis)*: An experimental analysis. *J. Exp. Biol.* 88 : 49–72.

Luecke, C., and W. J. O'Brien. 1981a. Phototoxicity and fish predation: Selective factors in color morphs in *Heterocope. Limnol. Oceanogr.* 26 : 454–60.

Luecke, C., and W. J. O'Brien. 1981b. Prey location volume of a planktivorous fish: A new measure of prey vulnerability. *Can. J. Fisheries Aquat. Sci.* 38 : 1264–70.

Luecke, C., and W. J. O'Brien. 1983a. The effect of *Heterocope* predation on zooplankton communities in arctic ponds. *Limnol. Oceanogr.* 28 : 367–77.

Luecke, C., and W. J. O'Brien. 1983b. Photoprotective pigments in a pond morph of *Daphnia middendorffiana. Arctic* 36 : 365–68.

Maly, E. J., S. Schoenholtz, and M. T. Arts. 1980. The influence of flatworm predation on zooplankton inhabiting small ponds. *Hydrobiologia* 76 : 233–40.

McCabe, G. D., and W. J. O'Brien. 1983. The effects of suspended silt on feeding and reproduction of *Daphnia pulex. Am. Midland Naturalist* 110(2) : 324–37.

Mellors, W. K. 1975. Selective predation on ephippial *Daphnia* and the resistance of ephippial eggs to digestion. *Ecology* 56 : 974–80.

Mittelbach, G. G. 1981. Foraging efficiency and body size. A study of optimal diet and habitat use by bluegills. *Ecology* 62 : 1370–86.

O'Brien, W. J. 1975. Some aspects of the limnology of the ponds and lakes of the Noatak drainage basin, Alaska. *Verh. Int. Verein. Theoret. Angew. Limnol.* 19 : 472–79.

O'Brien, W. J. 1979. The predator-prey interaction of planktivorous fish and zooplankton. *American Scientist* 67(5) : 572–81.

O'Brien, W. J., C. Buchanan, and J. F. Haney. 1979. Arctic zooplankton community structure: Exceptions to some general rules. *Arctic* 32(3) : 237–47.

O'Brien, W. J., B. Evans, and C. Luecke. 1984. Apparent size choice of zooplankton by planktivorous sunfish: Exceptions to the rule. *Environ. Biol. Fish.* 13 : 225–33.

O'Brien, W. J., B. I. Evans, and G. L. Howick. In press. A new view of the predation cycle of a planktivorous fish, White crappie *(Pomoxis annularis). Can. J. Fish. Aquat. Sci.*

O'Brien, W. J., D. Kettle, and H. Riessen. 1979. Helmets and invisible armor: structures reducing predation from tactile and visual planktivores. *Ecology* 60 : 287–94.

O'Brien, W. J., N. A. Slade, and G. L. Vinyard. 1976. Apparent size as the determinant of prey selection by bluegill sunfish *(Lepomis macrochirus). Ecology* 57(6) : 1304–10.

Riessen, H. P. 1982. Predatory behavior and prey selectivity of the pelagic water mite *Piona constricta. Can. J. Fisheries Aquat. Sci.* 39 : 1569–79.

Riessen, H. P. 1984. The other side of cyclomorphosis: Why *Daphnia* lose their helmets. *Limnol. Oceanogr.* 29 : 1123–27.

Schmidt, D. R., and W. J. O'Brien. 1982. Planktivorous feeding ecology of Arctic grayling *(Thymallus arcticus). Can. J. Fisheries Aquat. Sci.* 39(3) : 475–82.

Schwartz, S. S., and P. D. N. Hebert. 1982. A laboratory study of the feeding behavior of the rhabdocoel *Mesostoma ehrenbergii* on pond Cladocera. *Can. J. Zool.* 60 : 1305–1307.

Szlauer, L. 1965. The refuge ability of plankton animals before models of plankton-eating animals. *Pol. Arch. Hydrobiol.* 13 : 89–95.

Threlkeld, S. T. 1979. The midsummer dynamics of two *Daphnia* species in Wintergreen Lake, Michigan. *Ecology* 60 : 165–79.

Tucker, R. P., and S. P. Woolpy. 1984. The effect of parthenogenic eggs in *Daphnia magna* on prey location by the bluegill sunfish *(Lepomis macrochirus). Hydrobiologia* 109 : 215–17.

Vinyard, G. L. 1980. Differential prey vulnerability and predator selectivity: Effects of evasive prey on bluegill *(Lepomis macrochirus)* and pumpkinseed *(L. gibbosus)* predation. *Can. J. Fisheries Aquat. Sci.* 37 : 2294–99.

Vinyard, G. L. 1982. Variable kinematics of Sacramento perch *(Archoplites interruptus)* capturing evasive and nonevasive prey. *Can. J. Fisheries Aquat. Sci.* 39 : 208–211.

Vinyard, G. L., and W. J. O'Brien. 1975. Dorsal light response as an index of prey preference in bluegill sunfish *(Lepomis macrochirus). J. Fish. Res. Bd. Can.* 33 : 2845–49.

Vinyard, G. L., and W. J. O'Brien. 1976. Effects of light and turbidity on the reactive distance of bluegill sunfish *(Lepomis macrochirus). J. Fish. Res. Board Can.* 33 : 2845–49.

Von Holst, E. 1950. Quantitative messung von Stimmungen im Verhalten der Fische. *Symp. Soc. Exp. Biol.* 4 : 143–172.

Ware, D. M. 1973. Risk of epibenthic prey to predation by rainbow trout *(Salmo gairdneri). J. Fisheries Res. Bd. Can.* 30 : 787–97.

Werner, E. E., and D. J. Hall. 1974. Optimal foraging and the size selection of prey by the bluegill sunfish *(Lepomis macrochirus). Ecology* 55 : 1042–52.

Werner, E. E., G. G. Mittelbach, and D. J. Hall. 1981. The role of foraging profitability and experience in habitat use by the bluegill sunfish. *Ecology* 62 : 116–25.

Wetterer, J. K., and C. J. Bishop. 1985. Planktivore prey selection: The reactive field volume model versus the apparent size model. *Ecology* 66 : 457–64.

Wright, D. I. 1981. The planktivorous feeding behavior of white crappie *(Pomoxis annularis)*: Field testing a mechanistic model. Ph.D. diss., University of Kansas, Lawrence, Kan.

Wright, D. I., and W. J. O'Brien. 1982. Differential location of *Chaoborus* larvae and *Daphnia* by fish: The importance of motion and visible size. *Am. Midlands Naturalist* 108(1) : 68–73.

Wright, D. I., and W. J. O'Brien. 1984. The development and field test of a tactical model of the planktivorous feeding of white crappie *(Pomoxis annularis)*. *Ecol. Monogr.* 54 : 65–98.

Wright, D. I., W. J. O'Brien, and C. Luecke. 1983. A new estimate of zooplankton retention by gill rakers and its ecological significance. *Trans. Amer. Fish. Soc.* 112 : 638–46.

Wright, D. I., W. J. O'Brien, and G. L. Vinyard.

1980. Adaptive value of vertical migration: A simulation model argument for the predation hypothesis. In W. C. Kerfoot (ed.), *Evolution and ecology of zooplankton communities*, pp. 138–147. Hanover, N.H.: University Press of New England.

Zaret, T. M. 1972. Predators, invisible prey, and the nature of polymorphism in the *Cladocera* (class crustacea). *Limnol. Oceanogr.* 17 : 171–84.

Zaret, T. M. 1980a. *Predation and freshwater communities*. New Haven, Conn.: Yale University Press.

Zaret, T. M. 1980b. The effect of prey motion on planktivore choice. In W. C. Kerfoot (ed.), *Evolution and ecology of zooplankton communities*, pp. 594–603. Hanover, N.H.: University Press of New England.

Zaret, T. M., and W. C. Kerfoot. 1975. Fish predation on *Bosmina longirostris*: Body size selection versut visibility selection. *Ecology* 56 : 232–37.

Zaret, T. M., and J. S. Suffern. 1976. Vertical migration in zooplankton as a predation avoidance mechanism. *Limnol. Oceanogr.* 21 : 804–13.

2. General Predators and Unstable Prey Populations

William W. Murdoch

James Bence

Quite commonly, both fish and invertebrate generalist predators in freshwater systems drive their prey locally extinct. We illustrate this using our research on the biological control of mosquitoes by the mosquitofish *(Gambusia)* and the backswimmer *Notonecta*. We propose that such generalist predators are typically a source of instability for coupled predator–prey systems because of constraints on the ability of the predator to respond in a non-lagged density-dependent manner to fluctuations in the density of any particular prey species. Three characteristics determine the outcome with generalist predators: (1) These predators tend to have generation times much longer than those of their prey; (2) because they are generalists their dynamics are often weakly coupled to any particular prey population; and (3) these features combine to ensure that the numerical and developmental responses are at best neutral in their effects on prey stability and are likely to be destabilizing.

Only the short-term behavioral, i.e., functional, response remains as a potential source of stabilizing mortality. Satiation and handling time, however, are destabilizing forces on the functional response. They might in principle be overcome by changes in preference in response to changes in the densities of the various prey species, but we think this is unlikely to be common. Labile preferences seem to be uncommon among at least some predatory invertebrates. Furthermore, there appear to be significant time lags in the response of preference of fish to changing prey densities. Such lags detract from the predator's stabilizing capacity in general, and in the case of preferred prey species the lags might well account for the continued disproportionate predation, after they become scarce, that drives them extinct. We illustrate these arguments with data on *Notonecta* and *Gambusia*.

The literature suggests that local extinction of prey by predators is more common in aquatic than in terrestrial systems. We discuss whether this difference reflects real differences between the environments, in their physical structure, degree of isolation, or the nature of their predators, or whether it reflects the different interests of the ecologists who work in the two environments.

Terrestrial, freshwater, and marine ecologists often ask different sorts of questions. For example, ecologists working with terrestrial insects have spent decades attempting to determine whether and how such field populations are stabilized around equilibrial densities (e.g., Strong et al. 1984). By contrast, these issues have rarely been examined in field populations in freshwaters (but see Neill 1981; Murdoch et al. 1984; Murdoch and McCauley 1985), though classic laboratory studies of population stability have been done with freshwater organisms (e.g., Slobodkin 1954; Smith 1963; Neill 1975; Goulden et al. 1982). On the other hand, ecologists of freshwater environments have developed an impressive body of theory and evidence on the structural aspects of populations (e.g., distribution, cyclomorphology) and communities (e.g., the role of competition and predation) (Brooks and Dodson 1965; Werner and Hall 1976; Werner 1977; Kerfoot 1980; Tilman 1982; Mittelbach 1984; Werner et al. 1983).

In searching for possible explanations for this difference in approach, one is struck by an associated difference concerning the apparent effects of predators on population stability. Predators in freshwater systems (as in marine environments) have frequently been recorded as driving their prey extinct, whereas such a result is virtually nonexistent in the terrestrial literature. For example, in some lakes and ponds fish drive extinct some or all of the larger zooplankton species and other macroinvertebrates, such as predatory insects (e.g., Brooks and Dodson 1965; Hurlbert and Mulla 1981; Gliwicz and Rowan 1984). Cooper and Hemphill (personal communication) have recently extended this result to stream pools. Murdoch et al. (1984) show that an insect predator also can drive zooplankton populations extinct. By contrast, the only terrestrial example we have found of (probable) extinction of a local population is in a gall-wasp (Washburn and Cornell 1981), although Murdoch et al. (1985) argue that local extinctions may be common in biological control.

There is of course a problem of scale in discussing population extinction: at the smallest scale the death of an individual can be viewed as "local" extinction; at a sufficiently large scale, extinction is probably very rare in ecological time. The apparent difference between land and freshwater persists in spite of this difficulty. Zooplankton are often driven extinct from an entire habitat—a pond or lake—and insects frequently are driven extinct over the entire segment of stream in which large fish exist. Extinctions are simply not recorded from terrestrial systems, even from small patches of habitat.

Whether these reported patterns reflect real differences between the two environments (e.g., in their physical complexity or in the degree of isolation of habitats) or simply reflect differences in the world view and interests of the two groups of ecologists is not clear.

In this paper we explore properties of freshwater predators that influence whether or not they will tend to stabilize or destabilize populations of their prey. We begin with two examples of biological control of mosquitoes in freshwater systems because these provide both dramatic examples of the efficacy of predators in controlling their prey populations and evidence on the question of the stability of the interaction.

BIOLOGICAL CONTROL OF FRESHWATER MOSQUITOES

The theory of biological control, developed with terrestrial insects in mind, has as its central tenet that predators control pests by driving them to a low-equilibrium population size that is stable as a consequence of the characteristics of the predators (e.g., Hassell 1978, but

see Murdoch et al. 1985). The evidence from the control of mosquitoes does not appear to be consistent with this idea.

The mosquitofish, *Gambusia affinis*, is widely used as a biological control agent of pest mosquito larvae. It is the most extensively used biological control agent of mosquitoes in California (Gall et al. 1980) and has become the world's most widely distributed species of freshwater fish as a result of its use in control programs (Gerberich and Laird 1965). Experimental data showing that this fish can control mosquitoes abounds (Hildebrand 1919; Sokolov 1936; Krumholz 1948; Hoy and Reed 1970; Norland and Bowman 1976).

Local extinction of the mosquito pest is a frequent result when this fish is introduced (e.g., Emerick 1941; Green and Imber 1977; Coykendall 1980). In fact, successful control and extinction are often considered to be equivalent. In cases where extinction of mosquitoes did not result, the presence of vegetation as a refuge from predation has been implicated (Hildebrand 1919; Gerberich and Laird 1965; Coykendall 1980). The fish must be reintroduced regularly to temporary habitats, but in permanent bodies of water this is not always necessary (Emerick 1941; Coykendall 1980; personal observation).

The key to successful control, combined with pest extinction, is that mosquitofish populations can survive and grow when the pest is gone. The reason is that the mosquitofish is a very general predator. It consumes many different types of prey, including zooplankton, insects, other invertebrates, young fish, and in some cases even plant material (e.g., Hess and Tarzwell 1942; Harrington and Harrington 1961; Washino and Hokama 1967; Farley 1980). Thus, mosquitoes form only a minor part of this fish's diet; the predator can survive on other types of prey when mosquitoes are rare, and the population dynamics of the fish are at most weakly coupled to those of the mosquitoes. In fact, mosquitoes alone may not even be an adequate diet for the mosquitofish (Reddy and Pandian 1972).

Figure 2.1 shows an example of control by mosquitofish from our own work; mosquitoes were absent from rice paddies stocked with *Gambusia*, while at the same time mosquitoes became abundant in paddies from which the fish had been excluded. Superimposed on this figure is the average number of fish caught in traps in the paddies stocked with fish. Even though mosquitoes were absent throughout the season, mosquitofish survived, reproduced, and became increasingly more abundant over time.

Our second example from control of mosquitoes is the backswimming bug *Notonecta*, a naturally occurring predator of mosquitoes in small ponds and stock tanks in the ranch country of southern California. Murdoch et al. (1984) and Chesson (1984) showed by experimentally manipulating the predator in the field that *Notonecta* not only controls mosquitoes but also usually drives them extinct. *Notonecta* is a generalist predator and so is able to survive on other prey when mosquitoes are absent. Figure 2.2 shows an example of the experimental data.

These two control agents, although highly efficacious, are very general predators and are thus at the opposite extreme from the specialized parasitoid, which is the generally accepted ideal control agent in the biological control literature. In this regard they are, however, archetypal freshwater predators.

Below, we explore the components of predation to arrive at a possible explanation for the observed destabilizing effects seen in the control of mosquitoes. In addition, we ask whether these features may be widespread in freshwater predators. First, however, we need to provide a framework for determining the effect of various predation processes on the stability of a predator–prey interaction.

DENSITY DEPENDENCE, TIME DELAYS, AND STABILITY

We will argue below that freshwater predators, if they are to stabilize their prey populations, are most likely to do so via short-term behavioral responses to changes in prey density, and in particular via the functional response. Our argument rests on two broad assumptions: first, that stabilizing processes are those that have a density-dependent effect on the vital rates of the populations; second, that time delays in the *predator* population tend to destabilize the system. The first of these assumptions is supported by the entire body of theoretical work done on predator–prey systems (e.g., May 1974; Murdoch and Oaten 1975; Hassell 1978; Nisbet and Gurney 1982), with the important proviso that matters become complicated when time lags occur (e.g., May 1973).

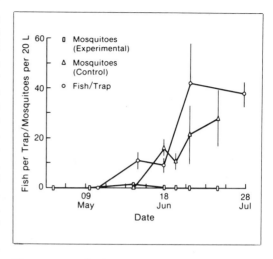

Figure 2.1. *The density of mosquitoes in control and experimental (stocked with mosquitofish) rice paddies, and the number of fish caught per trap in experimental paddies. Error bars indicate standard errors. For details see Bence (1985).*

The body of predator–prey theory concerning systems with time lags that is most relevant to freshwater systems is cast in delay-differential equations, and that also acknowledges the existence of age structure because numbers in freshwater populations change more or less continuously, generations overlap, and age structure is important. The dynamic outcomes of such models are not always simple, as Hastings (1983) has emphasized. Nevertheless, a recurrent conclusion in such studies is that delays in recruitment, or maturation, in the *predator* population tend to be destabilizing (Nisbet and Gurney 1982; Nunney 1985a, 1985b; Hastings 1984b; Murdoch et al., in press). This is the case even when the delay is short (Nunney 1985c). Nunney (1985b) points out that any undelayed aspects of the predator's numerical response (in his models, predator mortality) can help to counteract the destabilizing effect of the predator delay. But we will argue below that even mortality in freshwater predators will generally react with a time lag to changes in prey density. Furthermore, in a model in which only adult prey are eaten, Hastings (1984b) has shown that a long lifetime in the adult predator, relative to the length of the prey's juvenile period, is also strongly destabilizing. Murdoch et al. (in press) show that longer predator lifetimes are also de-

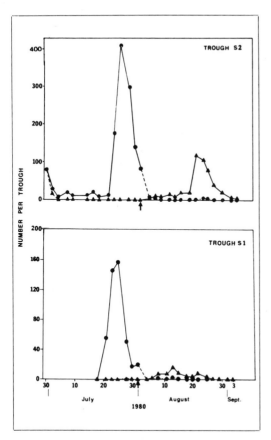

Figure 2.2. *The effect of the predator Notonecta on the number of large mosquito larvae in two stock tanks. Each tank was divided in half and predators were confined to the right side (triangles) of the tank. The vertical arrow marks the point at which predators were moved to the left side (circles). The solid line joins mosquito numbers in the absence of predators and the dashed line joins mosquito numbers (usually zero) in the presence of predators. From Murdoch et al. (1985).*

stabilizing when only juvenile prey are eaten. Long adult lifetimes in predators, relative to their prey species, are of course common in many freshwater systems.

Both Hastings (1983) and Nunney (1985c) have shown that time delays are not always destabilizing. This result needs to be interpreted with care, however. The general result is that delays in processes that are density-dependent on the prey population tend to destabilize (a very broadly based result derived in a wide variety of models); delays in processes that are inversely density-dependent on the prey (i.e., destabilizing in unlagged systems) can weaken

their destabilizing effect, but this is a much more narrowly based result. Nunney (1985a) has found one condition in which a *predator* delay can be stabilizing over some of its possible range of values, but this arises in rather narrowly defined conditions.

Another conclusion from these studies that is relevant to our discussion below is that Oaten and Murdoch's (1975) qualitative result on functional response in unlagged systems carries over to lagged systems: a type 3 functional response tends to be stabilizing, and a type 2 destabilizing, whether the system is lagged or not (Nunney 1980; Wollkind et al. 1982; Hastings 1983). Note that the response itself is assumed in these analyses to be unlagged. When a type 3 (stabilizing) functional response itself has a lag (e.g., due to learning) it tends to become destabilizing (Bence, unpublished data). The reason for this is straightforward. The functional response has its effect on the stability of a predator–prey model via the self-damping term of the prey population—it affects the prey death rate as a function of prey density; thus, lagging a type 3 response is like adding a lag to the prey's own feedback loop.

PREDATORY BEHAVIOR AND INSTABILITY

We return now to an evaluation of the likely effects on stability of various components of predation by freshwater organisms. Much of the literature on predation over the last two decades has sought to explain how predators can stabilize their prey populations (e.g., Murdoch and Oaten 1975; Hassell 1978). Here we arrive tentatively at an opposite conclusion, namely, that predators in freshwater environments are likely to be sources of instability even when these particular mechanisms are taken into account.

Predator populations might help stabilize their prey through one or more of several possible responses: the numerical response (Solomon 1949), which includes changes in the death rate as well as in the birth rate of predator; the functional response; the aggregative response (Hassell 1978); and the developmental response (Murdoch 1971). Among generalist predators of freshwater, the functional and aggregative responses are the most likely to be stabilizing, for reasons discussed below.

First, the numerical response is not likely to

be important in adding stability to prey populations because most freshwater predators change abundance much more slowly than do their prey (Hall et al. 1976). *Notonecta*, for example, has a generation time that is about 12 times longer than that of mosquitoes; fish also typically have generation times more than an order of magnitude longer than those of zooplankton. Changes in prey abundance are translated into changes in adult predator abundance only after long lags (relative to prey fluctuations) because of energy storage (which prevents starvation) or because of reproduction and the growth required to enter the predator population. In addition, the birth and death rates of generalist predators seem likely to respond weakly if at all to changes in the abundance of a single-prey species. As noted earlier, freshwater predators also tend to be long-lived relative to their prey (i.e., the death rate tends to be relatively low), and this also appears to be destabilizing.

The insensitivity of the density of a general predator to short-term changes in the abundance of its major food supply is demonstrated by Murdoch and Orr (unpublished data). They studied populations of *Notonecta* (with mixed age distributions) that were interacting in experimental stock tanks with zooplankton populations. In addition, each day they were exposed to a food supply (*Drosophila* adults) that fluctuated (from zero to highly abundant) on a 2-week period. The *Drosophila* were the major source of food. Nevertheless, *Notonecta* density fluctuated no more than did that of the controls that were exposed to a constant supply of food. *Notonecta* clearly has a large capacity to buffer fluctuations in its food supply on this time scale, yet a couple of weeks is the approximate time scale on which significant fluctuations in the prey can be expected to occur (e.g., Murdoch et al. 1984).

Second, the developmental response of freshwater predators also does not appear to be stabilizing under field conditions. Fish grow slowly relative to the numerical fluctuations in most of their prey species, as do most predatory insects in freshwater. As in the case of the numerical response, the growth rate of such generalist predators also is likely to respond weakly to change in density of any single prey species. In addition, as aquatic predators grow, typically they change the species included in their diets (Keast 1977, 1978; Helfman 1978;

Lemly and Dimmick 1982; Schmitt and Holbrook 1984; Scott and Murdoch 1983; Werner and Gilliam 1984). This will act against any stabilizing impact of the developmental response because, although larger predators may eat more total prey, they will often eat fewer of the prey species they were eating previously.

Murdoch and Sih (1978) explored the stabilizing potential of the developmental response in *Notonecta* feeding on mosquito larvae. The predator has a type 2 functional response. In searching for mechanisms that could be stabilizing, Murdoch and Sih asked whether there were feasible conditions in which the combined functional and developmental responses could yield density-dependent mortality. Groups of *Notonecta* were provided with mosquito larvae every day. The number was held constant for 2 or 3 days (depending on the experiment), then was increased during the next 2- or 3-day period, and so on. The predators were allowed to grow in response. One set of experiments was done in tubs in the laboratory, and prey were increased by 20 (fig. 2.3a) or 40 (fig. 2.3b) every 3 days. The second set was done in stock tanks on the campus of the University of California at Santa Barbara, and prey typically were increased by 150 every 2 days. In the second set of experiments, one treatment had six adults (fig. 2.3c) and the other had 12 (fig. 2.3d). In both treatments one first-instar *Notonecta* was added each day to simulate spring recruitment.

Fig. 2.3 reanalyzes the results and shows that the general result was inversely density-dependent, or density-independent, mortality. Murdoch and Sih plotted the results as the number eaten per day as a function of prey abundance. They noted that there was density dependence in two of the treatments over a narrow range of prey densities. This can be seen in figures 2.3a and 2.3c as increasing regions of the graphs.

Although these results show that density dependence is possible, the overall result is that the mortality is destabilizing. The only marked density dependence (fig. 2.3a at higher prey densities) is in the treatment that had an imposed rate of increase of the mosquito prey that was very low compared with rates of increase in the field (Murdoch et al. 1984; Chesson 1984). Furthermore, the overall mortality was density-independent even in this treatment: a regression of the data in figure 2.3a (mortality rate versus prey abundance)

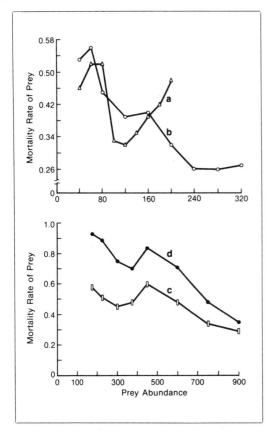

Figure 2.3. *The fraction of mosquito larvae killed by groups of* Notonecta *that grew throughout the experiment, as a function of the abundance of mosquitoes;* **a** *and* **b** *are from lab experiments,* **c** *and* **d** *from field experiments (see text for explanation).*

was not significant. In all other treatments mortality declined significantly with prey abundance (for fig. 2.3b: p, <0.001; for fig. 2.3c: p, <0.017; and for fig. 2.3d: p, <0.001, all based on linear regression). Thus, the combined developmental and functional responses of *Notonecta* are at best density-independent, with neutral stability effects, and are typically destabilizing.

Murdoch and Sih's experiments were designed to maximize the chances of finding potentially stabilizing mortality. Prey numbers increased steadily through time; density dependence would have been much less likely had prey abundance declined or fluctuated, because the predator would not have decreased in size

in response to declines in prey abundance. Also, unlike field situations, no alternative prey were present; they would have served to decouple *Notonecta*'s growth rate from mosquito abundance.

We are thus left with the predator's short-term behavior as the most plausible potential source of stabilizing mortality for the prey. What is the evidence on this issue for the predators we have been discussing?

In the aggregative response, individual predators spend longer in patches with more prey. Hassell and May (1973, 1974) show that this response can sometimes stabilize the otherwise unstable Nicholson–Bailey parasitoid–host model. It is not clear how general this result is, since Chesson and Kerans (personal communication) have shown that a similar model can also be made less stable by aggregation. The Nicholson–Bailey formulation, with its discrete and with nonoverlapping generations is not in any case appropriate for freshwater populations. The effect of aggregation on local host density is only now being examined in more appropriate models that have overlapping generations (Murdoch, unpublished data). Finally, there seem to be no data on the effect of aggregative responses on prey mortality rates in freshwater systems.

We turn, finally, to the functional response as the one most likely to be a source of stabilizing predation in freshwater systems. A fairly complete picture has been developed in our lab for *Notonecta*'s functional response, and is presented in Murdoch et al. (1984). The functional response in the presence of a single-prey species is type 2, i.e., inversely density-dependent and hence destabilizing. When given a mixture of prey species at varying relative and absolute densities, preference between species remains constant (i.e., there is no switching); indeed, behavior is very stereotyped, and preferences among new combinations of prey can be predicted from preferences between subsets of the species. The exception to this picture is the result of Lawton et al. (1974) that a species of *Notonecta* switched between two prey species. We were not able to repeat this result, and we suspect that such behavior is rare.

Our expectation that the overall mortality generated by *Notonecta* populations will be destabilizing is consistent with the facts that *Notonecta* (at constant density and size–structure) drove *Daphnia* populations extinct in the laboratory (Murdoch and Scott 1984) and that populations of *Notonecta* drove extinct both *Daphnia* and mosquitoes in the field (Murdoch et al. 1984). Coexistence between this predator and *Daphnia*, when it did occur, was the result of compensatory behavior of the zooplankter (Murdoch and Scott 1984).

To what extent is *Notonecta* an exemplar of the generalist freshwater predator? In particular, do fish also cause destabilizing or at best neutral (density-independent) mortality on their short-lived prey species, such as zooplankton and mosquito larvae? We have argued that the key lies in whether the functional response is stabilizing. Hence, to answer the question we next review results on the predation rates of individual fish in response to changes in the relative and absolute abundances of their prey species.

The basic functional response of fish in the presence of a single-prey species is typically type 2 (Murdoch and Oaten 1975; Townsend and Risebrow 1982). Murdoch and Oaten analyzed data showing that naive fish may have a type 3 response, which arises because the fish at low prey densities take longer to learn that prey are present than do those at high prey densities. Such a response is ephemeral, and once learning has occurred, the response is type 2.

In contrast with the situation for *Notonecta*, stereotyped behavior does not seem to be the rule for freshwater fish. There are many examples where the selectivity of a predatory fish has been found to depend on the fish's experience (e.g., Beukema 1968; LeBrasseur 1969; Bryan 1973; Murdoch et al. 1975; Milinski and Loewenstein 1980; Dill 1983; Bence, in press). This behavioral flexibility could lead to a potentially stabilizing (i.e., type 3) functional response. Whether this occurs, however, will depend in part on how rapidly the fish learns and adjusts its behavior: time lags in the behavioral response could turn a potentially stabilizing mechanism into a destabilizing one (see previous section). In addition, density- or frequency-dependent preferences do not guarantee, even in the absence of time lags, that the response to any particular prey will be type 3; that will depend also on how fast the total predation rate increases relative to the rate of increase of the density of the prey in question.

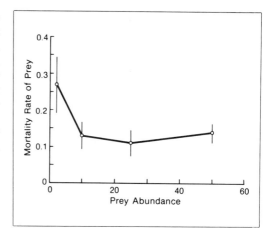

Figure 2.4. *The fraction of* Cerioadaphnia dubia *killed per minute by individual mosquitofish (standard length, 17–20 mm), as a function of* C. dubia *density. Error bars indicate standard errors. See text for details.*

Case Study: Poeciliid Fish

In an extensive series of experiments we found that the mosquitofish (*G. affinis*) did not change its preferences, over the short term, for different sizes (Bence and Murdoch 1986) or different taxa of prey (Bence 1985), as either the absolute or relative abundance of prey was varied. The consequence of these constant preferences was generally a type 2 functional response for each prey type. Figure 2.4 shows an example of the mortality rate suffered by the prey as a function of its density. In this experiment one prey type (cyclopoid copepods) was held at a constant density (50/L), and the target prey (*Ceriodaphnia dubia*, Cladocera) was varied in abundance. The trials were done in 2-L containers, and the methods follow those in Bence (in press). To avoid depletion effects, prey were replenished whenever their abundance had been reduced to 75% of the initial numbers. As in most of our experiments, there is no tendency for the mortality rate to increase as prey density increases.

We did find in one exceptional case that mosquitofish increasingly preferred mosquitoes over *Notonecta* as the abundance of mosquitoes was increased (with *Notonecta* density fixed) (Bence and Murdoch 1982). This exception is particularly interesting because it shows that density-dependent changes in preferences are not necessarily stabilizing. Based on the results

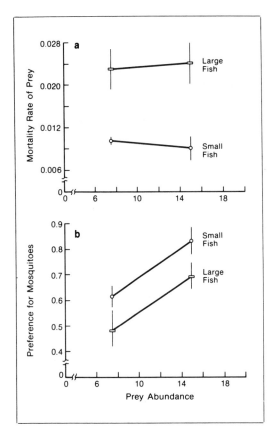

Figure 2.5. *The fraction of mosquito larvae killed per minute per fish:* **a**, *by groups of three large (standard length, 37–40 mm) or small (standard length, 27–30 mm) mosquitofish as function of the abundance of mosquitoes;* **b**, *preference for mosquitoes (over Notonecta) as a function of mosquito density. Small and large fish (as in a), preference measured by standard forage ratio (Chesson 1978). Error bars indicate standard errors. See text for details.*

in Bence and Murdoch (1986), it appears that this counterexample was due to a satiation effect and not to the fish's response to the absolute or relative abundance of mosquitoes per se. We found that as mosquitofish became more satiated they increasingly preferred small prey but fed at slower rates. The fact that mosquitoes were smaller than *Notonecta*, and fish consumed more total prey biomass at high than at low densities of mosquitoes, could explain the result. Of particular importance here is that the mosquito mortality rate was independent of mosquito abundance for both small and large fish (fig. 2.5).

Both the guppy (*Poecilia reticulata*) and the mosquitofish can modify their preferences in

response to their experiences over longer time spans (Murdoch et al. 1975; Bence 1985, in press). When total prey abundance was fixed, and the relative abundance of a prey type was increased or decreased gradually over 12 days, the guppy imposed density-dependent mortality on its prey (Murdoch et al. 1975). There are, however, at least three important caveats to this result. First, prey abundance was varied slowly. Even then, some individuals showed appreciable lags in their response to changes in prey density (and such a delay tends to be destabilizing). Second, the density-dependent result stemmed in part from the fact that total prey density was fixed. Attacks were concentrated on a prey type as it became more abundant because the other prey type was becoming less abundant, as well as because of changes in preference between prey. Thus, in order to be stabilizing, switching by guppies may need to act in concert with some mechanism (e.g., competition) that fixes total prey density. Finally, both guppies and the mosquitofish (Bence 1985, in press) appeared to respond more quickly to increases than to decreases in abundance of their more profitable (and preferred) prey types. This kind of lag may be particularly destabilizing because the fish will concentrate on those prey types even after they have been driven to low numbers.

Evidence from Other Freshwater Fish

There is surprisingly little information on whether fish feeding on a mixture of prey types causes density-dependent mortality on a given prey type. Sticklebacks apparently do not change their prey preferences in response to short-term changes in prey abundance (Gibson 1980; see also Butler and Bence 1984; Wetterer and Bishop 1985). Not surprisingly, the functional response is type 2, and the resulting mortality rates are inversely density dependent. This matches Beukema's (1968) results for this fish feeding on a single prey type. Sticklebacks can change their prey preferences over longer time periods (Beukema 1968), i.e., with a long time lag. Again, the delay is likely to be a source of instability.

Reed (1971) found that bluegill sunfish choosing between mosquitoes and chironomids exhibited anti-switching; they increasingly preferred a prey type as it became relatively (and absolutely) rare. The net effect of this behavior was that mortality rates of the prey were independent of prey density. Bohl (1982) also found anti-switching in the field for a cyprinid fish feeding on two types of zooplankton. He did not present data on the relationship between mortality rate of the prey and prey density.

To our knowledge, there are no examples of type 3 functional responses in freshwater fish to *short-term* changes in prey abundance that can be attributed to density-dependent changes in preferences among taxa of prey. Results for bluegill sunfish suggest that this fish sometimes causes density-dependent mortality, at intermediate densities, on the larger-size classes of its prey (Werner and Hall 1974; Wetterer, personal communication). The mortality of the population as a whole, however, remains inversely density-dependent. [Bohl (1982) found short-term switching in a cyprinid fish that was feeding on two sizes of *Daphnia*.] The theoretical consequences of these types of behaviors, which would require an analysis of stage-structured population dynamics, are virtually unknown. This is particularly curious given the concentration in optimal diet studies on preferences between different-size classes of prey.

In summary, the functional response is the only component of predation by fish that has been shown to have the potential to stabilize prey populations. That response, however, is likely to be destabilizing in nature whenever only one prey species is present. Fish in mixtures of prey species sometimes have preferences that vary as the prey densities vary, and this in principle could be stabilizing. For several reasons, however, varying preferences are not likely to be stabilizing. First, responses to change in prey density occur with an appreciable lag, which is likely to be destabilizing. Second, when fish do learn to prefer more profitable prey, they tend to continue to prefer them even as they decline in abundance, a behavior that is likely to lead to local extinction of that prey species; indeed that may be one reason why such predators often drive their most preferred prey extinct. Finally, the underlying destabilizing forces of satiation and handling time in predators remain even if preferences vary with prey density. (This is probably the explanation of the density-independent mortality in fig. 2.5). Only in the most restrictive conditions (e.g., total prey density con-

stant) will frequency-dependent predation (as in switching) lead to density-dependent predation. This was the case in the example of switching by damselfly naiads found by Akre and Johnson (1979). The presence of refuges, of course, may produce effectively type 3 functional responses (Murdoch and Oaten 1975), but then the refuge, not the predator, is the explanation for stability of the prey.

DISCUSSION

The short-term behavior of many aquatic predators is probably the only component of their response to changes in prey density that has the potential to be stabilizing. The dynamic consequences of an aggregative response in systems of this type are unknown; therefore, information on this response does not exist. The numerical and developmental responses of these predators are too slow relative to the time scale of the prey population's fluctuations. The evidence we have surveyed suggests that the functional response of the backswimmer *Notonecta*, of fish, and perhaps of most generalist freshwater predators, tends to be a destabilizing force on their prey populations. This may explain why extinction is often an outcome of the activity of *Notonecta* and fish. We suspect that stability is not likely to be an inherent feature of the predator–prey interaction per se.

Our conclusion that freshwater predators are generally forces for instability is tentative, and for various reasons needs to be viewed simply as a hypothesis to be tested. First, there is still remarkably little direct evidence on whether predators stabilize or destabilize the density of prey populations with which they coexist. Second, we have argued from detailed knowledge of a few predators—especially *Notonecta* and some fish—to freshwater predators in general, but we do not know if our exemplars truly are typical. Third, our analysis extends ideas (for example, of the relationship between density dependence and stability and between time lags and instability) from a set of mathematical theory to field populations. This transference may sometimes be inappropriate because the theory has not considered all possible circumstances. For example, theory has not considered predators that change their preferences among different sizes of prey as a function of total prey density.

We return to the comparison with terrestrial systems. There is more evidence for extinction of prey by predators in fresh water than on land. Three possible explanations for this difference suggest themselves. First, the difference between these habitats in frequency of extinction may be more apparent than real. As noted earlier, it has been argued that successful biological control of insect pests in long-lived agroecosystems may typically involve local extinctions, and may only rarely involve regulation of the pest around a stable equilibrium (Murdoch et al. 1985). It may similarly be the case that extinction by predation commonly occurs in natural terrestrial systems but has not been looked for.

As a corollary, we need also to consider the possibility that freshwater systems have been sampled very nonuniformly with respect to population phenomena. Like ecologists in the intertidal areas, those in freshwater environments have been more interested in the effects of predators on community structure than on population stability. Thus, whether predators in freshwater systems increase or decrease the stability of the prey populations with which they coexist remains an interesting area for investigation.

A second possible explanation for the apparent difference between land and fresh water is that the greater physical complexity of land communities, by providing refuges and spatial heterogeneity, may prevent extreme instability leading to local extinction. We know of no evidence bearing on this hypothesis, other than the obvious observation that many freshwater environments are physically simpler than most terrestrial environments. Alternatively, a crucial difference may be that freshwater systems tend to be more closed to immigration than terrestrial habitats, so that trends toward extinction run their course more frequently in the former than in the latter type of habitat.

A third possibility is that terrestrial and freshwater predators have different properties in general. One obvious difference is that specific predators are common on land, at least among insect parasitoids, whereas virtually all freshwater predators attack many prey species. Also, unlike most freshwater predators, the generation times of parasitoids are frequently similar to those of their prey. There is evidence that specific parasitoids can cause density-dependent mortality. For example Strong et al. (1984) found 7 insect life table studies, out of 31, in which parasitoids caused density-dependent mortality. On the other hand, they failed to find evidence of this sort in the majority of the studies. Dempster (1983), by contrast, found only 1 case in which a parasitoid was potentially stabilizing among 24 life table studies for Lepidoptera (an example also listed by Strong et al.). Dempster also found that parasitoids were destabilizing in one study. Reeve and Murdoch (in press) found that the parasitoid controlling red scale in California did not cause density-dependent mortality. Turning to terrestrial predators rather than parasitoids, both Strong et al. and Dempster found only one additional study (the same one) in which predators were the stabilizing mortality factor, whereas Dempster found between four and nine studies in which they were destabilizing.

The same diversity of outcomes pervades studies not using the life table approach, including those of generalist predators. One study of weasels in Wytham Woods in England showed that they caused density-dependent mortality on nestlings in nest boxes (Dunn 1977); another showed this was not the case for weasels preying on voles and mice (King 1980). Linden and Wickman (1983) found a density-dependent functional response in goshawks preying on grouse, whereas Erlinge et al. (1983) found strong inversely density-dependent responses (and overall predation) by a group of general predators of voles and mice. Thus, the extent to which terrestrial predators (specific or generalist) cause density-dependent mortality, and freshwater predators typically cause destabilizing mortality, remains both an open and an interesting question.

What does stabilize those freshwater prey populations that appear stable, if not predators? Obvious potential mechanisms include refuges in space and time and other aspects of physical heterogeneity; invulnerable age or size classes in the prey, although these are probably less common in freshwater than terrestrial systems (Murdoch et al., in press); and compensatory responses by the prey (e.g., increases in the rate of reproduction or survival in response to increased food per head when predation reduces prey abundance). Examples of refuges can be found in the mosquito control literature, as discussed above. Compensatory responses by the prey were shown in natural populations of zooplankton preyed on by *Chaoborus* (Neill 1981) and in experimental zooplankton populations exposed to destabilizing predation by *Notonecta* (Murdoch and Scott 1984). More field studies are needed to determine the relative importance of these and other stabilizing mechanisms.

A more fundamental question also remains. Are stable populations at all common in nature? This is a vigorously debated question (Roughgarden 1975; Murdoch 1979; Connell and Sousa 1983).

The problem of stability raised in this paper has been explored by ecologists for at least 50 years. It is perhaps surprising that we are not closer to a general conclusion. In part this is because the study of population stability has been somewhat out of vogue in recent years. Yet it is basic to much else in ecology, including community theory. Furthermore, the analytic tools for modeling and understanding the dynamics of real systems have recently been developing apace. Of special interest to the aquatic ecologist are substantial advances in modeling of size-structured systems (e.g., Hastings 1984a; Nisbet and Gurney 1983; Mittelbach and Chesson, this volume). We therefore end with the suggestion that population stability is a theme to which ecologists, including those working in freshwater habitats, might well return.

ACKNOWLEDGMENTS

The research reported here was done with the support of NSF grant BSR-83-15235 to WWM and grants to both authors from the University of California Mosquito Control Research Program and the Program for Appropriate Technology. We thank Scott Cooper, Ed McCauley, Roger Nisbet, Len Nunney, and Russ Schmitt for commenting on a draft of the manuscript.

REFERENCES

Akre, B. G., and D. M. Johnson. 1979. Switching and sigmoid functional response curves by damselfly naiads with alternative prey available. *J. Anim. Ecol.* 48 : 703–20.

Bence, J. R. 1985. Specificity in choice of prey and the predatory impact on invertebrates of the freshwater mosquitofish. Ph.D. thesis. University of California, Santa Barbara, Calif.

Bence, J. R. In press. Feeding rate and attack specialization: the roles of predator experience and energetic tradeoffs. *Environ. Biol. Fish.*

Bence, J. R., and W. W. Murdoch. 1982. *Gambusia* as a predator upon *Notonecta*: Laboratory experiments. *Proceedings of the California Mosquito and Vector Control Association* 50 : 51–53.

Bence, J. R., and W. W. Murdoch. 1986. Prey size selection by the mosquitofish and its relation to optimal diet theory. *Ecology* 67 : 324–36.

Beukema, J. J. 1968. Predation by the three spined stickleback (*Gasterosteus aculeatus* L.): The influence of hunger and experience. *Behavior* 30 : 1–126.

Bohl, E. 1982. Food supply and prey selection in planktivorous cyprinidae. *Oecologia* 53 : 134–38.

Brooks, J. L., and S. I. Dodson. 1965. Predation, body size, and composition of plankton. *Science* 150 : 28–35.

Bryan, J. E. 1973. Feeding history, parental stock, and food selection in rainbow trout. *Behavior* 45 : 123–53.

Butler, W., and J. R. Bence. 1984. A diet model for planktivores that follow density-independent rules for prey selection. *Ecology* 65 : 1885–94.

Chesson, J. 1978. Measuring preference in selective predation. *Ecology* 59 : 211–15.

———. 1984. Effect of notonectids (Hemiptera: Notonectidae) on mosquitoes (Diptera: Culicidae): predation or selective oviposition? *Environ. Entomol.* 13 : 531–38.

Connell, J. H., and W. P. Sousa. 1983. On the evidence needed to judge ecological stability or persistence. *Amer. Naturalist* 121 : 789–824.

Coykendall, R. L. 1980. *Fishes in California mosquito control.* Sacramento, Calif.: CMVCA Press.

Dempster, P. J. 1983. The natural control of populations of butterflies and moths. *Biol. Rev.* 58 : 461–81.

Dill, L. M. 1983. Adaptive flexibility in the foraging behavior of fishes. *Can. J. Fisheries Aquat. Sci.* 40 : 398–408.

Dunn, E. 1977. Predation by weasels (*Mustela nivalis*) on breeding tits (*Parus* spp.) in relation to the density of tits and rodents. *J. Anim. Ecol.* 46 : 633–52.

Emerick, A. M. 1941. Symposium on operating problems: Mosquitofish. *Proc. Calif. Mosq. Control Assoc.* 12 : 128–29.

Erlinge, S. 1983. Predation as a regulating factor on small rodent populations in southern Sweden. *Oikos* 40 : 36–52.

Farley, D. 1980. Prey selection by the mosquitofish *Gambusia affinis* in Fresno County rice fields. *Proc. Calif. Mosq. Control Assoc.* 48 : 51–55.

Gall, G., J. Chech, R. Garcia, V. Resh, and R. Washino. 1980. Mosquitofish—an established predator. *Calif. Agric.* 34 : 21–22.

Gerberich, J. B., and M. Laird. 1965. An annotated bibliography of papers relating to the control of mosquitoes by the use of fish (revised and enlarged to 1965). WHO/EBL/66.71. Geneva: World Health Organization.

Gibson, R. M. 1980. Optimal prey size selection by three-spined sticklebacks (Gasterosteus aculeatus): A test of the apparent size hypothesis. Z. Tierpsychol. 52 : 291–307.

Gliwicz, Z. M., and M. G. Rowan. 1984. Survival of Cyclops abyssorum tatricus (Copepoda, Crustacea) in alpine lakes stocked with planktivorous fish. Limnol. Oceanogr. 29 : 1290–99.

Goulden, C. E., L. H. Henry, and A. J. Tessier. 1982. Body size, energy reserves, and competitive ability in three species of Cladocera. Ecology 63 : 1780–89.

Green, M. F., and C. F. Imber. 1977. Applicability of Gambusia affinis to urban mosquito problems in Brulington County, New Jersey. Mosq. News 37 : 383–85.

Hall, D. J., S. T. Threlkheld, C. W. Burns, and P. H. Crowley. 1976. The size-efficiency hypothesis and the size structure of zooplankton communities. Annu. Rev. Ecol. Syst. 7 : 177–208.

Harrington, R. W., and E. S. Harrington. 1961. Food selection among fishes invading a high subtropical salt marsh from onset of flooding through the progress of a mosquito brood. Ecology 42 : 646–56.

Hassell, M. P. 1978. The dynamics of arthropod predator-prey systems. Princeton, N.J.: Princeton University Press.

Hassell, M. P., and R. M. May. 1973. Stability in insect host-parasite models. J. Anim. Ecol. 42 : 693–736.

Hassell, M. P., and R. M. May. 1974. Aggregation in predators and insect parasites and its effect on stability. J. Anim. Ecol. 43 : 567–94.

Hastings, A. 1983. Age-dependent predation is not a simple process. I: Continuous time models. Theor. Pop. Biol. 23 : 347–62.

Hastings, A. 1984a. Age-dependent predation is not a simple process. II: Wolves, ungulates, and a discrete time model for predation on juveniles with a stabilizing tail. Theor. Pop. Biol. 26 : 271–82.

Hastings, A. 1984b. Delays in recruitment at different trophic levels: Effects on stability. J. Math. Biol. 21 : 35–44.

Helfman, G. S. 1978. Patterns of community structure in fishes: Summary and overview. Environ. Biol. Fish. 3 : 129–48.

Hess, A. D., and C. M. Tarzwell. 1942. The feeding habits of Gambusia affinis affinis, with special reference to the malaria mosquito, Anopheles quadrimaculatus. Amer. J. Hygiene 35(1) : 142–51.

Hildebrand, S. F. 1919. Fishes in relation to mosquito control in ponds. Public Health Rep. 34 : 1113–28.

Hoy, J. B., and D. E. Reed. 1970. Biological control of Culex tarsalis in a California rice field. Mosq. News 30 : 222–30.

Hurlbert, S. H., and M. S. Mulla. 1981. Impacts of mosquitofish (Gambusia affinis) predation on plankton communities. Hydrobiology 83 : 125–51.

Keast, A. 1977. Feeding and food overlaps between the year classes relative to the resource base, in the yellow perch, Perca flavescens. Environ. Biol. Fish. 2 : 55–70.

Keast, A. 1978. Feeding interrelations between age-groups of pumpkinseed (Lepomis gibbosus) and comparison with bluegill (L. macrochirus). J. Fish. Res. Bd. Can. 35 : 12–27.

Kerfoot, W. C. 1980 (ed.). Evolution and ecology of zooplankton communities. Hanover, N.H.: University Press of New England.

King, C. M. 1980. The weasel Mustela nivalis and its prey in an English woodland. J. Anim. Ecol. 49 : 127–59.

Krumholz, L. A. 1948. Reproduction in the western mosquitofish Gambusia affinis (Baird and Girard), and its use in mosquito control. Ecol. Monogr. 18 : 1–48.

Lawton, J. J., J. R. Beddington, and R. Bonser. 1974. Switching in invertebrate predators. In M. B. Usher and M. H. Williamson (eds.), Ecological stability, pp. 141–58. London: Chapman and Hall.

LeBrasseur, R. J. 1969. Growth of juvenile chum salmon (Oncorhynchus keta) under different feeding regimes. J. Fish. Res. Bd. Can. 26 : 1631–45.

Lemly, A. D., and J. F. Dimmick. 1982. Growth of young-of-the-year and yearling centrarchids in relation to zooplankton in the littoral zone of lakes. Copeia 1982 : 305–21.

Linden, H., and M. Wikman. 1983. Goshawk predation on tetraonoids: Availability of prey and diet of the predator in the breeding season. J. Anim. Ecol. 52 : 953–68.

May, R. M. 1973. Time-delay versus stability in population models with two and three trophic levels. Ecology 54 : 315–25.

May, R. M. 1974. Stability and complexity in model ecosystems. (2nd ed.). Princeton, N.J.: Princeton University Press.

Milinski, M., and C. Loewenstein. 1980. On predator selection against abnormalities of movement: A test of an hypothesis. Tierpsychol. 53 : 323–40.

Mittelbach, G. G. 1984. Predation and resource partitioning in two sunfishes (Centrarchidae). Ecology 65 : 499–513.

Murdoch, W. W. 1971. The developmental response to predators to changes in prey density. Ecology 52 : 132–37.

Murdoch, W. W. 1979. Predation and the dynamics of prey populations. Fortschr. der Zool. 25 : 295–310.

Murdoch, W. W., S. Avery, and M. E. Smythe. 1975. Switching in a predatory fish. Ecology

56 : 1094–1105.

Murdoch, W. W., J. Chesson, and P. L. Chesson. 1985. Biological control in theory and practice. *Amer. Naturalist* 125 : 344–66.

Murdoch, W. W., and E. McCauley. 1985. Stability and cycles in planktonic systems: Implications for ecological theory. *Nature* 316 : 628–30.

Murdoch, W. W., R. M. Nisbet, S. P. Blythe, W. S. C. Gurney, J. D. Reeve. In press. An invulnerable age class and stability in delay-differential parasitoid-host models. *Am. Nat.*

Murdoch, W. W., and A. Oaten. 1975. Predation and population stability. *Adv. in Ecol. Res.* 9 : 1–131.

Murdoch, W. W., and M. A. Scott. 1984. Stability and extinction of laboratory populations of zooplankton preyed on by the backswimmer, *Notonecta. Ecology* 65 : 1231–48.

Murdoch, W. W., M. A. Scott, and P. Ebsworth. 1984. Effects of the general predator, *Notonecta* (Hemiptera), upon a freshwater community. *J. Anim. Ecol.* 53 : 791–808.

Murdoch, W. W., and A. Sih. 1978. Age-dependent interference in a predatory insect. *J. Anim. Ecol.* 47 : 581–92.

Neill, W. 1975. Experimental studies of microcrustacean competition, community composition and resource utilization. *Ecology* 56 : 809–26.

———. 1981. Impact of *Chaoborus* predation upon the structure and dynamics of a crustacean zooplankton community. *Oecologia* 48 : 164–77.

Nisbet, R. M., and W. S. C. Gurney. 1982. *Modelling fluctuating populations.* New York: John Wiley and Sons.

Nisbet, R. M., and W. S. C. Gurney. 1983. The systematic formulation of population models for insects with dynamically varying instar duration. *Theor. Pop. Biol.* 23 : 114–35.

Norland, R. L., and J. R. Bowman. 1976. Population studies of *Gambusia affinis* in rice fields: Sampling design, fish movement and distribution. *Proc. Calif. Mosq. Control Assoc.* 44 : 53–56.

Nunney, L. 1980. The influence of the type 3 (sigmoid) functional response upon the stability of predator-prey difference equations. *Theor. Pop. Biol.* 18 : 257–78.

Nunney, L. 1985a. Absolute stability in predator-prey models. *Theor. Pop. Biol.* 27 : 202–21.

Nunney, L. 1985b. The effects of long time delays in predator-prey systems. *Theor. Pop. Biol.* 27 : 202–21.

Nunney, L. 1985c. Short time delays in population models: A role in enhancing stability. *Ecology* 66 : 1849–58.

Oaten, A., and W. W. Murdoch. 1975. Functional response and stability in predator-prey systems. *Amer. Naturalist* 109 : 289–98.

Reddy, S. R., and T. J. Pandian. 1972. Heavy mortality of *Gambusia affinis* reared on diet restricted

to mosquito larvae. *Mosq. News* 32 : 108–10.

Reed, R. C. 1971. An experimental study of prey selection and regulatory capacity of bluegill sunfish *(Lepomis macrochirus).* M.A. thesis, University of California, Santa Barbara, Calif.

Reeve, J. R., and W. W. Murdoch. 1986. Biological control by the parasitoid *Aphytis melinus,* and population stability of the California red scale. *J. Anim. Ecol.* (in press).

Roughgarden, J. 1975. *Theory of population genetics and evolutionary ecology: An introduction.* New York: Macmillan.

Schmitt, R. J., and S. J. Holbrook. 1984. Ontogeny of prey selection by black surfperch *Embiotoca jacksoni* (Pisces: Embiotocidae): The roles of fish morphology, foraging behavior, and patch selection. *Mar. Ecol.* 18 : 225–39.

Scott, M. A., and W. W. Murdoch. 1983. Selective predation by the backswimmer, *Notonecta. Limnol. Oceanogr.* 28 : 352–66.

Slobodkin, L. B. 1954. Population dynamics in *Daphnia obtusa* Kurz. *Ecol. Monogr.* 24 : 69–88.

Smith, E. 1963. Population dynamics in *Daphnia magna* and a new model for population growth. *Ecology* 44 : 4.

Sokolov, N. P. 1936. L'acclimatisation du *Gambusia patruelis* en Asie centrale. *Riv. Malariol.* 15(5) : 325–44.

Solomon, M. E. 1949. The natural control of animal populations. *J. Anim. Ecol.* 18 : 1–35.

Strong, D. R., J. H. Lawton, and T. R. E. Southwood. 1984. *Insects on plants.* Cambridge, MA: Harvard University Press.

Tilman, D. 1982. Resource competition and community structure. Princeton, N.J.: Princeton University Press.

Townsend, C. R., and A. J. Risebrow. 1982. The influence of light level on the functional response of a zooplanktivorous fish. *Oecologia* 53 : 293–95.

Washburn, J. O., and H. V. Cornell. 1981. Parasitoids, patches, and pheonology: Their possible role in the local extinction of a cynipid gall wasp population. *Ecology* 62 : 1597–1607.

Washino, R. K., and Y. Hokama. 1967. Preliminary report on the feeding pattern of two species of fish in a rice field habitat. *Proc. Calif. Mosq. Control Assoc.* 35 : 687–713.

Werner, E. E. 1977. Competition and habitat shift in two sunfishes (Centrarchidae). *Ecology* 58 : 869–76.

Werner, E. E., and J. F. Gilliam. 1984. The ontogenetic niche and species interactions in size-structured populations. *Annu. Rev. Ecol. Syst.* 15 : 393–425.

Werner, E. E., and D. J. Hall. 1974. Optimal foraging and size selection of prey by the bluegill sunfish *Lepomis macrochirus. Ecology* 55 : 1042–52.

Werner, E. E., and D. J. Hall. 1976. Niche shifts in sunfishes: Experimental evidence and significance.

Science, N.Y. 191 : 404–406.

Werner, E. E., G. G. Mittelbach, D. J. Hall, and J. F. Gilliam. 1983. Experimental tests of optimal habitat use in fish: The role of relative habitat profitability. *Ecology* 64 : 1549–55.

Wetterer, J. K., and C. J. Bishop. 1985. Planktivore prey selection: The reactive field volume model versus the apparent size model. *Ecology.* 66 : 457–64.

Wollkind, D., A. Hastings, and J. Logan. 1982. Age structure in predator-prey systems. II. Functional response and stability and the paradox of enrichment. *Theor. Pop. Biol.* 21 : 57–68.

II. Types of Indirect Interactions

3. Redefining Indirect Effects

Thomas E. Miller

W. Charles Kerfoot

By means of a simple three-species situation plus some examples, we distinguish between three qualitatively different types of "indirect effect": trophic linkage, behavioral, and chemical response. If species C affects species A by altering the abundance of species B, then the effect is of the first type. In this type of effect, abundance changes act in series along trophic pathways. On the other hand, the second and third types need not involve changes in abundance. If species C affects species A by influencing the behavior of B, the effect is termed a behavioral effect. If species C alters the nature of A's response to B through chemically mediated actions, the effect is termed a chemical response effect. Examples of the second type are now quite common, whereas the third type includes many cases of induction or sequestering of noxious compounds.

There is a growing amount of literature that discusses multispecies interactions in communities. Many, if not all, of these papers deal with determining how and to what degree pairwise species interactions are influenced by the presence and density of other species in the community. Many different terms have been used to describe dependencies between species interactions, including indirect effect, higher-order interaction, keystone predator effect, apparent competition, apparent predation, and vaulting. Unfortunately, there is some confusion in the use of this melange of terms, with very similar interactions being described using different terms [e.g., Case and Bender 1981 (higher-order interaction); Abrams 1984 (indirect effect)] and very different interactions being described using the same term (e.g., indirect effect in Levine 1976; Schmitt et al. 1983).

We propose here a terminology for qualitatively different types of indirect effects, one that describes different ways in which dependencies in three-species interactions may occur. It is hoped that this will begin to provide an organization for understanding the similarities and differences between different types of indirect species interactions in communities.

We begin with a general definition of the indirect effect of a species as an effect that modifies the direct effect of one species on another.

In the simplest two-species system, the growth of a species (*focal species* A) can be represented as some direct function of the abundance of a second species (*associate species* B):

$$dN_A/dt = f_i(N_A, N_B) \qquad (3.1)$$

This simple representation gives the population growth of species A as a function of the abundances of species A (N_A) and species B (N_B). In our formulation, the specific functional form (f_i) is intentionally left unspecified. We propose that there are at least three distinct ways in which the inclusion of a third species, C, may indirectly affect focal species A by changing the impact of species B on species A.

First, the total abundance of species B may be a function of the third species, C. This produces a chain of direct effects in which N_C determines N_B and N_B determines N_A. This type of pathway, in which species C affects the abundance of species A by changing the total abundance of an intermediate species, species B, we will define as a *trophic linkage indirect effect*. Through trophic linkage effects, species C may have a positive (facilitation) or negative (detrimental) effect on the abundance of species A, depending on the type of direct effects involved in the chain. In formal analyses of communities, interactions among species (N_i) at equilibrium can be summarized in an *a*-matrix of pairwise and more complicated terms, representing all possible paths of influence. Direct effects would be described by the simplest linkage paths between two species (the elements a_{ji}, a_{ij}), whereas interactions along longer paths of influence (e.g., $a_{21}a_{32}a_{13}$) describe indirect effects [the "loops" of Levins (1975)].

A second way in which a third species can affect the two-species system given in equation 3.1 is for the function f_i to be influenced by species C independent of total abundance, per se. The function f_i describes precisely how the density of species B affects the growth rate of species A. This function is determined by characteristics of both the species having the effect (species B) and the species responding (species A). The contribution of species B to this function can be altered by its "behavior": for example, the presence of a predator may stimulate radically different prey activity patterns in time or space, without necessarily influencing total prey abundance. The response need not be restricted to prey. If species B is a predator, interactions with species C may influence the speed at which it forages, or if species B is a plant, the arrangement of leaves that shade the area beneath the plant. The contribution of focal species A to the function f_i is determined by how well it can adapt to changes created by other species: the ability of species A to avoid encounters with a predator or to tolerate low light conditions. A third species can affect either the behavior of B or the response of A. While there are many possibilities, for the contributions in this volume, two are especially important. If the presence of the third species affects the behavior of species B, it is a *behavioral indirect effect*; and if the presence of the third species affects the nature of the interaction between A and B through chemically mediated actions, it is a *chemical response indirect effect*. These types of indirect effects of species may be either facilitative or detrimental to the focal species.

The three classes of indirect effects (trophic linkage, behavioral, and chemical response) also can be distinguished by different pathways and different mechanisms. Trophic linkage and behavioral indirect effects involve two direct interactions in series, e.g., an interaction pathway in which species C affects species B, which in turn affects species A. Chemical response indirect effects can involve series or two

direct species interactions in parallel pathways, e.g., both species C and species B affecting species A. The three types of indirect effects also can be distinguished by differences in the mechanism by which the third species affects a two-species system. Trophic linkage indirect effects require species C to change the total abundance of species B. Behavioral and chemically mediated indirect effects involve species C changing the distribution, behavior, morphology, or physiology—not absolute total abundance—of associated species. Interestingly, behavioral and chemical response indirect effects are mathematically very similar, both changing f_i in equation 3.1, even though they can involve very different pathways of interactions. To illustrate these different types of indirect effects, we present an example of each type.

Lubchenco (1978) has demonstrated that the structure of an intertidal algal community is strongly influenced by both competition and predation. The ephemeral alga, *Enteromorpha*, the competitive dominant in tide pools, prevents the growth of perennial species such as *Chondrus crispis*. However, *Enteromorpha* is the preferred food of the grazing snail *Littorina littorea* (fig. 3.1A). Where *Littorina* is present and grazes on *Enteromorpha*, the competitive subordinate *Chondrus* is able to thrive. This is an abundance indirect effect of *Littorina* on *Chondrus*, where *Littorina* facilitates the persistence of *Chondrus* by consuming the competitive dominant *Enteromorpha* (fig. 3.1A).

Werner et al. (1983) investigated the effect of predation risk on the habitat use and growth of bluegill sunfish (*Lepomis macrochirus*). For small individuals of *L. macrochirus*, feeding in the open water habitat was more profitable than in the vegetated areas of a pond, in large part because of an abundant population of the zooplankter *Daphnia pulex* in the open water. The presence of the predatory largemouth bass (*Micropterus salmoides*) caused the smaller bluegills to forage in vegetation rather than open water, apparently to avoid predation (fig. 3.1B). Werner et al. demonstrated that when no bass predators were present in an experimental pond, the bluegills eliminated all *D. pulex* within 10 days. When bass were present, *D. pulex* were able to persist for > 20 days. The bass were experimentally prevented from having a large effect on bluegill densities by replacing the bluegills eaten by the bass. Werner et al. estimated the number of bluegills eaten by the bass and replaced those individuals with fish from a stock pond. So the bass prolonged the persistence of *D. pulex* by affecting the habitat use, rather than the total abundance, of bluegill sunfish. This is a facilitative behavioral indirect effect of the bass on the zooplankter *D. pulex* (fig. 3.1B).

Smiley et al. (1985) have investigated a system that demonstrates both behavioral and chemical response indirect effects. The beetle *Chrysomela aenicollis* was found to feed on willows (especially *Salix orestera*), which contain salicin, a toxic phenol glycoside (fig. 3.1C). The beetles use salicin as a substrate for producing defensive secretions (predators include ants and a number of arthropods). Smiley et al. were able to demonstrate that the beetles preferentially fed on trees with high leaf levels of salicin and that beetles on trees with high salicin levels had higher survivorship than beetles on trees with low salicin levels. The trees have a behavioral indirect effect on the predators in that the trees enabled the beetles to produce more secretion, thus changing the defensive behavior of the beetle. However, the trees have a chemical indirect effect on the beetles in that the trees change the nature of the predator–prey interaction from susceptible to resistant prey (fig. 3.1C). Note that the class of indirect effect of the willow tree changes depending on which of the other two species is the species of interest.

Although the terminology introduced here is new, these divisions of indirect effects are not entirely new. Trophic linkage indirect effects encompass what many studies have previously called simply indirect effects. Many theoretical studies have suggested that species interactions may occur through changes in the total abundance of some intermediate species (e.g., Levine 1976; Holt 1977, 1984; Lawlor 1979; Schaffer 1981), and there is a great deal of experimental evidence for the importance of abundance indirect effects in natural communities (e.g., Paine 1966; Harper 1969; Paine and Vadas 1969; Dodson 1970; Lubchenco 1978; Lubchenco and Menge 1978; Davidson 1980; Davidson et al. 1984; Sih et al. 1985).

Behavioral and chemical response indirect effects encompass what many studies have previously called higher-order interactions. *Higher-order* is a term that has caused a great deal of confusion, and we suggest that it be redefined along clear lines or be dropped entirely. Several

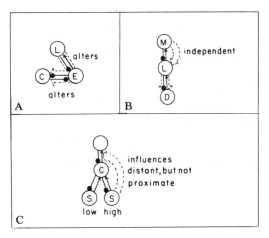

Figure 3.1. *Pathways of direct effects that lead to indirect effects in three experimental systems:* **A**, *the abundance indirect effect of the snail* Littorina littorea *on the alga* Chondrus crispis *investigated by Lubchenco (1978);* **B**, *the behavioral indirect effect of the bass* Micropterus salmoides *on the zooplankton* Daphnia pulex *(from Werner et al. 1983);* **C**, *the chemical response indirect effect of the willow* Salix orestera *on the herbivore* Chrysomela aenicollis *(from Smiley et al. 1985).*

studies have "tested for higher-order interactions" by testing for the significance of mathematically higher-order terms (any term more complex than aN_iN_j—e.g., $aN_iN_j^c$ or $aN_iN_jN_k$) in explaining the abundance or growth of a species and in doing so have confused a mathematical form with a biological process (e.g., Seifert and Seifert 1976, 1979; Rey and Stoner 1984). However, most of the previous studies have interpreted a higher-order interaction as occurring when a third species "changes the nature" (Abrams 1983) of the interaction between two others. This usually has been tested for by measuring the constancy of some coefficient of interaction, often a competition coefficient (e.g., Vandermeer 1969; Wilbur 1972; Neill 1974; Richmond et al. 1975; however, see Case and Bender 1981). The terminology proposed here recognizes that the value of the interaction coefficient can be influenced by both the effect of one species and the response of the other and that a third species can affect these two components separately (behavioral and chemical response indirect effects, respectively).

Behavioral and chemical response indirect effects have not been previously discussed as separate, distinct classes of indirect effects. However, recent work on predation risk has highlighted the importance of behavioral indirect effects. In many systems, the addition of predators seems to have a greater effect on prey behavior than on prey number (e.g., Sih 1982; Stamps 1983; Mittelbach 1984). Predator-induced behavioral changes often will lead to changes in interactions of the prey with other species (e.g., competition or predation; see Abrams 1984; Mittelbach and Chesson, this volume). Examples of chemical or response indirect effects can be found in a variety of different types of systems, including fish communities (Mittelbach 1984), marine invertebrate systems (Schmitt et al. 1983), plant–herbivore communities (Smiley et al. 1985), and plant communities (Miller 1985). We are certain that there are examples from other systems of which we are not aware.

The three types of indirect effects defined here may operate on very different time scales. Because trophic linkage indirect effects are mediated through changes in population size, they do not always need to be considered explicitly for us to understand the instantaneous or short-term dynamics of a system (Abrams 1984). Behavioral and chemical indirect effects can occur much more rapidly and may need to be incorporated directly into population dynamics models (see also Schaffer 1981; Bender et al. 1984).

It is also apparent that although the three types of indirect effects are defined by different interaction pathways and mechanisms, they may often act simultaneously. For example, in the bass–bluegill–*Daphnia* system (fig. 3.1B) investigated by Werner et al. (1983), the bass can potentially change both the behavior and the abundance of bluegills and so have both an abundance and a behavioral indirect effect on *Daphnia*. We don't believe that this lessens the value of the proposed terminology: even when different types of indirect effects are co-occurring, they operate on different time scales and through different mechanisms. Anyone attempting to understand or model the system would need to treat the different types of indirect effects separately.

This terminology of indirect effects allows a comparison of the role of complex species interactions in different systems, as well as the relative importance of different types of indirect effects in a single system. Although many studies have investigated specific types of indirect effects, it is telling that we can find only one paper that discusses both abundance indirect effects and "higher-order interactions" (behavioral and generalized response indirect effects) in any detail (Bender et al. 1984). It is hoped that a better organization will lead to a better understanding of both what we know about interactions between species effects and what we still need to know.

ACKNOWLEDGMENTS

The senior author thanks G. Mittelbach, M. Liebold, C. Osenberg, S. Gleeson, and E. Werner for comments. The junior author acknowledges NSF grants DEB80–04654 and 82–07007. We also thank E. Werner, J. Lubchenco, M. Liebold, C. Osenberg, P. Abrams, A. Winn, A. Sih, D. Wilson, and T. Case for reading earlier versions. This is W. K. Kellogg Biological Station contribution No. 565.

REFERENCES

Abrams, P. 1983. Arguments in favor of higher order interactions. *Amer. Naturalist* 121 : 887–91.

Abrams, P. 1984. Foraging time optimization and interactions in food webs. *Amer. Naturalist* 124 : 80–96.

Bender, E. A., T. J. Case, and M. E. Gilpin. 1984. Perturbation experiments in community ecology: theory and practice. *Ecology* 65 : 1–13.

Case, T. J., and E. A. Bender. 1981. Testing for higher order interactions. *Amer. Naturalist* 118 : 920–29.

Davidson, D. W. 1980. Some consequences of diffuse competition in a desert ant community. *Amer. Naturalist* 116 : 92–105.

Davidson, D. W., R. S. Inouye, and J. H. Brown. 1984. Granivory in a desert ecosystem: Experimental evidence for indirect facilitation of ants by rodents. *Ecology* 65 : 1780–86.

Dodson, S. I. 1970. Complementary feeding niches sustained by size-selective predation. *Limnol. Oceanogr.* 15 : 131–47.

Harper, J. L. 1969. The role of predation in vegetation diversity. *Brookhaven Symp. Biol.* 22 : 48–62.

Holt, R. D. 1977. Predation, apparent competition, and the structure of prey communities. *Theoretical Population Biology* 12 : 197–229.

Holt, R. D. 1984. Spatial heterogeneity, indirect interactions, and the coexistence of prey species. *Amer. Naturalist* 124 : 377–406.

Lawlor, L. R. 1979. Direct and indirect effects of n-species competition. *Oecologia* 43 : 355–64.

Levine, S. H. 1976. Competitive interactions in ecosystems. *Amer. Naturalist* 110 : 903–10.

Levins, R. 1975. Evolution in communities near equilibrium. In M. L. Cody and J. M. Diamond (eds.), *Ecology and evolution of communities*, pp. 16–50. Cambridge, MA: Harvard University Press.

Lubchenco, J. 1978. Plant species diversity in a marine rocky intertidal community: Importance of herbivore food preference and algal competitive abilities. *Amer. Naturalist* 112 : 23–39.

Lubchenco, J., and B. A. Menge. 1978. Community development and persistence in a low rocky intertidal zone. *Ecol. Monogr.* 48 : 67–94.

Miller, T. E. 1985. Competition and complex interactions among species: community structure in an early old-field plant community. Ph.D. diss., Michigan State University, Lansing, Mich.

Mittelbach, G. G. 1984. Predation and resource partitioning in two sunfishes (Centrarchidae). *Ecology* 65 : 499–513.

Neill, W. 1974. The community matrix and interdependence of the competition coefficients. *Amer. Naturalist* 108 : 399–408.

Paine, R. T. 1966. Food web complexity and species diversity. *Amer. Naturalist* 100 : 65–75.

Paine, R. T., and R. L. Vadas. 1969. The effects of grazing by sea urchins, *Stronglyocentrotus* spp., on benthic algal populations. *Limnol. Oceanogr.* 14 : 710–19.

Rey, J. R., and A. W. Stoner. 1984. Macroinvertebrate associations on the egg masses of the sea hare, *Aplysia brasiliana* Rang (Gastropoda: Opisthobranchia). *Estuaries* 7 : 158–64.

Richmond, R. C., M. E. Gilpin, S. P. Salas, and F. J. Ayala. 1975. A search for emergent competitive phenomena: The dynamics of multispecies *Drosophila* systems. *Ecology* 56 : 709–14.

Schaffer, W. M. 1981. Ecological abstractions: The consequences of reduced dimensionality in ecological models. *Ecol. Monogr.* 51 : 383–401.

Schmitt, R. J., C. W. Osenberg, and M. G. Bercovitch. 1983. Mechanisms and consequences of shell fouling in the kelp snail, *Norrisia Norrisi* (Sowerby) (Trochidae): Indirect effects of octopus drilling. *J. Exp. Mar. Biol. Ecol.* 69 : 267–81.

Seifert, R. P., and F. H. Seifert. 1976. A community matrix analysis of *Heliconia* insect communities. *Amer. Naturalist* 110 : 461–83.

Seifert, R. P., and F. H. Seifert. 1979. A *Heliconia* insect community in a Venezuelan cloud forest. *Ecology* 60 : 462–67.

Sih, A. 1982. Foraging strategies and the avoidance of predation by an aquatic insect, *Notonecta hoffmanni*. *Ecology* 63 : 786–96.

Sih, A., P. Crowley, M. McPeek, J. Petranka, and K. Strohmeier. 1985. Predation, competition, and prey communities: A review of field experiments. *Annu. Rev. Ecol. Syst.* 16 : 269–311.

Smiley, J. T., J. M. Horn, and N. E. Rank. 1985. Ecological effects of salicin at three trophic levels: New problems from old adaptations. *Science* 229 : 649–51.

Stamps, J. A. 1983. The relationship between ontogenetic habitat shifts, competition and predator avoidance in a juvenile lizard (*Anolis aeneus*). *Behav. Ecol. Sociobiol.* 12 : 19–33.

Vandermeer, J. H. 1969. The community structure of communities: an experimental approach with protozoa. *Ecology* 50 : 362–71.

Werner, E. E., J. F. Gilliam, D. J. Hall, and G. G. Mittelbach. 1983. An experimental test of the effects of predation risk on habitat use in fish. *Ecology* 64 : 1540–48.

Wilbur, H. M. 1972. Competition, predation, and the structure of the *Ambystoma-Rana sylvatica* community. *Ecology* 53 : 3–21.

4. Indirect Interactions between Species That Share a Predator: Varieties of Indirect Effects

Peter Abrams

A large number of types of indirect interactions may occur between two prey species that do not interact directly but that share a common predator. Changes in the population density of one prey species may cause changes in any of the following properties of the other prey: (1) the instantaneous population growth rate, (2) the equilibrium population size, (3) the existence and/or magnitude of population cycles, and (4) the evolution of traits related to predator avoidance or escape. All of these effects are discussed, and some are explored using mathematical models. In all cases, the qualitative nature of the indirect effects depends on the details of the population dynamics of the interacting species; an increase in the population size of one prey species may increase or decrease the population growth rate, equilibrium population size, or stability of the second prey. The results presented here differ from previous work in showing (1) prey that share a food-limited predator may increase each other's population size (contra Holt 1977), and (2) addition of a second prey usually makes it less likely that the first prey species will have stable population dynamics (contra Inouye 1980, and others). In predator-prey systems that have multiple stable equilibria, the qualitative nature of indirect effects between prey may differ at different equilibria; indirect effects in such systems may change in a discontinuous manner with a small change in the population density of one prey species. It is important that field ecologists who manipulate predator-prey systems be aware of the large range of possible indirect effects that may occur.

Species 1 may be said to have an indirect effect on species 2 under the following circumstances. A change in some property of species 1 causes a change in some property of species 3 or a set of other species 3, 4, 5, etc.). The change brought about in species 3 (4, 5, etc.) then causes a change in some property of species 2. The word *property* usually refers to per capita birth or death rates or population size, but the definition need not be restricted to these cases. The property need not be the same for the three species involved. More briefly, and less precisely, indirect effects may be said to occur when one species alters the effect that another species has on a third. An indirect effect thus involves a minimum of three species. I will refer to these as the donor, transmitter, and receiver, corresponding to species 1, 3, and 2, respectively, in the above definition.

In general, indirect effects have received less attention from ecologists than they deserve. However, one particular variety of indirect effect has a long history in the ecological literature. Exploitative competition involves effects that occur because of mutual depletion of a set of resources. Competition is a direct interaction only when it occurs via interference mechanisms such as allelopathy. Textbooks often refer to exploitative competition as a direct interaction, however, and types of indirect interactions other than exploitative competition seem to have been almost completely ignored by ecologists until the mid-1970s. At that time, a number of theoretical papers were published, all pointing out that the effect that one species had on the equilibrium density of a second species could depend on other species in the same food web. Levins (1975) and Levine (1976) developed this theme in a general way. Holt (1977) explored indirect effects between prey that share a common predator. Lawlor (1979) investigated indirect effects in guilds of competing species. Vandermeer (1980) analyzed the seeming mutualism that could arise between two consumers of competing prey species. Bender et al. (1984) emphasized the importance of studying indirect effects before interpreting population manipulation experiments. Recent field studies have confirmed the existence of indirect effects in several natural systems (Dethier and Duggins 1984; Davidson et al. 1984; Kerfoot and DeMott 1984; other chapters in this volume).

It is common for ecologists to speak loosely about "effects" of one species on another without specifying what the effects are. For example, several textbooks (e.g., Odum 1983, Pianka 1983) define interactions between species on the basis of "the sign of the effect that each species has on the other." Unfortunately, because of the many possible meanings of *effect*, this sort of definition is ambiguous. For example, the immediate effect on per capita population growth rate may differ in sign from the eventual effect on equilibrium population density. There also may be effects on the stability of population dynamics and on the existence of equilibrium points. Species may have an effect on the evolution of traits relevant to determining population dynamics in another species. It is unlikely that all of these effects are always qualitatively similar to each other or that they can always be meaningfully described by a + or − sign. The need to define *effect* applies to discussions of both direct and indirect effects. The need is probably more pressing in the case of indirect effects for two reasons:

1. Previous theory about indirect effects has concentrated almost exclusively on the effects that species have on each other's equilibrium population size. This is true of all of the articles discussed in the preceding paragraph. Abrams (1984) considers effects on per capita population growth rates. There seems to be a general lack of awareness that other types of effects are possible. Theory dealing with direct effects, on the other hand, has considered all of the types of effect listed above.

2. It is more common for the qualitative nature of different kinds of effects to differ when the effects are indirect. If an increase in one species directly causes a reduction in the instantaneous population growth rate of another, it will generally decrease that second species' equilibrium population size also. If the effects are indirect, however, it is more likely for a

species to have a negative effect on the instantaneous population growth rate and a positive effect on equilibrium population size (or vice versa). Examples of this point are provided below.

This article will present a (probably incomplete) catalogue of the varieties of indirect effects that may occur between two species that do not interact directly but do interact indirectly because they share a common predator. The reasons for choosing this particular system are: (1) it is impossible, for reasons of time, space and ability, to consider many systems; (2) this particular system has received considerable attention, largely because of Holt's excellent 1977 paper on the subject; and (3) this system exhibits a wide variety of possible indirect effects. Holt (1977) presented a very detailed examination of indirect effects of prey species on each other's equilibrium population densities, in which he argued that prey species that shared a predator would usually reduce each other's equilibrium population density, an effect that he labeled "apparent competition." Holt acknowledged that other effects on equilibrium density were possible if the predator's population density were determined by some factor other than food. This article will expand on Holt's theme by (1) examining a broader range of models of population dynamics, and (2) examining effects on the receiver other than changes in equilibrium population density. The goal of the analysis is to illustrate the diversity of indirect effects that are possible even in this very simple three-species system.

ABBREVIATED GUIDE TO THE CONTENTS

Indirect effects may be classified according to (1) what property of the donor species is changed, (2) what property of the transmitter species is changed, and (3) what property of the receiver is changed. In the one-predator-two-prey systems discussed here, each prey species functions as both donor and receiver, and the predator functions as the transmitter. The present analysis focuses on cases in which the donor changes its population size, and it examines the following properties of the receiver: (1) instantaneous per capita population growth rate, (2) equilibrium or average population size,

(3) stability of population dynamics, and (4) evolutionary equilibrium of ecologically important traits (e.g., predator avoidance). Each of these four types of effect is considered.

The range of biologically possible models illustrating any one of these four types of effects is essentially infinite. To keep this analysis to a readable length, I will restrict myself to differential equation models of monomorphic species. The large range of questions addressed here makes it impossible to examine any single model in great detail. Possible outcomes are emphasized rather than deriving exact necessary and sufficient conditions for those outcomes. A typical model has the following form:

$$dP/dt = P(a_1f_1(N_1,N_2) + a_2f_2(N_1,N_2) - D - I(P))$$

$$(4.1)$$

$$dN_1/dt = N_1g_1(N_1) - Pf_1(N_1,N_2)$$

$$dN_2/dt = N_2g_2(\dot{N}_2) - Pf_2(N_1,N_2),$$

where P, N_1, and N_2 are the population sizes of predator, first prey and second prey, respectively; g_i is the per capita growth rate of prey species i in the absence of the predator; f_i is the functional response of the predator on prey species i; a_i is the predator's efficiency of converting prey species i into new predators; D is a density-independent predator death rate, and $I(P)$ is a function describing the effect of predator numbers on its own per capita population growth for a given number of prey. The two prey species do not interact directly; if $P = 0$, dN_i/dt is independent of N_j.

Given such a model, some of the important questions that determine the nature of the four types of effect listed above are as follows:

1. What is the functional response of the predator while it is foraging?
2. Are the predator and/or prey capable of adaptation to each other, either behaviorally or evolutionarily? (Behavioral adaptation affects the functional response, so this overlaps with question 1.)
3. What determines the numerical response of the predator; i.e., is the equilibrium population size of the predator determined entirely by prey, entirely by some other factor, or by a combination of prey and another factor?

All of these questions are addressed to some extent within each of the four major sections that follow.

INDIRECT EFFECTS ON INSTANTANEOUS PER CAPITA POPULATION GROWTH RATE

The general model (4.1) shows that prey species i will only affect the instantaneous population growth rate of prey species j if the functional response of the predator on prey species i is a function of the density of prey j as well as prey i. If the functional response to prey density is linear, the number of prey j does not affect the rate at which prey i is eaten, and there is no indirect effect of one prey on the instantaneous growth rate of the other. Although theoretical works commonly assume that functional responses are linear, most measured functional responses are nonlinear (Hassell 1978), and it is physically impossible for functional responses to remain linear when prey density becomes sufficiently large. Nonlinear functional responses may have a variety of forms. Below are some possibilities, together with discussions of the indirect effects on population growth rate that result from them.

Holling Type 2

The Holling type 2 response (Holling 1965) assumes that prey species are consumed by a predator that (1) requires a finite handling time T_i to capture and consume prey species i, and (2) encounters prey in proportion to the prey's abundance while searching. This results in the following expression for the functional responses on each of two prey in the same habitat:

$$f_1 = C_1N_1/(1 + C_1T_1N_1 + C_2T_2N_2)$$

$$f_2 = C_2N_2/(1 + C_1T_1N_1 + C_2T_2N_2),$$

where C_i is the rate constant for capturing prey i while the predator is searching. It is clear that an increase in the population density of prey j will reduce the rate at which species i is consumed. Each prey therefore has a positive effect on the per capita population growth of the other. This phenomenon was noted by Holt (1977) and was discussed in Abrams (1983a). It should be noted that if the two prey types occur in different habitats and if the predator does not change its foraging effort in each habitat with changes in relative density of prey, the functional response on each prey type will be of the form $C_iN_i/(1 + C_iT_iN_i)$. In this case, the two prey will not affect each other's instantaneous per capita growth rate. This situation, however, seems rather unlikely; a well-adapted predator would adjust foraging effort between two habitats based on the returns from foraging in each. If satiation is a cause of the leveling off of the functional response, and there is frequent movement of the predator between habitats, then intake in one habitat will generally affect intake in a second habitat (Holt, *personal communication*).

Holling Type 3

Holling (1965) suggested that an S-shaped functional response could arise if a species had to learn how to capture prey successfully, and forgot when it did not encounter prey. Hassell (1978) suggests a number of other possible mechanisms; one possibility is that prey refuges from predation are in limited supply, so higher densities result in higher capture rates per prey. The most commonly used functional representation of a type 3 functional response curve for a single-prey type is $CN^2/(1 + CTN^2)$. If there are two functionally equivalent prey species, the functional response on prey i may be written as $C(N_i + N_j)N_i/(1 + CT(N_i + N_j)^2)$. More generally, the two prey species will not be completely equivalent with respect to the predator's functional response. Learning to catch an individual of species 1 may help the predator in its subsequent attempts to catch species 2 individuals, but it generally will not help as much as having learned on a species 2 individual. Alternatively, the prey may have only partial overlap in their use of refuges from predation. One might then represent the functional response on species i as

$$\frac{C_i(N_i + \beta_{ij}N_j)N_i}{1 + C_iT_i(N_i + \beta_{ij}N_j)N_i + C_jT_j(N_j + \beta_{ji}N_i)N_j}.$$

$$(4.2)$$

Here β_{ij} represents a conversion coefficient describing, for example, overlap in refuge use by the two prey species. Given this general form for the functional response, it is possible for a change in species j's population to have either a plus or a minus effect on the per capita growth rate of species i. The derivative of equation 4.2 with respect to N_j has the opposite sign of the effect of species j on the per capita growth rate of i. This derivative is positive provided

$$\beta_{ij} > T_jC_j\beta_{ij}N^2_j + 2T_jC_jN_iN_j + T_jC_j\beta_{ji}N^2_i.$$

$$(4.3)$$

Inequality (4.3) may or may not be satisfied. β values close to 1 (very similar prey species) and small prey population sizes make it more probable that equation 4.3 is satisfied and that larger numbers of prey j will therefore decrease prey i's population growth rate. It is also possible for larger numbers of prey j to decrease the per capita growth rate of i, even though larger numbers of i increase the growth rate of j's population. Thus, there may be $(++)$, $(--)$, or $(+-)$ indirect interactions between the two prey when the predator has a type 3 functional response. For the special case of equivalent prey, equation 4.3 reduces to the rule that each species will have a positive effect on the other's per capita population growth rate if $N_i + N_j > (CT)^{-\frac{1}{2}}$, and a negative effect if the inequality is reversed. At low prey densities, increased numbers of prey increase the predator's hunting efficiency, whereas at high prey densities, more prey decrease efficiency (defined as number caught per unit time divided by number available). The limiting case of $\beta_{ij} = \beta_{ji} = 0$ corresponds to the biological assumption that the density of any one prey does not alter the probability that a searching predator will catch a member of the other prey species. In this case, increased numbers of one prey simply decrease the search time available for exploiting the second; each prey has a positive effect on the per capita growth rate of the other.

There are many other ways of deriving a two-species version of a type 3 functional response that result in formulas other than equation 4.3. Most possibilities result in the same broad range of indirect effects.

Switching Functional Responses

Roughly speaking, switching functional responses occur when a predator eats proportionately more of the more abundant prey. Murdoch and Oaten (1975) discuss different ways of representing such responses mathematically. No explicit analysis will be given here. The range of possible switching responses is too broad to discuss within the limits of this paper. Switching may result from adaptive variation in the relative effort a predator devotes to capturing each of two prey types. It is clear that switching responses will generally result in two prey having positive effects on each other's population growth. Increasing numbers of one prey result in the predator increasing its relative capture rate for that particular prey. This would tend to result in increasing populations of each prey species having a positive effect on the population growth rate of the other. It is of course possible that the absolute level of the predator's consumption might change in such a way as to override the change in relative consumption levels that arose from switching.

Functional Responses
Incorporating Adaptive Behavior on the Part of Predator and/or Prey

With the exception of some switching responses, the functional responses discussed above do not incorporate any concept of adaptive behavior on the part of either predator or prey. In fact, many of the two species functional responses presented above are incompatible with notions of adaptive or optimal foraging for many possible values of parameters and variables. Much work has been done to determine what the adaptive behavior of individual predators should be (see reviews by Schoener 1971; Pyke et al. 1977; Pyke 1984), but very little of this work has been incorporated into predator–prey models. Abrams (1982, 1984) and Sih (1984) are exceptions to this generalization. The examples discussed below are some of the broader categories of adaptive behavior that can be incorporated into predator-prey models.

FORAGERS WITH ADAPTIVE PREY CHOICE ASSUMING FINITE HANDLING TIME. This is the traditional optimal diet choice problem (Pyke et al. 1977), which is usually analyzed using a type 2 functional response. The optimal behavior for the predator consists of ranking prey by energy content divided by handling time and eating the next lower ranked prey only if the top

ranked ones are sufficiently rare. If there are two prey, with species 1 having a higher E/T value than species 2, indirect effects between the prey on each other's per capita population growth rate are as follows. Increasing N_1 decreases the consumption rate of N_2, as in the strict type 2 functional response model, until a critical population density of species 1 is reached. At this density, species 2 is dropped from the diet entirely, and the per capita population growth rate of species 2 increases dramatically. Further increases in species 1 have no effect on species 2 because the latter is no longer preyed on. Thus, there is an abrupt shift from a (+ +) to a (0 0) interaction. Increasing the population density of the second-ranked species has a positive effect on the first provided that both are included in the diet; there is no drastic shift in the magnitude of the indirect effect. Combining a type 3 functional response with optimal diet choice may result in more complex patterns; e.g., if predator learning is involved, the ranks of the prey may change with changes in prey abundance.

ADAPTIVE VARIATION IN THE OVERALL FORAGING EFFORT OF THE PREDATOR. This is an issue I have discussed previously (Abrams 1982, 1984). Foraging is generally an activity that entails costs as well as benefits. A forager that maximizes its fitness should vary time spent foraging or some other measure of foraging effort as a function of total available prey. Foraging time or effort may either increase or decrease with prey density, depending on the shapes of cost and benefit curves. Increasing predator foraging effort with increasing prey numbers would tend to result in an increased population density of one prey having a negative effect on the other's per capita growth rate. Decreasing foraging effort with increasing prey numbers would generally result in the opposite effect. Abrams (1982) gives explicit formulas for functional responses incorporating variable foraging time in one-prey–one-predator systems; these may easily be modified to include two prey types. In any natural system, variation in foraging effort will occur together with factors such as finite handling time, learning by the predator, and the like. Indirect effects from these different factors may sometimes be opposite in sign, and there do not seem to be any general considerations for determining which effect will dominate.

ADAPTIVE VARIATION IN PREDATOR AVOID-
ANCE BY THE PREY. The functional response of
a predator depends on the behavior of its prey
as well as its own behavior. However, the effect
of adaptive behavior by the prey on predator
functional response has received very little attention (but see Abrams 1984). Prey behaviors
that decrease the chance of being captured by
a predator generally entail some cost. Often
prey are more vulnerable to predators when
the prey themselves are foraging (see Abrams
1984 for references), so avoiding predators
means decreasing food intake. Alternatively,
prey may expose themselves to other potential
mortality factors (e.g., harsh physical conditions) to avoid predators. If increasing the
numbers of a second prey species changes the
functional response of a predator on the first
prey species, this will alter the first prey's optimal tradeoff point for predator avoidance. In
most cases, the fact that the first prey adjusts
its behavior will not change the qualitative
nature of the indirect effect of the second prey
on the first. It will, however, generally reduce
the magnitude of negative effects and increase
the magnitude of positive effects. Consider, for
example, a predator with a type 2 functional
response. Addition of a second prey species
reduces the risk for the first prey species because the predator spends more time handling
prey, resulting in a positive indirect effect on
the per capita population growth rate. Consequent adaptive changes in the first prey's behavior relative to the predator will by definition increase its per capita growth rate further.
If the second prey species had a negative effect
on the first (e.g., type 3 predator response and
low prey density), adaptive behavior would by
definition reduce the magnitude of this effect.
For example, the adverse impact of increased
predator foraging might be partially offset if
the prey spent more time hiding in refuges.

The process of adaptation in one prey species
may itself cause indirect effects on the other
prey species. These are discussed further in
"Evolutionary Indirect Effects," below.

In summary, indirect effects of independent
prey species on each other's instantaneous per
capita population growth rates are determined
by the predator's functional response. The
foraging behavior of the transmitter species
determines the nature of the indirect interaction between the two prey, each of which functions as donor and recipient. The predator

functional response can have a large variety of
functional forms, and it may be affected by
adaptive behavior by both predator and prey.
There are consequently a variety of possible interactions; $(++)$, $(--)$, and $(+-)$ interactions
figured most prominently in the preceding
analysis; $(+0)$ and (-0) interactions are also
possible as limiting cases. For example, if one
species has an effectively zero handling time (if
capturing or consuming prey j did not decrease
the predator's ability to capture prey i), one
would observe a $(+0)$ interaction. Effects of
prey species on each other's per capita population growth rates can be very important in determining their population dynamics. If a
changing environment prevents the system
from reaching equilibrium, indirect effects
transmitted via the predator's functional response may be more important in determining
prey population dynamics than will the effects
via predator population density discussed in
the next section. Unfortunately, however, instantaneous population growth rates are difficult to measure in the field, and I know of no
examples in which any of the effects discussed
above have been demonstrated.

INDIRECT EFFECTS
ON EQUILIBRIUM OR
AVERAGE POPULATION SIZE

This topic has already been considered in some
detail by Holt (1977), and I will try to avoid
excessive repetition of results derived in that
work. The theoretically possible varieties of indirect effects include every possible combination of pairs of 0, +, and − signs. Holt,
however, argued that $(--)$ interactions
should be the only type observed if the predator were food-limited. Most of his analysis was
devoted to the case of food-limited predators
with linear functional responses. In such a system, the predator's per capita growth rate
would be given by an increasing function of
$C_1 N_1 + C_2 N_2$. The relationship between
equilibrium prey densities is defined by setting
this function equal to zero; it is easy to show
that the two prey densities are negatively related. Holt labeled this relationship "apparent
competition." Holt argued briefly that apparent competition would still be observed with
other types of functional response, but that
other types of interaction were possible for a
system in which the predator was not food-

limited. The remainder of this section will be subdivided according to the question of what limits the predator population.

Total Food Limitation

Holt (1977) is probably correct that "apparent competition" is the most likely indirect effect in this case. There are some exceptions which should be mentioned here.

DIFFERENT HABITATS. If the two prey species occur in different habitats and the predator population distributes itself optimally between the habitats, then changes in the density of one prey species will not change the equilibrium density of the second (Holt 1984).

ADAPTIVE PREY BEHAVIOR BALANCING RESOURCE EXPLOITATION AND PREDATOR AVOIDANCE. If the prey adjust their own behavior adaptively to balance conflicting demands of predator avoidance and resource exploitation, interactions other than $(--)$ may occur. The remainder of this section justifies this claim for one particular model.

Assume that the prey species are vulnerable to the predator only when the prey themselves are foraging and that prey adjust their own foraging time so as to maximize fitness. This leads to models of population interaction that have been described in Abrams (1984) for the case of one resource, one prey, and one predator species. An analogous system relevant to the present discussion consists of two prey, each exploiting a separate resource population and each being consumed by the same predator species. The analysis given below shows that there are conditions under which it is possible for a predator to increase the equilibrium density of its prey, by forcing the prey to exploit their own resources more prudently. In such a system, it is possible for an increase in the population size of a second prey species to indirectly force the first species to become more prudent in its resource exploitation, as the result of increased predator pressure. The end result is that greater numbers of the second prey cause an increase in the equilibrium density of the first prey. If the prey species have equivalent ecological parameters, each will increase the equilibrium density of the other. The following model supports this verbal analysis.

The general form of a version of the model with a single prey species is given by equations (1) in Abrams (1984). A particular realization of this general framework will be considered, so that quantitative results can be obtained. I will assume the following:

1. The resource obeys logistic population growth.
2. The consumer (prey) species has a linear functional response while foraging, and its birth rate is a linear function of its rate of resource consumption.
3. The prey can be caught by the predator only while the prey are foraging. The probability that a prey individual is caught is proportional to the square of the fraction of its available time spent foraging. (Available time is total time minus that required for mating and other essential activities.)

These assumptions result in the following model of population dynamics:

$$dR/dt = rR(1 - (R/K)) - C_0NRT_{op} \qquad (4.4)$$

$$dN/dt = b_1C_0RNT_{op} - C_1PNT_{op}^2 - D_1N$$

$$dP/dt = b_2C_1PNT_{op}^2 - D_2P.$$

Here r and K are the logistic growth parameters, the C_i are predation rate parameters, the b_i are conversion efficiencies of resource to prey or prey to predator, and the D_i are density-independent mortality rates. T_{op} is the optimal (evolutionarily stable) foraging time and is given by $T_{op} = b_1C_0R/(2C_1P)$, provided this quantity is less than 1; otherwise $T_{op} = 1$ (This expression is derived in Abrams 1984, p. 84). If there are no predators in the system, the prey will spend all of its available time foraging, and the equilibrium prey and resource densities (R) will be

$$R = D_1/(b_1C_0) \qquad (4.5a)$$

$$N = (r/C_0)(1 - (D_1/(b_1C_0K))). \qquad (4.5b)$$

It can be shown that the system given by equation 4.4 always has a single stable internal equilibrium point. (In determining stability T_{op} is replaced by the formula given above.) The equilibrium prey density in the system specified by equation 4.4 may be shown to be

$$N = \frac{r^2 K^2 b_1^2 b_2 C_0^2 C_1 D_2}{(b_1 C_0^2 D_2 K + 2 D_1 r C_1 b_2)^2}. \tag{4.6}$$

It is possible to show that there exist conditions under which the equilibrium prey density with a predator (eq. 4.6) will be greater than that without the predator (eq. 4.5b). A necessary condition for this situation to hold is that the right-hand side of equation 4.5a must be less than $K/2$. This condition implies that the prey is overexploiting its resource in the absence of the predator. Conditions for the predator to increase prey density will not be explored in detail here, however, because the goal of this analysis is to examine the effect of adding or increasing the population size of a second prey on the equilibrium population size of the first prey.

The equations describing the second resource and second prey species have parameters identical to those for the first. The predator equation becomes $dP/dt = b_2 C_1 P(N_a + N_b) T_{op}^2 - D_2 P$. The expression for the equilibrium density of either prey in the two-prey system is

$$N = \frac{2 r^2 K^2 b_1^2 b_2 C_0^2 C_1 D_2}{(b_1 C_0^2 D_2 K + 4 D_1 r C_1 b_2)^2}. \tag{4.7}$$

The condition for prey density in the two-prey system to be greater than that in the single-prey system is

$$D_2 b_1 C_0^2 K > \sqrt{2}\, (2 r C_1 D_1 b_2). \tag{4.8}$$

This condition is closely related to the formulas for equilibrium resource population size relative to $K/2$, which is the resource density that results in maximum resource productivity. It can be shown that in the single-prey system, $R < K/2$ if $D_2 b_1 C_0^2 K > 2 r C_1 D_1 b_2$. In the two-prey system, $R < K/2$ if $D_2 b_1 C_0^2 K > 4 r C_1 D_1 b_2$. Thus, the following cases are possible:

1. Overexploitation of resources by prey occurs in both one and two prey systems. Inequality (eq. 4.8) is always satisfied, and the prey have a $(++)$ interaction; an increase in the population density of either increases the equilibrium population density of the other.
2. Overexploitation of resources by prey occurs in neither one- nor two-prey systems. In this case, inequality (eq. 4.8) cannot be satisfied, and the prey will always have a $(--)$ interaction.
3. Overexploitation occurs in the single-prey system but not in the two-prey system. In this case, either a $(++)$ or a $(--)$ interaction may occur, depending on whether the parameter values satisfy inequality (4.8).

If a $(++)$ interaction does occur, the mechanism is as follows. Adding the second equivalent prey increases predator population size and thereby reduces the first prey's optimum foraging time from

$$T_{op} = \frac{2 D_1}{K C_0 b_1} + \frac{D_2 C_0}{b_2 C_1 r}$$

to

$$T_{op} = \frac{2 D_1}{K C_0 b_1} + \frac{D_2 C_0}{2 b_2 C_1 r}$$

When inequality (4.8) is satisfied, the resultant increased productivity of the resource more than compensates for the decreased foraging time and increased predator population that result from the addition of the second prey. The numbers of the two prey species could be reversed in this argument, showing that the interaction is $(++)$. It should be noted that there are several indirect effects in this example. The predator transmits effects of prey i on prey j and of prey i on the resource of prey j (where i and j are 1 or 2).

The above analysis shows that adaptive changes in the prey's foraging behavior in response to the predator can result in a $(++)$ indirect interaction between equilibrium densities of two prey species. The mechanism illustrated by this example may potentially operate whenever prey overexploit their own resources and when predator abundance affects the prey's resource exploitation. It does not depend on the specific functions used in this example. It does depend on the prey's resource being a reproducing entity; detritivores cannot overexploit their food supply.

There might be some objection to including this discussion in a section dealing with predators that are totally food-limited. When one substitutes for T_{op} in the dP/dt equation of equation 4.4, one obtains $dP/dt = P((b_2 b_1^2 C_0^2 R^2)/(4C_1 P^2 - D_2)$. The per capita rate of increase of the predator decreases as predator density goes up when food is held constant. This initially appears to be a case of predator self-limitation. However, the apparent self-limitation arises solely because of prey behavior. The number of prey eaten per predator is the only factor one need know to determine predator per capita population growth. These considerations make it appropriate to consider this as a case of pure food limitation.

A recent review of field studies of predation (Sih et al., 1985) has shown that the addition of a predator species often reduces their prey's population size by more than a factor of 2. Because the prey in the above model are also predators on their resource, this would suggest that resources are often reduced below half their carrying capacity. Thus, overexploitation may be a common phenomenon, unless resource growth departs significantly from the logistic model in the direction of greater competition at low population densities. Unfortunately, I am not aware of any field studies in which the mechanism described here has been shown to operate.

Predator Population Independent of Prey Population

The size of the predator population could be determined by some factor other than number of prey consumed: predation by a predator on a higher trophic level or a limited number of territories are two of the possibilities. Predator equilibrium population size may also be determined by a combination of food and a second factor; this is discussed in the following section.

If predator population density is independent of prey population size (above some minimum), the indirect effects between prey on each other's population size are determined by the predator's functional response. This principle was noted by Holt (1977). The analysis of the population level effects is therefore very similar to the analysis of indirect effects on population growth rate. If the predator has a linear (type 1) functional response, prey species will have no effect on each other's equilibrium population size. There will also be no indirect effects if the prey species occur in different habitats; this has the effect of making the functional responses on the two prey independent of each other.

Indirect effects in a system with type 2 functional responses are complicated by the fact that there can be more than one equilibrium point for certain types of prey growth functions. Noy Meir (1975) and May (1977) present graphic analyses of such systems. In any predator–prey system, the equilibrium prey density is determined by setting the predator functional response times the number of predators equal to the prey population growth rate. The equilibria may be illustrated by graphing each of these two quantities as a function of prey population size. If the functional response is type 2, and the prey has logistic growth modified so that there is a constant immigration rate (I) of prey into the system, this graphical analysis looks like figure 4.1. The actual prey population dynamics equation is

$$dN/dt = I + rN(1 - (N/K)) \qquad (4.9)$$

$$- CNP/(1 + CTN).$$

Each of the three intersection points in figure 4.1 represents an equilibrium, but the middle equilibrium is always unstable. By changing parameter values, one can modify the system so that there is only one equilibrium point; it may be either to the left or to the right of the hump in the prey growth curve. The effect of a second prey on the equilibrium density of the first may be examined as follows. In the presence of a second prey species, the predator's functional response on the first prey becomes $CN_1/(1 + C_1 T_1 N_1 + C_2 T_2 N_2)$. Thus, the functional response curve in figure 4.1 is lowered, but the population growth curve for the prey is unchanged. A large enough N_2 will eliminate the lower equilibrium point. Thus, if the one prey system is at its lower equilibrium point, gradually increasing the density of the second prey will gradually increase the equilibrium density of the first, until a critical density is reached, at which the lower equilibrium disappears and the density of the first species increases dramatically. Here the indirect effect is $(+ +)$, but the actual magnitude of the effect may have a discontinuity. It should be noted that it is necessary to have immigration or some constant-number type of refuge from predation if a model with a type 2 functional response is to have alternative stable equilibria.

If the predator has a type 3 functional response, however, refugia or immigration are not required for alternative equilibria. Ludwig et al. (1978) analyze such a system in which the population growth of the single prey is described by the logistic equation and the predator functional response is represented by an expression of the form $CN^2/(1 + CTN^2)$. Here I will examine the indirect interaction of two prey species in an analogous model. This is a special case of a two-competitor–one-predator model discussed briefly by Noy Meir (1981). The nature of the indirect effects on equilibrium density will be shown graphically. Figure 4.2 plots per capita population growth of a logistically growing prey as a function of prey density, and the predation rate per prey $(CPN/(1 + CTN^2))$, also as a function of prey density. These two quantities must be equal at an equilibrium point. The prey growth rate curve is not affected by the number of prey species. The addition of a second prey species changes the predation rate per prey curve for the first prey species. If the two prey are equivalent from the predator's viewpoint, the predation rate per prey for the first prey becomes $CP(N_1 + N_2)/(1 + CT(N_1 + N_2)^2)$. An analysis of this formula shows that its maximum occurs at lower N_1 than does the single prey curve; $(1/(CT)^{1/2}) - N_2$ as opposed to $1/(CT)^{1/2}$. However, the value of the predation rate curve at its maximum is given by $PC^{1/2}/2T^{1/2}$ independent of N_2. Qualitatively, then, the effect of increasing N_2 is to shift the predation rate curve to the right and to compress it. If the original system has two stable equilibria (as shown in fig. 4.2), and if the system is at the low prey density equilibrium point, then adding more individuals of a second species will have the following effect. Initially, as N_2 is increased from 0, the equilibrium point will move to the left, implying a smaller equilibrium density of species 1. At a critical number of N_2, the lower equilibrium point may disappear. It will disappear for some N_2 provided that $r > PC^{1/2}/2T^{1/2}$; this condition is satisfied in the system illustrated. N_1 will increase abruptly to the upper equilibrium at this critical point. Further increases in N_2 will then increase the equilibrium density of N_1. Because the prey are equivalent, the effect of 1 on 2 is the same as that of 2 on 1. There is a $(--)$ interaction at the lower equilibrium and a $(++)$

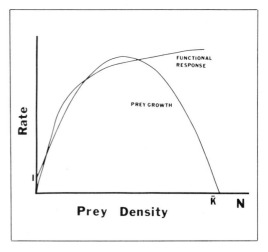

Figure 4.1. *Graphical analysis of equilibria in a predatory-prey system with (a) constant predator numbers, (b) type 2 predator functional response, (c) logistic prey growth with an immigration term. $\overline{K}$ is the equilibrium density in the absence of predation, and is given by $K(r + (r^2 + (4Ir/K))^{1/2}/2r$. I is the immigration rate. The curve labeled* functional response *is actually predator population density times the functional response.*

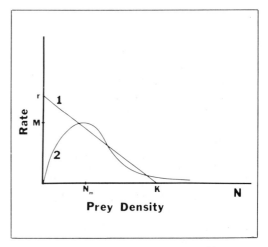

Figure 4.2. *Graphical analysis of equilibria in a predator-prey system with (a) constant predator numbers, (b) type 3 predator functional response, (c) logistic prey growth. The line labeled 1 gives the prey per capita growth rate (not including predation). The line labeled 2 is the predation rate per prey (predator numbers times functional response divided by prey numbers); r and K, parameters of the logistic growth curve; M, maximum predation rate per prey; N_m, prey density at which this maximum occurs.*

interaction at the upper equilibrium. Near the density at which the lower equilibrium point disappears, there may be a huge + effect of one species on the other. It is, of course, possible that the parameter values result in only the higher equilibrium being present in the single prey system; in that case there is only a (+ +) effect. It is also possible for both the one- and two-species systems to have only the lower equilibrium, resulting in a (− −) interaction. Finally, it is possible for the one prey system to have only the lower equilibrium, while the two prey system has two alternative equilibria. Which situation applies is determined by prey growth parameters, predator numbers, and predator functional response parameters. The above discussion has assumed that prey are equivalent and that both have logistic growth. Neither assumption is necessary for the existence of more than one equilibrium point. Allowing the more general type 3 response presented earlier (eq. 4.2) and considering prey with different growth functions may allow more complex patterns of indirect effects, but no further analysis will be presented here.

The discussion thus far has not considered indirect effects on population density when the functional responses of the predator are influenced by adaptive behavior of predator or prey. Most of the possible types of indirect effect in such situations can be understood from the analysis of indirect effects on population growth presented in the preceding section. A brief discussion of the types of adaptive variation considered in that section follows.

Adaptive Diet Choice with a Type 2 Functional Response. If prey i's per capita population growth rate is given by the nonincreasing function $g_i(N_i)$, then the equation determining equilibrium prey density is $g_i(N_i) = C_iP/(1 + C_iT_iN_i + C_jT_jN_j)$, assuming that both prey are included in the diet. It is easy to show that increasing N_j increases N_i. If N_j is the preferred prey, then at a certain density of N_j, N_i will be dropped from the diet, resulting in an abrupt increase in N_i to its carrying capacity and eliminating the indirect effect. Thus, the interaction can be either (+ +) or (00).

Predator Changes Foraging Effort Adaptively with Changes in Prey Density. If the predator has linear functional responses while foraging, adaptive variation in foraging effort can result in (+ +) or (− −) interactions between prey, depending on whether foraging effort decreases or increases (respectively) with increased total prey density. The combination of nonlinear functional responses while foraging and adaptive variation in foraging effort may result in functional responses with multiple changes in concavity (Abrams 1982). Such functional responses could result in essentially any type of indirect interaction between prey. There may be multiple equilibria in such a system also (Abrams 1983b), resulting in phenomena similar to those discussed under the type 3 functional response above.

Prey Changes Behavior Adaptively in Response to Changes in the Predator Functional Response. As noted above, in the section on adaptive variation in predator avoidance by prey, adaptive responses in the prey generally enhance positive indirect effects on the population growth rate and reduce the magnitude of negative indirect effects. The same applies to indirect effects on population density of prey when the predator population size is fixed. It is also possible for indirect interactions to result from changes in the prey's exploitation of their own resources when the predator functional response changes. Thus, phenomena analogous to those discussed above, in "Adaptive Prey Behavior Balancing Resource Exploitation and Predator Avoidance," may occur.

Predator Population Partially Determined by Prey Consumed and Partially Determined by a Second Factor

This scenario is intermediate between those discussed in the two preceding sections. It is also more likely to be true of any given natural system. Unfortunately, more general models are usually more difficult to analyze, and that rule holds in this case.

As pointed out by Holt (1977), the addition of a second source of predator density dependence (other than shortage of prey) does not qualitatively change the indirect effects on equilibrium population size in systems with linear functional responses; each competitor still has a negative effect on the per capita rate of increase of the other. Depending on the functional form of the density dependence, however, the quantitative effect may be greatly reduced.

The next simplest model is one in which the

predator has a type 2 functional response and the prey has logistic growth:

$$dN/dt = rN(1 - (N/K)) - CPN/(1 + CTN)$$

$$(4.10)$$

$$dP/dt = P(aCN/(1 + CTN) - D - IP).$$

This model assumes that the predator's per capita population growth rate decreases linearly with predator population size for a given per capita food intake, resulting in the -IP term. This could conceivably result from aggressive behavior increasing as population density increased, or from a limited supply of some other essential resource. This model exhibits a much wider range of possible dynamic behaviors than the analogous model without the interference (-IP) term where I is a measure of interference between predators. It has been analyzed by Bazykin (1974), who showed that all of the following were possible outcomes, depending on parameter values:

1. There is a single stable equilibrium point at relatively low prey densities.
2. There is an unstable equilibrium point at low prey density and a stable equilibrium point at high prey densities. Regardless of initial conditions, prey density reaches the higher equilibrium.
3. There are stable equilibria at a low and a high prey density. Initial conditions determine which point is reached.
4. There is a stable equilibrium at high prey density and an unstable point at low density. The low density point is surrounded by a stable limit cycle. Initial conditions determine the outcome.
5. There is a single stable (low prey) equilibrium with a small domain of attraction. This is surrounded by an unstable cycle that is in turn surrounded by a stable cycle. Depending on initial densities, the prey either reaches the equilibrium point or undergoes limit cycle oscillations.

It is beyond the scope of the present paper to provide a complete analysis of the statics or dynamics of the two-species version of this particular model. However, it is not difficult to show that this system can exhibit many of the same qualitative features as the type 3 func-

tional response, constant predator density system discussed earlier. In particular, changing the density of one prey can create or destroy equilibria, resulting in discontinuities in the indirect effects. Also, the sign of the effect that one species has on another can depend on the equilibrium point at which the prey population is located.

In a single-prey version (eq. 4.10), equilibrium prey density is determined by setting $dN/dt = dP/dt = 0$, and eliminating P from the resulting simultaneous equations. This results in the following cubic equation for prey density:

$$aCN/(1 + CTN) - D \qquad (4.11)$$

$$= (Ir/C)(1 + CTN)(1 - (N/K)).$$

To simplify the algebra, I will consider only the special case of $D \cong 0$. This means that predators can reproduce with very little food if the predator population is very low. Equation 4.11 can then be rewritten as

$$aCN/(1 + CTN)^2 = (Ir/C)(1 - (N/K)). \quad (4.12)$$

The two sides of equation (12) are very similar to the two functions shown in figure 4.2, and it is clear that two stable equilibria can arise in the same manner. The left-hand side of equation 4.12 has a maximum value of $a/4T$ at $N = 1/CT$. The right-hand side is a straight line with a negative slope. The left-hand side does not intersect the positive N axis. Two such curves can intersect one or three times.

If a second species with the same a, C, and T values is present at population density N_2, the right-hand side of equation 12 does not change, but the left hand side becomes

$$aC(N_1 + N_2)/(1 + CT(N_1 + N_2))^2. \qquad (4.13)$$

The maximum of equation 4.13 has the same value as the maximum of the one species version (left-hand side of eq. 4.12), but it occurs at $N_1 = (1/CT) - N_2$. Thus, the effect of adding a second species is to shift the curve to the right and compress it, just as in the model illustrated in figure 4.2. The varieties of indirect effects on equilibrium density are thus the same as for the type 3 functional response system with constant predator numbers. The population dynamics are more complex, however, because the lower prey density equilibrium point is not always stable. Allowing for type 3 functional responses, different prey growth functions, and adaptive behavior results in complicated models that will not be analyzed here.

EFFECTS ON STABILITY

Both predator–prey models and laboratory predator–prey systems are known for their propensities to exhibit cyclical population dynamics. If the addition of a second prey species changed a predator prey system from one with a stable equilibrium to one exhibiting limit cycles, this might be a much more dramatic effect than the indirect effects on per capita population growth rate or density at an equilibrium point. Limit cycles in predator–prey models of the variety considered here often result in fluctuations in population density over several orders of magnitude. Although the past decade witnessed the publication of dozens of stability analyses of predator–prey systems, very few have addressed the topic of how the addition of one or more prey species affect the stability of a predator–prey interaction. This section will begin by analyzing the local stability of equilibria in models of one predator and one prey and models of one predator and two equivalent prey species. The conclusion is that the addition of a second prey is more often than not a destabilizing force. I close with a review of previous work on this topic.

The model analyzed in detail by Holt (1977) assumed linear predator functional responses and logistic resource growth. The equilibrium point of this system is locally stable in both one- and multiple-prey versions. The models analyzed below are all simple systems in which one-predator–one-prey versions can exhibit either a stable equilibrium or a limit cycle.

Predator Has a Linear Functional Response; Prey Exhibits an Allee Effect

An Allee Effect occurs when a population's per capita growth rate increases with population size at relatively low population densities. A simple model for a predator–prey system in which the prey has an Allee Effect is the following:

$$dN/dt = N(r + \alpha N - \beta N^2) \tag{4.14}$$

$$dP/dt = P(aCN - D).$$

The criterion for a stable equilibrium in this system may be found by standard methods (e.g., May 1973), and it works out to be

$$D > \alpha aC/(2\beta). \tag{4.15}$$

If a second competitor with an identical population growth equation is added to the system, the stability criterion becomes

$$D > \alpha aC/\beta. \tag{4.16}$$

Clearly, the range of parameter values allowing stability is smaller in the two-prey system. The addition of a second equivalent species will result in population cycles for any value of D (the predator density-independent death rate) that lies between the right-hand sides of equations 4.15 and 4.16. Addition of a second equivalent prey cannot stabilize an unstable one-prey system. Even though the effect of adding a second prey species is to reduce the average population density of the first species, simulation results show that the maximum density that occurs during a cycle may be significantly greater than the equilibrium density in the absence of the second prey.

Predator Has a Type 2 Functional Response; Prey Have Logistic Growth

The one-prey version of the model is

$$dN/dt = rN(1 - (N/K)) - CNP/(1 + CTN) \tag{4.17}$$

$$dP/dt = P(aCN/(1 + CTN) - D).$$

The conditions for stability are given in numerous works:

$$D > (CKT - 1)a/(T + CKT^2) \tag{4.18}$$

The dynamics of the system following the addition of a second equivalent prey may be understood by considering the dynamics of total prey density. Denoting total prey density by N_t, one obtains

$$dN_t/dt = rN_t - \frac{r(N_1^2 + N_2^2)}{K} \qquad (4.19)$$

$$- CN_tP/(1 + CN_tT)$$

$$dP/dt = P(aCN_t/(1 + CN_tT) - D).$$

If the two-prey system is near its equilibrium point, N_1 is approximately equal to N_2, so $N_1^2 + N_2^2 \cong N_t^2/2$. Making this substitution in equation 4.19 results in a model that is identical to the original one-species system with a doubled carrying capacity. The criterion for stability in the two-prey case is therefore

$$D > a(2CKT - 1)/(T + 2CKT^2). \qquad (4.20)$$

Equation 4.20 is a more stringent criterion than equation 4.19. Thus, it is possible to choose parameter values resulting in a stable one-prey system but an unstable two-prey system. The destabilizing effects of adding a second prey species are analogous to the destabilizing effect of increasing prey-carrying capacity in a one-prey system, the well-known "paradox of enrichment" (Rosenzweig 1971).

If the two prey occur in separate habitats and if the predator does not adjust its use of the habitats depending on relative prey density, the functional responses on the two prey will be independent (see "Holling Type 2," above). A stability analysis for such a model again supports the conclusion that more prey destabilize the system. In this case, the exact stability criterion is

$$D > 2a(CKT - 1)/(T + CKT^2), \qquad (4.21)$$

which is clearly more stringent than equation 4.18.

One can analyze a system in which the predator has a type 3 functional response in a similar manner; again, addition of an equivalent prey species is analogous to doubling the carrying capacity. The addition of the second prey makes it less probable that the system will be stable.

The above discussion has assumed that the two prey are identical but independent.

Models with two prey having different population dynamics may exhibit a wider range of behavior. In particular, it is not difficult to construct systems in which (1) one of the two possible one-predator–one-prey systems is unstable and one is stable, and (2) the two-prey system is stable. This might be viewed as stabilization of the system as a result of adding a second prey. However, the results obtained above for two equivalent prey argue that the stabilization should be attributed to a specific feature of the prey. Addition of prey species per se is destabilizing.

It is worth noting that the impact of destabilization on the population biology of a prey species is quite often much greater in magnitude than effects on equilibrium or average population size. Consider the system given by equation 4.17 with a value of D close to but larger than the minimum required for stability. Adding a second prey species that differs from the first in being 20 times more difficult to capture ($C_2 = 0.05C_1$) changes the average population size of the first species by only a few percent, but it results in population cycles in which the minimum and maximum sizes of the first prey species differ by a factor of 2.3. These are results of simulations with a particular set of parameter values; although the amplitude of the limit cycles depends on the values of T, C_1, D, r, K, and a, the amplitude of the cycles generally increases rapidly with destabilizing changes in any of the parameter values (Gilpin 1975). It is not unusual for a single-prey system to be stable but for the analogous system with two similar prey species to exhibit population cycles whose maximum and minimum values differ by an order of magnitude.

The conclusion reached above contrasts with the findings of the few previous papers that have considered the question of how additional prey affect the stability of a predator–prey system. Inouye (1980) suggested that a predator might have to reduce its capture rate of each prey if it were to search for both simultaneously, and argued that this could stabilize the predator–prey interaction. Using the criteria for stability derived for model 4.17, for example, one can see that the capture rate, C, would have to be reduced by more than one-half to make the two-species system more stable. This is biologically unlikely, however, because predator individuals could then achieve higher fitness by ignoring one or the other type of prey. Freedman (1980) presents a brief analysis of a predator–prey system with two prey that occur in different habitats. He points out that such a system can have a stable equilibrium point even when one of the prey, if present alone, would have an unstable interaction. Although this is possible, as noted above, it says little about the general stabilizing or destabilizing effect of adding a second prey. Powell (1980) discussed a simulation model of a particular one-predator–three-prey community, and showed that one of the prey and the predator persisted over a broader range of parameter space in the three-prey system. He concluded that the additional prey stabilized the system. This is a very different definition of stability than is generally used, and it is not clear to what extent the results depend on the particular functions built into the model. Almost all of the interest in one-predator–several-prey models has been centered on comparing systems of competing prey with and without predators (e.g., Cramer and May 1972; Comins and Hassell 1976; Vance 1978; Hutson and Vickers 1983). Relatively little attention has been directed to noninteracting prey.

EVOLUTIONARY INDIRECT EFFECTS

Even when there are no indirect effects on instantaneous population growth rate or equilibrium population size on an ecological time scale, prey that share a predator may have indirect effects on each other through their direct effect on the predator's evolution. This type of indirect effect differs from those discussed above in that the property of the transmitter species that is changed is its genetic composition. However, the nature of the effects should be similar to those that occur when the predator changes its behavior adaptively. If the constraints of the genetic or behavioral systems do not interfere, both types of change in the predator will result in the same indirect effects of prey on each other's equilibrium population densities. A second category of evolutionary indirect effects consists of those cases in which the property of the donor species that changes is its genetic composition. If such a change has an effect on the predator's population density or an effect on the predator's traits related to prey capture, it will lead

to an indirect effect on the other prey species. For example, one prey species may evolve a better means of avoiding its predator. If this reduces the population density of the predator, it should thereby increase the equilibrium population density of the second prey species. The altered predator density also may create a selection pressure for the second prey to alter traits affecting predator escape. In general, the evolution of any prey trait affecting equilibrium population size or loss rate to the predator is likely to have indirect effects on both the population size and predation-related traits of other prey that share the predator. Holt (1977) contains additional discussion of evolutionary indirect effects.

CONCLUSIONS

Holt (1977) concluded that unless factors other than prey eaten limited the predator's population size, prey that shared a predator would cause decreases in each other's equilibrium population size. The analysis presented above suggests that this generalization was too broad; (+ +) interactions are also possible in this situation. The example presented here exhibited (+ +) effects because of indirect interactions involving the prey's resources as well as their predator. More significantly, a consideration of predators that are not strictly food-limited and a consideration of effects on quantities other than equilibrium population size suggests that indirect effects between prey may be characterized by essentially any combination of +, −, and 0 signs. In addition, some of the more important indirect effects between prey, changing the number of equilibrium states or their stability, cannot be characterized by plus or minus signs.

A number of the indirect effects that were discussed above lie completely within the realm of speculation. In my opinion, they represent plausible speculation, but there are admittedly no examples of natural systems exhibiting many of the possibilities I have discussed. I still believe that it is worthwhile to call attention to the possibilities. Theory plays a major role in determining which of the infinity of possible observations of nature are made or quantified. I hope that drawing attention to the possibilities will stimulate appropriate observations. This whole volume is evidence of the fact that many instances of indirect effects on equilibrium population density were "out there," waiting to be studied. It is certainly more than coincidence that the field studies reported here were preceded by a spate of theoretical articles on indirect effects in the late 1970s and early 1980s. Future work may result in other kinds of indirect effects receiving their fair share of attention.

A knowledge of possible types of interactions is also important in interpreting the results of field experiments. Bender et al. (1984) have recently emphasized the need to consider indirect effects in interpreting the results of population manipulation experiments. Their analysis considered only indirect effects on equilibrium population size. The present work illustrates the fact that many other types of indirect effects are possible, even in relatively simple three-species systems. Any field biologist who is interested in either predicting or interpreting the results of a population introduction or manipulation must be aware of the full range of indirect effects that may occur.

I do not want to close by suggesting that the ball is entirely in the field worker's court. Writing this article has made me aware of how little attention theorists have devoted to systems with numbers of species between two and many. These are the sorts of systems in which indirect effects might be expected to play an observable role. Unfortunately, many mathematical techniques that are available in two-dimensional systems are not applicable to those with three or more dimensions. As a result, it is only in the last few years that we have begun to understand the population dynamics that are possible in Volterra-type models of, for example, three competitors (Hallam et al. 1979; Coste et al. 1978). Such work is a prerequisite for understanding indirect effects in this system.

This article has treated the case of prey that do not interact directly. A more general analysis would have allowed for the possibility of competition between prey. Competition between prey has been shown to be an important determinant of indirect effects in at least two marine systems (Dethier and Duggins 1984; Dungan, *this volume*). Allowing for competition between prey species greatly increases the range of population dynamics that are possible in the system. Noy Meir (1981) has shown that multiple alternate equilibria may occur in such systems. As Vance (1978) and Gilpin (1979)

first noted, it is possible for chaotic population dynamics to occur in a Lotka Volterra–type model with one predator and two competing prey. It is only very recently that adequate conditions for the persistence of systems with one predator and two competing prey have been derived for the Volterra-type model (Hutson and Vickers 1983). In addition to having more complex population dynamics, systems with competing prey have a much wider range of possible indirect effects. Many of these arise because predation pressure can affect the magnitude of competitive interactions between the prey (Case 1982; Abrams 1986). There is clearly much theoretical work that needs to be done before we understand all of the indirect effects that are possible in this system. An even wider variety of indirect effects are possible in systems with more than two trophic levels; Abrams (1984) illustrates a few of the possibilities. It would not be surprising if we eventually found that indirect effects play at least as large a role as direct effects in determining the distribution and abundance of species.

ACKNOWLEDGMENTS

I would like to thank Don DeAngelis, Bob Holt, and Bob Sterner for their comments on an earlier draft of this paper.

REFERENCES

Abrams, P. A. 1982. Functional responses of optimal foragers. *Amer. Naturalist* 120 : 382–90.

Abrams, P. A. 1983a. Alternative forms for links in trophic webs. In D. L. DeAngelis, W. M. Post, and G. Sugihara (eds.), *Current trends in food web theory*, pp. 91–94. Technical Report 5983. Oak Ridge, Tenn.: Oak Ridge National Laboratory.

Abrams, P. A. 1983b. Life history strategies of optimal foragers. *Theoretical Population Biology* 24 : 22–38.

Abrams, P. A. 1984. Foraging time optimization and interactions in food webs. *Amer. Naturalist* 124 : 80–96.

Abrams, P. A. 1986. Character displacement and niche shift analyzed using consumer-resource models of competition. *Theoretical Population Biology* 29 : 107–60.

Bazykin, A. D. 1974. Volterra's system and the Michaelis-Menton equation. In V. A. Ratner (ed.), *Problems in mathematical genetics*. USSR Academy of Science. (In Russian.)

Bender, E. A., T. Case, and M. Gilpin. 1984. Perturbation experiments in community ecology:

Theory and practice. *Ecology* 65 : 1–13.

Case, T. J. 1982. Coevolution in resource-limited competition communities. *Theoretical Population Biology* 21 : 69–91.

Comins, H. N., and M. P. Hassell. 1976. Predation in multi-prey communities. *J. Theoretical Biology* 62 : 93–114.

Coste, J., J. Peyraud, P. Coullet, and A. Chenciner. 1978. About the theory of competing species. *Theoretical Population Biology* 14 : 165–84.

Cramer, N. F., and R. M. May. 1972. Interspecific competition, predation, and species diversity: a comment. *J. Theoretical Biology* 34 : 289–93.

Davidson, D. W., R. S. Inouye, and J. H. Brown. 1984. Granivory in a desert ecosystem: Experimental evidence for indirect facilitation of ants by rodents. *Ecology* 65 : 1780–86.

Dethier, M. N., and D. O. Duggins. 1984. An "indirect commensalism" between marine herbivores, and the importance of competitive hierarchies. *Amer. Naturalist* 124 : 205–19.

Freedman, H. I. 1980. *Deterministic mathematical models in population ecology*. New York: Marcel Dekker.

Gilpin, M. E. 1975. Group selection in predator-prey communities. Princeton, N.J.: Princeton University Press.

Gilpin, M. E. 1979. Spiral chaos in a predator-prey model. *Amer. Naturalist* 113 : 306–308.

Hallam, T. G., L. J. Svoboda, and T. C. Gard. 1979. Persistence and extinction in three species Lotka-Volterra competitive systems. *Mathematical Biosciences* 46 : 117–24.

Hassell, M. P. 1978. *The dynamics of arthropod predator-prey systems*. Princeton, N.J.: Princeton University Press.

Holling, C. S. 1965. The functional response of predators to prey density and its role in mimicry and population regulation. *Mem. Entomol. Soc. Can.* 45 : 1–60.

Holt, R. D. 1977. Predation, apparent competition, and the structure of prey communities. *Theoretical Population Biology* 12 : 197–229.

Holt, R. D. 1983. Optimal foraging and the form of the predator isocline. *Amer. Naturalist* 122 : 521–41.

Holt, R. D. 1984. Spatial heterogeneity, indirect effects, and the coexistence of prey species. *Amer. Naturalist* 124 : 377–406.

Hutson, V., and G. T. Vickers. 1983. A criterion for the permanent coexistence of species, with an application to a two-prey, one-predator system. *Mathematical Biosciences*. 63 : 253–69.

Inouye, R. S. 1980. Stabilization of a predator-prey equilibrium by the addition of a second "keystone" victim species. *Amer. Naturalist* 115 : 300–305.

Kerfoot, W. C., and W. R. DeMott. 1984. Food web dynamics: Dependent chains and vaulting. In D. G. Meyers and J. R. Strickler (eds.), *Trophic interactions within aquatic ecosystems*, pp. 347–82, AAAS Selected Symposium #85. Washington, D.C.: Westview Press.

Lawlor, L. R. 1979. Direct and indirect effects of n-species competition. *Oecologia* 43 : 355–64.

Levine, S. H. 1976. Competitive interactions in ecosystems. *Amer. Naturalist* 110 : 903–10.

Levins, R. 1975. Evolution in communities near equilibrium. In M. L. Cody and J. M. Diamond (eds.), *Ecology and evolution of communities*, pp. 16–50. Cambridge, Mass.: Belknap Press.

Ludwig, D., D. D. Jones, and C. S. Holling. 1978. Qualitative analysis of insect outbreak systems: spruce budworm and forest. *J. Animal Ecol.* 47 : 315–32.

May, R. M. 1973. Stability and complexity in model ecosystems. Princeton, N.J.: Princeton University Press.

May, R. M. 1977. Thresholds and breakpoints in ecosystems with a multiplicity of stable states. *Nature* 269 : 471–77.

Murdoch, W. W., and A. Oaten. 1975. Predation and population stability. *Advances in Ecological Research* 9 : 2–131.

Noy Meir, I. 1975. Stability of grazing systems: an application of predator-prey graphs. *J. Ecol.* 63 : 459–81.

Noy Meir, I. 1981. Theoretical dynamics of competitors under predation. *Oecologia* 50 : 277–84.

Odum, E. P. 1983. *Basic ecology*. Philadelphia: W. B. Saunders.

Pianka, E. R. 1983. *Evolutionary ecology*. 3rd ed. New York: Harper & Row.

Powell, R. A. 1980. Stability in a one-predator, 3-prey community. *Amer. Naturalist* 115 : 567–79.

Pyke, G. H. 1984. Optimal foraging theory; a critical review. *Annu. Rev. Ecol. Syst.* 15 : 523–575.

Pyke, G. H., H. R. Pulliam, and E. L. Charnov. 1977. Optimal foraging: A selective review of theory and tests. *Quart. Rev. Biol.* 52 : 137–54.

Rosenzweig, M. L. 1971. Paradox of enrichment: Destabilization of exploitation ecosystems in ecological time. *Science* 171 : 385–87.

Schoener, T. W. 1971. Theory of feeding strategies. *Annu. Rev. Ecol. Syst.* 2 : 369–404.

Sih, A. 1984. Optimal behavior and density-dependent predation. *Amer. Naturalist* 123 : 314–26.

Sih, A., P. Crowley, M. McPeek, J. Petranka, and K. Strohmeier. 1985. Predation, competition, and prey communities: A review of field experiments. *Annu. Rev. Ecol. Syst.* 16 : 269–311.

Vance, R. R. 1978. Predation and resource partitioning in one-predator, two prey model communities. *Amer. Naturalist* 112 : 797–813.

Vandermeer, J. H. 1980. Indirect mutualism: Variations on a theme by Steven Levine. *Amer. Naturalist* 116 : 441–48.

III. Food Web Dynamics

5. Cascading Effects and Indirect Pathways

W. Charles Kerfoot

In food webs, path lengths permit a clear distinction between indirect and direct effects. For example, for a species, N_i, direct interactions are characterized by the dual elements a_{ji}, a_{ij} of the interaction matrix A, whereas effects along paths of greater length (the "loops" of Levins 1975) describe indirect interactions. Thus in a complex planktonic community, scramble or exploitative competition is an indirect interaction transmitted through loops of at least two links.

Although the paths can be distinguished, the repercussions of indirect effects are still difficult to predict. The potential impact of perturbations up food webs (the so-called bottom-up effects) are known to be severely constrained by energetic considerations and greatly influenced by time lags. Corresponding impacts down food webs (the so-called cascading or top-down effects) are depletion waves subject to a variety of variables (branching pattern, ecological efficiencies, the potential impact of targeted prey on lower trophic levels). Impacts down food webs can produce cascading effects if interaction strengths are large, targeted species monopolize resources, and generation times shorten along routes.

To illustrate cascading and indirect (facilitation) effects in lake food webs, I discuss examples drawn from 7 years of enclosure experiments in Lake Mitchell, Vermont. Perturbations of fish density are shown to have potentially strong repercussions on herbivore structure, invertebrate predator densities, and energy flow patterns. The magnitude of the disturbance is dependent on two properties: (1) the ability of a single herbivore genus (*Daphnia*) to monopolize resources even when primary production is stimulated considerably beyond natural levels and (2) the inclusive nature of dietary overlaps between this genus and most co-occurring grazers. *Daphnia* depress co-occurring herbivores by lowering the absolute level of highly edible phytoplankton (principally flagellates), placing many of these subordinate grazers at marginal maintenance levels. Phytoplankton heterogeneity, especially average changes in quality associated with the balance between highly edible flagellates and "resistant" taxa (hard-coated greens, gelatinous greens, and blue-greens), appears important in stabilizing community interactions by dampening potential oscillations in *Daphnia*.

Perturbation experiments demonstrate that certain invertebrate predators (e.g., *Mesocyclops*) benefit from modest increases in fish density because effects along indirect pathways are greater than those along direct pathways. The balance between the two path strengths, however, is sensitive enough to fish density that the overall interaction between predators can easily shift from positive to negative at high fish density.

DIRECT AND INDIRECT EFFECTS

One step to organizing natural complexity utilizes dietary information to group species into food chains and webs. Although the resulting linkage arrangement neglects some important spatial and temporal dynamics, the dual components of a typical predator–prey interaction provide paths along which effects travel through the community. Given that a differential equation can describe the rate of change of the ith species abundance (N_i) as a function f_i of the abundance levels of all other species (N_1, N_2, N_3, . . . , N_n) in the community,

$$dN_1/dt = f_i (N_i, N_2, N_3, . . . , N_n), \qquad (5.1)$$

we can distinguish easily between direct and indirect effects. Assume that in the neighborhood of any equilibrium point the behavior of the system will depend on the properties of the interaction matrix

$$\mathbf{A} = \begin{bmatrix} a_{11}\,a_{21} \cdots\cdots a_{n1} \\ a_{12}\,a_{22} \cdot \\ \cdot \ddots \cdot \\ \cdot \ddots \cdot \\ a_{1j} \cdots\cdots\cdots a_{nn} \end{bmatrix}, \tag{5.2}$$

where the elements are the coefficients of the N_{ij} in the equations for dN_i/dt. For a completely specified food web (i.e., one where resources are explicitly identified), consumption of prey promotes the growth of the consumer; hence in fig. 5.1a, coefficients that describe bottom-up interactions are positive. In contrast, removal of prey by the predator depresses the growth of the prey population; hence, top-down coefficients are negative. The oriented paths along the links further permit the definition of interaction loops of different lengths. This kind of "loop analysis" allows a clear distinction between direct and indirect effects by specifying the paths through which the interactions travel (Levins 1975; Hutchinson 1979; Bender et al. 1984). In both the diagrams and matrix, we can recognize three kinds of loops: (1) loops that connect species to themselves (the elements a_{ii}, a_{jj}; self-loops, e.g., cannibalism in a simple consumer-resource web), (2) loops that connect species to other species through a single link (a_{ij}, a_{ji}), and loops that go beyond single links and return to the focal species (e.g., $a_{21}\,a_{42}\,a_{34}\,a_{13}$. For species N_i, direct effects are simply loops described by the elements a_{ij}, a_{ji} in the $\mathbf{A}$ matrix, whereas indirect interactions are the effects transmitted along loops of greater length.

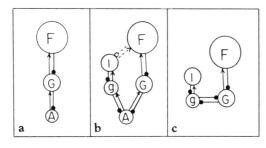

Figure 5.1. *Indirect interactions between grazers (3,g):* **(a)** *cascading,* **(b)** *completely specified,* **(c)** *as depicted in Lotka–Volterra (Gaussian) equations.*

Cascading Effects

Although the interaction paths allow a formal distinction between direct and indirect effects, the overall repercussions of indirect effects are still difficult to predict. The potential impact of perturbations that travel up food webs (the so-called bottom-up effects) are known to be severely constrained by energetic considerations and greatly influenced by time lags (May 1973). Corresponding impacts down food webs (the so-called cascading or top-down effects) (Paine 1980) are depletion waves subject to a variety of variables (branching pattern, ecological efficiencies, the ability of predators to limit key competitors). Impacts down food webs are likely to produce cascading effects if interaction strengths are large, targeted species monopolize resources, and generation times shorten along routes (Paine 1980). Cascading effects seem particularly important in intertidal and pelagic communities, principally because so many species are exposed to predators.

The chapters in this section (Food Web Dynamics) treat top-down influences of predators. Contributions by Adams and DeAngelis, Mills et al., and Kitchell and Carpenter examine the cascading effects of top predators down food webs, effects largely driven by oscillations in piscivorous fishes. The second set of contributions deals with actual experimental manipulations of communities, from lake and pond assemblages (Vanni, Threlkeld, and Morin) to intertidal species (Dungan). The first three touch on some of the indirect effects on primary producers, nutrient cycling, and benthic grazers. Chapter 13 (Dungan) considers an example of positive feedback (facilitation) brought about through indirect pathways.

In the remainder of this chapter, I will use my own investigations and those of my graduate students, in Lake Mitchell, to illustrate the nature of direct (cascading) and indirect (facilitative) effects in pelagic food webs. A companion chapter by Levitan attempts a formal stability analysis of the Lake Mitchell community.

Lake Mitchell: Pelagic Food Web

Consider the pelagic community described in figure 5.2, one composed of fish, invertebrate predators (copepods), a suite of subordinate grazers (rotifers, small microcrustaceans), and the herbivore's resources (algae, bacteria, detritus). This community is typical for many New England lakes undisturbed by eutrophication, and it characterizes the essential elements in Lake Mitchell, Vermont, site of our enclosure studies. Here the dual nature of consumer interactions is indicated by arrows (+) and closed circles (−), signifying the benefit delivered to the consumer and the simultaneous loss suffered by the resource.

As an example of opposing feedbacks transmitted through the food web, consider that under increasing fish predation, two alternative pathways influence the abundance of invertebrate predators (*Mesocyclops*): (1) a direct predator–prey link between fish and the invertebrate predator (interaction 6), and (2) an indirect pathway through algae and the suite of subordinate grazers (interactions 1–3, 5). The first is primarily suppressive, whereas the second is potentially facilitative. Note that this characterization goes beyond merely saying that *Mesocyclops* density is determined by a simple combination of food (resource) and predation because the immediate availability of high-quality resources (e.g., soft-bodied rotifers) is also indirectly related to top predator dynamics.

In Lake Mitchell, the net effect of increasing fish abundance on the planktonic community will depend on the balance between direct and indirect pathways, on the nature and strength of the interactions, and on how faithfully the simplified food web specifies the actual interactions (e.g., the crucial importance of the heterogeneous assemblage labeled algae). Our research has focused on two aspects of food web interactions: cascading effects and facilitative feedbacks between top predators.

BACKGROUND AND METHODS

Lake Mitchell is a small, shallow, and mesotrophic lake near Sharon, Vermont (11.2 ha; z_{max}, 5 m; chlorophyll *a* range, 1.1–15.5 μg/L; total alkalinity 1.9 meg/L). Basin morphology and basic physical and chemical data are given in DeMott and Kerfoot (1982). For the 1978–1980 manipulations, natural assemblages of phytoplankton were enclosed in two types of enclosures: 4-L glass jars and large rectangular plastic bags (12,000 L; 2 × 2 × 3 m deep; 0.2-mm-thick polyethylene). The 4-L glass jars received various combinations of grazers in pairwise competition experiments (*Daphnia pulicaria, D.*

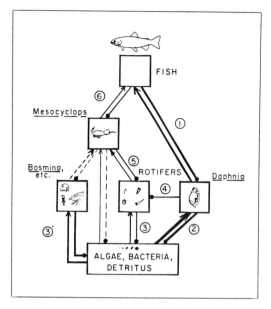

Figure 5.2. *Basic elements in the Lake Mitchell pelagic food web. Positive effects indicated by arrows, negative by closed circles; strengths of interactions crudely indicated by thickness of links. Numbers denote specific interactions noted in text.*

rosea, Bosmina longirostris). The jars were covered by Nitex netting and were suspended in wire baskets at 0.5- to 1.5-m depths for 2 to 4 weeks (Kerfoot and DeMott 1980; DeMott 1983). Treatments included responses to natural algal assemblages and to resources artificially simplified by the addition of an equal weight of *Chlamydomonas* or *Aerobacter*, or both supplements. The intent here was to guage the intensity of competition when grazers fed on the natural phytoplankton assemblage as opposed to when they shared a single resource, either a highly edible flagellate or a relatively large bacterium. Details of the resource supplement experiments are discussed in Kerfoot and DeMott (1980). Large enclosures were open at their tops, supported by wooden frames, and anchored in Lake Mitchell at a water depth of 0 to 3 m. Large-enclosure manipulations involved density manipulations of fish, *Daphnia*, and nutrients.

During placement, large enclosures were tipped on their sides, surrounding natural assemblages of plankton, then rotated into an upright position. No enclosures initially included fish. Treatments included addition of fish, netting and addition of *Daphnia*, and addition of nutrients. Sampling procedures are detailed in DeMott and Kerfoot (1982) and Levitan et al. (1985). To place the fish introductions in perspective, only three species of fish have been recorded in Lake Mitchell since 1977: rainbow trout (*Salmo gairdneri*), brook trout (*Salvelinus fontinalis*), and the northern creek chub (*Semotilus atromaculatus*). The usual stocking levels at Lake Mitchell are 36 to 268 rainbow trout per hectare and 170 to 268 brook trout per hectare, or approximately 40 to 100 kg/ha. Only rainbow trout and creek chubs were used in fish density experiments.

The netting manipulations included *Daphnia* addition and removal. *Daphnia* were added to two enclosures using a 280-μm-mesh (Nitex) plankton net. Additional tows came from the lake at a time (July 21, 1979) when *Daphnia* species comprised about 80% of the total number of cladocerans and 90% of the microcrustacean biomass. *Daphnia* were removed from two other enclosures, using the same plankton net. On the day following manipulations, additions showed the following average absolute densities: *D. rosea*, 74.2/L; *D. pulicaria*, 25.1/L; *Diaphanosoma*, 4.4/L; and *Bosmina*, 3.0/L. Removals lowered densities to the following levels: *D. rosea*, 1.7/L; *D. pulicaria*, 0.6/L; *Diaphanosoma*, 0.8/L; and *Bosmina* 1.3/L. Although the netting procedure lowered *Diaphanosoma* and *Bosmina* densities somewhat, their densities still fell within typical summer values.

Nutrient additions involved measured amounts of phosphates and nitrates. In the 1979 experiments, three of five enclosures were left as unenriched controls, while two were enriched with 25.0 μM NaNO$_3$-N and 1.6 μM KH$_2$PO$_4$-P (N : P = 16 : 1), roughly a fivefold increase over control and lake levels (total P was 0.2–0.5 μM during the summers of 1979 and 1980). Prior to enrichment, an additional enclosure had six creek chubs added to simulate the effects of high fish stock. In 1980, two of the eight enclosures were unenriched controls, two received a large nutrient spike with a low N : P ratio (125 μM N : 16 μM P), and four received a large spike with a high N : P ratio (250 μM N : 16 μM P). Two of the latter enclosures also received eight rainbow trout fingerlings apiece to simulate high fish stock. The introduced trout came from the hatchery as the fish stocked in Lake Mitchell. Additional experiments involving more continuous nutrient inputs were done in 1978 and are reported elsewhere (DeMott and Kerfoot 1982; Kerfoot and DeMott 1984).

Each year we recorded the effects of various treatments on the grazer, algal, and invertebrate predator communities. In addition, we regularly monitored temperature, oxygen concentration, transparency, soluble reactive phosphorus, total phosphorus, and chlorophyll a (see DeMott and Kerfoot 1982; DeMott 1983; Levitan et al. 1985). In retrospect, we were fortunate to have had a relatively simple zooplankton community composed of few microcrustaceans. During 3 years of nearly weekly sampling, 1977 to 1979, the following species reached maximum densities: *D. pulicaria* (30/L), *D. rosea* (30/L), *Bosmina longirostris* (20/L), and *Mesocyclops edax* (4 adults per liter; 25 copepodids per liter), *Diaphanosoma brachyurum* (7/L), and *Tropocyclops prasinus* (12 total copepodids per liter).

SEASONAL DYNAMICS: PHYTOPLANKTON AND ZOOPLANKTON

Over both seasons that phytoplankton species were monitored in detail, assemblages consisted of (1) early spring dominance of flagellates (mainly μ-flagellates, *Rhodomonas*, *Cryptomonas*, and *Chlamydomonas*), (2) summer dominance of gelatinous greens and other small protected species, (3) late spring blooms of *Dinobryon* with occasional fall reappearances, and (4) scattered spring and fall pulses of diatoms. The summer assemblages were especially rich in species that were accessible on a size basis to *Daphnia* and yet were very resistant to mastication and digestion. In many samples, these small gelatinous greens (*Asterococcus*, *Sphaerocystis*, *Planktosphaeria*, *Quadrigula*), small thick-walled greens (*Oocystis*, *Crucigenia*), and small blue-greens (*Chroococcus*, *Aphanocapsa*) accounted for 70 to 90% of the total cell count. Based on their ability to withstand gut passage, these genera were collectively termed the "resistant algae" assemblage (fig. 5.3). Some other genera, e.g., *Scenedesmus*, which are normally considered highly edible, also showed surprising resistance to grazing (see Horn 1981 for a similar circumstance in Saidenbach Reservoir, East Germany).

Zooplankton succession was also basically similar during 1978–1979. Each year *Bosmina* increased exponentially during early June and became the most abundant cladoceran during

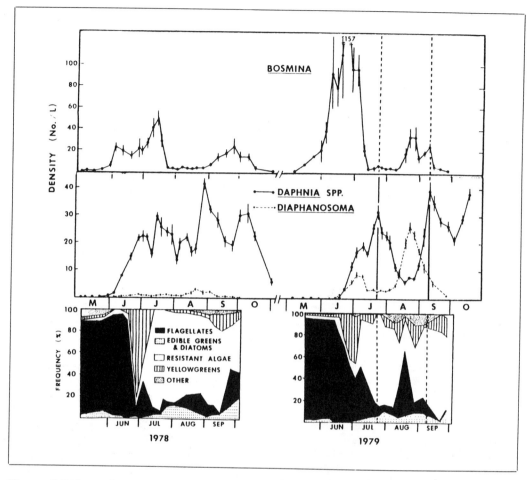

Figure 5.3. *Seasonal succession of cladocerans and algal groups in Lake Mitchell. Only total Daphnia densities indicated. Vertical lines indicate interval of* the flagellate period *Daphnia depression in 1979 and associated responses in flagellates and smaller-bodied cladocerans.*

the flagellate period (fig. 5.3). *Daphnia* and *Diaphanosoma* began to increase somewhat later than *Bosmina*, during late June. *Daphnia* increased steadily until it dominated the summer community, constituting 70 to 90% of the total microcrustacean biomass, whereas *Diaphanosoma* never achieved substantial densities except during the 1979 *Daphnia* depression. *Bosmina* declined dramatically to low densities during midsummer, then recovered somewhat during the 1978 fall resurgence in flagellates and during the 1979 *Daphnia* depression.

The 1979 *Daphnia* depression coincided with a rapid temperature fluctuation and subsequent reproductive failure of both *Daphnia* (*D. pulicaria* and *D. rosea*). See DeMott (1983) for a detailed treatment of this mortality event in relation to *Daphnia* dynamics. The subsequent events were interesting. Flagellates rose to abnormally high levels, both as percentages of total phytoplankton (fig. 5.3) and as absolute numbers (Kerfoot, DeMott, and DeAngelis, 1985). Both *Diaphanosoma* and *Bosmina* increased in abundance, then declined as *Daphnia* rebounded to typical late-summer densities. These fluctuations suggested evidence for interspecific interactions between the co-occurring grazers.

Demographic evidence for general food limitation during the summer was strong. At the beginning of the *Daphnia* increase in late May and June of both years, brood sizes started at

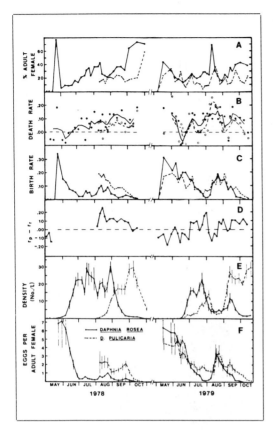

Figure 5.4. *Details of* Daphnia *dynamics in Lake Mitchell during 1978–1979, documenting seasonal succession of two* Daphnia *species (D. rosea, spring dominant; D. pulicaria, late summer and fall dominant) in terms of fecundity, density, age structure, instantaneous birth and death rates, and differences in specific growth rates (r_p = D. pulicaria growth rate; r_r = D. rosea growth rate). From DeMott (1983).*

seven eggs per adult female (fig. 5.4). Brood size then declined steadily throughout June, dropping below one egg per adult instar in July of both years after populations attained typical summer densities. By the middle of July, *Daphnia* fluctuated moderately between 10 and 40 organisms per liter. The approach of *Daphnia* to its summer density range coincided with the summer depression of flagellates and the onset of the "resistant" algal period. The very low levels of reproduction for all species, but especially for *Daphnia*, suggest that the resistant algae were relatively low in food value (Kerfoot and DeMott 1980; DeMott 1983). This was confirmed by enriching small enclosures with the flagellate *Chlamydomonas* (Kerfoot and De-

Mott 1980). Although total algal density ranged between 2 and 9 × 10^3 cells • ml⁻¹, and chlorophyll *a* remained relatively constant throughout the season (DeMott and Kerfoot 1982), the readily digestible fraction of the phytoplankton (i.e., the flagellates) often fell below 1 × 10^3 cells • ml⁻¹, and amounted to an even smaller fraction of the total biomass.

Within the genus *Daphnia*, there was a pronounced seasonal succession. *D. rosea* had an early advantage but was rapidly replaced by *D. pulicaria* from late summer to autumn (fig. 5.4). In both years, brood sizes of adult *D. rosea* declined to very low values by early to late July. During these intervals, many *D. rosea* carried no eggs nor embryos at all, an indication of severe food shortage. In contrast, *D. pulicaria* was able to sustain active reproduction during the summer months, often carrying one to three eggs per adult (fig. 5.4). In 1978, *D. pulicaria*'s population growth began in early August, with replacement occurring during September. In 1979, population growth began in June–July, was interrupted by the *Daphnia* depression in August, then continued with species replacement in September. During the *Daphnia* depression event in 1979, both species responded to lower total densities with a burst of reproductive activity (fig. 5.4).

If the seasonal shifts in grazer demographics are plotted as a function of total *Daphnia* density, some strong patterns are evident. Brood size shows a negative exponential decline in both *Daphnia* species, with *D. rosea* much more depressed than *D. pulicaria* (Kerfoot et al. 1985a, 1985b). Responses for *Diaphanosoma* brood size are also negative. The relationship of population growth (*r*) is much more linear, but is negative for both *Daphnia* and *Diaphanosoma* (DeMott and Kerfoot 1982; Kerfoot et al. 1985a, 1985b). The relationship between *Bosmina* and *Daphnia* is weak for summer fluctuations of *Daphnia*, showing very low correlations (DeMott and Kerfoot 1982; Kerfoot and DeMott 1984). The nature of the interspecific conflict between these two genera is evident only in perturbation experiments.

CASCADING EFFECTS
Planktivore-Zooplankton Repercussions

Increased densities of either rainbow trout or creek chubs beyond 100 kg/ha led to depres-

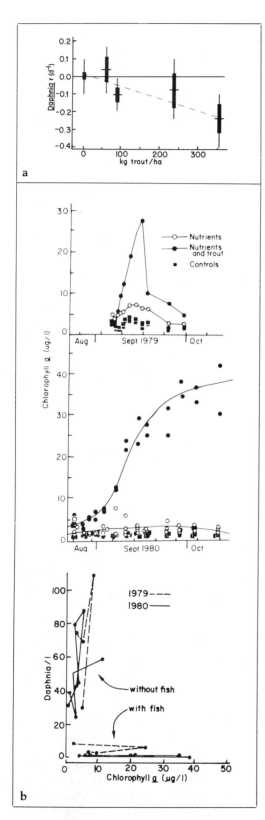

Figure 5.5. *Responses of* Daphnia *and algae to manipulations.* **(a)** *decline in* Daphnia *related to increasing densities of trout;* **(b)** *influence of nutrient additions in the absence and presence of high fish stocks (1979, six creek chubs; 1980, eight rainbow trout fingerlings; see Background and Methods for details). From Levitan et al. (1985).*

sion or elimination of *Daphnia* within large enclosures (Kerfoot and DeMott 1980; DeMott and Kerfoot 1982; Levitan et al. 1985). The susceptibility of large-bodied grazers such as *Daphnia* to fish predation is well established (Hrbàček 1962; Hrbàček and Novotná-Dvořáková 1965; Brooks and Dodson 1965; Galbraith 1967). Particulate feeding fishes differentially remove the larger, more conspicuous instars of *Daphnia* (O'Brien 1979). As these instars often contribute disproportionately to population birth rates (Hall et al. 1976, Jacobs 1978, Taylor 1980) and filtration rates (Paloheimo et al. 1982), the size bias of selectivity strikes at the very heart of this grazer's ability to monopolize algal resources, influencing both functional and numerical responses. If the difference between treatments and controls is calculated, *Daphnia* declined at a maximum per capita rate of -0.24/day. Given the poor summer reproduction of *Daphnia* on resistant algal assemblages, mortality from fish predation can have substantial effects (fig. 5.5a).

When fish manipulations depress *Daphnia* within enclosures, smaller-bodied grazers show elevated reproduction and increased population levels (fig. 5.6). Species responding include *Diaphanosoma brachyurum, Bosmina longirostris, Tropocyclops prasinus,* and various rotifers (Kerfoot and DeMott 1980, 1984; DeMott and Kerfoot 1982). *Diaphanosoma* is especially sensitive to *Daphnia* density.

Nutrient enrichment treatments with and without fish produce even more dramatic results (Levitan et al. 1985). When fish are absent from enclosures, *Daphnia* respond quickly to nutrient additions and subsequent increases in primary productivity, achieving densities near or over 100 per liter within 10 to 14 days, a 3- to 10-fold increase above typical lake densities (fig. 5.5b). The increased densities of *Daphnia* control the population growth of algae, funneling primary productivity to higher trophic levels. In contrast, where introduced fish stocks suppressed *Daphnia*, nutrients stimulated ele-

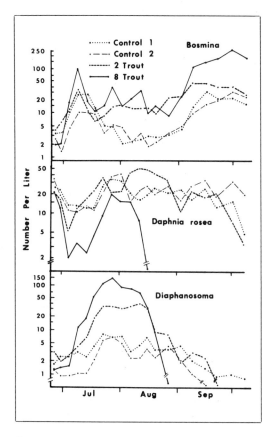

Figure 5.6. *Responses of small-bodied grazers to fish-mediated declines in* Daphnia. *Data from 1978 manipulations (from DeMott and Kerfoot 1982). Note log scale.*

vated phytoplankton standing crops, as measured by chlorophyll *a* concentrations. A plot of chlorophyll *a* concentration versus *Daphnia* density (fig. 5.5) illustrates the sharp dichotomy between the two responses.

A few simple calculations help explain the effectiveness of *Daphnia* grazing pressure when populations of the grazer respond quickly to additions. Field estimates of clearance rates, run at the time of the enclosure experiments, showed average filtration rates of approximately 0.85 ml per *Daphnia* per hour. This is certainly a conservative estimate of potential filtering rate, perhaps reduced by the low quality of ambient food. Other investigators report much higher filtering rates for similar-size *Daphnia*. For example, Lampert (1981) gives 1.6 ml per *Daphnia* per hour for a 1.4-mm *D. pulicaria*. Assuming the 0.85 conservative esti-

mate, at a density of 50 animals per liter the *Daphnia* population within enclosures would filter slightly more than the equivalent of the entire water column once every day (1.02/day); whereas at the maximum density of 100 per liter, that amount would increase to more than twice per day (2.04 at 100 *Daphnia* per liter. In order to withstand this grazing rate, algae would have to achieve a doubling time of 0.69 days at 50 *Daphnia* per liter and 0.35 days at 100 *Daphnia* per liter. These mortality rates approach or exceed the maximum intrinsic reproductive capacity of most algal groups (Lehman et al. 1975).

In simple fish density manipulations, i.e., those without nutrient additions, algal populations showed modest changes in biomass and species composition (DeMott and Kerfoot 1982; Kerfoot and DeMott 1984). Shifts involved a two- to threefold increase in biomass and an increase in flagellates, small diatoms, and some green algae. However, interpretations were complicated by the fact that introduced fish were excreting nutrients into enclosures. For this reason, we conducted two types of experiments. In the first, trout were fed *Daphnia* in aquaria, then the filtered aquarium water was added to some enclosures (Levitan 1986). In the second series, *Daphnia* were suppressed by a procedure (netting) that did not involve nutrient release. The second series of experiments is discussed here briefly.

Zooplankton–Algal Interactions

To investigate the influence of *Daphnia* on co-occuring grazers and algae independent of fish presence, a pulse perturbation was performed. *Daphnia* were selectively enriched in two enclosures, depressed in two others, and left undisturbed in the remaining two (controls). The manipulations were begun before the onset of the midsummer *Daphnia* depression on July 21, 1979. On the day following manipulations, total *Daphnia* were at 99.3 organisms per liter in additions, 2.3 per liter in removals, and 29.3 to 37.6 per liter in controls.

Of all species, *D. rosea* appeared the most sensitive to treatments. In additions, the elevated *Daphnia* densities had the overall effect of accelerating the seasonal replacement that was well underway. At the onset of the manipulation and throughout the duration of the experiment, *D. rosea* had much less reproduction

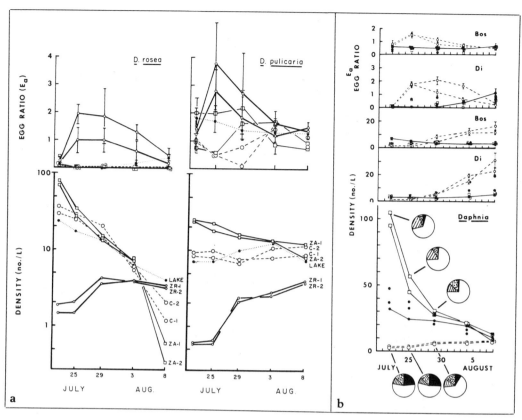

Figure 5.7. *General cladoceran and phytoplankton responses during netting experiment.* **(a)** *Adult egg ratios and densities for D. rosea and D. pulicaria during pulse manipulations (closed circles, lake; open circles, controls; open squares, zooplankton addition; open circles connected by thick solid line, zooplankton removal). In egg ratio estimates, 95% confidence limits given for lake and removal replicates. Notice log scale in bottom panel.* **(b)** *Top four panels give adult egg ratios and densities for* Bosmina *and* Diaphanosoma *(connected closed circles, lake; unconnected closed circles, controls; open squares, zooplankton addition; open circles, zooplankton removal). 95% confidence intervals set around adult egg ratios, ±1 SE given about density estimates. Total* Daphnia *densities shown in bottom panel along with pie diagrams for algal relative abundances (clockwise: solid, flagellates; unshaded, gelatinous algae; crosshatched, Oocystis; stipled, other).*

in controls and lake populations than *D. pulicaria*. Virtually no female *D. rosea* carried eggs in either controls or addition treatments. Addition treatment accelerated population declines, increasing mortality. In contrast, removals stimulated the fecundity of *D. rosea* and allowed nearly stable populations during the period of lake and control decline.

The seasonally competitive dominant, *D. pulicaria*, showed its superiority. Reproductive output remained moderately high even in addition treatments (fig. 5.7a). Densities declined in addition treatments but remained fairly stable in controls and the lake. Removal treatments caused a modest increase in fecundity,

which allowed *D. pulicaria* to increase rapidly throughout the experimental interval.

Reduction of *Daphnia* led to immediate increases in highly edible flagellates (fig. 5.7b). Subsequently, both *Diaphanosoma* and *Bosmina* showed an increase in brood sizes and population growth. *Diaphanosoma* reached a maximum density of 22 to 31 organisms per liter, whereas *Bosmina* increased to 12 to 16 organisms per liter. Additions of *Daphnia* had less effect because edible resource levels were so low initially.

Thus, the responses to perturbations conclusively verified the importance of strong density-dependent interactions between *D. rosea* and *D. pulicaria* during seasonal succession. Moreover, the experiments strongly confirmed the late-summer competitive superiority of *D. pulicaria*. The removal experiments led to a sequence of changes that anticipated those during the natural *Daphnia* depression in mid-August. Release of highly edible flagellates promoted reproduction and increased biomass in previously suppressed small-bodied grazers.

Summary of Grazer Competition Experiments

Direct tests of competition within suspended jars and introductions in large-scale enclosures have shown the following:

1. Strong competitive interactions occur between congeneric *Daphnia* (*D. rosea, D. pulicaria, D. galeata*). The two most abundant species, *D. rosea* and *D. pulicaria*, will seasonally replace each other in the natural environment or in large-scale enclosures, given that adequate seed populations are present (DeMott 1983). A normally replaced species can persist at high densities if the competitive dominant is removed from the water column or not seeded into enclosures at midyear. When *D. pulicaria* replaces *D. rosea* during late summer, the filtering and assimilation rates of the latter are depressed relative to the former species (DeMott 1983; Kerfoot et al., 1985). If, however, *D. rosea* are removed from the lake and placed in a suspension of highly edible algae (e.g., *Chlamydomonas*), their filtration and assimilation rates rebound to healthy levels (DeMott 1983).

2. Moderately strong competition occurs between *Daphnia* and *Diaphanosoma*. *Diaphanosoma* show demographic fluctuations strongly correlated with the density of *Daphnia* (DeMott and Kerfoot 1982, Kerfoot et al. 1985). Dual-label experiments and observations of ingested algae, however, suggest that *Diaphanosoma* prefer slightly smaller particles than do *Daphnia*, although they are also generalized filter feeders (DeMott and Kerfoot 1982; DeMott 1985, Kerfoot et al. 1985b).

3. Competition occurs between *Daphnia* and *Bosmina*, but the interaction is complicated by *Bosmina*'s radically different feeding behavior and penchant for flagellates. In Lake Mitchell, demographic or numerical fluctuations in *Bosmina* populations are poorly correlated with *Daphnia* oscillations (DeMott and Kerfoot 1982; Kerfoot and DeMott 1984). Yet removal of *Daphnia* stimulates *Bosmina*, presumably through release of flagellates. *Bosmina*'s dual feeding mechanism and ability to selectively graze flagellates when the latter are at low densities seems to aid persistence in *Daphnia*-dominated waters (DeMott and Kerfoot 1982; DeMott 1982, 1985; Bogden and Gilbert 1982, 1985).

4. *Daphnia* primarily depress rotifers by lowering resources below sustaining levels, although there are some interference effects. For example, the food level at which populations of *Brachionus calyciflorus* just maintain themselves is about 0.2 to 1 mg carbon per liter, while comparable values for *Daphnia* are about 0.02 to 0.04 mg carbon per liter (Lampert 1977; Porter et al. 1983). Both field and laboratory experiments document severe depression of rotifers when *Daphnia* are abundant (Kerfoot and DeMott 1984; Neill 1984; Gilbert 1985). *Daphnia*'s ability to injure rotifers, by buffeting individuals drawn into the brachial chamber, however, is a new discovery (Gilbert and Stemberger 1985) (fig. 5.2, interaction 4).

Dietary Observations

Because the summer algal assemblages were so dominated by "resistant" species, ingested algae could be quantified in grazer guts (Kerfoot et al. 1985b; Kerfoot, unpublished data). Matching of ingested algae with natural assemblages revealed interesting differences between cladoceran genera. In general, *Daphnia* drew their resources from a much broader variety of algae, whereas the diets of *Diaphanosoma* and *Bosmina* were considerably more restricted, largely on a size basis. All grazers showed relatively poor correspondence to available algae, the best match being *Daphnia* at correlations of 0.4 to 0.6 (Kerfoot et al. 1985b). These differences were partly attributable to the ingestion of moderately scarce, yet large-bodied, phytoplankton species (*Dinobryon, Pediastrum, Cosmarium, Coelastrum, Mougeotia, Aphanothece, Aphanocapsa, Synedra*) by *Daphnia*, species relatively uncommon in the guts of smaller-bodied grazers. Moreover, some rather abundant moderate-size algae were reasonably common in the guts of *Daphnia* (e.g., *Sphaerocystis*,

Planktosphaeria, Chroococcus, Crucigenia, and *Quadrigula*) but only occasional in the guts of *Diaphanosoma* and *Bosmina*. A plot of electivities (fig. 5.8) emphasizes the nature of size preferences among three cladoceran genera. Electivities derived from inert particles (glass beads) are superimposed on the algal values. Both of these data indicate that *Bosmina* and *Diaphanosoma* have difficulty with compact items larger than 15 μm, whereas *Daphnia* show positive electivities over a broad range of larger-size algal species. Thus, the broad, opportunistic feeding range of *Daphnia* explains in part this genera's ability to monopolize algal resources in Lake Mitchell. The correspondingly restricted feeding ranges of smaller-bodied grazers also provide a major clue to their inability to control total phytoplankton abundance.

Dietary similarities were strongest among different species of *Daphnia*. Dietary matching between *D. rosea* and *D. pulicaria* gave high correlations (r, .92–.97) (Kerfoot and DeMott 1984; Kerfoot et al. 1985b) irrespective of time, site, or treatment. Yet gut analyses revealed markedly fewer ingested cells in the guts of *D. rosea* than in those of *D. pulicaria* during summer (Kerfoot, 1985b).

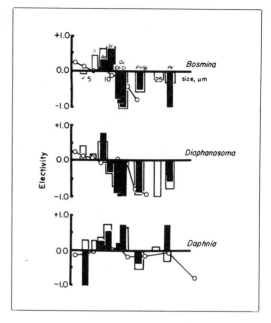

Figure 5.8. *Electivities for various algae compared to values from glass bead experiments. (O, Oocystis; As, Asterococcus; Sc, Scenedesmus; Ch, Chroococcus; Qu, Quadrigula; Cr, Crucigenia; Pl + Sp, Planktosphaeria and Sphaerocystis; Pe, Pediastrum.) Solid bars based on "window" technique; open bars indicate values from the more accurate "squash" technique (see Kerfoot et al., 1985). Circles and connecting lines trace electivities from the glass bead trials.*

FACILITATION: POSITIVE FEEDBACKS

In general, for many large-bodied invertebrate predators (e.g., *Mysis, Chaoborus*), fish predation is usually suppressive, diminishing densities and selecting for transparency or diurnal vertical migrations. However, in the case of intermediate-size species, it might actually be facilitative by indirectly increasing the food supply through modification of potential competitors (Kerfoot and DeMott 1984). This form of facilitation, originally discussed by Levine (1976) and Vandemeer (1980), relies on the net positive interaction along links 1–3 and 5 in figure 5.2.

When fish depressed *Daphnia* in enclosures, the increase in small-bodied grazers and algae also coincided with a longer-term increase in *Mesocyclops* (Kerfoot and DeMott 1984). The exact combination of factors (i.e., whether increased phytoplankton for nauplii or increased prey for advanced copepodids and adults) that vaulted *Mesocyclops* into prominence deserves careful attention. Detailed data on adult body and brood sizes showed that moderate levels of

fish predation (two rainbow trout fingerlings per enclosure) increased adult size and fecundity in the second generation but failed to stimulate density responses during 1979. Presumably, these would have followed in the third generation if monitoring had continued in the enclosure during the subsequent year. In contrast, higher levels of fish predation (eight fingerlings per enclosure) resulted in greatly increased adult body and brood sizes in the first generation, which then led to a major numerical response in the second generation (fig. 5.9). Addition of nutrients alone, which increased primary production, produced only minor responses in adult length and brood sizes, hardly above control levels (Kerfoot and DeMott 1984). These results suggest that the increased availability of small, fragile prey (rotifers, *Tropocyclops* nauplii, juvenile small-bodied cladocerans) is largely responsible for the immediate *Mesocyclops* demographic response, rather than

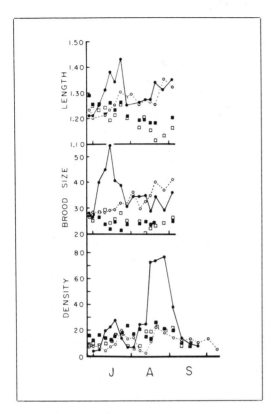

Figure 5.9. *Response of* Mesocyclops *during the 1978 trout treatments. Variables include total length and brood size of adult females; densities are for late instars (total copepodids + adults). See Kerfoot and DeMott (1984) for confidence limits around estimates. Symbols for inset:* ○, *addition of 2 trout fingerlings;* ●, *addition of 8 trout fingerlings;* □, *control (no trout);* ■, *lake populations.*

just relief from a juvenile "bottleneck" (Neill and Peacock 1980). The small prey provided a larger forage base for the omnivore.

The relationship between fish and *Mesocyclops* would not appear strongly reciprocal but does appear to be an interesting example of positive feedbacks. Dietary items in the stomachs of adult *Mesocyclops* verified the independence of the two top consumers (fig. 5.2, interactions 5 and 1). Fish were primarily consuming *Daphnia* and initially took no *Mesocyclops*. *Mesocyclops*, on the other hand, were consuming dinoflagellates (*Peridinium*), rotifers (trophi of *Polyarthra, Keratella, Synchaeta,* and *Conochilus*), and small cladocerans (Williamson 1981). Laboratory studies of *Mesocyclops* feed-

ing preferences further corroborated these findings (Gilbert and Williamson 1978; Williamson and Gilbert 1980; Williamson 1981). *Polyarthra*, small *Asplanchna, Brachionus,* and *Keratella* are all vulnerable to attack and ingestion by advanced *Mesocyclops* instars. No evidence for *Daphnia* in the guts of *Mesocyclops* was found, and laboratory ingestion rates were very low.

Thus, increased fish predation depressed *Daphnia* and encouraged the construction of a subordinate food chain, one that delivered a benefit to *Mesocyclops*. However, do the developing subordinate chains deliver a reciprocal benefit back to *Daphnia* and hence to fish? Preliminary experiments discussed in Kerfoot and DeMott (1984) suggest that the interaction is asymmetrical, with a much weaker tie from invertebrate predators back to *Daphnia* and fish. The reasons lie in the dietary specializations of the subordinate grazing suite. Although *Daphnia* are capable of monopolizing algal resources in Lake Mitchell, the subordinate grazers are individually much more restricted in their dietary preferences. *Mesocyclops* feeding is spread over a variety of prey, each of which has only a restricted impact on the phytoplankton assemblage. Thus, transmission of a signal is weakened by branching (multiple trophic links) and inefficiency.

In one perturbation experiment, artificial enrichment of *Mesocyclops* in non-fish enclosures resulted in a rapid return (within 3–10 days) to initial copepod densities. Either the particular enrichment protocol, rapid response to low resources, or cannibalism could have contributed to responses. There was no indication of a transmitted benefit to *Daphnia*, but the duration of the perturbation was too short for adequate assessment of transmitted effects.

DISCUSSION

The food web perturbations in Lake Mitchell illustrate the potential for strong top-down (cascading) interactions and for positive (facilitative) indirect effects among invertebrate and vertebrate predators, despite the lengthy pathways over which certain signals must pass. However, given the absolute densities of both fish and invertebrate predators in Lake Mitchell, realized top-down effects are probably rare (Levitan et al. 1985).

Some of the most intriguing interactions in the Lake Mitchell food web involve the grazer-phytoplankton link. Given the historical reliance of community ecology on Lotka–Volterra (Gaussian) competition equations (May 1973), it is ironic that exploitative competition really is an indirect interaction (fig. 5.1b). That is, intraspecific (self-limitation) and interspecific conflicts are mediated through the suspended resources (phytoplankton, bacteria, detritus) without direct contact of grazers (but see interference mechanisms discussed in Folt, this volume). Competition involves quantitative and qualitative changes in the resource assemblage that, in turn, are transmitted back to co-occurring herbivores. The central message here is that indirect effects not only are likely but include some of the central interactions. Of course, the linkage effect is only one of a number of qualitatively different "indirect" effects (Miller and Kerfoot, this volume).

Additional aspects of the grazer–phytoplankton interaction include (1) the evolution of resistance among phytoplankton species (not treated here), (2) asymmetries during the pulse experiments, and (3) the dampening effects of resource quality on *Daphnia* oscillations. In the netting experiments, *Daphnia* removal influenced co-occurring grazers far more than *Daphnia* addition. Removal coincided with immediate release of edible flagellates, resources then consumed by *Diaphanosoma* and *Bosmina*; whereas addition only slightly depressed edible flagellates.

The dampening ability of resource heterogeneity seems relatively straightforward. If *Daphnia* are fed on a monospecific culture of highly edible algae (e.g., *Chlamydomonas*), severe population oscillations are usually commonplace, as energetic and reproductive time lags ensure overshoots. However, if the resource includes a heterogeneous mixture of high- to low-quality species, dampening is possible. Approach of populations to their carrying capacity would coincide with resource shifts toward lower average quality, curbing the customary time-delayed overshoots and allowing a more gradual approach to equilibrium conditions (Kerfoot et al. 1985b). Overall community stability, however, depends on much more than just dampening of *Daphnia* oscillations. Stability calculations must include the whole array of nutrient–phytoplankton–herbivore–predator

interactions. Such an analysis is offered by Levitan (this volume).

ACKNOWLEDGMENTS

This work supported by NSF Grants DEB 76-20238 and 80-04654.

REFERENCES

Bender, E. A., T. J. Case, and M. E. Gilpin. 1984. Perturbation experiments in community ecology: Theory and practice. *Ecology* 65 : 1–13.

Bogden, K. G., and J. J. Gilbert. 1982. Seasonal patterns of feeding by natural populations of *Keratella, Polyarthra*, and *Bosmina*: Clearance rates, selectivities and contributions to community grazing. *Limnol. Oceanogr.* 27 : 918–34.

———. 1985. Body size and food size in freshwater zooplankton. *Proc. Natl. Acad. Sci., U.S.A.* 81 : 6427–31.

Brooks, J. L., and S. I. Dodson. 1965. Predation, body size, and composition of plankton. *Science* 150 : 28–35.

DeMott, W. R. 1982. Feeding selectivities and relative ingestion rates of *Daphnia* and *Bosmina*. *Limnol. Oceanogr.* 27 : 518–27.

———. 1983. Seasonal succession in a natural *Daphnia* assemblage. *Ecol. Monogr.* 53 : 321–40.

———. 1985. Relations between filter mesh-size, feeding mode, and capture efficiency for cladocerans feeding on ultrafine particles. *Arch. Hydrobiol. Beih., Ergebn. Limnol.* 21 : 125–34.

DeMott, W. R., and W. C. Kerfoot. 1982. Competition among cladocerans: Nature of the interaction between *Bosmina* and *Daphnia*. *Ecology* 63 : 1949–66.

Galbraith, M. G., Jr. 1967. Size-selective predation on *Daphnia* by rainbow trout and yellow perch. *Trans. Am. Fish. Soc.* 96 : 1–10.

Gilbert, J. J. 1985. Competition between rotifers and *Daphnia*. *Ecology* 66 : 1943–50.

Gilbert, J. J., and R. S. Stemberger. 1985. Control of *Keratella* populations by interference competition from *Daphnia*. *Limnol. Oceanogr.* 30 : 180–88.

Gilbert, J. J., and C. E. Williamson. 1978. Predator-prey behavior and its effect on rotifer survival in associations of *Mesocyclops edax, Asplanchna girodi, Polyarthra vulgaris*, and *Keratella cochlearis*. *Oecologia* 37 : 13–22.

Hall, D. J., S. T. Threlkeld, C. W. Burns, and P. H. Crowley. 1976. The size-efficiency hypothesis and the size structure of zooplankton communities. *Annu. Rev. Ecol. Syst.* 7 : 177–208.

Horn, W. 1981. Phytoplankton losses due to zooplankton grazing in a drinking water reservoir. *Int. Revue ges. Hydrobiol.* 66 : 787–810.

Hrbáček, J. 1962. Species composition and the amount of zooplankton in relation to the fish stock. *Rozpr. Cesk. Akad. Ved. Rapa Mat. Prir. Yed* 72 : 1–116.

Hrbáček, J., and M. Novotná-Dvořáková. 1965. Plankton of four backwaters related to their size and fish stock. *Rozpr. Cesk Akad. Yed Rada Mat. Prir. Ved.* 75 : 3–10.

Hutchinson, G. E. 1979. *An introduction to population ecology.* New Haven, Conn.: Yale University Press.

Jacobs, J. 1978. Coexistence of similar zooplankton species by differential adaptation to reproduction and escape in an environment with fluctuating food and enemy densities. III. Laboratory experiments. *Oecologia* 35 : 35–54.

Kerfoot, W. C., and W. R. DeMott. 1980. Foundations for evaluating community interactions: The use of enclosures to investigate coexistence of *Daphnia* and *Bosmina.* In Kerfoot, W. C. (ed.), *Evolution and ecology of zooplankton communities,* pp. 726–741. Hanover, N.H.: University Press of New England.

———. 1984. Food web dynamics: Dependent chains and vaulting. In D. G. Meyers and J. R. Strickler (eds.), *Trophic interactions within aquatic ecosystems,* pp. 347–82. AAAS Selected Symposium #85. Washington, D.C.: Westview Press.

Kerfoot, W. C., W. R. DeMott, and D. L. DeAngelis. 1985b. Interactions among cladocerans: Food limitation and exploitative competition. *Arch. Hydrobiol. Beih. Ergebn. Limnol.* 21 : 431–51.

Kerfoot, W. C., W. R. DeMott, and C. Levitan. 1985a. Nonlinearities in competitive interactions: component variables or system response? *Ecology* 66 : 959–65.

Lampert, W. 1977. Studies on the carbon balance of *Daphnia pulex* as related to environmental conditions. I. Methodological problems of the use of ^{14}C for the measurement of carbon assimilation. *Archiv für Hydrobiol.* (Suppl.) 48 : 287–309.

———. 1981. Inhibitory and toxic effects of blue-green algae on *Daphnia. Int. Rev. ges. Hydrobiol.* 66 : 289–99.

Lehman, J. T., Botkin, D. B., and G. E. Likens. 1975. The assumptions and rationales of a computer model of phytoplankton population dynamics. *Limnol. Oceanogr.* 20 : 343–64.

Levine, S. H. 1976. Competitive interactions in ecosystems. *Am. Nat.* 110 : 903–10.

Levins, R. 1975. Evolution in communities near equilibrium. In M. L. Cody and J. M. Diamond (eds.), *Ecology and evolution of communities,* pp. 16–50. Cambridge, Mass.: Belknap Press, Harvard University Press.

Levitan, C. 1986. The structure and stability of a freshwater pelagic community. Ph.D. thesis, Dartmouth College, Hanover, N.H.

Levitan, C., W. C. Kerfoot, and W. R. DeMott. 1985. Ability of *Daphnia* to buffer trout lakes against periodic nutrient inputs. *Verh. Internat. Verein. Limnol.* 22 : 3076–82.

May, R. M. 1973. *Stability and complexity in model ecosystems.* Princeton, N.J.: Princeton University Press.

Neill, W. E. 1984. Regulation of rotifer densities by crustacean zooplankton in an oligotrophic montane lake in British Columbia. *Oecologia* 61 : 175–81.

Neill, W. E., and A. Peacock. 1980. Breaking the bottleneck: interactions of invertebrate predators and nutrients in oligotrophic lakes. In W. C. Kerfoot (ed.), *The evolution and ecology of zooplankton communities,* pp. 715–24. Hanover, N.H.: University Press of New England.

O'Brien, W. J. 1979. The predator-prey interaction of planktivorous fish and zooplankton. *Amer. Sci.* 67 : 572–81.

Paine, R. T. 1980. Food webs: Linkage, interaction strength and community infrastructure. *J. Anim. Ecol.* 49 : 667–85.

Paloheimo, J. E., S. J. Crabtree, and W. D. Taylor. 1982. Growth model of *Daphnia. Can. J. Fish. Aquat. Sci.* 39 : 598–606.

Porter, K. G., J. D. Orcutt, Jr., and J. Gerritsen. 1983. Functional response and fitness in a generalist filter feeder, *Daphnia magna* (Cladocera: Crustacea). *Ecology* 64 : 735–42.

Stemberger, R. S., and J. J. Gilbert. 1985. Body size, food concentration, and population growth in planktonic rotifers. *Ecology* 66 : 1151–59.

Taylor, B. E. 1980. Size-selective predation on zooplankton. In W. C. Kerfoot (ed.), *Evolution and ecology of zooplankton communities,* pp. 377–87. Hanover, N.H.: University Press of New England.

Vandermeer, J. 1980. Indirect mutualism: Variation on a theme by Stephen Levine. *Am. Nat.* 116(3) : 441–48.

Williamson, C. E. 1981. The feeding ecology of the freshwater cyclopoid copepod *Mesocyclops edax.* Ph.D. thesis, Dartmouth College, Hanover, N.H.

Williamson, C. E., and J. J. Gilbert. 1980. Variation among zooplankton predators: The potential of *Asplanchna, Mesocyclops,* and *Cyclops* to attack, capture, and eat various rotifer prey. In W. C. Kerfoot (ed.), *The evolution and ecology of zooplankton communities,* pp. 509–17. Hanover, N.H.: University Press of New England.

6. Formal Stability Analysis of a Planktonic Freshwater Community

Charles Levitan

The Lake Mitchell midsummer planktonic community is shown to be stable. The summer phytoplankton and zooplankton populations of the lake were significantly more constant than those in spring or in manipulated enclosures. The community was resilient after reversible perturbations and was shown to be stable through a formal stability analysis, the latter based on Lyapunov analysis of the community interaction matrix. Elements of that matrix were required to satisfy the equation $dX_i/dt = a_{ij} \cdot x_j$, which describes the direct effect of changes in density of one population on the growth rate of another. Equations were formulated for interactions between each of six compartments: Digestion-resistant and digestion-susceptible phytoplankton, *Daphnia* sp., *Bosmina*, *Mesocyclops*, and fish. All equations were based on measurable functional relationships, such as type 2 feeding by grazers and Michaelis–Menton nutrient uptake dynamics by phytoplankton. The source of community stability was established through a sensitivity analysis of the community matrix. Stability was found to be sustained primarily by the interactions among grazers and phytoplankton, with predators having little influence. Enrichment of the community with limiting nutrients was predicted to have a destabilizing effect.

Planktonic communities are extremely dynamic, containing organisms with generation times of days, nutrients whose residence times are measured in minutes, and populations that are capable of consuming the equivalent of their entire resource base in just days. It is not surprising, therefore, that lake communities often feature compositions that change rapidly, as well as strong patterns of patchiness. Under such conditions a community with a constant composition would be unexpected. The planktonic community of Lake Mitchell, Vermont, however, does persist throughout the summer with a fairly constant composition and is also consistent between summers (DeMott and Kerfoot 1982; DeMott 1983). This community (fig. 6.1) features the herbivores *Daphnia* and the less abundant *Bosmina* feeding on a phytoplankton assemblage that consists primarily of "resistant" algae and a smaller number of flagellates and edible green algae, with periodic appearances of *Dinobryon*. The grazers are in turn preyed on by the invertebrate omnivore *Mesocyclops edax*, the creek chub, and stocked trout. The constancy of the community, even in the face of such perturbations as occasional storms and fish stockings, suggests that the community may be stable in the formal sense; that is, its internal dynamics cause its own composition to revert toward an equilibrium condition whenever perturbed away from that state. Because formal analyses of the stability of natural communities are extremely scarce (Pimm 1982) and because much is already known about the Lake Mitchell community, an analysis testing its community stability was attempted. Stability was assayed through three separate methods. First, the constancy of the community was documented. The resiliency of the community was then observed following reversible perturbations. Finally, the community's stability was mathematically documented through an analysis of interactions between populations.

CONSTANCY OF THE LAKE MITCHELL COMMUNITY

The biotic community of Lake Mitchell is very consistent throughout the summer and between years during the summer (DeMott and Kerfoot 1982; DeMott 1983; Kerfoot et al. 1985a). This constancy, or persistence (Pimm

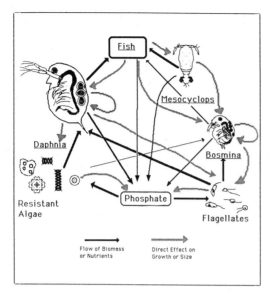

Figure 6.1. *The most common zooplankton in the Lake Mitchell ecosystem are three species of* Daphnia, *one* Bosmina, *and one* Mesocyclops. *Phytoplankton consist of those highly susceptible to predation by grazers and those resistant to digestion after ingestion. Fish populations include native creek chub (*Semotilus atromaculus*) and stocked rainbow (*Salmo gairdnerii*) and brook (*Salvelinus frontinalis*) trout. Solid arrows indicate all major paths of energy and nutrient flow; while dashed arrows indicate other interactions affecting the growth rate of compartments of the model. Competition is not indicated directly in the model, as it is interpreted through the pattern and magnitude of other interactions.*

1982), is evidence of and a prerequisite for stability in a system, although not sufficient proof of stability.

Methods

Constancy may be quantified in several ways. Alternative measures of variability are based on the coefficient of variation (CV) or the coefficient of fluctuation (CF). The coefficient of variation is the standard deviation of the population densities at different times normalized by the mean population size:

$$CV = [\Sigma(N_t - \overline{N})^2/(n - 1)]^{0.5}/\overline{N}, \qquad (6.1)$$

where $\overline{N}$ represents the mean population density and n the number of estimates. The coefficient of fluctuation is a similar measure but treats population densities on a logarithmic scale, which may be preferable because of the

nature of population growth (Whittaker 1975, p. 13):

$$CF = \text{antilog } D_l, \qquad (6.2)$$

$$D_l = [\Sigma(\log \overline{N} - \log N)^2/(t - 1)]^{0.5}, \qquad (6.3)$$

$$\overline{N} = [\Pi(N_i)]^{1/t} = \text{antilog } [(\Sigma \log N_i)/t] \qquad (6.4)$$

The coefficients of fluctuation and variation were calculated for lake phytoplankton populations in the spring and summer. These values are compared to those for lake populations that have been experimentally manipulated by the addition of nutrients and/or fish.

Results and Discussion

Coefficients of fluctuation and variation for phytoplankton are listed in table 6.1 for the sets of experiments performed during the early and late summer of 1980 (see Levitan 1986 for individual species calculations). For both early and late summer, the coefficients of variation and coefficients of fluctuation were lower for control enclosures and lake samples than for experimental enclosures. Table 6.1, showing the distributions of mean coefficients of fluctuations, also shows that the spring phytoplankton community had significantly higher mean coefficients of fluctuation than the midsummer lake or control enclosures. These results indicate that, relative to other times during the year and to modified enclosures, the phytoplankton communities of Lake Mitchell and of unperturbed control enclosures during midsummer are much less variable, indicating a persistent and possibly stable community.

PERTURBATION STUDIES OF THE LAKE MITCHELL ECOSYSTEM

The stability of a system may be assayed by perturbing it and observing the subsequent changes of the community. To test fairly for stability, the perturbation should be reversible; that is, it should not fundamentally and permanently change the nature of the interactions of the system. Implicit in this type of analysis is that the system in question has one stable equilibrium point and that no species will become extinct during the recovery of the system. The addition of large amounts of nutrients to a sys-

Table 6.1. *Mean coefficients of fluctuation and coefficients of variation of the densities of dominant phytoplankton species*

Measure of variation (average)	Spring lake	Early summer manipulated enclosures	Early summer lake, control enclosures	Late summer manipulated enclosures	Late summer lake, control enclosures
Coefficient of fluctuation	4.656	1.405	1.309	1.349	1.156
	3.413	1.577	1.016	1.533	1.035
		1.580	1.124	1.491	
		1.473		1.150	
		1.267		1.144	
		1.384		1.137	
		1.466			
		1.464			
Mean	4.034	1.452	1.105	1.295	1.095
		$p < 0.05$	$p < 0.05$	$p < 0.05$	
Coefficient of variation	1.372	1.373	1.152	1.170	1.066
	1.253	1.293	1.011	1.161	0.947
		1.514	0.993	1.521	
		1.155		1.415	
		1.209		1.164	
		1.337		1.080	
		1.497			
		1.350			
Mean	1.312	1.341	1.052	1.252	1.006
		N.S.	$p < 0.05$	$p < 0.05$	

Samples are from Lake Mitchell during the spring (May–June) and summer (July–September) and enclosures that had either been perturbed with additions of nutrients and fish (manipulated) or left unmanipulated (control). All differences between spring and summer conditions and between summer manipulated and unmanipulated communities were significant.

tem would not be a reversible perturbation on a short time scale. Those nutrients are conserved in the system due to fundamental mass-balance considerations; their abundance would continue (for some time) to be some function of the total amount introduced during the perturbation. Further, the introduction of nutrients would fundamentally change the nature of the interactions within and between populations of phytoplankton, eliminating any interaction based on the concentration or supply of nutrients. In contrast to this, a reversible perturbation is one that involves the change in density of some population, which can then recover or diverge further, according to the particular dynamics of the system.

Procedure

The perturbation performed involved the depletion of *Daphnia* from the midsummer system. During the summer of 1981, two large (12,000-L) enclosures were depleted of *Daphnia* by repeatedly towing a 130-μm, 0.25-m-diameter net through the water, removing 98% of all net plankton. The enclosures were then enriched with 5.0 μg K_2HPO_4-P and 35 μg $NaNO_3$-N/L, replacing those nutrients sequestered in the zooplankton. The enclosures were re-sieved after 7 days. After 13 days six small 75-L microenclosures were suspended in one of the netted enclosures. The microenclosures, fabricated of clear polyethylene can liners (White River Paper Co., White River Junction, Vt.) sealed with rubber stoppers in Tygon plastic sleeves, were filled with a modified Schindler trap. The trap, which had an impermeable spout mounted around the net and outlet, collected water at three depths—0 to 0.5 m, 1.0 to 1.5 m, and 2.0 to 2.5 m—to fill the microenclosures. The water was sieved during collection. Approximately 500 *Daphnia* freshly collected from the lake were introduced into three of the six microenclosures. The bags

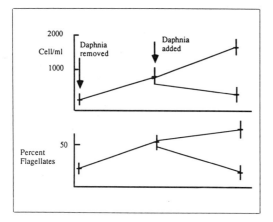

Figure 6.2. *Cells per milliliter and composition, as percentage of flagellates, for six microenclosures depleted of Daphnia; Daphnia were reintroduced into three of those enclosures after 13 days. Lines around each point indicate one standard error of the mean.*

were all then suspended from a beam within the large enclosure at a depth of 1.0 to 1.5 m below the surface. After 12 days (25 days after the beginning of the experiment) the bags were sampled for *Daphnia* densities, chlorophyll concentration, and phytoplankton densities. The phytoplankton were enumerated into only two categories: flagellates, including *Cryptomonas* and *Rhodomonas* plus small greens, and those cells judged to have thick cell walls or a gelatinous sheath, the so-called resistant phytoplankton. In addition to other species, the latter included *Scenedesmus*, *Pediastrum*, and *Oocystis*.

Results and Discussion

Phytoplankton cell densities are graphed for the experiments in figure 6.2, along with their percentage composition. After the grazers were removed from the large enclosures, flagellates began growing, stimulated by the nutrient pulse and released from predation, rising in density from a mean of 450 to 2,030 per milliliter in 13 days, for a realized growth rate of 0.11 per day (fig. 6.2). Resistant phytoplankton also increased in numbers, rising from 1,560 to 2,200 per milliliter for a realized growth rate of 0.026 per day. After introduction of the *Daphnia* the phytoplankton densities in the microenclosures fell to near-original levels, and flagellates made up a far smaller proportion of the

cells than in the released and enriched treatments.

The compositions of the microenclosures, in terms of percentage of flagellates, were compared using the Difference Between Proportions test (Zar, 1974, p. 296), with Z being the proportion of the area under the normal curve. All tests were done with one-tailed hypotheses.

$$Z = \frac{(p_1 - p_2)}{\sqrt{(pq/n_1 + pq/n_2)}} \tag{6.5}$$

where

p_1 = proportion of members of a sample in a given category,

p_2 = proportion of members of the other sample in the same category

p = proportion of pooled samples in the same category,

q = $1 - p$,

n_1 and n_2 = size of samples 1 and 2.

It is clear that the composition of the enclosures did diverge after the original perturbation. That difference was significant and continued to be significant in those enclosures not reinoculated with *Daphnia*. In those enclosures with *Daphnia* reintroduced, the difference between experimental and unmanipulated controls decreased until the mean Z fell from Z = 4.51 ($p < 0.001$) for enclosures without *Daphnia* to Z = 1.04 ($p = 0.15$). This indicates that the community is stable in the sense that it is resilient; i.e., it returned to near the point of origin after some perturbation.

STABILITY ANALYSIS OF THE LAKE MITCHELL ECOSYSTEM

Having shown that the Lake Mitchell plankton community was both persistent and resilient, a formal stability analysis was performed. This analysis was done both to confirm the existence of stability and to clarify which components and interactions in the community were most responsible for the maintenance of stability. This analysis is based on the Lyapunov criteria for stability of interactive systems (Ashby 1960), involving the calculations of eigenvalues of the community interaction matrix.

The correct derivation of the community interaction coefficients a_{ij} relies on the proper interpretation of the basic community dynamics equation (Gardner and Ashby 1970):

$$\frac{dx(t)}{d(t)} = \mathbf{A} \cdot x(t). \tag{6.6}$$

This equation represents a linearization of growth equations around the equilibrium point, using just the first term of a Taylor series (see Levitan 1986, chapter 2). At equilibrium the left side of the equation must be zero, and the community dynamics equation reduces to

$$\mathbf{A} \cdot x(t) = 0. \tag{6.7}$$

If stable, nontrivial solutions (i.e., $x(t) \neq 0$) must also satisfy the equation

$$\det [\mathbf{A} - \lambda \cdot \mathbf{I}] = 0 \tag{6.8}$$

where $\mathbf{I}$ is the identity matrix, λ are the eigenvalues of the community interaction matrix $\mathbf{A}$, and all eigenvalues λ must have negative real parts. The basic community-dynamics equation may be linearized about the point of stability ($x = (x_1, x_2, \ldots, x_n)$), with the rate of change of each member of the community described by the equation

$$dX_i/dt = (a_{i1}x_1 + a_{i2}x_2 + \ldots a_{in}x_n) \tag{6.9}$$

(Ashby 1960). Each a_{ij} is calculated at equilibrium (Ashby 1960) as

$$a_{ij} = \partial(dx_i/dt)/\partial x_j, \tag{6.10}$$

which when linearized about the point of stability becomes

$$d(X_i)/(dt) = a_{ij}x_j + K, \tag{6.11}$$

similar in form to the Lotka–Volterra predation equation (May 1981, p. 78). This derivation is similar to that of May (1973, p. 21), who uses

$$dX_i/dt = (a_{i1} \cdot (X_{1s} - X_1) + \tag{6.12}$$

$$a_{i2} \cdot (X_{2s} - X_2) + \ldots a_{in} \cdot (X_{ns} - X_n),$$

which is also a Taylor expansion around the point of equilibrium. In this treatment, X_{ns} is the size of population n at equilibrium, and

each dX_i/dt is influenced by the deviation of other species from equilibrium, rather than by their absolute numbers.

Using equation 6.10 the value of a_{ij} is calculated from the change in the growth rate of population i with the change in size of population j. The value of a_{ij} may be estimated either by examining the relationship of X_j and (dX_i/dt) in field populations or by actually observing the response of population i to different levels of population j in controlled laboratory experiments. Both approaches are taken in this study. The simulation of field conditions in these laboratory experiments is especially important, as all functional and numerical responses are specifically linearized about the point of assumed stability (i.e., average field conditions, see table 6.2).

Most troublesome among these calculations is the derivation of the values of all a_{ii}, a problem addressed more generally in Levitan (1986, chap. 2). Small changes in the abundance of X_i may not affect per capita birth rates but may affect the population growth rate. This is due to constant per capita reproduction and growth coupled to altered rates of predation by predators that switch or whose feeding rates have begun to saturate. The response of a given population's growth rate to changes in its own population size will depend on at least three factors: the release of resources coupled to the population decrement; uniform changes in birth, growth, respiration, and death rates; and disproportionate changes in loss to predation. The mathematics of this approach are detailed below for each population. The formal analysis of the stability of a system therefore requires the definition of compartments making up the system, the measurement of the interactions between all compartments, and the calculation of eigenvalues of the resulting community matrix. As the choice of compartments in the food web model largely determines the nature and strength of all interactions, their correct assignment is vital to the accurate assessment of stability.

Phosphorus Concentrations and Flows in Lake Mitchell

The pool of inorganic nutrients, and in particular of phosphate, is of paramount importance in the structure of the Lake Mitchell ecosys-

Table 6.2. *The Lake Mitchell ecosystem during midsummer*

Group	Density	Excretion rate, ng P/ind./hr	Clearance rate, ml/ind./hr (I)	(II)	Mortality after ingestion (I)	(II)	Total mortality by population, d[1] (I)	(II)
Daphnia	20/L	3.7	0.85	0.85	1.00	0.33	0.135	0.400
Bosmina	5/L	0.11	0.08	0.00	1.00	0.00	0.000	0.01

Group	Density no./ml	Fraction of total P	[P] µg/L	Mortality d[-1]	Phosphorus loss, µg/L/d	Phosphorus gain, µg/L/d	Gut P uptake µg/L/d	Ambient P uptake µg/L/d
Flagellates	1000	15	1.50	0.400	0.60	0.60	0.0	0.60
Resistants	9000	85	8.50	0.133	1.15	1.15	0.56	0.59
Totals	10,000	100	10.00	——	1.75	1.75	0.56	1.19

(I), flagellates; (II), resistant cells.
Values are from this chapter, personal observations, DeMott and Kerfoot (1982), Kerfoot and DeMott (1984), Levitan, Kerfoot, and DeMott (1985), and Peters and Rigler (1973).

tem. Phosphate concentration affects, or is affected by, virtually all other compartments in the food web. It is distinct from other compartments in that it is abiotic and consequently has no "birth" or "death" rate. The actual size of the pool is also very difficult to estimate.

The growth of phytoplankton in Lake Mitchell is phosphorus-limited during the summer. This limitation has been demonstrated over several years using nutrient-supplementation assays. In all assays, phosphorus caused significant phytoplankton growth, either alone or in combination with nitrogen (see Levitan 1986, chap. 2, table 2). The addition of nitrogen alone caused a very small increase in the concentrations of phytoplankton carbon and chlorophyll. These results indicated that the dynamics of phosphorus were most important in determining the growth of nutrient-limited primary producers, and other nutrients such as nitrogen were not considered.

During the summer months concentrations of phosphate were estimated with a variety of methods. These included the molybdate blue method (Murphy and Riley 1962) with an ascorbic acid reductant (Strickland and Parsons 1968); the inhibitive effect on the action of the enzyme alkaline phosphatase by phosphate ions (Petterson 1979); and in an indirect manner, i.e. the multiplication of the turnover time of the free phosphorus pool by the rate of uptake of phosphates. Assuming that inorganic carbon and phosphorus are incorporated in direct proportion to their relative abundance in cells, phosphate uptake was calculated through the measurement of primary production and knowledge of the elemental composition of phytoplankton (Levitan 1986).

Using the molybdate-blue method, phosphate concentrations between 0 and 3 µg P/L were routinely seen. Phosphate measured through enzyme inhibition was uniformly lower than that measured by the traditional molybdate-blue method, often below the limit of detection (0.2 µg P/L). This discrepancy of results between methods is not surprising (Rigler 1975) and probably represents the concentration of some form of phosphate made soluble by the highly acidic chemical conditions of the Murphy and Riley assay (pH 0.4). Similar discrepancies, often of comparable magnitude, also were documented by Petterson. Phosphate turnover times were short, with mean phosphorus residence times as low as 2.4 min. Concentrations calculated from uptake rates were extremely low, always below 1.0 µg/PO$_4$-P/L.

Because of its limiting role in phytoplankton dynamics, the actual ambient concentration of free phosphate in Lake Mitchell is extremely low, and the "lifetime" of any phosphate ion in the water column is orders of magnitude shorter than any other component in the food web. The concentration of phosphate responds to changes in populations so rapidly that indirect interactions between organisms affecting the phosphate pool would appear to be direct.

These fast dynamics are used to justify elimination of the phosphate pool from the food web model and treatment of nutrient competition as a direct interaction.

Compartments of the Lake Mitchell Food Web

The stability analysis of the Lake Mitchell community was based on a six-compartment food web. The grouping of populations into food web compartments is a well-recognized practice (Sprules and Holtby 1980; Sprules and Knoechel 1985; but see Polis 1984). As community interaction models are based on matrices describing direct interactions, the sole requirement for being included in a compartment is that a population interact with members of other compartments in a similar fashion, i.e., that it be similar. The clumping of species into compartments also allows for the gradual replacement of similar species within compartments while the dynamics of the entire system are maintained.

Lake Mitchell contains three daphnids: *Daphnia rosea*, *D. pulicaria*, and *D. galeata mendotae*. The first two, which are most common, are ecologically very similar (Kerfoot 1985) in such features as feeding electivity and susceptibility to predation. During the summer they exhibit competitive replacement (DeMott 1983). The third *Daphnia* is relatively rare. *Bosmina*, the other principal grazer in the lake, differs from all of the *Daphnia* in its feeding ecology and size (DeMott and Kerfoot 1982; DeMott 1985) and in its reproductive response to food availability.

More than 50 species of phytoplankton have been reported in Lake Mitchell (Zelazny 1983), covering a wide range of sizes and phyletic divisions. To try to separate those species into meaningful categories, a least-squares dendrogram study was performed on the correlation matrix of their abundances under a wide variety of conditions. The distance metric employed was D = 1 − r, on the assumption that dissimilar species would be negatively correlated. It should be noted that there is some danger in this procedure if extensive competitive replacement actually causes negative correlations between the abundances of ecological equivalents. Linkage was average rather than single to avoid chaining of points in clusters. Analysis was performed using the Biomedical

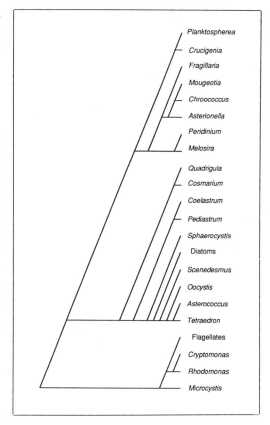

Figure 6.3. *Dendrogram of the occurrence of 22 species of phytoplankton in three separate treatments, with 2 or 4 replicates per treatment, based on cluster analysis of variables. The phytoplankton grouped broadly into three categories, as detailed in the text.*

Data Processing program 2M, cluster analysis of variables (Dixon 1981). The dendrogram for populations present during the summer of 1980 is shown in figure 6.3. Three clear groups are apparent. The first consists of flagellates, including *Rhodomonas* and the blue-green *Microcystis*. This grouping is not surprising; all of those species are fast-growing and are extremely susceptible to predation by both *Daphnia* and *Bosmina*. *Microcystis* in Lake Mitchell occurs as unicells or rare small colonies, and probably is too rare (≪2%) to cause any toxic effects (DeBernardi et al. 1981). The second group of phytoplankton consists of species whose members are ingestible by *Daphnia* but have some structure giving resistance to digestion and mortality from ingestion. These structures include gelatinous coats (*Sphaerocystis*) and thick cell walls (*Scenedesmus*). Such cells may have access to nutrients at high concentrations while inside grazer guts (Porter 1976). Many of these cells are not accessible to *Bosmina* because of their size (Kerfoot et al. 1985a). The third group of cells includes *Peridinium* and *Asterionella*, which are generally too large to be efficiently grazed even by *Daphnia*. These cells have slow growth rates (Banse 1976) and are exposed to neither grazer mortality nor grazer-gut nutrient pools. Because members of this group are extremely rare in the midsummer Lake Mitchell ecosystem, they have been omitted from consideration in the following analysis.

Uptake of Ambient Nutrients by Phytoplankton

The nutrient cycle in Lake Mitchell was analyzed following the set of interactions diagrammed in figure 6.4. Briefly, grazers assimilate energy and nutrients from all digested phytoplankton, the latter of which are subsequently excreted. The nutrients excreted by grazers and predators are then absorbed through active uptake by all phytoplankton. In addition, those digestion-resistant phytoplankton not killed after ingestion are able to absorb nutrients while still in grazer guts. Although direct interactions among phytoplank-

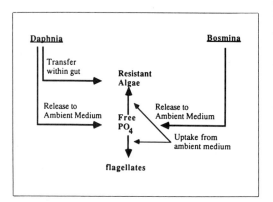

Figure 6.4. *Diagram of the cycling of nutrients, primarily phosphate, within the Lake Mitchell ecosystem. Although benthic, littoral, and atmospheric inputs exist, they are much lower than internal cycling. In this model only* Daphnia *releases nutrients to phytoplankton that it has ingested.*

ton are limited to such processes as allelopathy, shading, and occasionally predation (some flagellates), phytoplankton populations must be interpreted as interacting directly in any description of a system in which their shared limiting resources are not included. This *de facto* interaction of phytoplankton can be modeled only through an understanding of the transport and the control of transport of unbound (inorganic) nutrients through the community.

The primary source of inorganic nutrients, particularly phosphate, in the Lake Mitchell summer community is the *in situ* excretion of ions by grazers. Zooplankton excretion, as measured directly and confirmed through literature values, adequately accounts for support of the total primary production by phytoplankton. Phosphate excretion rates by zooplankton are far greater than the contributions of air input (Peters 1977; Rigler 1975) and stream input. Likewise, the contribution of phosphorus by the benthos is probably minimal, because of the well-oxygenated nature of the bottom. The sediments are capped with an oxidized zone extending at least 10 cm deep (personal observation), so high rates of release of phosphate are unlikely (Harter 1968, Shukla et al. 1971, Rigler 1975). The exclusion of the benthos from this nutrient flow model is further justified by the similarity of nutrient dynamics in the lake and in control enclosures. These enclosures had sealed bottoms that completely isolated the water column from the benthos. In midsummer the bacterial populations in Lake Mitchell are small (< 1,000/ml), as measured by the direct-count epifluorescence technique. The bacteria were small (< 0.5 μm) and seemed to be mainly inactive, as judged by their predominant binding of acridine orange to DNA (glowing green) as opposed to RNA (found in faster-growing cells, giving an orange glow). These bacteria also may be nutrient sinks rather than nutrient sources (Mayfield and Innis 1978).

Phosphate released by grazers (predominantly *Daphnia*) is either taken up directly by digestion-resistant algae (see below) or is released into the water and diluted to ambient concentrations. Phosphate release is therefore partitioned into two separate pools, with removal by one (gut nutrients) directly reducing input to the other (see fig. 6.4). The uptake dynamics of the resistant algae are discussed elsewhere but are noted here as decreasing the release of nutrients to the bulk medium.

This analysis considers the bulk medium to have a uniform distribution of phosphate. Lehman and Scavia (1982a,b) showed that large grazers may leave "patches" of high nutrients in their wakes, especially for phosphorus (Scavia and McFarland 1982). This phenomenon is not considered for several reasons. First, the phosphate concentrations expected in most of the patches, approximately 125 nM above ambient concentrations, are still within the linear portion of all phytoplankters' uptake capacity; hence, the relative rates of nutrient uptake by different species will be the same within and outside the patches (although the absolute rate of uptake will differ between individuals). Further, phosphate patches in Lake Mitchell will be small, rare, and short-lived. The densities and sizes of *Daphnia* present during the summer would create patches occupying no more than 0.002% of the water column at any time, and persist for only seconds.

Phosphate is absorbed by phytoplankton from the bulk medium through active transport. The rate of uptake follows Michaelis–Menton enzyme dynamics, and the uptake per cell is quantified as

$$V = V_m \cdot (P/(K_m + P)), \tag{6.13}$$

where

V = the rate of uptake of phosphate per cell per time,
V_m = the maximum rate of phosphate uptake per cell per time,
P = the concentration of phosphate in the water, and
K_m = the Michaelis constant for nutrient uptake, being the concentration at which $V = V_m/2$.

This equation is identical in form to the Monod population growth equation (Monod 1950, in Tilman 1982, p. 46). However, the Michaelis–Menton equation describes only nutrient uptake dynamics for cells growing at a given rate, and presumably at steady state, at a given nutrient concentration. The confusion of these two equations can lead to the estimation of unreasonably high growth rates at high concen-

trations of nutrients (McCarthy and Goldman 1979; Goldman et al. 1979). Both approaches also ignore the fact that the nutrient uptake kinetics of phytoplankton change under different nutrient regimes (Jackson 1980).

At the extremely low phosphate concentrations seen in Lake Mitchell during the summer, nutrient concentrations are far smaller than the Michaelis constant K_m, and equation 6.13 may be simplified to

$$V = V_m \cdot (P/(K_m)) \tag{6.14}$$

$$= (V_m/K_m) \cdot P. \tag{6.15}$$

Although the concentrations of free phosphate may be below the limits of detection, measures of V_m/K_m for each population may be used to quantify relative uptake rates (Crowley 1975). Because several distinct nutrient-uptake enzyme systems may exist in a single phytoplankter, the measurement of uptake kinetics must be done only at low phosphate concentrations. Rather than estimate the parameters separately, using Burke–Lineweaver (Lehninger 1975) calculations or other methods of independently estimating V_m and K_m, the ratio of these parameters may be calculated directly, through the estimation of nutrient turnover times in low-nutrient environments:

$$(V_m/K_m) \cdot N = (V/P) \cdot N = K, \tag{6.16}$$

where N is the cell density and K is the rate of depletion of phosphate from a dilute solution:

$$P_{(t)} = P_{(o)} \cdot e^{-Kt}. \tag{6.17}$$

K is also equal to the inverse of the turnover time; the half-life of a phosphate ion in the water column is $[\ln(2)/K]$. Values for V_m/K_m or K may therefore be estimated from turnover times of phosphate in the presence of a known density of cells (as the number of enzyme sites, proportional to V_m, will be directly proportional to the cell density):

$$V_m/K_m = -\ln(P_{(t)}/P_{(o)})/(t \cdot N) = K/N. \tag{6.18}$$

The nutrient uptake from the ambient pool into each population *i* is

$$U_i = dP_i/dt = (V_{mi}/K_{mi}) \bullet P \bullet N_i, \qquad (6.19)$$

where

$U_i = dP_i/dt$ = the phosphate input to population i,

V_{mi}/K_{mi} = the *per capita* phosphate uptake capability of population i,

P = the phosphate concentration, and

N_i = the population density of population i.

Since phosphate concentrations in the lake are constant over long periods of time, the rate of phosphate release by grazers to the bulk medium will be equal to the sum of the rates of phosphate uptake by all populations of phytoplankton, which is

$$U_T = \sum_{i=1}^{n}(dP_i/dt) = \sum_{i=1}^{n}(V_{mi}/K_{mi} \bullet P \bullet N_i), \quad (6.20)$$

where

dP_i/dt = rate of phosphate uptake per volume by each population.

Unless otherwise specified, all summations $[\Sigma(x)]$ represent summations across all populations. The condition of nutrients (i.e., resources) being taken up at the same rate as they are supplied has been discussed by Tilman (1982, p. 67). As the concentration of phosphate is constant in a uniform medium, equation 6.20 may be simplified to

$$dP_T/dt = \Sigma(dP_i/dt) = P \bullet \Sigma(V_{mi}/K_{mi} \bullet N_i),$$

$$(6.21)$$

where

dP_T/dt = rate of total phosphate supply to the system, per volume.

The phosphate uptake by any population is equal to the total nutrient input rate times the relative uptake potential (based on population size and uptake enzyme characteristics), as in equation 6.19, and may then be calculated from equations 6.19 and 6.21 to be

$$U_i = dP_T/dt \bullet \frac{(V_{mi}/K_{mi} \bullet N_i)}{\Sigma(V_{mi}/K_{mi} \bullet N_i)}. \qquad (6.22)$$

Interaction coefficients in the Jacobian community interaction matrix measure the influence of the numbers of one population on the population growth rate of another. In a nutrient-limited system the control of nutrient uptake is equivalent to the control of growth. The influence of population k on the uptake rate of population j in a stable system then can be expressed as

$$\frac{dU_j}{dN_k} \qquad (6.23)$$

$$= \frac{d}{dN_k} \frac{(dP_T/dt \bullet (V_{mj}/K_{mj} \bullet N_j))}{(\sum_{i \neq k}((V_{mi}/K_{mi}) \bullet N_i) + (V_{mk}/K_{mk}) \bullet N_k)}.$$

All calculations of the derivatives of functions were performed using the mathematical package MACSYMA (Bogen 1977) running on a Digital VAX 11/780 computer under the UNIX operating system at Dartmouth College. The first derivative of U_j with respect to N_k is expanded, through differentiation, to

$$\frac{dU_j}{dN_k} = \frac{-(dP_T/dt) \bullet (V_{mj}/K_{mj} \bullet N_j) \bullet (V_{mk}/K_{mk})}{(\sum_{i \neq k}((V_{mi}/K_{mi}) \bullet N_i) + (V_{mk}/K_{mk}) \bullet N_k)^2}$$

$$(6.24)$$

$$= \frac{-(dP_T/dt) \bullet (V_{mj}/K_{mj} \bullet N_j) \bullet (V_{mk}/K_{mk})}{(\Sigma((V_{mi}/K_{mi}) \bullet N_i))^2}.$$

$$(6.25)$$

Equation 6.25 may be simplified by substituting equation 6.21 into the denominator:

$$\frac{dU_j}{dN_k} = \frac{-P \bullet (V_{mj}/K_{mj} \bullet N_j) \bullet (V_{mk}/K_{mk})}{\Sigma((V_{mi}/K_{mi}) \bullet N_i)},$$

$$(6.26)$$

which in turn may be simplified through substitution of equations 6.19 or 6.22, to

$$\frac{dU_j}{dN_k} = \frac{- U_j \bullet (V_{mk}/K_{mk})}{\Sigma((V_{mi}/K_{mi}) \bullet N_i)} . \qquad (6.27)$$

or, alternatively, noting from equations 19 and 21 that

$$U_k/U_T = [(V_{mk}/K_{mk}) \bullet N_k]/[\Sigma((V_{mi}/K_{mi}) \bullet N_i)], \qquad (6.28)$$

$$\frac{dU_j}{dN_k} = \frac{U_j \bullet U_k}{U_T \bullet N_k} = - U_j/U_T \bullet u_k,$$

where u_k is the per capita uptake rate of population k; $u_k = U_k/N_k$. The "effect" of the presence of a competitor is proportional to its ability to take up a renewed resource, relative to the total community capacity to take up that resource. This is similar to the Lotka–Volterra equations, where the populations are "weighted" by their ability to occupy a given portion of the "carrying capacity." In this case, there is no carrying capacity per se but rather a collection of populations, each of which absorbs nutrients at a rate to offset its mortality.

Uptake of Free Nutrients: Self-Regulation

The influence of the size of a population j on its own uptake rate is

$$\frac{dU_j}{dN_j} \qquad (6.29)$$

$$= \frac{d}{dN_j} \frac{(dP_T/dt \bullet (V_{mj}/K_{mj}) \bullet N_j)}{(\Sigma((V_{mi}/K_{mi}) \bullet N_i) + (V_{mj}/K_{mj}) \bullet N_j)}$$

$$= \frac{(dP_T/dt) \bullet (V_{mj}/K_{mj}) \bullet (\Sigma((V_{mi}/K_{mi}) \bullet N_i))}{(\Sigma(V_{mi}/K_{mi}) \bullet N_i)^2}, \qquad (6.30)$$

which is the partial derivative of equation 6.22, with respect to N_j. As the phosphorus concentration at equilibrium will be equal to the input rate divided by the phosphate uptake potential ($\Sigma((V_{mi}/K_{mi}) \bullet N_i)$; see eq. 6.21),

$$\frac{dU_j}{dN_j} = \frac{P \bullet (V_{mj}/K_{mj}) \bullet (\underset{i \neq j}{\Sigma}((V_{mi}/K_{mi}) \bullet N_i))}{(\Sigma(V_{mi}/K_{mi}) \bullet N_i)}$$

$$\qquad (6.31)$$

Expanding the left half of equation 6.18, and recalling the uptake rate for a population is $U_i = P \bullet (V_{mi}/K_{mi}) \bullet N_i$ (eq. 6.19):

$$\frac{dU_j}{dN_j} = U_j/N_j - \frac{U_j \bullet (V_{mj}/K_{mj}) \bullet N_j}{(\Sigma(V_{mi}/K_{mi}) \bullet N_i) \bullet N_j} \qquad (6.32)$$

$$\frac{dU_j}{dN_j} = U_j/N_j - \frac{U_j \bullet U_j}{U_T \bullet N_j} \qquad (6.33)$$

$$= U_j/N_j \bullet (1 - U_j/U_T). \qquad (6.34)$$

The effect of changes in population size are minimal if U_j makes up a large portion of the nutrient uptake by phytoplankton. If U_j is small ($U_j \ll U_T$), then dU_j/dN_j will be proportional to the population's per capita uptake rate (U_j/N_j), as a change in the population size will only marginally affect the total community uptake potential.

It is noted that the effect of population j on the per capita nutrient uptake rate of population i (which is directly proportional to the per capita growth rate) is

$$\frac{du_i}{dN_j} = \frac{- (V_{mj}/K_{mj})}{\Sigma((V_{mi}/K_{mi}) \bullet N_i)} . \qquad (6.35)$$

The effect of the presence of a competitor is proportional to its ability to take up a renewed resource, relative to the total community capacity to take up that resource. This is similar to the Lotka–Volterra equations, where the populations are weighted by their ability to occupy a given portion of the carrying capacity. In this case, there is no carrying capacity per se but rather a collection of populations, each of which absorbs nutrients at a rate to offset its mortality.

Uptake of Nutrients in Grazer Guts

The uptake of phosphate by phytoplankton as described above assumed equal access to grazer-released phosphate by all exploiters. In fact, this may not be the case. Phytoplankton which can be ingested without dying ("resistant" phytoplankton) may take up phosphate while in the grazer's gut (Porter 1973, 1976, 1977). The dynamics of this uptake are difficult to quantify precisely but are vastly different from uptake dynamics from the bulk medium. This uptake of nutrients from within the grazers' guts also will reduce the excretion rate of phosphate into the bulk medium.

The uptake rate of phosphorus by phytoplankton while in the grazer gut is difficult to characterize using traditional chemical methodology. It is probable that the concentration of phosphate in the gut is high; as phosphate excretion is proportional to food intake, at least at high food concentrations (Peters 1975), it is reasonable to assume that phosphate ion production is partly associated with digestion. If the phytoplankter takes up phosphate in the gut with the aid of active transport, Michaelis–Menton kinetics may again be used to model uptake:

$$V = V_m \cdot ([P]/(K_m + [P])). \qquad (6.36)$$

The phosphate concentration in the gut is certainly much greater than K_m, at least for enzymes associated with uptake from the bulk medium, so equation 6.19 may be restated as:

$$V = V_m \cdot ([P]/[P])) = V_m. \qquad (6.37)$$

A similar formulation applies if the rate-limiting step is determined by a finite storage capacity for nutrients. The rate of uptake for the entire resistant phytoplankton population is

$$U_r = N_r \cdot V_m \cdot C(N_r) \cdot t \qquad (6.38)$$

where

U_r = the population nutrient uptake rate,

N_r = the population density,

$C(N_r)$ = the per capita clearance rate of resistants by grazers, also equal to the reciprocal of the turnover time of water being filtered or searched, and

t = the average time a cell is in the grazer's gut.

The clearance rate $[C(N_r)]$ is independent of phytoplankton density at the low densities

found in Lake Mitchell during the summer. A different V_m may also be involved in the uptake of nutrients at high concentrations. The time spent in a grazer's gut is not known and is not clearly dependent on the density of phytoplankton. By reducing all terms independent of N, equation 6.38 becomes:

$$U_r = N_r \cdot V_{Rr}, \qquad (6.39)$$

in which V_{Rr}, the uptake rate by resistant algae in guts, may be calculated directly.

Competition between Digestion-Resistant and -Susceptible Cells

The uptake of phosphate from within the gut of a grazer reduces the grazer's rate of phosphorus release to the bulk medium. Since the total uptake of phosphate from the bulk medium by phytoplankton from the water is equal to the release rate by grazers, this uptake will proportionately decrease the uptake of phosphate from the water by all phytoplankton:

$$U_A = U_T - \Sigma(N_r \cdot V_{Rr}), \qquad (6.40)$$

where

U_A = the total release of nutrients to the bulk medium,

U_T = the total release of nutrients by the grazer,

and

$\Sigma(N_r \cdot V_{Rr})$ = the sum of nutrient uptake by individuals of all resistant species r while in the grazer's gut.

To maintain the constancy of units and to relate population sizes to nutrient requirements and nutrient uptake rates, populations may be measured as their nutrient-content equivalent. The rate of nutrient uptake for phytoplankton without access to the nutrients in grazer guts is still described by equation 6.22, with the total nutrient released by the grazer to the medium reduced by gut uptake by resistants:

$$U_j(\text{flagellates}) = \frac{U_A \cdot (V_{mj}/K_{mj}) \cdot N_j}{\Sigma((V_{mi}/K_{mi}) \cdot N_i)} \qquad (6.41)$$

$$= \frac{(U_T - \Sigma(N_r \cdot V_{Rr})) \cdot (V_{mj}/K_{mj}) \cdot N_j}{\Sigma((V_{mi}/K_{mi}) \cdot N_i)}, \qquad (6.42)$$

and for those capable of taking up nutrients in the gut is

$$U_j(\text{resistants}) = \frac{(U_T - \Sigma(N_r \cdot V_{Rr})) \cdot (V_{mj}/K_{mj}) \cdot N_j}{\Sigma((V_{mi}/K_{mi}) \cdot N_i)} + N_j \cdot V_{Rj} \tag{6.43}$$

(see eq. 6.19).

The presence of phytoplankton in the gut will affect the potential growth rates of other phytoplankton populations because fewer nutrients are released to the bulk medium, so equations 6.42 and 6.43 must reflect that change. To calculate the effect of densities of different types of phytoplankton, the derivatives of equations 6.43 and 6.42 are calculated for resistant and susceptible algae, respectively. For digestion-susceptible cells, the influence of the change in abundance of digestion-resistant cells is:

$$\frac{\partial U_j}{\partial N_k} = \frac{\partial}{\partial N_k} \frac{[((U_T - \Sigma(N_r \cdot V_{Rr})) - (N_k \cdot V_{Rk})) \cdot (V_{mj}/K_{mj}) \cdot N_j]}{\Sigma_{i \neq k}((V_{mi}/K_{mi}) \cdot N_i) + (V_{mk}/K_{mk}) \cdot N_k} + N_j V_{Rj} \tag{6.44}$$

$$= \frac{[- \Sigma((V_{mi}/K_{mi}) \cdot N_i) \cdot V_{Rk} - U_A \cdot (V_{mk}/K_{mk})] \cdot (V_{mj}/K_{mj}) \cdot N_j}{(\Sigma(V_{mi}/K_{mi}) \cdot N_i)^2}. \tag{6.45}$$

or, alternatively, recalling that $U_k/U_A = [(V_{mk}/K_{mk}) \cdot N_k]/[\Sigma((V_{mi}/K_{mi}) \cdot N_i)]$,

$$\frac{dU_j}{dN_k} = \frac{-[V_{Rk} + (V_{mk}/K_{mk})]}{\Sigma((V_{mi}/K_{mi}) \cdot N_i)}. \tag{6.46}$$

The influence of the density of phytoplankton on any population with access to gut nutrients does not change from the previous model. The basal growth rate, however, will probably be higher, with members of other populations causing smaller relative changes in its growth rate. This is in part due to the increase in rate of nutrients released.

The influence of the density of a phytoplankton population on its own growth rate also will change if it has access to gut nutrients:

$$\frac{\partial U_j}{\partial N_j} = \frac{\partial}{\partial N_j} \frac{[(U_T - \Sigma(N_r \cdot V_{Rr}) - (N_j \cdot V_{Rj}) \cdot (V_{mj}/K_{mj}) \cdot N_j + N_j \cdot V_{Rj}]}{\Sigma((V_{mi}/K_{mi}) \cdot N_i)} \tag{6.47}$$

$$= [U_j/N_j \cdot (1 - U_j/U_T)] + V_{Rj}. \tag{6.48}$$

As U_j approaches U_T, the effect of changes of a population's density on its own growth rate becomes small. Not surprisingly, the effect is also small for small U_j/N_j, that is, for cells with low per-capita nutrient uptake rates while in the bulk medium. Populations that (hypothetically) rely only on nutrient sources within grazer guts have no effect of their density on their per capita uptake rate:

$$\partial U_j/\partial N_j = V_{Rj}. \tag{6.49}$$

This condition will persist only near the point of equilibrium. In summary, although phytoplankton are not interpreted as directly interacting, the effect of a change in one's population size on the density of the other may be modeled by assuming that nutrient concentrations will respond extremely rapidly to changes in the populations of exploiters. By excluding nutrients from direct consideration while still realistically modeling the sources and sinks for those nutrients, the competitive effects of the density of one species of phytoplankton on the growth rate of another may be predicted.

Partitioning of Phosphate Flow in Lake Mitchell

The derivation of all interaction coefficients describing nutrient transfers requires the estimation of nutrient flows from grazers to phytoplankton. These transfer rates are difficult to assay directly, as *Daphnia* release nutrients simultaneously to resistant cells in their gut and to the ambient medium, and resistant cells acquire nutrients both from the bulk medium and from grazers' guts. Those flows may be quantified, however, by partitioning the total nutrient flow out of *Daphnia* into two separable pools of phytoplankton and by studying the nutrient uptake capability of the two pools in the open water. With knowledge of the total rate of phosphate excretion by the *Daphnia*, it is possible to calculate the magnitude of transfers from the grazer to the two phytoplankton groups.

PROCEDURE. Only the transfer of phosphorus was measured in these nutrient partitioning experiments, because of both the availability of radioisotopes of phosphorus and the status of phosphorus as by far the most limiting nutrient in the lake. *D. rosea* were fed a diet of ^{32}P-labeled *Cryptomonus erosa* for 2 weeks to label them uniformly with the radiophosphorus. Body loads were approximately 0.02 μCi (650 CPM) ^{32}P per animal. It is not known whether such a body load can affect the physiology of a *Daphnia*; loads in excess of this were reached by Lehman and Scavia (1982a,b). Phytoplank-

ton of two size ranges were used as nutrient sinks. This caused the two species to be differentially ingested and also allowed for mechanical separation of the cell populations after exposure to nutrients. The two species of phytoplankton used were *Scenedesmus* sp., isolated from Lake Mitchell *Daphnia* feces as the ingestible fraction, and a large *Pediastrum* (mean diameter = 52 μm) as the inaccessible fraction. The experimental design required careful selection of both the phytoplankton separation procedure and the size of the *Daphnia* used as the phosphorus source. *Pediastrum* were originally prepared by filtering a whole culture twice through 37-μm Nitex netting using low-vacuum ($\approx$ 1 mm Hg) filtration, and only those fractions retained twice by the filter were used in the experiment. *Scenedesmus* sp. ($\bar{x}$ = 22 μm, sd = 3 μm) were sieved through the 37-μm Nitex to eliminate large clumps of cells, but their sizes were otherwise fairly uniform. Separation of the two fractions was fairly efficient, with less than 15% overlap. To confirm this, single-species cultures were presorted as described above, mixed, and labeled with 20,000 CPM of ^{32}P, incubated, and refractionated as though they had been mixed with the other culture. All *Daphnia* used in this experiment were shorter than 1.5 mm (Burns 1968), to further ensure that only the smaller fraction would be ingested. This selection excluded only the largest adult *Daphnia*.

During the nutrient transfer experiments both species of phytoplankton were exposed to phosphorus from 50 radioactively tagged *Daphnia* for 12 hours in a 500-ml flask. *Daphnia* either were stored in a flask for 12 hours, then removed and the water (and excreted ^{32}P) filtered into the phytoplankton cultures ("*Daphnia* isolated"), or the *Daphnia* were placed directly in contact with the two phytoplankton species ("*Daphnia* contact"; see fig. 6.5). Four replicates of each treatment were performed. After 12 hours the experiments were then filtered through 500-μm Nitex netting to remove *Daphnia*, and the water containing phytoplankton was sieved through a 37-μm Nitex mesh. Each phytoplankton fraction was then filtered onto Whatman GF/C glass filters, and the activity of the two fractions was measured by liquid scintillation in a Beckman 101 liquid scintillation counter.

The two phytoplankton species' rates of phosphate uptake from the ambient medium

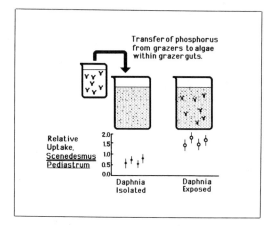

Figure 6.5. *Diagram of the experimental design for fractionating the flow of phosphorus from* Daphnia *to phytoplankton. Grazers in one design were isolated from all phytoplankton, ensuring that all phosphate excretions were released for input from the ambient medium. In the other design, grazers were placed in intimate contact with nutrient exploiters, allowing uptake while inside grazer guts as well as from the ambient medium.*

were measured by exposing both species of phytoplankton to the excretion of isotopically labeled *Daphnia* isolated from the phytoplankton. This prevented differential ingestion of the two species and ensured that no uptake was from the grazer guts. The phytoplankton mixture was then fractionated using the same procedures as in the grazer contact experiments.

RESULTS. The efficiency of sieves at segregating the two species of phytoplankton approached 90% (Levitan 1986). Assays also confirmed that the selected *Daphnia* grazed predominantly on the smaller fraction (Levitan 1986). In the following analyses all fractionation efficiencies are assumed to be ideal to simplify calculations.

The ^{32}P content of *Pediastrum* and *Scenedesmus* either exposed to *D. pulicaria*'s radioactive excretions or in contact with radioactive *D. pulicaria* are shown in table 6.3. Raw counts shown represent counts of samples minus 120 CPM background from the scintillation counter. No corrections for quench were made, as quench was comparable for all samples. The means of those counts are shown in lines 1 and 2 of table 6.4. Approximately 16% of the *Scenedesmus* died during the experiment, reducing that population's capacity to absorb nutrients from grazer guts. The derivation of all phosphate flows is outlined below.

The uptake of ^{32}P by the *Scenedesmus* while inside the *Daphnia* gut was calculated by subtracting uptake while outside the *Daphnia* from the total uptake. The uptake of phosphate from the ambient medium by *Scenedesmus* while exposed to *Daphnia* grazing was calculated in line 3 of table 6.3 by multiplying the uptake of the *Pediastrum*, essentially an "internal standard," by the relative uptake rates of *Scenedesmus* and *Pediastrum* when isolated from *Daphnia*, as calculated in line 1. It is estimated, however, that the densities of *Scenedesmus* fell (due to mortality) so that its average density was only 84% of the original density. Its uptake rates from the ambient medium are shown corrected in line 4. The uptake by *Scenedesmus* while in the gut is calculated by difference in line 5. The uptake of ^{32}P is converted into phosphate uptake in line 6, using Peters and Rigler's (1973) equation for excretion as a function of *Daphnia* length (p. 835) for poorly fed *Daphnia*. At this concentration of *Daphnia* fully half of the *Scenedesmus*' phosphorus supply is acquired while in the *Daphnia*'s gut. This fraction could be expected to decrease with decreasing *Daphnia* density and decreasing time in the grazer's gut.

Although these experiments revealed only the relative phosphate transfer rates, the total phosphate excretion rate for *Daphnia* is a well-established parameter (Peters and Rigler 1973; Scavia and Gardner 1982), from which the actual phosphate transfer rates to the two phytoplankton species could be estimated. These experiments measured only phosphorus released by the *Daphnia* itself, rather than that released by other phytoplankton being digested. However, Scavia and Gardner found that phosphorus excretion rates by *Daphnia* remained high for up to 11 days after the *Daphnia* was

Table 6.3. *Counts of the two size fractions of phytoplankton exposed to excretion of 50 radioactively labeled (1300 CPM^{32}P/individual) Daphnia in 500 ml filtered lake water for 12 hours, either isolated from the grazer (excretion of grazer filtered) or in contact with the grazer*

| | | Activity (counts per minute) | |
| | | *Scenedesmus* ($<37\ \mu m$) | *Pediastrum* ($\geq 37 \mu m$) |
Calculation	Treatment		
Raw counts for three replicates (CPM)	*Daphnia* isolated	7551	9211
		6690	10510
		8765	12108
		6914	11235
	Daphnia contact	9662	6452
		10676	7387
		11333	6660
		13001	7820

Table 6.4. *Computation of the partitioning of the excreted phosphate by grazers to the two phytoplankton pools*

Calculation	Species: Sizes: Cell treatment/ml:	Scenedesmus (< 37 μm) 4500	Pediastrum (≥ 37 μm) 1000	Source of PO$_4$
		Activity (counts per minute)		
Mean values (S.D.) of uptake by fractions in two treatments (CPM)	Daphnia isolated	7482 (1431)	10766 (1243)	Ambient
	Daphnia contact	11168 (1350)	7079 (628)	Ambient and gut
Uptake from ambient medium (CPM)	Daphnia contact	4919 = 7079. • (7482/10766)	7079	Ambient
Uptake from ambient medium, correct for 0.16 mortality (CPM)	Daphnia contact	4132 = 4919. • 0.84	7079	Ambient
Uptake from grazer gut (CPM)	Daphnia contact	7036 = 11168. − 4132	0	Gut
		Phosphate Flux Summary		
Uptake of phosphate[a] (assume 1.72 μg P released total as 12,990 CPM ^{32}P)	Daphnia contact	0.41 μg P 0.70 μg P	0.62 μg P 0	Ambient Gut

$$\frac{\text{Phosphate uptake/cell}}{(Daphnia/L).hr} = \frac{0.70/4.5 \cdot 10^6}{100 \cdot 12} = 1.3 \cdot 10^{-10} \mu g \; P/cell/hr \; Gut^{b}$$

[a]In phosphate units.
[b]At 1.0 Daphnia/L.

isolated from a food source. As digested phytoplankton would probably contribute proportionately more phosphate to other cells than the whole Daphnia, the estimation of the gut nutrient uptake is probably conservative. The experiments also were based on the assumption that both phytoplankton fractions absorbed phosphorus while suspended and that the ingested phytoplankton absorbed additional phosphorus while in the Daphnia's gut. The uptake of phosphorus by the inaccessible fraction was therefore a measure of the phosphorus taken up by the ingestible cells while not inside the Daphnia.

The flux of phosphate from grazers to phytoplankton may be estimated from a knowledge of the community composition and of the types of interactions between the populations present. The Lake Mitchell ecosystem is considered here as containing two phytoplankton groups and two zooplankton groups. Principal interactions consist of consumption and excretion by grazers, the subsequent mortality of phytoplankton groups, and the uptake of nutrients by the two phytoplankton groups.

The components and their densities in the model community, based on averages of the Mitchell midsummer community, are outlined in the left halves of table 6.2. Also tabulated are measured or literature values for phytoplankton mortality and phosphate excretion by grazers, as well as other descriptors of interactions. Phosphate uptake by each phytoplankton group was calculated by setting it equal to the loss from predation, that is, the mortality rate times the population phosphorus content. The uptake of phosphate from grazer guts was calculated from the uptake per passage through grazer guts by resistant cells times the clearance rate of Daphnia, less mortality to the resistant phytoplankton, and the uptake in the bulk medium by difference. It is interesting to note that the per capita uptake of phosphate by flagellates in the bulk medium is much higher than that of resistants; almost eight (9,000/1,000 • .52/.59) times as high. This range of uptake capacities is large, but less than that previously described (Nalewajko and Lean 1980). It is also remarkable that fully 40% of the resistants' phosphorus is absorbed after ingestion, representing a 180-fold increase in phosphorus uptake rates over that in the ambient medium (assuming a 15-minute gut residence period 0.34 times per day).

Phytoplankton Mortality

Population dynamics of phytoplankton are strongly influenced by grazing pressure. The mortality to flagellates caused by grazers was equivalent to ingestion rates; no flagellates were observed intact after ingestion. The mortality of flagellates was directly proportional to grazer densities. Phytoplankton densities during the summer were below those causing saturation of feeding rates by grazers. DeMott (1982) reported *Daphnia* beginning feeding saturation at 10,000 *Chlamydomonas* per milliliter and *Bosmina* at 4,000 per milliliter. Midsummer concentrations were always below this.

Mortality to Resistant Algae

As outlined above, phytoplankton may survive ingestion by a zooplankter, and even receive some benefit from ingestion (Porter 1976) through access to limiting nutrients while in transit through the gut. As nutrient availability often limits phytoplankton growth, and ambient nutrients may be immeasurably low, this benefit can be considerable. Any gain in nutrients after ingestion is offset, however, by the possibility of mortality during ingestion and digestion. Any measure of the benefits of ingestion must balance this mortality. Two approaches were used to quantify the effect of the grazers' density on the mortality of digestion-resistant algae. The first measured the change in population growth rates of phytoplankton in field enclosures as grazers were eliminated. The second was based on mortality to phytoplankton under laboratory conditions.

PROCEDURE. During the summer of 1980, paired 12,000-L enclosures in Lake Mitchell were enriched with 1,000 μg/L nitrate–nitrogen and 150 μg/L phosphate–phosphorus to remove any nutrient limitation to phytoplankton. Zooplankton were then eliminated from half of the enclosures by introducing zooplanktivorous creek chub (*Semotilus atromaculatus*). Phytoplankton densities were monitored using Lugols-preserved samples. The differences in phytoplankton population growth rates were attributed to differences in grazing pressure, since saturating nutrient concentrations eliminated any advantage to being ingested attributable to access to gut nutrients.

Mortality to resistant cells following ingestion also was measured using cultured resistant cells in the laboratory. *Scenedesmus* sp. and *Pediastrum* sp. (not the strain used for the phosphate-segregation study above) were isolated by L. Zelazny and myself from the guts of *Daphnia* collected from Lake Mitchell. Freshly collected *Daphnia* were rinsed in 0.45-μm-Millipore-filtered lake water to rinse out algae caught within their carapace and were then transferred and fed glass beads. The ejected feces containing algae were then diluted and cultured in MBL f/2 medium. Cultures of *Scenedesmus* and *Pediastrum* were prepared for experiments by being grown in 0.45-μm-Millipore-filtered lake water supplemented with 75 μg PO_4–P/L and 850 μg NO_3–N/L. Those cultures were subdivided and incubated under five levels of *Daphnia* herbivory, from controls (no *Daphnia*) to 100 *Daphnia* per liter. Four replicates were run for each density of *Daphnia*. Cultures were examined each day to ensure that all *Daphnia* were alive. Dead animals were removed by pipette and counted, and an equal number of live animals was replaced. At the end of the experiment, generally 4 days, the phytoplankton were counted. A parallel set of experiments measured the effect of nutrient concentration on the mortality caused by the grazing of *Daphnia*. In these experiments *Scenedesmus* was mixed with *Cryptomonas* in cultures and exposed to high (75 μg PO_4–P/L and 850 μg NO_3–N/L) or low (ambient, ≤ 5 μg PO_4–P/L) nutrient levels and high or low (30 or 0 *Daphnia* per liter) herbivore densities and allowed to grow for 4 days. All cultures were preserved after 4 days and counted using the Utermohl technique.

RESULTS. The field experiment was marred by rapid regrowth of *Daphnia* and *Bosmina* in the depleted enclosure, as well as increased densities in the undepleted enclosure. There were significantly higher densities of flagellates in the depleted enclosure, but the confounding conditions make interpretation difficult. The results from laboratory experiments are listed in tables 6.5 and 6.6 as population sizes at the end of the experiment and as calculated growth rates. Growth rates were assumed constant for the duration of the experiment and were calculated from final concentrations of phytoplankton:

$$r = \ln(N_t/N_0)/t = \qquad (6.50)$$

$$\ln ((\text{final cell/ml})/1,000)/\text{days}$$

The *Pediastrum* experiment with the highest density of *Daphnia* was omitted because of the small number of cells and the difficulty in recognizing those remaining; other experiments were terminated early as indicated. Those experiments revealed two important phenomena. First, higher grazer densities were associated with declines in phytoplankton population growth rates. The suppression was generally much less than the calculated grazing rate, however. Second, relative death rates increased with increasing grazing rates. A polynomial regression analysis of growth rates as functions of grazer densities confirmed significant first- and second-order relationships (table 6.7) for grazing on *Scenedesmus*, and significant first-order relationships for *Pediastrum*. *Pediastrum's* regression coefficient was higher than that expected from grazing, and its Y-intercept (growth free of grazers) larger than expected, indicating similarities to *Scenedesmus'* second-order relationship. This increase in mortality per ingestion may have been influenced by the change in average times between ingestions, with a loss of recovery time leading to decreased survival. Even under rather low grazing pressure, some individual phytoplankters will be reingested soon after egestion. As ingestion is a rare discrete event and ingestion episodes are independent (*Daphnia* release loose uncompacted feces, the contents of which are probably immediately ingestible by *Daphnia*), the number of ingestions per cell per time interval should follow a Poisson distribution. Repeated ingestions in short time intervals may be a source of mortality not represented by extrapolation from rare ingestion events. A frequency distribution of times between ingestion events may be derived through computer simulation by assuming ingestion episodes are rare and unassociated (Fig. 6.6). The distribution is identical, in fact, to an age distribution of a population with constant age-independent mortality, except that it is a horizontal, rather than a vertical, life table. The mortality caused by grazing seems not to be affected by the nutritional state of the phytoplankton. The difference in growth rates of grazed and ungrazed populations falls from 0.12/day to 0.21/day with the addition of nutrients; any increase in cell death expected

Table 6.5. *The cell densities (cells/ml) and calculated population growth rates (days⁻¹) of* Scenedesmus *and* Cryptomonas *together in each of four replicates of each treatment at the end of 4 days*

Phytoplankton treatment species	Density (growth rate, days^{-1})				Geometric mean density (mean r)
Scenedesmus					
CC	1491 (0.10)	1821 (0.15)	1433 (0.09)	1174 (0.04)	1462 (0.10)
CN	3553 (0.317)	3089 (0.282)	2509 (0.23)	4087 (0.35)	3257 (0.29)
ZC	842 (−.043)	662 (−.103)	1597 (0.117)	852 (−.04)	918 (−.02)
ZN	1416 (0.087)	1492 (0.10)	1297 (0.065)	1323 (0.07)	1378 (0.08)
Cryptomonas					
CC	4953 (0.40)	3597 (0.32)	6050 (0.45)	5365 (0.42)	4903 (0.40)
CN	36161 (0.897)	16510 (0.701)	66686 (1.05)	29964 (0.85)	33051 (0.87)
ZC	1008 (0.002)	662 (−.103)	1682 (0.13)	818 (−.05)	979 (−.01)
ZN	7690 (0.51)	5366 (0.42)	11473 (0.61)	6553 (0.47)	7463 (.50)

The original density of each species was 1000 cells/ml. Treatments included growth in lake water with no zooplankton (CC), lake water enriched with 75 μg PO$_4$-P/L and 850 μg NO$_3$-N/L (CN), unenriched lake water plus 20 adult *Daphnia*/L (ZC), and enriched lake water plus *Daphnia* (ZN).

Table 6.6. *The cell densities (cells/ml) and calculated population growth rates (days^{-1}) of* Scenedesmus *and* Pediastrum, *each grown with* Cryptomonas *in four replicates of each* Daphnia *density, at the end of 2 to 4 days*

Phytoplankton species and *Daphnia*/L (days)	Density (growth rate, days^{-1})				Geometric mean density (mean *r*)
Scenedesmus					
0 (4)	3553 (0.317)	3089 (0.282)	2509 (0.230)	4055 (0.35)	3251 (0.29)
15 (4)	2225 (0.20)	1973 (0.17)	1377 (0.08)	3189 (0.29)	2096 (0.18)
30 (4)	1416 (0.087)	1492 (0.10)	1297 (0.065)	1323 (0.07)	1380 (0.08)
50 (3)	280 (−0.42)	330 (−0.37)	430 (−0.28)	215 (−0.51)	305 (−0.39)
80 (2)	120 (−1.06)	150 (−0.95)	60 (−1.41)	100 (−1.14)	102 (−1.14)
Pediastrum					
0 (4)	1320 (0.07)	2710 (0.25)	1822 (0.15)	1973 (0.17)	1895 (0.16)
15 (4)	564 (−0.143)	1635 (0.123)	1280 (0.062)	930 (−0.02)	1022 (0.005)
30 (3)	330 (−0.37)	740 (−0.10)	470 (−0.25)	515 (−0.22)	492 (−0.24)
50 (3)	55 (−0.96)	110 (−0.74)	75 (−0.85)	205 (−0.525)	100 (−0.77)

The original density of each species was 1000 cells/ml. All treatments were run in lake water enriched with 75 μg PO$_4$-P/L and 850 μg NO$_3$-N/L.

Table 6.7. *Polynomial analysis of growth rates (days^{-1}) of the phytoplankton* Scenedesmus *and* Pediastrum *under varying* Daphnia *densities*

Degree	Regression coefficient	Standard error	t-Value	Sum of squares	Degrees of freedom	F	Tail prob.
Scenedesmus growth rate (*r*) against *Daphnia* density							
0	0.301	.049	6.12	5.55	3	154.18	<0.001
1	−0.0041	.0030	−1.36	0.290	2	12.09	<0.001
2	−.000175	.000036	−4.87	0.009	1	0.76	0.396
Pediastrum growth rate (*r*) against *Daphnia* density							
0	.233	.057	4.04	1.98	3	40.42	<0.001
1	−0.018	0.002	−9.72	0.08	2	2.59	0.116

Tail probability is for test of hypothesis that higher-order polynomial should be included.

from a decline in nutrient availability was counterbalanced by the possible supplementation of nutrients released from grazing *Daphnia*. Table 6.8 lists the results of a three-way full factorial ANOVA on the three treatments (nutrients, phytoplankton species, and *Daphnia* grazing) as run on the computer program ANOVAR*** on Dartmouth College Time Share. The analysis revea 1 no interaction of nutrients and grazing ($p > 0.50$) nor of nutrients, grazing, and phytoplankton species ($p > 0.25$).

The suppression of both edible and resistant cells has been detected in Lake Mitchell. Flagellate densities are suppressed by grazing *Daphnia*, as has been demonstrated, and both their absolute densities and proportional representation in the phytoplankton decline with increases in grazer densities. At higher densities of grazers this phenomenon reverses as flagellates once again make up a larger portion of the phytoplankton. Figure 6.7 graphs the percent composition of flagellates for samples from the lake during 1977 and 1978, as well as from two unmanipulated enclosures in 1977. A polynomial regression was performed, regress-

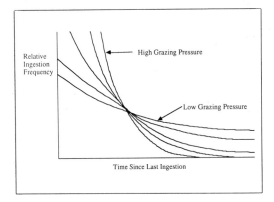

Figure 6.6. *Distribution of times since last ingestions for phytoplankton exposed to a range of grazing pressures. Model assumes probability of ingestion independent of the time since last ingestion, as well as no mortality during ingestion.*

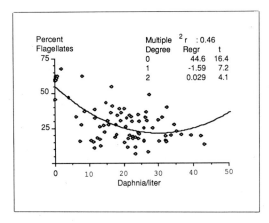

Figure 6.7. *Polynomial analysis of the composition of the phytoplankton in lake and unenriched enclosures as a function of the density of* Daphnia. *Composition is measured as the arcsin-transformed percentage of flagellates in the phytoplankton. Second-order (f(density Daphnia²)) effects account for the upward-open parabola shape of the curve.*

ing the percentage of the phytoplankton assemblage as flagellates (after arcsin transformation; see Zar 1974, p. 220) against the density of *Daphnia*. Regression coefficients also are listed in fig. 6.7. The proportion of the phytoplankton composed of flagellates did rise at the highest *Daphnia* concentrations encountered. The phenomenon of increased mortality to resistant algae is probably compounded by the alleviation of nutrient competition due to rap-

id regeneration by grazers. This phenomenon also may explain the absence of large numbers of resistant algae in more productive lakes, which also have high densities of grazers as well as the complete dominance of flagellates in enriched experimental enclosures with high *Daphnia* densities (Levitan et al. 1985).

Effect of Phytoplankton on Growth of Grazers

Daphnia's rate of egg production and population growth were found to be proportional to the density of phytoplankton, especially flagellates, in Lake Mitchell. Its highest average egg numbers occurred while the density of flagellates was also at its highest, on August 18, 1979 (DeMott 1983). *Daphnia*'s response to phytoplankton numbers is known to be highly dependent on the species of phytoplankton present. *Daphnia*'s ingestion (Lampert 1977a; Kerfoot et al. 1985a) and assimilation rates (Lampert 1977b), and consequently its rates of growth and egg production (Weglenska 1971), are all dependent on the species of phytoplankton present. Those species previously found to be assimilated by *Daphnia* with the highest efficiencies have included green alga and flagellates; diatoms, gelatinous or thick-walled greens, and gelatinous blue-greens are taken up less efficiently (Schindler 1970). To confirm that Lake Mitchell *Daphnia* exhibit this differential response to food type, the demographics of the Lake Mitchell *Daphnia* were correlated with the phytoplankton present, and controlled laboratory experiments were conducted to assay *Daphnia*'s performance while feeding on different phytoplankton.

The demographics of the Lake Mitchell *Daphnia* populations have been extensively analyzed (DeMott 1983; Kerfoot et al. 1985b) from the perspective of possible competitive interactions. Their performance can be explained best by the food resources available to them. For this study the egg ratio (eggs per adult female) of the *Daphnia* species present were regressed stepwise against the densities of highly digestible phytoplankton, phytoplankton resistant to digestion (and of lower food quality), and all *Daphnia* species. Data were derived from two summers' native lake populations and from populations in two unenriched enclosures. Statistical analysis was performed using the stepwise regression package (program

Table 6.8. *Three-way full factorial ANOVA (model 1) on growth rate (days^{-1}) of phytoplankton under specified conditions*

Term	Sum of squares	D.F.		F	Prob
Nutrient	0.8221	1	24	121.79	<.0001
Daphnia	0.6064	1	24	89.83	<.0001
Phytoplankton sp.	0.8662	1	24	128.34	<.0001
Nutrient X *Daphnia*	0.0025	1	24	0.38	0.5523
Nutrient X Phytoplankton	0.2362	1	24	34.99	<.0001
Daphnia X Phytoplankton	0.1004	1	24	14.88	<.001
Nutrient X *Daphnia* X. Phytoplankton	0.0088	1	24	1.31	.2639
Within-cell error	0.162		24		
Total (corrected for mean)	2.805		31		

Table 6.9. *Summary statistics of the egg production rates of* Daphnia pulicaria *and* D. rosea *during the summers of 1978 and 1979 in Lake Mitchell and in two unenriched fish exclosures in Lake Mitchell*

		Regression coefficient (S.E.)					
Step	Zero intercept	Density of flagellates	Density of resistants	Density of *Daphnia* sp.	F-to-enter	r^2	p
Daphnia pulicaria eggs per female							
0	2.09	.00086 (.00013)			42.2	0.563	<0.0001
1	1.03	.00086 (.00013)			42.2	0.563	>0.0001
2	0.93	.00087 (.00014)	.00002 (.00005)		0.1	0.551	>.25
3	1.99	.00054 (.00017)	.00004 (.00005)	−.038 (.012)	9.3	0.648	<0.0005
Daphnia rosea eggs per female							
0	1.26						
1	−.297	.00136 (.00018)			56.8	.559	<.0001
2	1.245	.00121 (.00018)	−.00016 (.00007)		5.5	.601	<.01
3	2.39	.00058 (.00018)	−.00011 (.00005)	−.077 (.014)	29.1	.761	<.0001

[a] See Kerfoot et al. 1985b.

E_A is regressed stepwise against the total density of flagellates (*Cryptomonas*, *Chlamydomonas*, *Rhodomonas*, plus other flagellates), other algae (mainly resistant algae), and the total density of *Daphnia*. Regressions are based on 33 samples for *Daphnia pulicaria* and 45 for *D. rosea*.

P2R) of the Bio-Medical Data Processing package (Dixon 1981). Table 6.9 lists the coefficients and tests of significance for each independent variable. In these regressions the highly digestible phytoplankton were strongly correlated with egg production by all *Daphnia*, whereas phytoplankton resistant to digestion correlated only weakly, and occasionally negatively, with grazer egg numbers. The significance of the negative correlation between *Daphnia* densities and *Daphnia* egg production is discussed below.

Interactions among Grazers

Direct interactions between the grazers in Lake Mitchell were not specifically assayed in this study. Exploitative competition, the suppression of the supply of shared limiting resources by populations, is not considered to be a direct interaction. Such interactions are described in the analysis of food webs only through coupled interactions with compartments containing shared food resources. The only exception in this model is the omission of inorganic phosphorus from the model, and the resulting inclusion of competition between phytoplankton. Whereas copepod populations may interact directly through allelopathy or mechanical interference (Folt and Goldman 1982) and reduce a competitor's feeding rate, such allelopathy has not been documented for *Daphnia* (Seitz 1984). *Daphnia* may interfere with small grazers (Gilbert personal communication; Durand 1984) by physically battering them during feeding but cannot damage a grazer as large as a *Bosmina* or a juvenile *Daphnia*.

The population growth rates and egg numbers (E_A) of *Daphnia* species in Lake Mitchell have both been shown to be inversely proportional to total *Daphnia* population density (Kerfoot et al. 1985b), indicating a possible negative interaction. Exploitative competition, which is considered to be an indirect interaction in this study, has been well documented between both *Daphnia* species and *Bosmina* in Lake Mitchell (DeMott 1983; DeMott and Kerfoot 1982). The suppression of birth rates and population growth rates documented in those studies (Kerfoot et al., 1985a, 1985b) could be ascribed to changes in the density or quality of the phytoplankton on which the grazers feed. Total phytoplankton density in unenriched enclosures and in the lake was inversely proportional to the density of *Daphnia* species, as was the fraction of phytoplankton that were flagellates (fig. 6.7). Much of the variance in the number of eggs carried by *Daphnia* could be attributed (using stepwise regression techniques) to the density of flagellates and to other more resistant algae (table 6.9), with very little additional variance accounted for by differences in density of *Daphnia* per se (additional r^2 < 0.09 for *D. pulicaria* and < 0.16 for *D. rosea*). *Daphnia pulicaria* has maintained reproduction ($E_A \approx 0.15$) in Lake Mitchell enclosure enrichment experiments even when its population density rose to more than 100 per liter. Those experiments, performed in 1980, 1981, and 1982, also featured some of the lowest (< 800 cells per milliliter) phytoplankton densities seen in any field experiment. All of this evidence, although circumstantial, suggests that increasing densities of grazers in Lake Mitchell do not directly affect the growth rate of any grazer population but do so indirectly through the resource pool. The small amount of "inhibition" noted here may have resulted from some measure of food quality not measured in the analysis, although it must be noted in the community analysis.

Predation on Bosmina and Daphnia

The cyclopoid copepod *Mesocyclops* preys on smaller zooplankton as well as large algae. The functional, and especially the numeric, responses of this predator to the densities of prey have not been well explored. Brandl and Fernando (1978) found that the daily ration of *Cyclops* was closely proportional to the density of *Bosmina* at up to 5,000 *Bosmina* per liter. Predation rates were approximately 0.002 to 0.003 *Bosmina* per *Cyclops* per day. There was no evidence of saturation; in fact, predation rates rose disproportionately with prey density. *Mesocyclops* seemed to have negative electivities for the large *Daphnia* and *Ceriodaphnia*, as confirmed by Confer (1971) and Williamson (1982). Brandl and Fernando (1978) found that *Mesocyclops* could consume almost 10% of the crustacean population of a lake per day, although those statistics were derived from eutrophic ponds in which 18 to 58% of the zooplankton were *Mesocyclops*. Williamson (1982) found that *Mesocyclops* preyed on *Bosmina* in Lake Mitchell extremely inefficiently, being repelled by the hard carapace of *Bosmina*, although *Bosmina* were susceptible to predation while molting. Laboratory experiments indicated that *Mesocyclops* would consume 0.2 to 0.8 *Bosmina* per day under field densities, which is very close (0.2–0.8 vs. 0.2–0.3 *Bosmina* per *Cyclops* per day) to the findings of Brandl and Fernando (1978). Thus, *Daphnia* and *Bosmina* are presumed to have a negligible effect on *Mesocyclops* reproduction in Lake Mitchell.

Copepods are in turn preyed on by fish. For the purposes of this analysis, that predation rate is set to be a fixed fraction of the rate of fish predation on *Daphnia*, a zooplankter of comparable visibility. That predation rate is discussed below.

Fish Predation on Grazers

All fish in Lake Mitchell prey on *Daphnia* during the summer. Fish predation inflicts a mortality directly proportional to fish density (Levitan et al. 1985). That mortality was calculated by comparing *Daphnia* population growth rates (r, d^{-1}) in enclosures with and without trout. The suppression of r was approximately 0.000587 • (d^{-1}) • kg trout • ha^{-1}. The density of trout in Lake Mitchell has never been estimated directly. One may calculate crude biomass estimates from stocking rates and survival times of trout. Less than 20 kg of trout per hectare exist in Lake Mitchell, causing about 0.01/day mortality to the *Daphnia*. Although fish predation may lead to increased grazer productivity by causing changes in the phytoplankton or nutrient pools (Kitchell et al. 1975, 1978), those increases are all due to secondary effects and are explained elsewhere in the analysis of the food web.

Whereas the influence of fish populations on zooplankton assemblages is well known (cf. Brooks and Dodson 1965), especially acting through size-selective predation, the influence of zooplankton on the fish populations exploiting them is less well documented. Positive relationships between phytoplankton or zooplankton density and fish production have been assumed (Sheldon et al. 1977), although such relationships have rarely been demonstrated. For example, Mills and Forney (1982a, 1982b) showed no correlation between the growth of yellow perch and *Daphnia* density at a time that perch fed exclusively on *Daphnia*.

In Lake Mitchell, the dependency of fish growth on *Daphnia* density is reduced by several additional factors. Trout, which constitute a large proportion of the total fish population, are recruited primarily through artificial stocking. *Daphnia* are consistently found in trout stomachs during the summer, but their biomass contribution is small compared to that of crayfish and terrestrial insects (personal observation). Native recruitment seems most directly affected by events upstream in Mitchell Brook (R. Doyle, personal communication). The stomachs of creek chub (*Semiotilus atromaculatus*) in Lake Mitchell contain about equal proportions of *Daphnia* and amorphous material during the summer, but these small fish are probably less abundant than trout.

FORMAL STABILITY ANALYSIS OF THE LAKE MITCHELL ECOSYSTEM

The stability of the Lake Mitchell summer ecosystem was analyzed based on a six-compartment model: two groups of phytoplankton, two groups of zooplankton grazers, an invertebrate omnivore, and planktivorous fish (see fig. 6.1). Growth rates of all components ($f(X)$) were determined to be functions of the densities of other components, as influencing population growth ($g(X')$) or death ($d(X')$) rates:

f(flagellates) = g(free nutrients, flagellates)—d(*Daphnia*, *Bosmina*)

f(resistants) = g(free nutrients, gut nutrients, resistants)—d(*Daphnia*)

f(*Daphnia*) = g(flagellates, resistants)—d(fish, *Daphnia*)

f(*Bosmina*) = g(flagellates)—d(copepods, fish, *Bosmina*)

f(*Mesocyclops*) = g(*Bosmina*)—d(fish, *Mesocyclops*)

f(fish) = g(*Daphnia*, *Mesocyclops*)—d(fish)

Nutrients, both those free in the water and those concentrated in the alimentary canal of the grazers, are accounted for in the descriptions of the interactions of nutrient excreters and nutrient exploiters. The interaction coefficients of that ecosystem are derived from the equations formalizing those relationships, based on the "model ecosystem" detailed in Fig. 6.1., with mean densities listed in table 6.2. Once the interaction coefficients are derived, the model neither requires nor predicts the sizes and growth rates of those populations (although those parameters are used in the derivation of the interaction coefficients); it predicts the stability and magnitude of the stability (as return time from perturbations) using only the influence of the change of one population's size on another's growth rate.

This analysis was done in two stages. To appreciate the complex interactions between plants and herbivores, those four components (two grazer and two phytoplankton compartments) were analyzed first apart from the predators. The predators, whose effects on other parts of the system were suspected to be small, were then added to the system, and further analysis of the system's dynamics was performed. The portion of the community interaction matrix describing plant–herbivore interactions may be diagrammed as shown in Table 6.10.

Using the functional definitions of interaction coefficients developed in the equations above and solved with the empirically derived

Table 6.10. *Portion of the community interaction matrix that describes plant-herbivore interactions*

	Effect of			
	Flagellates	Resistants	*Daphnia*	*Bosmina*
Effect on				
Flagellates	a_{11}	a_{12}	a_{13}	a_{14}
Resistants	a_{21}	a_{22}	a_{23}	a_{24}
Daphnia	a_{31}	a_{32}	a_{33}	a_{34}
Bosmina	a_{41}	a_{42}	a_{43}	a_{44}
	Flagellates	Resistants	*Daphnia*	*Bosmina*
Flagellates	−0.137	−0.046	−0.307	−01.07
Resistants	−0.088	−0.087	−0.102	−0.00
Daphnia	+0.2	+0.066	−0.02	0.00
Bosmina	+0.06	0.00	0.00	0.00

Elements are identified in top matrix, while values for Lake Mitchell are given in bottom matrix.

6.11. *Coefficients of the plant-grazer portion of the community matrix, with values listed for all coefficients as well as the equation, table, or formula used to calculate the coefficient*

Alpha coefficient	Value	Derivation
a_{11}	−0.137	$\partial U_i/\partial N_i$ from equation 6.34: $U_i/N_i \cdot (1 - U_i/U_T)$ minus derivative of density-dependent predation: $U_r \cdot C(N_r)$
		$= 0.6/1.5 \cdot (1 - .6/1.75) - 0.4 = -.137$
a_{12}	−0.046	$\partial U_j/\partial N_k$ from equation 6.28: $-(U_j \cdot U_k)/(U_T \cdot N_k)$
		$= -(0.6 \cdot 1.15)/(1.75 \cdot 8.5)$
a_{13}	−0.307	1000 flagellates/ml $\cdot$ 8.10^{-7} mg/flag $\cdot$ 10^3 ml/l = 8 mg/L
		.0008 L/*Daphnia*/hr $\cdot$ 20 D./mg $\cdot$ 24 hr/d = .384 L/mg/d
		(Table 6.2; DeMott and Kerfoot 1982; DuMont et al. 1975)
		.384 L/mg/d $\cdot$ .8 mg/L = .307 d^{-1}
a_{14}	−1.07	1000 flagellates/ml $\cdot$ 8.10^{-7} mg/flag $\cdot$ 10^3 ml/L = .8 mg/L
		.00039 L/*Bosmina*/hr $\cdot$ 144 B./mg $\cdot$ 24 hr/d = 1.34 L/mg/d
		(DeMott and Kerfoot 1982)
		1.34/mg/d $\cdot$ .8 mg/L = 1.07 d^{-1}
a_{21}	−0.088	$\partial U_i/\partial N_j$ from equation 6.46
		$= -[(0.56/8.5)/(1.75/0.3) + (1.15/1.75)/8.5] = -.011 - .077 = -.088$
a_{22}	−0.087	$\partial U_i/\partial N_i$ from equation 6.48:$[U_j/N_i \cdot (1 - U_j/U_T)] + V_R$, less grazing
		$= 1.15/8.5 \cdot (1 - 1.15/1.75) + -0.133 = 0.046 - 0.133 = -.087$
a_{23}	−0.102	$0.33 \cdot a_{13}$ (table 6.2)
a_{24}	0.0	Low electivity (Kerfoot 1985)
a_{31}	0.20	$E_a = 0.0008 \cdot$ (flagellates/ml) (table 6.2)
		Flagellates/ml $\cdot$ $8.10^{-4} = $ mg (flagellate)/L
		Daphnia/L $\cdot$ .05 = mg (*Daphnia*)/L
		$dN/dt = N \cdot E_a/D$
		$= 20$ D./L $\cdot$ .05 mg/D. $\cdot$ 0.0008 $\cdot$ [(flag./ml)/(8.10^{-4})]/5 d
		$= 0.2$ d^{-1}
a_{32}	0.066	$= 0.33 \cdot a_{31}$ (table 6.2)
a_{33}	−0.02	Table 18 of Levitan 1986; possible allelopathy
a_{34}	0.00	No direct interaction between *Bosmina* and *Daphnia*
a_{41}	0.06	Table 18 of Levitan 1986
a_{42}	0.00	No benefit to *Bosmina* from resistants
a_{43}	0.00	No direct interaction between *Bosmina* and *Daphnia*
a_{44}	0.00	No direct interaction among *Bosmina*

values for those coefficients, the community interaction matrix may be established. For the purpose of maintaining uniform units between populations, all populations and growth rates were based on biomass per volume, as milligrams per liter, or the biomass equivalent for inorganic nutrients, with all rate measures converted to days^{-1}. The derivation of the first 16 coefficients is briefly outlined with individual comments in table 6.11. The resulting submatrix is shown in table 6.10.

The submatrix was then analyzed for Lyapunov stability through the method of Hurwitz. All analyses were done with the program ROOT written in BASIC8 (Levitan 1986, Appendix 1). Occasional checks were performed with the computer package MACSYMA (Bogen 1977).

Analysis shows this community to be stable. A few features of this community matrix bear inspection. First, competition between phytoplankton types is recognized, is modeled as being a direct interaction, and is described as being strong. Generally speaking, each phytoplankton type specializes on one primary source of nutrients (bulk medium vs. in grazers' guts), so intraspecific competition is slightly stronger than interspecific competition. The model describes both *Daphnia* and *Bosmina* preying

strongly on flagellates but only *Daphnia* having a strong effect on the density of resistant phytoplankton, approximately one-third of its effect on the susceptible phytoplankton. Grazers do not interact directly, except for the weak intraspecific interaction of *Daphnia* possibly describing allelopathy. Both grazers benefit from the presence of flagellates, with *Bosmina* the more specialized feeder. Only *Daphnia* benefits from resistant phytoplankton.

The origin and the limits of the stability of the system may be outlined by slowly changing parameters in the system and monitoring the effect of the change on the stability of the system. Interpreting the biological significance of each change was fairly straightforward, using the equations functionally describing the interaction coefficients. Selected changes in the interaction matrix, the interpretation of those changes, and the consequences to the community are listed in table 6.12. Generally, any change that directly affects the balances between differential growth and death rates for the two populations of phytoplankton acts to destabilize the community.

The interactions between predators and the components of the grazer–phytoplankton group are fairly weak, due to both the externally limited nature of the fish populations and

Table 6.12. *Stability analysis of the phytoplankton-grazer portion of the community interaction matrix*

Interaction coefficient	Change[a]	Biological interpretation	Stability status
a_{11}	−0.1 to −0.08	Reduction in competition for nutrients in bulk medium, increase in nutrient levels	Destabilize
a_{12}	−0.08 to 0.12	Varying interactions between the phytoplankton while free in water has no effect	Stability maintained
a_{13}	−0.12 to −0.06	Reduction in efficiency of *Daphnia* as grazer, refuge for flagellates	Destabilize
a_{14}	−0.04 to 0.00	Reduction in feeding efficiency by *Bosmina*, loss of control over flagellates	Destabilize
a_{22}	−0.10 to −0.08	Reduction in competition for external nutrients, loss of source from gut	Destabilize
a_{23}	−0.04 to −1.0	Increase in mortality of resistant algae, increased digestion by the *Daphnia*	Destabilize
a_{31}	0.04 to 0.10	Increased feeding or growth efficiency of *Daphnia* feeding on flagellates.	Destabilize
a_{32}	0.01 to 0.03	Increased ability of *Daphnia* to digest and assimilate resistant algae	Destabilize
a_{41}	0.06 to 0.1	*Bosmina* lose ability to effectively exploit flagellates	Destabilize

Stability was determined through the use of Routh–Hurwitz criteria as each element was individually varied.
[a]Describes approximately the minimal change in each element required to cause a shift in stability.

the inefficiency of the predator *Mesocyclops*. The entire community interaction matrix may be diagrammed as shown in Table 6.13.

Interaction coefficients reflect the large size of members of the added population (fish), reflecting both predatory interactions and nutrient excretion (hence, positive interactions of fish with both phytoplankton populations). The coefficients added at this stage were derived from experimental data (Levitan et al. 1985), literature estimates, or best estimates based on relative predation rates. The derivation of these coefficients is outlined below and individual comments given in table 6.14. As described below, there is wide latitude in the values of many coefficients that permit stability.

These coefficients reflect the omnivorous diet of *Mesocyclops*. They also reflect its poor performance in capturing and assimilating the cladocera *Bosmina* and *Daphnia*. Fish are described as effective predators on both *Daphnia* and *Mesocyclops*. Much of their diet comes from nonplanktonic biomass, such as terrestrial insects, crayfish, and other fish (personal observation). In addition, fish recruitment is not strongly affected by conditions in the planktonic community, being influenced by watershed events (R. Doyle, personal communication) and management practices. Fish dynamics are therefore initially described as self-limiting (negative

value for a_{66}), with this condition relaxed during further analysis.

The community interaction matrix was subject to another sensitivity analysis by varying the coefficients involving the two predators within biologically realistic ranges and observing the changes in the stability of the system (see table 6.15). The entire system remained stable as each of the interaction coefficients was varied severalfold in magnitude. This continued stability implies that the interactions between phytoplankton groups and the two grazers establish a community with robust stability. Although perturbations in the populations of fish and copepods might conceivably lead to fluctuations (all eigenvalues of the community matrix had non-zero imaginary portions, indicating a certain amount of oscillatory behavior of the system during recovery), the system should tend to return to near the original equilibrium state after a perturbation.

DISCUSSION AND SUMMARY

A stable community will perpetuate itself through simple persistence and resilience after perturbations. A stable community will not be at its point of equilibrium at all times but will tend to return to it, directly or with oscillations, over time. Properties of communities

Table 6.13. *Complete community interaction matrix*

Effect on	Effect of					
	Flagellates	Resistants	*Daphnia*	*Bosmina*	*Mesocyclops*	Fish
Flagellates	a_{11}	a_{12}	a_{13}	a_{14}	a_{15}	a_{16}
Resistants	a_{21}	a_{22}	a_{23}	a_{24}	a_{25}	a_{26}
Daphnia	a_{31}	a_{32}	a_{33}	a_{34}	a_{35}	a_{36}
Bosmina	a_{41}	a_{42}	a_{43}	a_{44}	a_{45}	a_{46}
Mesocyclops	a_{51}	a_{52}	a_{53}	a_{54}	a_{55}	a_{56}
Fish	a_{61}	a_{62}	a_{63}	a_{64}	a_{65}	a_{66}
	Flagellates	Resistants	*Daphnia*	*Bosmina*	*Mesocyclops*	Fish
Flagellates	−0.137	−0.046	−0.307	−1.07	−0.02	0.00
Resistants	−0.088	−0.087	−0.102	−0.00	−0.01	0.00
Daphnia	+0.20	+0.066	−0.02	0.00	0.00	−0.006
Bosmina	+0.06	+0.00	0.00	0.00	−0.05	0.00
Mesocyclops	+0.02	+0.01	0.00	+0.002	0.00	−0.006
Fish	0.00	0.00	+0.0012	0.000	+0.0012	−0.002

Elements are identified in top matrix, while values for Lake Mitchell are given in bottom matrix.

Table 6.14. *Coefficients of the community matrix describing interactions between and among the predators Mesocyclops and trout, with values listed for all coefficients as well as the equation, table, or formula used to calculate the coefficient; as described in the text, the values of these coefficients do not have a major influence on the stability of the Lake Mitchell community*

Alpha coefficient	Value	Derivation
a_{15}	0.00	Low feeding rate of *Mesocyclops* (approx.)
a_{16}	0.00	Fish do not feed on flagellates; no effect of anoxia.
a_{25}	0.00	Low feeding rate of *Mesocyclops* (approx.)
a_{26}	0.00	Fish do not feed on resistants; no effect of anoxia.
a_{35}	0.00	No predation of *Mesocyclops* on *Daphnia*
a_{36}	−0.006	20 *Daphnia*/L • 50μg/*Daphnia* • (1 mg/1000 μg) = 1 mg/L
		20 kg fish/ha • 10^6 mg/kg • 1 ha lake/3.10^7 L = .6 mg fish/ha
		mortality of − 0.01/d • (.6 mg/L fish)/(1 mg D/L) = − .006/d
a_{45}	−0.05	Williamson (1982)
a_{46}	0.00	Fish do not prey on *Bosmina*
a_{51}	0.00	Low feeding rate of *Mesocyclops* (approx.)
a_{52}	0.00	Low feeding rate of *Mesocyclops* (approx.)
a_{53}	0.00	No predation by *Mesocyclops* on *Daphnia*
a_{54}	0.00	Low predation by *Mesocyclops* on *Bosmina*
a_{55}	0.00	No direct interaction among *Mesocyclops*
a_{56}	−0.006	Same as for *Daphnia* (a_{36}) (approx.)
a_{61}	0.00	Fish do not feed on flagellates.
a_{62}	0.00	Fish do not feed on flagellates.
a_{63}	0.0012	20% assimilation efficiency (approx.)
a_{64}	0.00	Fish do not feed on *Bosmina*.
a_{65}	0.0012	20% assimilation efficiency (approx.)
a_{66}	0.002	$(1-0.5^{(1/356)})$; 50% loss per year to outside influence; density dependent.

Table 6.15. *Stability analysis of the entire community interaction matrix; only elements of the last two rows and columns are subject to analysis*

Interaction coefficient	Change[a]	Biological interpretation	Stability status
a_{26}	−0.01 to 0.05	Increased excretion by fish	Stability maintained
a_{36}	−0.2 to −0.4	Increased mortality of *Daphnia* in the presence of fish	Stability maintained
a_{45}	−0.01 to −.05	Increased predation and growth by	Stability maintained
a_{54}	−0.02 to 0	*Mesocyclops* fed *Bosmina*	
a_{55}	−0.01 to 0.0	Eliminate self-regulation by *Mesocyclops*	Stability maintained
a_{56}	−0.15 to −0.30	Increased death rate of *Mesocyclops* in presence of fish	Stability maintained
a_{63}	0.02 to 0.20	Increased growth of fish fed *Daphnia*	Stability maintained
a_{65}	0.02 to 0.15	Increased fish growth of fish fed *Mesocyclops*	Stability maintained
a_{66}	−0.01 to 0.0	Eliminate self-regulation by fish	Stability maintained

Stability was determined through the use of Routh–Hurwitz criteria as each element was individually varied.
[a]Describes the limits to which individual elements were varied while assaying community stability.

conferring stability on that community are of interest for that reason—those features will tend to persist along with the community, and therefore those features should become relatively common.

The stability of the summer Lake Mitchell community is not easily ascribed to any one feature of the community. The basic "structure" that most strongly contributes to stability consists of two grazers with differential access to two phytoplankton groups, which in turn have differential access to nutrient sources. Prominent in this structure is the change in composition (and thus in food quality) of the phytoplankton as the zooplankton change in abundance. The change in phytoplankton composition in turn affects the relative success of the two grazers (DeMott and Kerfoot 1982). The sensitivity analysis of the community interaction matrix indicates that all of these features contribute to the stability of the system.

Studies on the stability of food webs (see Pimm 1982 for summary) have been restricted to just a few areas: the effects of species additions and deletions, the importance of food web size, and the role of food web complexity. The origin of stability has been investigated for small food webs, often through simultaneous solutions of growth equations for all populations included in the food web (see May 1981 for examples). Many of those studies have utilized Lotka–Volterra or other idealized growth equations. In doing so, none has adequately addressed the problem of self-limitation and the definition of coefficients of the principal diagonal of the community interaction matrix. Few adequately differentiate between direct effects such as predation and indirect effects such as competition, and few correctly include inorganic components (light, nutrients) in their description.

Formal analyses of the stability of natural ecological systems (communities) are virtually nonexistent. The analyses require an understanding of the interactions between all members of the community. Further, the dynamics of the constituent populations must be understood well enough to predict the effects of changes in their own densities on their population growth rates, both through changes in their production and their loss to predation. Finally all of these relationships must be quantified. Practical requirements dictate that the analyzed communities have few populations or be reducible to a small number of "compartments." Having little physical structure and fairly simple food webs, planktonic communities are particularly amenable to such analysis. Those of fresh water are especially appropriate, as they contain relatively few phylogenetic groups. This study presents such an analysis, demonstrates that a stability analysis of a natural community is feasible, and suggests possible changes in that ecosystem that could lead to destabilization. This type of analysis therefore provides a simple way of predicting the effects of changes in a community and provides an alternative to the study of each individual interaction (e.g., Rosenzweig 1971).

Other lake ecosystems differ from the Lake Mitchell ecosystem in ways that might influence the stability of those systems. Eutrophic systems, for example, may lack many of the competitive relationships among phytoplankton that maintain the stability seen here. Large populations of fish, supported by nonplanktonic sources of food (i.e., benthic and littoral sources) may be present in eutrophic lakes and can eliminate *Daphnia* and other large grazers. Those grazers are capable of harvesting a wide spectrum of phytoplankton types, which might remain unharvested in their absence. In the absence of fish, *Daphnia* may grow to such large numbers that they completely eliminate all but the fastest-growing phytoplankton, flagellates, and blue-greens (Levitan et al. 1985). The stability of such a system has not been explored but also seems dependent on the geometry of the lake in question and the presence of refugia.

While this analysis has focussed on summer dynamics, it is important to note that the midsummer Lake Mitchell community is not permanent, for each autumn it changes into a different community. Climatic changes include reductions of temperature and light intensity; there are local increases in nutrient inputs and lake basin washout and biological events such as decomposition of macrophytes and the formation of plankton resting eggs. Such events alter interaction coefficients, and the basic definition of important processes, beyond the scope of the present investigation.

ACKNOWLEDGMENTS

I thank W. Charles Kerfoot and Donald DeAngelis for invaluable guidance. This study supported by NSF Grant DEB 80-04654.

REFERENCES

Ashby, W. R. 1960. *Design for a brain: The origin of adaptive behaviour.* 2nd ed. New York: John Wiley & Sons.

Banse, K. 1976. Rates of growth, respiration, and photosynthesis of unicellular algae as related to cell size—a review. *J. Phycol.* 12 : 135–40.

Bogen, R. 1977. MACSYMA *reference manual.* Boston: Mathlab Group, MIT.

Brandl, Z., and C. H. Fernando. 1978. Prey selection by the cyclopoid copepods *Mesocyclops edax* and *Cyclops vicinus. Verh. Internat. Verein. Limnol.* 20 : 2505–10.

Brooks, J. L., and S. I. Dodson. 1965. Predation, body size, and composition of plankton. *Science* 150 : 28–35.

Burns, C. W. 1968. The relationship between body size of filter-feeding Cladocera and the maximum size of particle ingested. *Limnol. Oceanogr.* 14 : 675–78.

Confer, J. L. 1971. Intrazooplankton predation by *Mesocyclops edax* at natural prey densities. *Limnol. Oceanogr.* 16 : 663–66.

Crowley, P. H. 1975. Natural selection and the Michaelis constant. *J. Theor. Biol.* 50 : 461–75.

DeBernardi, R., G. Grussani, and E. L. Pedretti. 1981. The significance of blue-green algae as food for filter-feeding zooplankton: Experimental studies on *Daphnia* spp. fed by *Mycrocystis aeruginosa. Verh. Int. Verein. Limnol.* 21 : 477–83.

DeMott, W. R. 1982. Feeding selectivities and relative ingestion rates of *Daphnia* and *Bosmina. Limnol. Oceanogr.* 27 : 518–27.

———. 1983. Seasonal succession in a *Daphnia* assemblage. *Ecol. Monogr.* 53 : 321–40.

———. 1985. Relations between filter mesh-size, feeding mode, and capture efficiency for cladocerans feeding on ultrafine particles. *Arch. Hydrobiol. Beih. Ergebn. Limnol.* 21 : 125–34.

DeMott, W. R., and W. C. Kerfoot. 1982. Competition among cladocerans: Nature of the interaction between *Bosmina* and *Daphnia. Ecology* 63 : 1949–66.

Dixon, W. J., ed. 1981. BMDP statistical software. Berkeley, Calif.: University of California Press.

Dumont, H. J., I. V. de Velde, and S. Dumont. 1975. The dry-weight estimate of biomass in a selection of cladocera, copepoda, and rotifera from the plankton, periphyton, and benthos of continental waters. *Oecologia* 19 : 75–97.

Durand, M. A. 1984. The impact of filamentous cyanophytes on the population biology of *Keratella* and *Daphnia.* Master's thesis, Dartmouth College, Hanover, NH.

Folt, C. L., and C. R. Goldman. 1982. Allelopathy between zooplankton: A mechanism for interference competition. *Science* 213 : 1133–35.

Gardner, M. R., and W. R. Ashby. 1970. The connectance of large dynamic systems: Critical values for stability. *Nature* 288 : 784.

Goldman, J. C., J. J. McCarthy, and D. G. Peavey. 1979. Growth rate influence on the chemical composition of phytoplankton in oceanic waters. *Nature* 279 : 210–15.

Harter. 1968. Adsorption of phosphorus by lake sediment. *Soil Sci. Soc. Amer. Proc.* 32 : 514–18.

Jackson, G. A. 1980. Phytoplankton growth and zooplankton grazing in oligotrophic oceans. *Nature* 284 : 439–41.

Kerfoot, W. C. 1985. Zooplankton diets: Food limitation and inclusive overlaps. Abstracts, ASLO Annual Meeting, Minneapolis, MN.

Kerfoot, W. C., and W. R. DeMott. 1984. Food web dynamics: Dependent chains and vaulting. In D. G. Meyers and J. R. Strickler (eds.), *Trophic interactions within aquatic ecosystems,* pp. 347–82. AAAS Selected Symposium #85. Washington, D.C.: Westview Press.

Kerfoot, W. C., W. R. DeMott, and D. DeAngelis. 1985a. Interactions among cladocerans: Food limitation and exploitative competition. *Arch. Hydrobiol. Beih. Ergebn. Limnol.* 21 : 431–51.

Kerfoot, W. C., W. R. DeMott, and C. Levitan. 1985b. Nonlinearities in competitive interactions: Component variables or system response? *Ecology* 66 : 959–65.

Kitchell, J. F., J. F. Koonce, and P. S. Tennis. 1975. Phosphorus flux through fishes. *Verh. Int. Verein. Limnol* 19 : 2478–84.

Kitchell, J. F., R. V. O'Neill, D. Webb, G. W. Gallepp, S. M. Bartell, J. F. Koonce, and B. S. Ausmus. 1978. Consumer regulation of nutrient cycling. *Bioscience* 29 : 28–34.

Lampert, W. 1977a. Studies on the carbon balance of *Daphnia pulex* as related to environmental conditions. II. The dependence of carbon assimilation on animal size, temperature, food concentration, and diet species. *Arch. Hydrobiol.* (Suppl.) 48 : 336–60.

———. 1977b. Studies on the carbon balance of Daphnia pulex as related to environmental conditions. IV. Determination of the "threshold" concentration as a factor controlling the abundance of zooplankton species. *Arch. Hydrobiol.* (Suppl.) 48 : 361–68.

Lehman, J. T., and D. Scavia. 1982a. Microscale nutrient patches produced by zooplankton. *P.N.A.S.* 79 : 5001–5005.

———. 1982b. Microscale patchiness of nutrients in plankton communities. *Science* 216 : 729–30.

Lehninger, A. L. 1975. Biochemistry. *The molecular basis of cell structure and function* (2nd ed.). Worth Publishers.

Levitan, C. 1986. The structure and stability of a freshwater pelagic community. Ph.D. thesis, Dartmouth College, Hanover, N.H.

Levitan, C., W. C. Kerfoot, and W. R. DeMott. 1985. Ability of *Daphnia* to buffer trout lakes against periodic nutrients. *Verh. Internat. Verein. Limnol.* 22 : 3076–82.

May, R. M. 1973. Stability and complexity in model ecosystems. Princeton, N.J.: Princeton University Press.

——. (ed.). 1981. *Theoretical ecology* (2nd ed.) W. B. Saunders.

Mayfield, C. I., and W. E. Innis. 1978. Interactions between freshwater bacteria and *Ankistrodesmus braunii* in batch and continuous culture. *Microbial. Ecol.* 4 : 331–44.

McCarthy, J. J., and J. C. Goldman. 1979. Nitrogenous nutrition of marine phytoplankton in nutrient-depleted waters. *Science* 203 : 439–41.

Mills. E. L., and J. L. Forney. 1982a. Impact on *Daphnia pulex* of predation by young yellow perch in Oneida Lake, New York. *Trans. Amer. Fish. Soc.* 112 : 154–61.

——. 1982b. Regulation of Daphnid abundance by young yellow perch. *Trans. Amer. Fish. Soc.* 111 : 153–61.

Monod, S. 1950. La technique de culture continue; theorie et applications. *Ann. Inst. Pasteur* 79 : 390–410.

Murphy, T., and S. Riley. 1962. *Anal. Chim. Acta.* 27 : 31.

Nalewajko, C., and D. R. S. Lean. 1980. Phosphorus. In I. Morris (ed.), *The physiological ecology of phytoplankton*, pp. 235–58. *Studies in ecology*, Vol. 7. Berkeley, CA: University of California Press.

Peters, R. H. 1975. Phosphorus regeneration by natural populations of limnetic zooplankton. *Verh. Internat. Verein. Limnol.* 19 : 273–79.

——. 1977. Availability of atmospheric orthophosphate. *J. Fish. Res. Bd. Canada* 34 : 918–24.

Peters, R. H., and F. H. Rigler. 1973. Phosphorus release by *Daphnia*. *Limnol. Oceanogr.* 18 : 821–39.

Petterson, K. 1979. Enzymatic determination of orthophosphate in natural waters. *Int. Rev. ges. Hydrobiol.* 64 : 585–607.

Pimm, S. L. 1982. *Food webs*. London: Chapman and Hall.

——. 1984. The complexity and stability of ecosystems. *Nature* 307 : 321–26.

Polis, G. A. 1984. Age structure component of niche width and intraspecific resource partitioning: can age groups function as ecological species? *Amer. Naturalist* 123 : 541–64.

Porter, K. G. 1973. Selective grazing and differential digestion of algae by zooplankton. *Nature* 244 : 179–80.

——. 1976. Enhancement of algal growth and productivity by grazing zooplankton. *Science* 192 : 1332–34.

——. 1977. The plant-animal interface in freshwater ecosystems. *Amer. Sci.* 65 : 159–70.

Rigler, F. H. 1975. Phosphorus cycling in lakes. In F. H. Ruttner (ed.), *Fundamentals of limnology*, pp. 263–73. Toronto: University of Toronto Press.

Rosenzweig, M. L. 1971. Paradox of enrichment: destabilization of exploitation ecosystems in ecological time. *Science* 171 : 385–87.

Scavia, D., and W. S. Gardner. 1982. Kinetics of nitrogen and phosphorus release in varying food supplies by *Daphnia magna*. *Hydrobiology* 96 : 105–11.

Scavia, D., and M. J. McFarland. 1982. Phosphorus release patterns and the effect of reproductive stage and ecdysis in *Daphnia magna*. *Can. J. Fisheries Aquat. Sci.* 39 : 1310–14.

Schindler, J. E. 1970. Food quality and zooplankton nutrition. *J. Fish.* 89 : 589–95.

Seitz, A. 1984. Are there allelopathic interactions in zooplankton? Laboratory experiments with *Daphnia*. *Oecologia* 62 : 94–96.

Sheldon, R. W., W. H. Sutcliffe, and M. A. Paranjape. 1977. Structure of pelagic food chain and relationship between plankton and fish production. *J. Fish. Res. Bd. Can.* 34 : 2344–53.

Shukla, S. S., J. K. Syers, J. D. H. Williams, D. A. Armstrong, and R. F. Harris. 1971. Sorption of inorganic phosphorus by lake sediments. *Soil Sci. Soc. Amer. Proc.* 35 : 244–49.

Sprules, W. G., and L. B. Holtby. 1980. Body size and feeding ecology as alternatives to taxonomy for the study of limnetic zooplankton community structure. *J. Fish. Res. Bd. Canada* 36 : 1354–63.

Sprules, G., and R. Knoechel. 1985. Lake ecosystem dynamics based on functional representations of trophic components. In D. G. Meyers and J. R. Strickler (eds.), *Trophic interactions within aquatic ecosystems*, pp. 383–403. AAAS Selected Symposium 85. Washington, D.C.: Westview Press.

Strickland, J. D. H., and T. R. Parsons. 1968. *A practical handbook of seawater analysis*. Ottawa: Fisheries Research Board of Canada.

Tilman, D. 1982. *Resource competition and community structure*. Princeton, N.J.: Princeton University Press.

Weglenska, T. 1971. The influence of various concentrations of natural food on the development, fecundity, and production of planktonic crustacean filtrators. *Ekol. Pol.* 19 : 427–73.

Williamson, C. E. 1982. Behavioral interactions between a cyclopoid copepod predatoor and its prey. *Freshwater Biol.* 7 : 701–11.

Zar, J. H. 1974. *Biostatistical analysis*. Englewood Cliffs, N.J.: Prentice-Hall.

Zelazny, L. 1983. A coast of armor: Interactions between *Daphnia* and the green alga *Scenedesmus bijuga*. Master's thesis, Dartmouth College, Hanover, N.H.

A. Cascading Trophic Responses

7. Indirect Effects of Early Bass–Shad Interactions on Predator Population Structure and Food Web Dynamics

S. M. Adams

D. L. DeAngelis

Fluctuations in top predators are key elements in models of cascading effects, yet the influence of biotic and abiotic variables, particularly those that determine recruitment, are poorly understood.

Annual variability in environmental conditions such as water temperature can have dramatic effects on certain aquatic predator–prey relations, which in turn have repercussions on predator and prey populations and the remainder of the food web. We have investigated the predation of young-of-the-year largemouth bass (Micropterus salmoides) on shad [threadfin shad (Dorosoma petenense) and gizzard shad (D. cepedianum)] as a function of environmental conditions that affect spawning times of both bass and shad. A predation–growth model incorporating field data was developed to simulate bass size distributions through the first year of growth. The importance of timing of spawning and other environmental factors on predation and growth of bass was examined through analysis of the model. Predation and bass growth were extremely sensitive to the timing of spawning. This timing determines the initial size distributions of bass and shad and therefore the percentage of the shad population that is vulnerable to predation throughout the growing season. Because overwinter mortality of bass is dependent on the sizes attained by the end of the first growing season, recruitment of both bass and shad into adult stocks depend heavily on the degree of predation. Two general ecological implications of this study are: (1) components of the food web affected by the bass/shad predator–prey relationships may be subjected to variability induced by annual variations in the environmental conditions that affect timing of spawning, and (2) timing of bass and shad spawning may be subjected to natural selection feedbacks.

Many factors, operating independently or concurrently, can affect year-class strength in fish populations. These factors include (1) the environmental conditions that occur within a few weeks after spawning (Kramer and Smith 1962; Miller and Kramer 1971; Eipper 1975; Summerfelt 1975; Aggus 1979), (2) quality and size of food available during both the early life stages and throughout the first growing season (Pasch 1975; Jenkins and Morais 1976; Shelton et al. 1979; Timmons et al. 1980), (3) predation and competition for food (Pasch 1975; Mullan and Applegate 1965; Von Geldern and Mitchell 1975), (4) availability of adequate habitat (Aggus and Elliott 1975; Coutant 1975), and (5) time of predator–prey spawning (Pasch 1975; Shelton et al. 1979). In general, year-class strength will depend on each of these factors, although the relative contribution of each can vary from year to year. Without minimizing the importance of other controlling factors, we will emphasize in this study one particular combination of factors, or causal chain, that may often play a crucial role in determining predator year-class strength. This combination of factors may influence the predation and growth rate of a predator and, ultimately, regulate the strength of the year class by determining the vulnerability of the predator to overwinter mortality.

The interactions between largemouth bass (Micropterus salmoides) and either of its principal prey species, threadfin (Dorosoma petenense) or gizzard (D. cepedianum) shad, may often satisfy the condition of close ecological coupling. Mutual adaptations may occur in response to this predation interaction. In the present study we explore the influence of timing of spawning of bass and shad, as well as other factors, on the predation rate and the growth of bass through its first growing season. Our purpose is to determine the relative importance of the timing of spawning to the success of largemouth bass in reaching adulthood. Success in reaching maturity and reproducing may feed back and cause further adjustments in the timing of spawning by the two species.

In some predators, such as largemouth bass, the size distribution of the population at the end of the first growing season is a major determinant influencing further year-class strength and regulating recruitment into adult stocks the following spring (Shelton et al. 1979; Isely 1981; Gutreuter and Anderson 1985). Several studies have demonstrated that smaller individuals in largemouth and smallmouth bass populations suffer mortality over their first winter, which results in a relative increase in the number of larger individuals in a cohort (Pasch 1975; Von Geldern and Mitchell 1975; Timmons et al. 1980; Shelton et al. 1979; Adams et al. 1982; Aggus and Elliott 1975; Oliver et al. 1979). The most generally accepted explanation for this size-specific mortality of bass over the winter nonfeeding period is related to exhaustion of energy reserves in smaller fish, culminating in starvation (Saiki and Tash 1978; Oliver et al. 1979; Isely 1981; Adams et al. 1982). Small bass not only store proportionally less lipids than larger fish (Shuter et al. 1980; Adams et al. 1982) but also utilize lipids more rapidly because of relatively higher weight-specific metabolic rates (Oliver et al. 1979; Shuter et al. 1980). Starvation due to exhaustion of energy reserves renders fish more susceptible to mortality from numerous causes (Adams et al. 1985; Shul'man 1974), including breakdown in cellular metabolism when lipid levels reach physiologically critical levels (Wilkins 1967; Love 1970), predisposition to disease (Shul'man 1974; Glebe and Leggett 1981), increased vulnerability to predation (Herting and Witt 1967; Minton and McLean 1982), and increased susceptibility to environmental stress (Kwain et al. 1984; Wedemeyer and McLeay 1981). In addition to disproportionately smaller lipid stores in smaller bass, scarcity of adequately sized prey for smaller predators in the spring further extends their dependence on limited energy reserves (Adams et al. 1982).

The hypothesized chain of influences leading from timing of bass-shad spawning to recruitment of second-year bass is shown in figure 7.1. The survival or mortality of both largemouth bass and shad individuals, and thus their probabilities of eventual successful reproduction, is influenced by the predator–prey relationship during the first year. A potential long-term evolutionary consequence of survival versus mortality is the existence of natural selection feedback loops to the physiological and behavioral characteristics of the species that determine at what water temperatures the two species spawn. The interaction of these inherent biological factors with the particular environmental conditions of a given year regulate the largemouth bass and shad population for that year and influence other components of the food chain indirectly. For example, low shad survival due to a high predation rate of bass could reduce the grazing of shad on zooplankton, whereas increased intraspecific competition could occur with low predation and high shad survival (fig. 7.1).

METHODS AND APPROACHES

To assess the relative importance of timing of predator–prey spawning and other environmental factors on the size distribution and future recruitment success of a piscivorous fish population, a model was developed that could represent the complex details of size-structured predator–prey relationships. We chose a stochastic computer approach to simulate the predation and growth of a large number of largemouth bass through their first growing season. This modeling approach allowed us to study a wide variety of scenarios relative to factors influencing the size distribution of a predator population at the end of the first growing season. From simulations of a large number (500) of model predators, statistics were generated on the size distributions of the model cohort for each set of potentially influential factors. The model was used to predict the final size distributions of young-of-the-year (YOY) predators at the end of their first growing season, given both the initial predator size distribution and the changing prey size distribution through the growing season (fig. 7.2). The major deterministic and stochastic components designed into the model were those associated with the initial and temporal size distributions of the prey and the monthly growth in size of the predators. These components are discussed below.

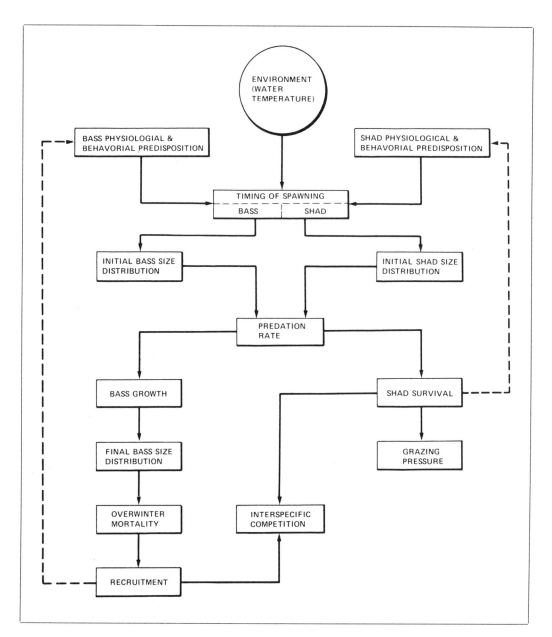

Figure 7.1. *Flow diagram showing the propagation of effects of environmental conditions and timing of predator (largemouth bass) and prey (threadfin and gizzard shad) spawning on the population dynamics of these fish through their first year. The dashed lines represent possible natural selection feedbacks on the timing of spawning. Some food web implications of bass–shad interactions also are shown.*

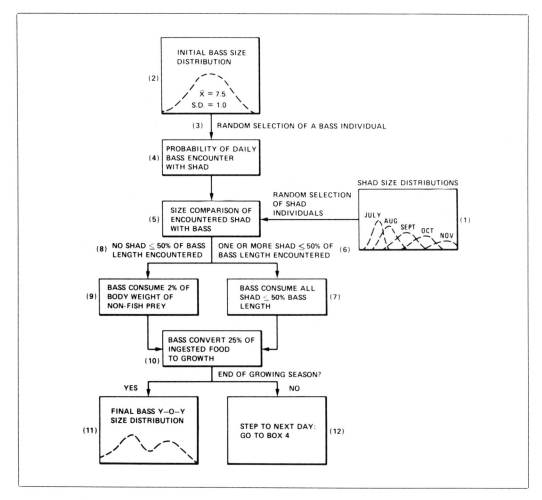

Figure 7.2. *Flow diagram showing the major stepwise components of the predation–growth model. Each of the numbered components (1–12) are discussed in the text.*

Prey Size Distribution

In many aquatic ecosystems of the southern United States, threadfin and gizzard shad are the primary prey of many predators, such as largemouth bass (Jenkins 1979; Nobel 1981). Shad are not normally available as prey until YOY bass reach their piscivorous stage at approximately 5 to 7 cm (Heidinger 1976). Bass smaller than this size consume mainly zooplankton and insects but ingest increasing quantities of fish prey as they become larger. Because there is a differential growth rate between YOY largemouth bass and YOY shad (Pasch 1975; Shelton et al. 1979), shad can outgrow the size range over which they are vulner-

able to many of their bass predators.

The size distribution of YOY shad potentially available as prey to YOY bass was incorporated into the model by using empirical data on YOY shad size distributions from several southern U.S. reservoirs (Johnson 1970; Pasch 1975; S. M. Adams, unpublished data, Oak Ridge National Laboratory). From the time of initiation of our simulated growing season on July 1, both the mean and standard deviation of shad sizes (lengths) were increased throughout the growing season to reflect the rate of monthly shad growth observed in the field (Johnson 1970; Pasch 1975; S. M. Adams, unpublished data, Oak Ridge National Laboratory) (fig. 7.2, no. 1). Although we did not explicitly model the feeding dynamics of the shad, we simulated the observed length change in the shad population through the growing season by allowing the shad to grow approximately 1 cm/month, with the mean size variability also increasing. For example, the initial size of a shad YOY cohort was 4.0 ± 0.5 cm, and the distribution of sizes follows a monthly growth pattern shown in figure 7.2 (no. 1). The population abundance constant, or the daily probability that an individual bass encounters prey, was incorporated into the model as an integral part of the predation rate, as discussed below.

Predator Growth and Population Distribution

INITIAL SIZE DISTRIBUTION. The model's initial size distributions of YOY bass at the start of the growing season on July 1 were based, as were the shad size distributions, on empirical field data. Information from Summerfelt (1975), Olmsted (1974), and Timmons et al. (1980) indicated that the mean size of YOY largemouth bass in southeastern reservoirs in July is 7 to 8 cm, with a standard deviation of 0.5 to 1.0 cm. At the beginning of the simulated growing season (July 1), a bass was chosen randomly (fig. 7.2, no. 3) from a size-frequency distribution that had a mean of 7.5 cm and a standard deviation of 1.0 cm (fig. 7.2, no. 2).

PREDATION ON SHAD. A bass selected at random from the initial size distribution shown in figure 7.2 (no. 2) has a daily probability, λ, of encountering prey in the environment (fig. 7.2, no. 4). This encounter rate was assumed to be the coefficient of a Poisson process (Paloheimo 1971; DeAngelis et al. 1984). On a given day, a bass could encounter a particular number, m, of shad with a probability,

$$\text{Prob } (n = m) = \frac{\lambda^m e^{-\lambda}}{m!} \quad (m = 0, 1, 2, \ldots).$$

The value of the encounter rate, λ, was determined by adjusting this coefficient in the nominal model case to yield end-of-the-year YOY bass size distributions consistent with distributions observed in the field (e.g., Shelton et al. 1979; Timmons et al. 1980; Adams et

al. 1982). In addition, DeAngelis et al. (1984) found that an encounter rate coefficient of 0.30, similar to the 0.31 value used in our model, gave good agreement between model simulations and field data on the density function of shad numbers consumed by largemouth bass in Watts Bar Reservoir, Tennessee. When a prey encounter occurred, the particular length of the shad encountered was chosen by a random process from the shad distributions shown in figure 7.2 (no. 1). Shad were randomly selected from this monthly distribution of sizes, depending on the appropriate month of the growing season being simulated in the model.

Given that bass encountered shad of a specific length, whether a particular shad was consumed depended on the length of the bass relative to the length of the encountered shad (prey length/predator length) (fig. 7.2, no. 5). Ingestion of prey occurred if the length of at least one encountered shad was ≤ 50% of bass length (fig. 7.2, nos. 6, 7). If bass encountered no shad during the day that were ≤ 50% of their lengths, then no shad were consumed; these bass, however, were allowed to ingest 2% of their body weight per day (fig. 7.2, nos. 8, 9) in nonfish prey in order to realize a minimum amount of growth. This percentage will be referred to as the nonfish feeding rate (α). A 2% daily ration is greater than the maintenance ration at 18°C (Niimi and Beamish 1974) but lower than consumption rates of bass on fish prey during the growing season (Cochran and Adelman 1982) (table 7.1). The prey-length/predator-length ingestion ratio of 50% appears to be the most acceptable value for this parameter based on extensive bass food habits data (Lawrence 1958; Shelton et al. 1979; Jenkins and Morais 1976).

Following consumption of fish prey (fig. 7.2, no. 7), prey length (L) in centimeters was converted to prey weight (W) in grams by the shad length–weight regression of Minton and McLean (1982): log W = 3.0 • log L − 2.07.

GROWTH EFFICIENCY AND BASS GROWTH. Weight of consumed shad was converted to weight of bass gained by the use of a gross growth efficiency (k_1) of 0.25 (fig. 7.2, no. 10). This efficiency is consistent with conversion ratios reported for YOY largemouth bass in the laboratory (Prather 1951) and field (Adams

Table 7.1. *Model predictions and field observations of average lengths [cm (± SD)] and quantity of shad consumed by young-of-the-year largemouth bass each month during the growing season at Watts Bar Reservoir*

Month	Size of shad consumed ($\overline{X}$ ± SD)[a]		Mean consumption (% body wt/day)	
	Model	Field	Model	Field
July	3.4 ± 0.5	4.0 ± 0.7	5.6	3.4
August	4.2 ± 0.7	5.2 ± 0.4	4.6	3.3
September	4.8 ± 1.0	6.2 ± 1.4	4.5	2.9
October	6.2 ± 1.2	7.3 ± 1.4	4.1	3.1
November	7.2 ± 1.5	7.5 ± 1.8	4.0	2.7
December	7.8 ± 1.6	8.4 ± 2.3	3.8	2.1

[a]No pairs of monthly means were significantly different at $p = 0.05$.

et al. 1982). Based on length–weight relationships of largemouth bass collected from Watts Bar Reservoir from 1980 to 1983, daily weight gain of bass (W) in grams was converted to daily length increments (L) in centimeters by applying the following regression: log L = 0.69 + 0.31 log W.

The sequence of bass predation and growth shown in figure 7.1 is for a single bass chosen at random from the initial bass size distribution shown in figure 7.2 (no. 2). This computer program or sequence was run 500 separate times to represent 500 bass and to generate the final distribution of bass sizes (fig. 7.2, no. 11). For each of the 500 bass, growth terminated on the last day (day 150 in the nominal case) of the growing season (fig. 7.2, no. 11). The sequence 4 to 12 in fig. 7.1 was repeated 150 times for each of the 500 bass to represent each of the 150 days in the growing season.

Field Studies

YOY largemouth bass were collected monthly during the growing season from July 1980 to December 1983 by electroshocking in Watts Bar Reservoir, a 13,300-ha impoundment of the Tennessee River in eastern Tennessee. For each monthly sample of bass, the total length in centimeters, weight in grams, and stomach contents of all fish were recorded. Total lengths and weight were measured for all shad consumed by bass. If prey were partially digested, original weights of consumed shad were calculated from a series of backbone-, standard-, and total-length regressions (Minton and McLean 1982). Monthly consumption by largemouth bass was

determined by the method of Adams et al. (1982).

RESULTS AND DISCUSSION

Because a variety of factors, operating concurrently or at different times, can affect year-class strength in fish populations (Kramer and Smith 1962), it is unrealistic, even with predator–prey models, to address in one study all of the hypotheses relative to factors that affect year-class strength. Assuming, however, that the size distribution of largemouth bass at the end of the first growing season dictates, to a large degree, the percentage of the bass population that survives the winter, then a model can be used to compare the relative importance of factors that indirectly affect the survival of bass and shape the size distribution during the first year. We have used the model to compare (1) the timing of predator and prey spawning; (2) factors affecting growth through metabolic and physiological pathways such as temperature, disease, or other environmental stresses; and (3) environmental conditions, such as temperature, that could decrease or extend the length of the growing season.

Internal Consistency of the Model

Before using the model to examine the relative importance of the three factors listed above on largemouth bass size distribution and future recruitment, we performed simulations to test the model's internal consistency. Patten et al. (1975) suggested three criteria that can be used to examine model consistency and credibility:

(1) the behavior of the model, if it lies within realistic ranges of measured initial states and if temporal sequences are reasonable; (2) the postulated internal organization, if it is consistent as evidenced by the model's ability to generate reasonable output dynamics not expressly incorporated into the model; and (3) responses to small perturbations, if they are reasonable.

To evaluate the first criterion related to realistic ranges of model parameters and temporal compartmental sequences, the model was used to simulate largemouth bass population distributions throughout the growing season, using field data on initial sizes of predator (bass) and prey (shad) at the beginning of the growing season (July 1 in our model). Final distributions of YOY bass populations generated by this model were similar to YOY bass populations observed in various southeastern reservoirs. For example, Adams et al. (1982) reported that the mean length and standard deviation of YOY bass in December in Watts Bar Reservoir were 16.1 ± 4.4 cm. These statistics are similar to the population means and standard deviations of 13.9 ± 4.4 and 15.5 ± 2.5 cm generated by the model for the two nominal cases. Shelton et al. (1979) and Timmons et al. (1980) found that these statistics for YOY largemouth bass in West Point Reservoir, Georgia, were 17.1 ± 7.4 and 19.2 ± 7.2 for the 1975 and 1977 year classes, respectively. The most convincing evidence supporting the first criteria for model credibility, however, was the generation of a bimodal population distribution by the model (see fig. 7.3 for bimodality of the nominal population). Bimodal size-frequency distributions have been observed in several largemouth bass populations, including West Point Reservoir, Georgia (Shelton et al. 1979; Timmons et al. 1980), Lake Carl Blackshear, Georgia (Pasch 1975), and Lake Fort Smith, Arkansas (Olmsted 1974). Differential growth caused primarily by disjunctive spawning within a year class and availability of appropriate-size prey during the first year of growth have been suggested as the principal causes of bimodality in YOY largemouth bass populations (Pasch 1975; Summerfelt 1975; Gutreuter and Anderson, 1985).

The internal consistency of the model can also be established by examining various outputs of the model that were not expressly incorporated in the model structure. For exam-

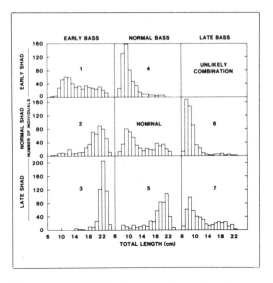

Figure 7.3. *Model predictions of the number of young-of-the-year largemouth bass in various length classes for seven different scenarios of possible predator-prey spawning situations: (1) early bass–early shad spawning; (2) early bass–normal shad spawning; (3) early bass–late shad spawning; (4) normal bass–early shad spawning, nominal case; (5) normal bass–late shad spawning; (6) late bass–normal shad spawning; and (7) late bass–late shad spawning. A late bass–early shad spawning would be unlikely (see text).*

ple, the average size (and size variability) of shad consumed by bass and the daily consumption rates of bass predicted by the model can be compared with actual field data (table 7.1). Predicted quality (sizes) and quantity of consumed shad were similar to those observed for bass feeding on shad in Watts Bar Reservoir (table 7.1). Sizes of shad consumed by bass in the field were slightly larger than those predicted by the model, mainly because field-collected bass were 10 to 20% larger on the average than those bass simulated in the model, and larger fish would be expected to consume larger prey. Given the high inherent variability involved in estimating daily consumption of predators in the field (Cochran and Adelman 1982; Rice and Cochran 1984), consumption rates predicted by the model and observed in field populations also were comparable. Model predictions also were consistent with these field observations and with published information relative to decreasing levels of weight-specific consumption as fish grow larger (Niimi and Beamish 1974; Lewis et al. 1974).

The third criterion for examining model credibility—sensitivity of the model to small perturbations—is addressed in the following section.

Sensitivity Analysis for Factors Affecting Size Distribution

Sensitivity analysis was performed on the major input parameters of the model to determine the extent of their influence on estimates of YOY bass growth and population distributions at the end of the growing season. Parameters such as the encounter rate coefficient, prey-length/predator-length ingestion ratio, and feeding rate coefficient were varied by ±10% and resultant changes in bass growth and size distribution compared with nominal estimates of growth and size distributions. Sensitivity of a model output estimate to a ±10% deviation of an input parameter was calculated using the method of Kitchell et al. (1977): $s_x(p) = (p \cdot \Delta x) / (x \cdot \Delta p)$; where, $s_x(p)$ = sensitivity of output parameter x to deviations of input parameter p; p = nominal value of input parameter p; Δp = input deviation of input pa-

Table 7.2. *Predicted mean lengths (standard deviations) of young-of-the-year largemouth bass populations at the end of the growing season for various values of the encounter rate coefficient, the prey-length/predation-length ingestion ratio, and the nonfish feeding rate coefficient*

Parameter	Model value	Simulated bass length (cm) Mean	±SD
ENCOUNTER RATE COEFFICIENT (λ)			
Nominal case	0.31	13.9[a]	4.4
+10%	0.34	17.4[b]	5.3
−10%	0.28	11.0[c]	2.5
PREY-LENGTH/PREDATOR-LENGTH			
Nominal case	0.50	13.9[a]	4.4
+10%	0.55	21.0[b]	3.0
−10%	0.45	9.3[c]	1.0
NONFISH FEEDING RATE (α)			
Nominal case	2%	13.9[ab]	4.4
Increase	3%	13.4[a]	3.8
Decrease	1%	12.8[ab]	4.3

Means with same letters for each parameter are not significantly different.

rameter p; x = nominal value of output parameter x from a standard simulation; and Δx = deviation of output parameter x due to Δp. A sensitivity of 1.0 indicates that a 10% change in an input parameter will yield a 10% change in an output estimate, such as the mean size of a largemouth bass population at the end of the growing season. The higher the absolute sensitivity value, the higher the corresponding change in an output parameter.

End-of-the-growing-season population statistics for YOY bass experiencing a 10% variation in encounter rate (λ), prey/predator ingestion ratio, and nonfish feeding rate (α) are shown in table 7.2. Variation in the encounter rate coefficient and the prey/predator ingestion ratio had a large effect on the mean sizes of the resulting YOY population, whereas the nonfish feeding rate parameter had little effect on the length distribution of the end-of-the-growing-season bass populations. Even though small (10%) changes in the encounter rate and the prey/predator ingestion ratio had a significant effect on the structure of the YOY bass distributions, this effect does not change the qualitative validity of our results for using this model to examine the relative importance of factors influencing the size distribution of YOY bass populations. In addition, we believe that the critical model assumptions for both encounter rate and the prey-length/predator-ingestion ratio are realistic representations of these two values based on field data and literature information (see Methods and Approaches).

In summary, based on the similarities between various outputs of the model and actual field observations (criteria 1 and 2) and the results of the sensitivity analysis (criterion 3 for model credibility), we conclude that this predation–growth model satisfies the criteria for model credibility and internal consistency. Therefore, we can apply the model to investigate the relative importance of factors influencing the size distributions and future recruitment success of YOY largemouth bass populations.

Importance of Factors Influencing Size Distributions

A series of three simulation experiments was performed with the model to evaluate the relative importance of the following factors in influencing the size distribution and therefore the future recruitment success of largemouth bass at the end of their first growing season: (1) timing of predator–prey spawning; (2) factors that affect growth through metabolic and physiological pathways such as temperature, disease, or other environmental stressors (e.g., pollution); and (3) environmental conditions, such as temperature, that could decrease or extend the length of the growing season.

TIMING OF PREDATOR-PREY SPAWNING. Variations in timing of predator–prey spawning were simulated by varying the initial nominal sizes of predator and prey by 10%. An earlier than normal (nominal) bass or shad spawning was reflected in these simulations by a 10% increase in the initial (nominal) size of a bass or shad at the beginning of the simulated growing season (July 1 in our model), while a late spawning for either predator or prey was indicated by a 10% decrease in their nominal initial sizes. The case where bass and shad both spawn early or both spawn late was represented by a 10% increase or 10% decrease, respectively, in the initial sizes of both bass and shad. Largemouth bass normally spawn at cooler water temperatures than do either gizzard or threadfin shad. Peak spawning of largemouth bass typically occurs at 16 to 18 °C (Kramer and Smith 1962; Heidinger

1976), gizzard shad at 16 to 21 °C (Shelton et al. 1982), and at 23 to 24 °C in threadfin shad (Johnson 1969).

Influence of predator–prey spawning on end-of-the-year YOY bass populations was investigated using two different initial bass and shad population mean lengths as the standard or nominal cases. The first nominal population (case 1) had an initial bass population mean length of 8.0 ± 1.0 cm and an initial shad population mean length of 4.5 ± 1.0 cm, whereas the other nominal population (case 2) represented an initial bass population with mean length of 7.5 ± 1.0 cm and an initial shad population mean length of 4.0 ± 1.0 cm. Both of these populations were considered nominal because they met the criteria standards for model credibility as discussed in the previous sections.

Percentages of YOY bass in the population that grew to be less than or greater than 16 cm at the end of the growing season under seven different scenarios of possible bass–shad spawning situations are shown in table 7.3. Because bass typically spawn at lower temperatures than do shad (Heidinger 1976; Shelton et al. 1982; Johnson 1969), the combination of an early shad spawn and a late bass spawn would

Table 7.3. *Model predictions for the percentage of young-of-the-year largemouth bass in two size categories at the end of the growing season under seven different possible bass-shad spawning situations*

No.	Possible spawning situations	Parameter variation	Case 1[a] 0–15 cm	Case 1[a] 16–25 cm	Case 2[b] 0–15 cm	Case 2[b] 16–25 cm
	Nominal case (Normal bass–normal shad)	None	67	33	54	46
1	Normal bass–early shad	+10% Shad IL	96	4	92	8
2	Normal bass–late shad	−10% Shad IL	15	85	9	91
3	Early bass–normal shad	+10% Bass IL	12	88	6	94
4	Late bass–normal shad	−10% Bass IL	97	3	94	6
5	Early bass–early shad	+10% Bass IL +10% Shad IL	63	37	44	56
6	Early bass–late shad	+10% Bass IL −10% Shad IL	1	99	1	99
7	Late bass–late shad	−10% Bass IL −10% Shad IL	77	23	56	44

The initial lengths (IL) of bass and shad were varied by ±10% of their nominal initial mean lengths to simulate various possible combinations of early and/or late predator–prey spawning.
Cases 1 and 2 represent two different nominal values for predator and prey initial lengths.
[a]Bass initial length, 8.0 ± 1.0 cm; shad initial length, 4.5 ± 1.0 cm.
[b]Bass initial length, 7.5 ± 1.0 cm; shad initial length, 4.0 ± 1.0 cm.

be unlikely; therefore, that spawning combination was not addressed in this study.

We have designated 16 cm as the critical size that YOY bass must attain by the end of their first growing season or their probability of overwinter survival and recruitment into the next age class will be drastically reduced. This critical survival size of approximately 16 cm has a sound biological basis. Adams et al. (1982) found that most YOY largemouth bass smaller than approximately 20 cm in Watts Bar Reservoir did not survive the winter because of the inability of these smaller bass to store sufficient energy reserves for overwintering and because of the unavailability of appropriate-size prey the following spring. Small bass not only store proportionally less energy (lipid) reserves than do large bass (Shuter et al. 1980; Adams et al. 1982) but utilize lipids more rapidly because of relatively higher weight-specific metabolic rates (Oliver et al. 1979; Shuter et al. 1980). Data on length-frequency distributions of YOY largemouth bass at the end of their first growing season in West Point Reservoir, Georgia, show that many of those bass smaller than approximately 16 cm did not survive the winter, as indicated by a large decrease in the number of smaller bass the following spring (Shelton et al. 1979; Timmons et al. 1980).

In situations where shad spawn early or bass spawn late relative to each other, YOY bass smaller than 16 cm dominate the populations at the end of the growing season (fig. 7.3; table 7.3, nos. 1 and 4). Conversely, in situations where shad spawn late or bass spawn early relative to each other, at least 80% of all individuals in the bass population at the end of the growing season are larger than 16 cm (fig. 7.3; table 7.3, nos. 2, 3). In those cases where bass and shad both spawn early (fig. 7.3; table 7.3, no. 5) or both spawn late (fig. 7.3; table 7.3, no. 7), the proportion of individuals in the two size categories are more evenly divided and approximate the nominal case (fig. 7.3, table 7.3). These results indicate that environmental conditions during the spawning season that accentuate the spawning time between bass and shad favor increased food availability and growth for bass, which leads to an increased number of fish in the population that reach 16 cm or greater by the end of the first growing season. An extreme example of this scenario is the case when bass spawn early and shad spawn late (table 7.3, no. 6). In contrast, environmental conditions during the spawning season that decrease the amount of time between bass and shad spawning tend to decrease the number of prey of adequate size available, reducing growth over the growing season and ultimately resulting in a smaller number of individuals longer than 16 cm in the end-of-the-year population.

Because YOY shad in most southeastern aquatic systems normally outgrow their YOY predators (Pasch 1975; Shelton et al. 1979), factors that tend to maximize the size of bass before shad become available as prey or minimize the size or growth of shad would tend to increase the period of time that shad would be vulnerable to predation. During the early growing period, conditions that maximize bass size and minimize prey size should increase the amount of shad consumed by a predator because a larger percentage of the shad encountered by a bass would be less than the critical 0.5 prey-length/predator-length ingestion ratio. In contrast, factors during the spawning season that minimize bass size relative to shad size will decrease the amount of shad consumed during the first growing season. For example, in our model simulations, 41 to 70% of shad encountered by bass were consumed under those spawning season conditions that maximized predator size relative to prey size (table 7.4, nos. 2, 3, 6), whereas only 6 to 17% of shad encountered by bass were consumed under those spawning conditions that minimized bass size relative to shad size (table 7.4, nos. 1, 4, 7).

Sensitivity analysis also indicates that the particular spawning conditions that maximize bass size relative to shad size are the most sensitive in terms of producing bass in the larger (16 to 25 cm) size group (table 7.5, nos. 2, 3, 6). Conditions in the spawning season that minimize bass size relative to shad size are less sensitive to producing bass in the smaller (0 to 15 cm) size group. These two relationships hold for both cases presented in table 7.5; each case represents different nominal initial lengths at the start of the simulated growing season. Based on the results of our modeling simulations, it appears that environmental conditions during the spring spawning period that maximize the time between bass and shad spawning are critical factors influencing prey availability and predator growth and therefore the length distribution of YOY largemouth

Table 7.4. *Model simulation of the number of shad encountered and amount of shad consumed during an average day during the growing season by 500 young-of-the-year largemouth bass under various possible spawning situations*

No.	Spawning situations	Number of shad encountered	Number of shad consumed	Percentage consumed	Consumption (% body wt/d)
	Nominal case	2032	380	19	4.8
1	Normal bass–early shad	2042	148	7	2.8
2	Normal bass–late shad	2111	947	45	8.7
3	Early bass–normal shad	2121	874	41	8.1
4	Late bass–normal shad	2000	122	6	2.4
5	Early bass–early shad	2051	380	19	5.1
6	Early bass–late shad	2088	1472	70	10.6
7	Late bass–late shad	2046	347	17	4.5

Table 7.5. *Normalized sensitivities of young-of-the-year largemouth bass in two size categories at the end of the growing season to 10% variations in the initial mean lengths (IL) of shad and bass populations*

			Sensitivity			
			Case 1[a]		Case 2[b]	
No.	Possible spawning situations	Parameter variation	0–15 cm	16–25 cm	0–15 cm	16–25 cm
1	Normal bass–early shad	+10% Shad IL	4.5	9.0	6.8	8.2
2	Normal bass–late shad	−10% Shad IL	7.7	15.4	8.6	10.3
3	Early bass–normal shad	+10% Bass IL	8.2	16.6	8.9	10.6
4	Late bass–normal shad	−10% Bass IL	4.5	9.0	7.2	8.6
5	Early bass–early shad	+10% Bass IL +10% Shad IL	0.6	1.3	2.0	2.4
6	Early bass–late shad	+10% Bass IL −10% Shad IL	9.5	19.7	10.0	11.9
7	Late bass–late shad	−10% Bass IL −10% Shad IL	1.5	3.0	0.3	0.3

Variation of initial mean sizes represents possible bass–shad spawning situations, with cases 1 and 2 representing two different nominal values for shad and bass initial lengths.
[a]Bass initial length, 8.0 ± 1.0 cm; shad initial length, 4.5 ± 1.0 cm.
[b]Bass initial length, 7.5 ± 1.0 cm; shad initial length, 4.0 ± 1.0 cm.

bass populations at the end of their first growing season.

GROWTH EFFICIENCY. To investigate the relative importance of those potential factors that can influence growth through metabolic pathways (such as temperature regimes, disease, and pollutant stress), we decreased the growth efficiency in the model by 10% from a nominal value of 25 to 22.5%. Stress of almost any nature generally increases the metabolism of an organism, thereby decreasing the amount of consumed energy available for growth. This 10% reduction in growth efficiency over the summer decreased the average size of the bass population from 13.9 to 12.2 cm with a notice-able increase in the numbers of bass smaller than 16 cm and disappearance of the bimodal distribution characteristic of the nominal population (fig. 7.4, table 7.6). In the standard case represented by a 25% growth efficiency, 67% of the bass at the end of the growing season were <16 cm; with a 10% reduction in growth efficiency, approximately 84% were <16 cm (table 7.6). Sensitivity of the number of bass in the two size categories to a 10% change in growth efficiency was relatively small compared with the sensitivities resulting from a 10% change in predator–prey spawning conditions (tables 7.5, 7.6). For example, the sensitivities of the 16-to-25-cm size group to a 10% change in either bass or shad spawning

Table 7.6. *Percentage of individuals and normalized sensitivities of young-of-the-year largemouth bass in two size categories and predicted end-of-the-growing-season population statistics under a 10% decrease in nominal growth efficiency and a ±10% variation in growing season length*

Parameter	Model value	Percent of bass		Sensitivity		Simulated bass length (cm)	
		0–15 cm	16–25 cm	0–15 cm	16–25 cm	X̄	±SD
Growth efficiency							
Nominal case[a]	25.0%	67	33	—	—	13.0	4.4
−10%	22.5%	84	16	2.5	5.1	12.2	3.4
Growing season length							
Nominal case[a]	150 d	67	33	—	—	13.9	4.4
+10%	165 d	66	34	0.03	0.06	14.4	5.0
−10%	135 d	71	29	0.57	1.10	12.0	3.9

[a]Nominal case is for bass with an initial size of 7.5 ± 1.0 cm and shad with an initial size of 4.0 ± 1.0 cm.

conditions varied between 9 and 16% (table 7.5), whereas the sensitivity of this size group to a 10% decrease in growth efficiency was only 5.1% (table 7.6).

LENGTH OF GROWING SEASON. The relative importance of an extended growing season (warmer than nominal temperatures) and a short growing season (cooler than normal temperatures) on the end-of-the-year YOY bass populations was also investigated (fig. 7.4). Under these conditions, the 135- and 165-day final population mean lengths were significantly different from each other, although neither the 135- or 165-day means were significantly different from the nominal 150-day case (fig. 7.4; table 7.6). The sensitivity of the number of fish in the two size groups to a 10% change in growing season length was very small, varying between 0.03 and 1.1 (table 7.6). Therefore, as was also found for the situation involving a 10% variation in growth efficiency, a 10% change in growing season length does not have as large a relative effect on the structure and distribution of end-of-the-year YOY bass populations as does a 10% variation in timing of predator and prey spawning.

CONCLUSIONS AND IMPLICATIONS

Our results demonstrate that environmental conditions can have a relatively large indirect effect on predator–prey relationships in aquatic ecosystems. For example, water temperature affects the timing of predator and

prey spawning, which in turn influences prey availability and the growth of YOY largemouth bass and, ultimately, the size distribution of bass populations at the end of the first growing season. Because population structure appears to be a major determinant regulating future recruitment of bass into adult stocks (see above), timing of predator and prey spawning, as influenced by water temperature, can be a major factor regulating future recruitment and reproductive success of a predator.

Environmental conditions during the spawning season that maximize the time period between predator (bass) and prey (shad) spawning are important factors influencing the nature of the end-of-the-year YOY bass population. An early bass spawn relative to the shad spawn allows bass to gain some additional size advantage on shad, rendering the prey population vulnerable to predation over a longer period of time compared with the situations in which time between predator and prey spawning is minimized. For example, an extended cool period following a major bass spawn might delay shad spawning, allowing a larger segment of the YOY bass population to gain a length advantage over shad while feeding on nonfish prey.

The sensitivity of YOY bass predation on shad to the timing of spawning of both species can have both ecological and coevolutionary implications for the aquatic community. The ecological implications stem from the amplification of variations in year-to-year environmental conditions to predator–prey dynamics. Small variations in the relative spawning times of bass and shad can lead to order-of-magnitude differ-

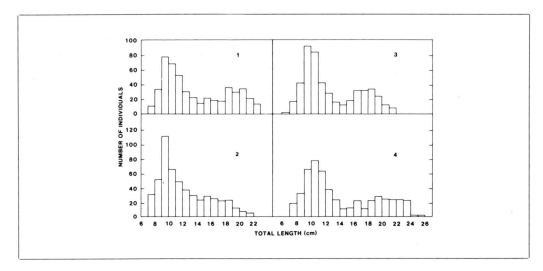

Figure 7.4. *Model predictions of the number of young-of-the-year largemouth bass in various length classes at the end of the growing season: (1) nominal population with a growth efficiency of 25% and a growing season length of 150 days, (2) −10% growth efficiency (22.5%), (3) −10% growing season length (135 days), and (4) +10% growing season length (165 days).*

ences in prey availability and the amount of shad biomass consumed, bass growth rate, and the number of bass large enough to avoid overwinter mortality and reproduce. Over evolutionary time periods—that is, over time periods of many generations of bass and shad—it is reasonable to expect that natural selection should act on the two species to time their spawnings to the advantage of each, insofar as other environmental constraints permit. The possibility of such an evolutionary trend is complicated by the yearly climatic variations, and thus water temperature that the fish would behaviorally key on for spawning, and by the possible significance of other environmental conditions at spawning time.

The interaction of controlling environmental variables, such as temperature, with the individual biological components of the food chain (bass and shad in the present case) can result in a variety of indirect effects on the remainder of the food web. A specific case of such an indirect effect relates to the position that YOY bass occupy in the food web. Food selection by these bass depends on their sizes. If, for example, the lack of available shad prey leads to a large number of small bass (e.g., fig. 7.3, nos. 4 and 6), these bass would continue to feed on zooplankton and insects, which are the primary foods for fry and fingerling bass less

than 5 to 7 cm. They would thus potentially compete with shad or other fish for such food (Von Geldern and Mitchell 1975). Intraspecific competition also will be influenced. Bass that are members of a population with a large variability in size distribution (such as fig. 7.3, no. 1) should experience less intraspecific competition than bass belonging to population size distributions with small variances. Environmental variables, therefore, not only can have direct effects on predator–prey interactions but also have indirect influences on other components of the food chain.

ACKNOWLEDGMENTS

This research was sponsored by the Office of Health and Environmental Research, U.S. Department of Energy, and in part by the National Science Foundation's Ecosystems Studies Program under Interagency Agreement BSR-831 under Contract No. DE-AC05-840R21400 with Martin Marietta Energy Systems, Inc. Publication No. 2617, Environmental Sciences Division, ORNL.

REFERENCES

Adams, S. M., J. E. Breck, and R. B. McLean. 1985. Cumulative stress-induced mortality of gizzard shad in a southeastern U.S. reservoir. *Environ. Biol. Fish.* 13 : 103–12.

Adams, S. M., R. B. McLean, and M. M. Huffman. 1982. Structuring of a predator population through temperature-mediated effects on prey availability. *Can. J. Fisheries Aquat. Sci.* 39 : 1175–84.

Aggus, L. R. 1979. Effects of weather on freshwater fish predator–prey dynamics. In H. Clepper (ed.), *Predator–prey systems in fisheries management*, pp. 47–56. Washington, D.C.: Sport Fishing Institute.

Aggus, L. R., and G. V. Elliott. 1975. Effects of cover and food on year-class strength of largemouth bass. In H. Clepper (ed.), *Black bass biology and management*, pp. 317–22. Washington, D.C.: Sport Fishing Institute.

Cochran, P. A., and I. R. Adelman. 1982. Seasonal aspects of daily ration and diet of largemouth bass, *Micropterus salmoides*, with an evaluation of gastric evacuation rates. *Environ. Biol. Fish.* 7 : 265–75.

Coutant, C. C. 1975. Responses of bass to natural and artificial temperature regimes. In H. Clepper (ed.), *Black bass biology and management*, pp. 272–85. Washington, D.C.: Sport Fishing Institute.

DeAngelis, D. L., S. M. Adams, J. E. Breck, and L. J. Gross. 1984. A stochastic predation model: Application to largemouth bass observations. *Ecological Modelling* 24 : 2–41.

Eipper, A. W. 1975. Environmental influences on the mortality of bass embryos and larvae. In R. H. Stroud and H. Clepper (eds.), *Black bass biology and management*, pp. 295–305, Washington, D.C.: Sport Fishing Institute.

Glebe, B. D., and W. C. Leggett. 1981. Temporal, intra-population differences in energy allocation and use by American shad (*Alosa sapidissima*) during the spawning migration. *Can. J. Fisheries Aquat. Sci.* 38 : 795–805.

Gutreuter, S. J., and R. O. Anderson. 1985. Importance of body size to the recruitment process in largemouth bass populations. *Trans. Amer. Fish. Soc.* 114 : 317–27.

Heidinger, R. C. 1976. Synopsis of biological data on the largemouth bass *Micropterus salmoides* (Lacepede) 1802 (Food and Agriculture Organization of the United Nations). *Fisheries Synopsis* 115.

Herting, G. E., and A. Witt, Jr. 1967. The role of physical fitness of forage fishes in relation to their vulnerability to predation by bowfin (*Amia calva*). *Trans. Amer. Fish. Soc.* 96 : 427–30.

Isely, J. J. 1981. Effects of water temperature and energy reserves on overwinter mortality in young-of-the-year largemouth bass (*Micropterus salmoides*). M.S. thesis, Southern Illinois University at Carbondale.

Jenkins, R. M. 1979. Predator–prey relations in reservoirs. In H. Clepper (ed.), *Black bass biology*

and management, pp. 123–34. Washington, D.C.: Sport Fishing Institute.

Jenkins, R. M., and D. I. Morais. 1976. Prey–predator relations in the predator-stocking-evaluation reservoirs. *Proceedings of the Annual Conference of the Southeastern Association of Game Fish Commission* 30 : 141–57.

Johnson, J. E. 1969. Reproduction, growth, and population dynamics of the threadfin shad, *Dorosoma petenense* (Gunther), in central Arizona reservoirs. Ph.D. diss., Arizona State University, Tempe.

———. 1970. Age, growth, and population dynamics of threadfin shad, *Dorosoma petenense* (Gunther), in central Arizona reservoirs. *Trans. Amer. Fish. Soc.* 99 : 739–53.

Kitchell, J. F., D. J. Stewart, and D. Weininger. 1977. Applications of a bioenergetics model to yellow perch (*Perca flavescens*) and walleye (*Stizostedion vitreum vitreum*). *J. Fish. Res. Board Can.* 34 : 1922–35.

Kramer, R. H., and L. L. Smith, Jr. 1962. Formation of year classes in largemouth bass. *Trans. Amer. Fish. Soc.* 91 : 29–41.

Kwain, W., R. W. McCauley, and J. A. McLean. 1984. Susceptibility of starved, juvenile smallmouth bass, *Micropterus dolomieui* (Lacepede) to low pH. *J. Fish Biol.* 25 : 501–504.

Lawrence, J. M. 1958. Estimated sizes of various forage fishes largemouth bass can swallow. *Proceedings of the Annual Conference at Southeastern Association of Game Fish Commission* 11 : 220–25.

Lewis, W. M., R. Heidinger, W. Kirk, W. Chapman, and D. Johnson. 1974. Food intake of largemouth bass. *Trans. Amer. Fish. Soc.* 103 : 277–80.

Love, R. M. 1970. *The chemical biology of fishes*. New York: Academic Press.

Miller, K. D., and R. H. Kramer. 1971. Spawning and early life history of largemouth bass (*Micropterus Salmoides*) in Lake Powell. In G. E. Hall (ed.), *Reservoir limnology and fisheries*, pp. 73–83. Washington, D.C.: American Fisheries Society. Special Publ. No. 8.

Minton, J. W., and R. B. McLean. 1982. Measurements of growth and consumption of sauger (*Stizostedion canadense*): Implications of fish energetics studies. *Can. J. Fisheries Aquat. Sci.* 39 : 1396–1404.

Mullan, J. W., and R. L. Applegate. 1965. The physical-chemical limnology of a new reservoir (Beaver) and a fourteen-year-old reservoir (Bull Shoals) located on the White River, Arkansas and Missouri. *Proceedings of the Annual Conference of Southeastern Association of Game and Fish Commission* 19 : 413–21.

Niimi, A. J., and F. W. H. Beamish. 1974. Bioenergetics and growth of largemouth bass (*Micropterus salmoides*) in relation to body weight and temperature. *Can. J. Zool.* 52 : 447–56.

Nobel, R. L. 1981. Management of forage fishes in impoundments of the southern United States. *Trans. Amer. Fish. Soc.* 110 : 738–50.

Oliver, J. D., G. F. Holeton, and K. E. Chua. 1979. Overwinter mortality of fingerling smallmouth bass in relation to size, relative energy stores, and environmental temperature. *Trans. Amer. Fish. Soc.* 108 : 130–36.

Olmsted, L. L. 1974. The ecology of largemouth bass (*Micropterus salmoides*) and spotted bass (*Micropterus punctulatus*) in Lake Fort Smith, Arkansas. Ph.D. diss., University of Arkansas, Fayetteville.

Paloheimo, J. E. 1971. A stochastic theory of search: Implications for predator–prey situations. *Math. Biosci.* 12 : 105–32.

Pasch, R. W. 1975. Some relationships between food habits and growth of largemouth bass in Lake Blackshear, Georgia. *Proceedings of the Annual Conference of Southeastern Association of Game Fish Commission* 28 : 307–21.

Patten, B. C., B. A. Egloff, and T. H. Richardson. 1975. Total ecosystem model for a cove in Lake Texoma. In B. C. Patten (ed.), *Systems analysis and simulation in ecology*, vol. 3, pp. 205–421. New York: Academic Press.

Prather, E. E. 1951. Efficiency of food conversion by young largemouth bass *Micropterus salmoides* (Lacepede). *Trans. Amer. Fish. Soc.* 80 : 154–57.

Rice, J. A., and P. A. Cochran. 1984. Independent evaluation of a biogenetics model for largemouth bass. *Ecology* 65 : 732–739.

Saiki, M. K., and J. C. Tash. 1978. Unusual population dynamics in largemouth bass, *Micropterus salmoides* (Lacepede), caused by a seasonally fluctuating food supply. *American Midland Naturalist* 100 : 116–25.

Shelton, W. L., W. D. Davies, T. A. King, and T. J. Timmons. 1979. Variation in the growth of the initial year class of largemouth bass in West Point Reservoir, Alabama and Georgia. *Trans. Amer. Fish. Soc.* 108 : 142–49.

Shelton, W. L., C. D. Riggs, and L. D. Hill. 1982. Comparative reproductive biology of the threadfin and gizzard shad in Lake Texoma, Oklahoma-Texas. In C. F. Bryan, J. V. Conner, and F. M. Truesdale (eds.), *Fifth Annual Larval Fish Conference*, pp. 47–51. Baton Rouge, La.: Louisiana Cooperative Fish Research Unit.

Shul'man, G. E. 1974. *Life cycles of fish*: Physiology and biochemistry. New York: John Wiley and Sons.

Shuter, B. J., J. A. MacLean, F. E. J. Fry, and H. A. Regier. 1980. Stochastic simulation of temperature effects on first-year survival of smallmouth bass. *Trans. Amer. Fish. Soc.* 109 : 1–34.

Summerfelt, R. C. 1975. Relationship between weather and year-class strength of largemouth bass. In H. Clepper (ed.), *Black bass biology and management*, pp. 166–74. Washington, D.C.: Sport Fishing Institute.

Timmons, T. J., W. L. Shelton, and W. D. Davies. 1980. Differential growth of largemouth bass in West Point Reservoir, Alabama-Georgia. *Trans. Amer. Fish Soc.* 109 : 176–86.

Von Geldern, C., and D. F. Mitchell. 1975. Largemouth bass and threadfin shad in California. In H. Clepper (ed.), *Black bass biology and management*, pp. 436–49. Washington, D.C.: Sport Fishing Institute.

Wedemeyer, G. A., and D. J. McLeay. 1981. Methods for determining the tolerance of fishes to environmental stressors. In A. D. Pickering (ed.), *Stress and fish*, pp. 247–75. London: Academic Press.

Wilkins, N. P. 1967. Starvation of the herring, *Clupea harengus* L.: Survival and some gross biochemical changes. *Comp. Biochem. Physiol.* 23 : 503–18.

8. Fish Predation and Its Cascading Effect on the Oneida Lake Food Chain

Edward L. Mills

John L. Forney

Kenneth J. Wagner

Over the last three decades, research on the Oneida Lake food web has revealed the importance of trophic-level interactions. Studies have illustrated functional relationships between predator and prey and have shown how the effects of fish predation are propagated through the food chain. Walleye and yellow perch are the predominant fish species and comprise a simple predator-prey association, with the walleye feeding on young and yearling yellow perch. Young yellow perch play a dual role, serving as prey for the walleye and as a predator on *Daphnia* and other zooplankton. Annual and seasonal variation in young yellow perch abundance affects growth, production, and recruitment of walleye and density of *Daphnia pulex*.

Age-0 yellow perch are selective predators, and in years when their biomass exceeds 20 kg/ha or their density exceeds 14,000/ha, *D. pulex* disappears and smaller herbivorous zooplankters increase. Loss of *D. pulex* results in elevated algal abundance and a shift in phytoplankton composition. Zooplankton-mediated alteration of algal biomass and composition affects nutrients, extending the impact of predation throughout the trophic hierarchy.

Predation serves as the chief source of density-dependent regulation in many animal associations. Ability of predatory fish to control prey populations is well documented (see Clepper 1979 for review), and suppression of forage species can impact other components of the aquatic community. Most notable in aquatic systems has been the impact of fish predation on the composition and size structure of zooplankton communities (Brooks and Dodson 1965; Galbraith 1967; Wells 1970; Sprules 1972; Threlkeld 1979; Mills and Forney 1983). Alterations of the zooplankton community structure can in turn influence the algal community. Early studies by Hrbàcek (Hrbàcek et al. 1961; Hrbàcek 1962) showed that highest algal concentrations occurred in ponds where abundant stocks of fish had eliminated large-bodied daphnids. More recent studies have confirmed that fish predation can indirectly impact phytoplankton through predation on zooplankton (Henrikson et al. 1980; Hurlbert et al. 1972; Hurlbert and Mulla 1981; O'Brien and De Noyelles 1974; Shapiro et al. 1975, 1982; Lynch and Shapiro 1981; Shapiro and Wright 1984).

THE SYSTEM

Oneida Lake is a shallow eutrophic lake (mean depth, 6.8 m) covering 20,700 ha on the Ontario Lake Plain of central New York (fig. 8.1). Numerous shoals are present, and approximately 26% of the lake bottom is shallower than 4.3 m. The lake is alkaline, pH is usually above 8.0 (Mills et al. 1978), estimated phosphorus loading is 0.72 gP $\cdot$ m^{-2} $\cdot$ yr^{-1} (Oglesby and Schaffner 1976), and retention time averages 235 days. Water chemistry is strongly influenced by nutrient-rich streams that flow from the south over outcroppings of Onondaga limestone and through the fertile and highly populated Ontario Lake Plain. The lake is usually homothermal, but temporary stratification may develop during extended periods of calm weather. Although more than 60 species of fish have been reported from Oneida Lake, species diversity is low. Walleye and yellow perch share the limnetic region with two serranids, white perch (*Morone americana*) and white bass (*Morone chrysops*), and a few large-bodied benthivores. Centrarchids are mostly restricted to a narrow vegetated zone, and cyprinids seldom venture into open water.

A HISTORICAL PERSPECTIVE

During its recorded history, the biological structure of Oneida Lake has undergone substantial changes. Most notable have been changes in the fishery. To early settlers, Oneida Lake was an important water route to the western frontier, and the fish provided an important food source for an expanding population. Salmon migrated from Lake Ontario, and as they entered streams to spawn, they were seined by early settlers. Ciscos were common, and the American eel, chain pickerel, northern pike, and walleye were the most important piscivorous species in the 1800s and early 1900s. Of these, only the walleye has remained abundant. Cultural changes contributed to the increase of the walleye population and the decline of other native species. Salmon were doomed to early extinction by lumbering and farming practices that degraded spawning streams. The decline in eels followed construction of the Barge Canal with its system of locks and dams, which may have restricted upstream migration of elvers from Lake Ontario. The fate of pickerel and northern pike was closely tied to conditions in the marshes where both species spawned. As wetlands were drained for agriculture and filled for urban development in the 1920s and 1930s, pickerel and pike populations declined. The walleye was left as the dominant piscivorous fish, and it prospered.

Trends in abundance of prey fish species indicate yellow perch has been the predominant species since the early 1900s (Forney 1977a), and invaders have fared rather poorly. For example, white perch invaded Oneida Lake from the Mohawk–Hudson river system in the late 1940s; they declined to low levels in the late 1960s and early 1970s but have shown signs of resurgence since 1978. Gizzard shad (*Dorosoma cepedianum*) were first reported in Oneida Lake in the early 1950s (Dence and Jackson 1959). Wide oscillations in young gizzard shad have been evident with population explosions documented in 1954 and again in 1984. Predation by walleye has played an important role in limiting the expansion of both white perch and gizzard shad.

Compositional changes in the zooplankton community, particularly daphnids, have oc-

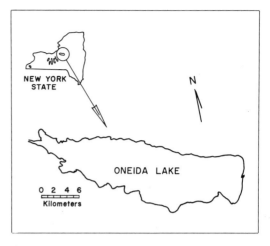

Figure 8.1. *Location of Oneida Lake, New York.*

curred over the last two decades. Prior to 1968, *Daphnia pulex* was seldom observed in Oneida Lake, and *D. galeata mendotae* and *D. retrocurva* predominated. During the 1970s, *D. pulex* gained dominance over other daphnid species, relinquishing its dominance only in 1972 and 1977. Since 1980, however, the *D. pulex* population has collapsed in 3 of 5 years and has been replaced by other daphnids and/or other smaller cladocera. This change in species composition appears to be tied to a resurgence in planktivore abundance.

One of the more dramatic biological changes in Oneida Lake has occurred in the benthic community. Prior to the mid-1960s, the mayfly *Hexagenia limbata* was the dominant benthic organism and emerged from Oneida Lake in vast numbers. However, mayfly abundance began to decline in 1959 (Jacobsen 1966), and the last nymph was seen in 1968 (Clady and Hutchinson 1976). Coincident with the mayfly decline were widespread oxygen shortages that occurred throughout most of the lake's deeper waters and probably led to the demise of the mayfly. Chironomids tolerant of low oxygen levels increased during the mayfly decline, and in subsequent years this population stabilized to become the dominant benthic organism in Oneida Lake.

METHODS

Information on interactions between walleye and yellow perch evolved from studies of population dynamics begun in 1957 (Forney 1980).

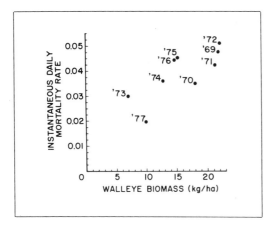

Figure 8.2. *Biomass of age-3 and older walleye versus daily instantaneous mortality rates for yellow perch cohorts in July–October, 1969–1977.*

Adult stocks of both species were estimated by mark–recapture, and, beginning in 1966, cohorts were followed through successive stages of development from egg to adult. Both species spawned in April, and larvae were pelagic through late June. Annual egg production was calculated from data on age-dependent maturity and fecundity (Clady 1976) and density of larvae estimated from the catch and volume strained by high-speed samplers (Noble 1968, 1970). After young switched to a demersal mode in July, trawls were fished at 10 to 17 sites at weekly intervals through mid-October, and abundance of cohorts were approximated from the catch and area swept (Forney 1971).

Stomachs of walleye collected at weekly or more frequent intervals were examined to measure frequency of cannibalism (Chevalier 1973) and rates of predation on yellow perch (Nielsen 1980). Reconstruction of yellow perch cohorts from numbers consumed by walleye (Forney 1977b) and comparison of calories ingested by walleye with energy requirements of the walleye population (Tarby 1977) provided some internal evidence that estimates of predator and prey abundance were reasonable.

Zooplankton was collected at weekly intervals from May to October, 1975–1984, and samples were processed either manually or by computer (Mills and Confer 1986). Zooplankton density was estimated from single vertical hauls with a 0.5-m diameter, 153-μm mesh net at up to five stations. Zooplankton were preserved in either 100% isopropyl alcohol or sugar formalin solution, identified to species, and enumerated in 3 1-ml aliquots. Zooplankters were identified from Edmondson (1963), and dry-weight conversion factors for *Daphnia* and other species (Chamberlain 1975) were used to estimate zooplankton biomass.

Chemical analyses were made on integrated water samples collected using 2-cm i.d. tygon tubing stretched from the surface to within 0.5 m from the bottom. Water was filtered through 934 AH glass fiber filters; the filters were used for pigment analysis, including phaeophytin (Strickland and Parsons 1972). Aliquots of unfiltered water were assayed for chemical parameters, including biologically available phosphorus (BAP) (Menzel and Corwin 1965), and a subsample was preserved for phytoplankton analysis. Phytoplankters were enumerated by the inverted microscope method (Lund et al. 1958) and the cell volume of individual species was converted to biomass assuming a specific gravity of 1.0 (Wellen 1959; Nauwerck 1963).

RESULTS
Walleye–Yellow Perch Interactions
Walleye and yellow perch comprise a demographically simple predator–prey association with walleye feeding mostly on age-0 yellow perch. Predation begins in June when young reach a length of 16 to 18 mm and continues through fall. Daily instantaneous mortality rates for yellow perch cohorts between June and October, 1969–1977, ranged from about 0.02 to 0.05 (fig. 8.2). Mortality rates in these years were roughly proportional to biomass of age-3 and older walleye, which implied most mortality was attributable to walleye predation. Comparison of numbers of yellow perch consumed and density of young in June dispelled any doubts about the efficiency of the walleye as a predator. Numbers of young eaten by walleye during summer and fall approximated estimates of number present in June (Forney 1977b).

In most years, more than 80% of the fish in walleye stomachs were age-0 yellow perch (Forney 1974). Consequently, annual growth and production by walleye was closely linked to first-year production by yellow perch cohorts (table 8.1). Production by age-0 yellow perch from June to October approximated the biomass of forage fish available to walleye. For

Table 8.1. *Comparison of age-3 and older walleye with age-0 yellow perch production (kilograms per hectare) from June to October, 1971–1977*

| Year | Production[a] | | Trophic level efficiency |
	Walleye	Yellow perch	
1971	13.3	129.8	0.103
1972	3.9	20.1	0.195
1973	5.6	17.8	0.315
1974	4.0	23.3	0.171
1975	11.7	94.9	0.123
1976	2.6	22.3	0.114
1977	11.1	97.4	0.114

[a]Sum of weekly estimates of production by the method of Chapman (1967).

example, the weight increment for age-3 walleye between June and October was 245 g in 1971, when young yellow perch production was 130 kg/ha, and 8.7 g in 1974, when yellow perch production was only 23 kg/ha. Annual walleye production closely matched fluctuations in production by yellow perch for all years except 1973, when age-0 white perch were more abundant than yellow perch both in trawls and in walleye stomachs. In other years, the narrow range in trophic efficiencies suggest few alternate prey were available.

Numerical response of walleye to fluctuations in abundance of age-0 yellow perch can be inferred from the contribution of walleye year classes to the adult stock. Strong year classes of walleye developed in years when prey were abundant and growth of age-0 walleye was rapid. As independent variables in a linear multiple regression, prey biomass (primarily age-0 yellow perch) and length of age-0 walleye on October 1 explained about half ($r^2 = 0.51$) the variability in survival of the 1966–1973 cohorts of walleye between age 0 and age 1 (Forney 1976). Inclusion of more recent year classes increased the r^2 to 0.76. High prey biomass and rapid first-year growth probably enhanced survival of young walleye by reducing intensity of cannibalism (Chevalier 1973).

The relation between prey abundance and survival of age-0 walleye suggests an equilibrium should exist between prey production and walleye biomass. High prey production will enhance growth of walleye and reduce cannibalism; but walleye are long-lived, and the increase in both numbers and biomass of walleye will suppress

growth and recruitment in subsequent years. The stable size composition of the adult stock and absence of any long-term trends in growth implies that the walleye has been in balance with its food supply for more than three decades (Forney 1977a).

Young yellow perch together with lesser numbers of age-0 serranids serve as the primary link in the transfer of energy from secondary to higher trophic levels. Yellow perch hatch in mid-May and concentrate in open water at depths of 0-5 m until midsummer when they attain lengths of 30 to 40 mm and become demersal (Forney 1980). Like most fishes, yellow perch are highly fecund, and temporal distribution of mortality during early life can have a profound effect on cohort biomass and production (LeCren 1962). Because abundance of young fish is often difficult to quantify and mortality rates difficult to measure, the importance of age-0 fish in trophic-level interactions has often been overlooked.

Abundance of yellow perch cohorts in Oneida Lake was measured at successive stages of development from egg to age 3 (table 8.2). Cohorts produced in 1971–1977 averaged 1 g on August 1 and 150 at age 3. Biomass calculated from these mean weights and the densities shown in table 8.2 indicate that biomass was higher at age 0 than age 3 for five of seven cohorts. More relevant to those concerned with energy flow are differences in production. Mean first-year production by the 1971–1977 year classes was 58 kg/ha and production by individual year classes ranged from approximately 20 to 130 kg/ha (table 8.1). Production estimates for older age groups are not available for the same time period, but combined somatic and gonadal production by age-3 and older perch in 1969–1971 averaged 17 kg/ha (Vashro 1975). These results suggest that both biomass and production of young yellow perch in Oneida Lake is substantial and may equal or exceed biomass and production of older yellow perch in most years.

Analysis of mechanisms controlling abundance of yellow perch is beyond the scope of this paper. It is sufficient to note here that the number of 8-mm larvae present in May was strongly correlated with climatic conditions during the period of egg incubation (Clady 1976). Mortality during the 6- to 8-week pelagic stage led to further divergence in cohort abundance, but the exact cause was ob-

Table 8.2. *Density (number per hectare) of the 1971–1977 year classes of yellow perch estimated at successive life stages*

Year class	Eggs (millions)	Age 0				Age 3
		8 mm	18 mm	Aug 1	Oct 15	
1971	2.16	210,000	218,000	37,700	3,520	7
1972	2.63	143,000	110,000	2,500	100	1
1973	3.26	39,000	18,000	1,200	510	52
1974	3.28	77,000	30,000	3,100	320	13
1975	2.46	362,000	152,000	21,400	450	7
1976	2.33	126,000	37,000	6,500	190	17
1977	2.08	220,000	65,000	14,700	4,220	243
Mean	2.60	168,000	90,000	12,400	1,330	48

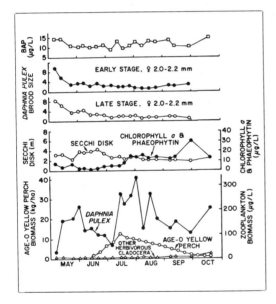

Figure 8.3. *Comparison of seasonal changes in zooplankton biomass, chlorophyll a plus phaeophytin, early- and late-stage* Daphnia pulex *brood size, age-0 yellow perch biomass, Secchi disc transparency, and biologically available phosphorus in Oneida Lake during 1976.*

scure. After cohorts attained a length of 18 mm, most mortality could be accounted for by examining stomachs of walleye and older yellow perch (Forney 1977b). Cohorts of yellow perch, which were exceptionally abundant in August at age 0, often contributed few fish to the adult stock at age 3 (table 8.2). Neilsen (1980) attributed density-dependent mortality of yellow perch between age 0 and age 3 to predation by walleye and observed that mortality was a function of body size. Large body size served

as a refuge from predation, and rapid first-year growth shortened the time cohorts were vulnerable to walleye predation. Subsequent studies demonstrated first-year growth of yellow perch was density-dependent, and body length attained by fall was governed by abundance of *D. pulex* (Mills and Forney 1981).

Yellow Perch–Zooplankton Interactions

Seasonal trends in abundance of *D. pulex* and age-0 yellow perch 1975 to 1979 indicated predation by large year classes of yellow perch could deplete *Daphnia* populations (Mills and Forney 1983). Whether daphnids persisted through the summer or disappeared was probably determined by abundance of yellow perch as illustrated by seasonal trends in *Daphnia* density and yellow perch abundance in 1976 and 1977 (figs. 8.3 and 8.4). In 1976, daphnid biomass rose in May, declined through early July, recovered in late July, and remained high through the summer. Biomass of young yellow perch was low in June, and their biomass never exceeded 12 kg/ha. In 1977, *D. pulex* peaked in mid-June, declined precipitously in July, and could not be detected in August through October. In contrast to 1976, yellow perch biomass exceeded 30 kg/ha and remained above 20 kg/ha into October. *D. pulex* did not reappear until June 1978. The assumption that predation was responsible for the disappearance of *D. pulex* in 1977 was supported by estimates of prey consumption. Daily consumption of *D. pulex* by age-0 yellow perch exceeded estimates of *D. pulex* production prior to col-

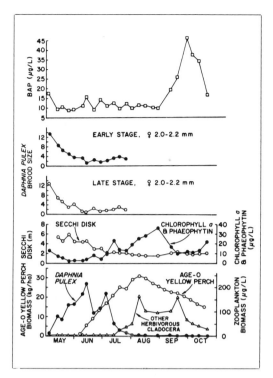

Figure 8.4. *Comparison of seasonal changes in zoo-plankton biomass, chlorophyll a plus phaeohytin, early- and late-stage Daphnia pulex brood size, age-0 yellow perch biomass, Secchi disc transparency, and biologically available phosphorus in Oneida Lake during 1977.*

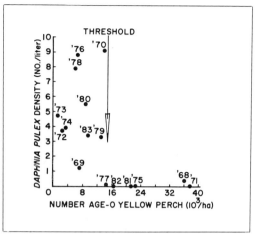

Figure 8.5. *Abundance of age-0 yellow perch determined from catches on August 1, compared with density of* Daphnia pulex, *averaged August–October in Oneida Lake, 1968–1983.*

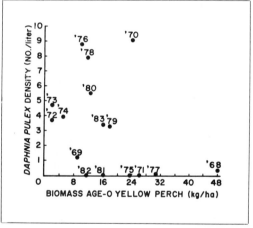

Figure 8.6. *Biomass of age-0 yellow perch determined from trawl catches on August 1, compared with density of* Daphnia pulex, *averaged August–October in Oneida Lake, 1968–1983.*

lapse of the population in 1977, but consumption by yellow perch was consistently less than one-third *Daphnia* production in 1976.

Density of *D. pulex* in late summer was consistently low in years when density of young yellow perch exceeded a threshold of 14,400/ha (fig. 8.5). More specifically, in 6 of 16 years when yellow perch density exceeded 14,400/ha, the mean density of *D. pulex* during August through October was 0.4/L or less. The relation between yellow perch biomass and *Daphnia* density was not as well defined, but in 4 out of 5 years when the August 1 biomass of yellow perch exceeded 20 kg/ha, mean density of *D. pulex* was 0.4/L or less (fig. 8.6). The only exception was 1970, when predation by young perch was not sufficient to collapse the daphnid population. In all 6 years when the biomass of young yellow perch was less than 10 kg/ha, *D. pulex* were abundant into late fall. When age-0 yellow perch biomass ranged between 10 and 20 kg/ha, however, *D. pulex*

were able to tolerate predation from yellow perch in 3 of 5 years. The 2 years (1981 and 1982) when *D. pulex* collapsed by fall were years when young yellow perch densities in midsummer exceeded 14,400/ha. These findings contrast with those of Levitan et al. 1985, who found in enclosure experiments that *D. pulicaria* could tolerate 100 to 150 kg/ha of fingerling rainbow trout (*Salmo gairdneri*). However, unlike fingerling trout, which are larger and can utilize an assortment of prey items, young yellow perch are dependent on zooplankton for food. Furthermore, because of size-dependent differences between fingerling trout and age-0 yellow perch, it is likely that differences in food intake, growth, and production also may impact on the tolerance level at which fish predation influences the zooplankton community, and this level (20 vs. 150 kg/ha) may go up as the dependence of planktivores on zooplankton declines.

Young yellow perch are selective predators, and prey size increases with fish size (Hansen and Wahl 1981). Although yellow perch over 35 mm are able to capture and ingest adult *D. pulex*, smaller immature *D. pulex* are preferred. Consequently, the impact of predation by young yellow perch in midsummer was concentrated on immature and early maturing adult *Daphnia*. Total lengths of young yellow perch on August 1, 1975–1983 averaged 41 to 55.5 mm. Based on a relationship between yellow perch size and the minimum–maximum size of *D. pulex* selected (Mills, Confer, and Ready 1984), the size of prey selected by 41- to 55.5-mm young yellow perch has ranged between 0.9 and 1.9 mm (table 8.3). Comparison of *Daphnia* density for the size range selected by young yellow perch with yellow perch density indicates that in years when the density of age-0 yellow perch exceeded 14,400/ha on August 1, the density of *D. pulex* on this date was very low (<0.02/ha). In these years (1975, 1977, 1981, and 1982), size-selective predation

by young yellow perch on intermediate-size prey in July most likely had a severe impact on daphnid recruitment because the *Daphnia* population collapsed by August of each year. In other years, young yellow perch density was less than 14,400/ha, and supplies of preferred-size *Daphnia* remained abundant in late summer. Apparently, reduced predation pressure by young yellow perch on intermediate-size *Daphnia* and their recruitment into the pool of reproducing adults (usually >1.9 mm) was sufficient to avoid population collapse.

In Oneida Lake, young yellow perch act as a keystone species, and their ability to depress populations of large-bodied daphnids may explain how competition and predation interact to regulate yellow perch recruitment. Positive selection for *D. pulex* occurs when the yellow perch cohort reaches 25 to 30 mm total length; and if yellow perch are sufficiently abundant, *D. pulex* will disappear and stocks of other herbivorous cladocerans will dramatically increase (figs. 8.3, 8.4, 8.7). Although age-0 yellow perch can subsist on small zooplankters, growth in late summer is substantially reduced (Mills and Forney 1981). Slow first-year growth increases losses from cannibalism by adult yellow perch (Tarby 1974) and prolongs the period of vulnerability to predation by walleye (Neilsen 1980). Consequently, cohorts of yellow perch that were exceptionally abundant in July and collapsed the *D. pulex* population should experience higher mortality than smaller cohorts. In support of this hypothesis, the 1968, 1971, 1975,

Table 8.3. *Comparison of age-0 yellow perch density with the density of intermediate-size prey preferred by young perch on August 1, 1975–1983*

	Age-0 yellow perch		*Daphnia pulex*	
Year	Mean total length (mm)	Density (no./ha)	Density (no./L)	Prey size range selected[a] (mm)
1975	43.5	23,690	0.01	1.0–1.8
1976	49.3	6,115	7.20	1.3–1.9
1977	55.5	14,435	0.02	1.4–1.9
1978	51.6	6,064	1.20	1.2–1.9
1979	48.0	14,175	1.20	1.1–1.9
1980	47.5	9,980	4.20	1.1–1.9
1981	41.0	20,020	0.00	0.9–1.8
1982	41.0	15,505	0.02	0.9–1.8
1983	50.7	9,625	1.60	0.9–1.8

[a]Based on mean total length of yellow perch size and the minimum and maximum lengths of *D. pulex* selected (Mills, Confer, and Ready 1984).

and 1981 cohorts that were dominant in early summer contributed fewer age-3 fish to the adult stock than initially smaller cohorts (Forney 1980; unpublished data).

Zooplankton–Phytoplankton

During its recorded history Oneida Lake has exhibited algal blooms characteristic of nutrient-enriched conditions, and seasonal patterns of species succession have been well defined. Following ice-out there is a pulse of diatoms and small flagellates leading to a spring algal bloom. Subsequent decline in diatom populations results in a period of maximum transparency in June, when small flagellates such as cryptophytes dominate the phytoplankton. Cyanophyte biomass begins to rise in late June through early July, culminating in one or more summer blooms, sometimes coincident with pulses of diatoms. Algal standing stock declines during the fall and remains low through winter.

The zooplankton community, particularly daphnids, contributes to the regularity of observed phytoplankton and water clarity patterns. Peak spring biomass of *Daphnia* coincides with the June period of low algal biomass and maximum water transparency. Loss of the *D. pulex* population resulting from predation by young yellow perch will trigger a dramatic increase in other herbivorous cladocera (fig. 8.7) which are both smaller and less efficient grazers of algae (Haney and Hall 1975). These smaller species include *Bosmina longirostris*, *Chydorus sphaericus*, *D. galeata mendotae*, *D. retrocurva*, and *Sida crystallina*. Such an event leads to an increase in the ratio of nanno- to net-phytoplankton, and subsequently to a combined decrease in lake water clarity and increase in algal biomass. Consequently, young yellow perch predation, mediating grazing by *Daphnia*, is largely responsible for structural shifts in the Oneida Lake phytoplankton community (Wagner 1983). More specifically, as *Daphnia* biomass increases, slower growing and digestible diatom species are intensively grazed on, and by June diatoms give way to faster growing small flagellates. These flagellates provide ideal food for *Daphnia*, are intensely grazed, and fail to reach a large biomass. By July, the available niche space is filled by less desirable and poorly consumed blue-green algae, which usually maintain dominance into fall.

The most critical period for *Daphnia* gener-

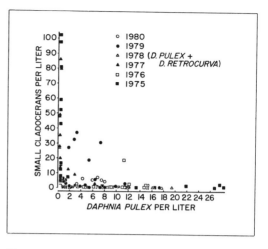

Figure 8.7. *Relationship between abundance of* Daphnia pulex *and density of other small cladocerans in Oneida Lake, New York, 1975–1980. (Points represent a combined average of two to five sites for each date and includes the period January–December of each year.)*

ally occurs in late June and early July. At this time *Daphnia* biomass is at a peak, and grazing on algae is intense. As a result, phytoplankton levels are greatly reduced and water transparency is high. The *Daphnia* population responds to declining food resources with an increase in ephippial egg production, a decline in the fraction of mature females bearing eggs, and a trend toward smaller gravid females (Wagner 1983). While *Daphnia* is responding to declining stocks of food, intense predation by young yellow perch begins. At this time, if predation by young yellow perch is intense, recruitment of young daphnids will not be sufficient to prevent collapse of the *Daphnia* population. On the other hand, if predation pressure by young yellow perch on *Daphnia* is low, recruitment of young daphnids will exceed losses from predation.

Reproduction by *D. pulex* during the critical June–July period is likely submaximal because of a variety of factors, including limited supplies of edible nannophytoplankton and inability of *Daphnia* to exploit efficiently a bountiful midsummer supply of large, nonconsumable algal species composed primarily of blue-greens. For *D. pulex* averaging 1.7 mm in length a concentration of 50 μg C/L is needed to sustain body weight, 200 μg C/L will allow parthenogenic reproduction, and maximal reproduction begins at about 700 μg C/L (Lampert 1978). Assuming that carbon comprised 10% of total phytoplankton biomass, only 55% of all total phytoplankton carbon estimates have exceeded the reproductive threshold of 200 μg C/L, and most of these values occurred when the phytoplankton assemblage was dominated by large nonconsumable and/or indigestible filamentous diatoms and cyanophytes (Wagner 1983). Consequently, many of these values represent an overestimate of carbon available to D. pulex. This raises the potential of food limitation for Oneida Lake Daphnia, which is somewhat surprising, considering the relatively high productivty of this eutrophic system.

Reproductive limitation of Oneida Lake Daphnia also can be seen in the fraction of mature D. pulex actually gravid. The fraction of mature, gravid (FMG) Daphnia should exceed 0.8 to 0.9 under conditions of optimal food resources (Hebert 1978). For Daphnia collected weekly from Oneida Lake during May through October of 1975–1981, less than 2% of FMG values were greater than 0.9, and only 10% exceeded 0.8. Reproductive limitation also was evidenced by early- and late-stage brood sizes, which clearly declined following the spring diatom pulse and occurred during late May through June when available food resources were low (figs. 8.3 and 8.4). Under conditions of low food resources, Daphnia respond by producing ephippia as a long-term maintenance measure, but this strategy diverts energy from parthenogenic egg production and reduces the effective birth rate. Consequently, reduced gravidity and brood sizes lead to submaximal birth rates and diminish the ability of Daphnia to overcome the pressures of predation from young yellow perch.

Response of Daphnia and algae to varying predation intensity by young yellow perch in Oneida Lake is illustrated by data for 1976 and 1977 (figs. 8.3 and 8.4). In 1976, D. pulex biomass rose in May and declined through early July. Yet the D. pulex population recovered in late July, remained high through late summer, and was never challenged by other herbivorous cladocerans. Early- and late-stage mean brood sizes were highest (9.8 and 8.3, respectively) in early May and declined to 1 and 3 during June through September. Predation by young yellow perch on D. pulex was modest because the density of these fish was low and their biomass never exceeded 12 kg/ha. Consequently, Daphnia were abundant, and grazing

by these herbivores on algae was presumably intense. Actual surrogates of algal abundance would support this observation because chlorophyll concentrations were generally less than 15 μg/L, and lake water was relatively clear with lakewide Secchi disk transparencies usually exceeding 2 m. In 1977, peak mean brood size occurred in May and preceded the period when D. pulex abundance was highest. Brood size declined in June to generally less than 3 eggs per mature female. Apparently reproduction was unable to compensate for losses due to predation by young yellow perch, which was sufficiently intense to collapse the D. pulex population. Loss of D. pulex triggered both an increase in smaller and less efficient herbivorous cladocera and an increase in algal density. In the absence of Daphnia, maximum chlorophyll concentrations in August were nearly two times greater than those for the same period in 1976 when D. pulex was present. Despite changes in the daphnid community between years, Secchi disc depths in August were not so different as might be expected, possibly due to the annual predominance of blue-greens in midsummer.

Nutrient Considerations

If young yellow perch structure the zooplankton community and Daphnia are largely responsible for shifts in the Oneida Lake phytoplankton community, it is possible such modifications in biological structure could influence cycling of nutrients, particularly phosphorus. For example, when small-bodied zooplankton replace Daphnia, standing stocks of algae increase dramatically, and more phosphorus could be incorporated in algae and remineralized in the water column (Carpenter and Kitchell 1984). In support of this statement, annual peak total phosphorus concentrations between 1975 and 1983 have generally occurred in years when Daphnia biomass was low (Spearman Rank correlation, −0.53; p < 0.10). Furthermore, a direct seasonal relationship exists between total phytoplankton biomass and total phosphorus, with as much as 71% of the variance in total phytoplankton biomass attributed to changes in total phosphorus concentration (table 8.4). Also examined was the seasonal response of BAP to changes in both daphnid and algal abundance (figs. 8.3 and 8.4). It might be expected that when Daphnia disappear, phytoplankton bio-

Table 8.4. *Dependence of phytoplankton biomass (Y) on total phosphorus concentration (X) in Oneida Lake, 1976–1980, based on June-through-October averages of phytoplankton and total phosphorus and only years when Daphnia pulex averaged >15% of total zooplankton biomass*

Year	Relationship	r^2	Significance level (p)
1976	Y = −301.0 + 156.3(X)	0.34	<0.05
1977	Y = −5909 + 404.8(X)	0.71	<0.001
1978	Y = −1102 + 209.8(X)	0.62	<0.25
1979	Y = 1944 + 60.8(X)	0.50	<0.01
1980	Y = −1125 + 129.2(X)	0.42	<0.025

mass would increase, and BAP levels would decline until an algal die-off occurred, at which time a BAP increase might be observed. In effect, young yellow perch, through *Daphnia* and phytoplankton, could produce a "ripple effect" on phosphorus. In 1976, *D. pulex* were present through the summer, chlorophyll concentrations generally remained below 20 µg/L, and BAP exhibited no dramatic pulse during the period. In 1977, however, *D. pulex* collapsed, total chlorophyll peaked to nearly 40 µg/L, and BAP levels increased dramatically following a decline in algae abundance. However, elevated BAP levels were not accompanied by increased Secchi depths, suggesting possible changes in chlorophyll-to-carbon ratios as the phytoplankton community changes (Carpenter and Kitchell 1984).

Community Level Interactions

The seasonal sequence and cycling of biological events follows distinct patterns that vary with planktivore abundance. A generalized model of annual biomass cycles of algae and *Daphnia* in years when maximum biomass of age-0 yellow perch ranged from 20 to 40 kg/ha is shown in fig. 8.8. The response of lower trophic levels to an early summer increase in yellow perch biomass can be likened to the "domino effect." As young yellow perch approach their maximum summer standing crop, predation depresses stocks of *D. pulex*, smaller herbivorous zooplankton appear, and reduced grazing allows an increase in biomass of both nanno- and netphytoplankton. Conversely, in years when young yellow perch biomass is less than 10 to 15 kg/ha, large-bodied daphnids dominate and stocks of nannophytoplankton are depressed as a consequence of intense grazing.

So far, this paper has described trophic level

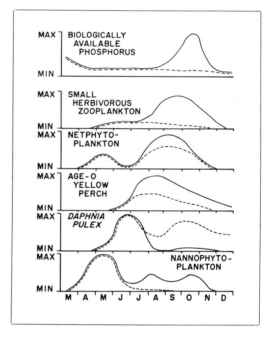

Figure 8.8. *Generalized annual cycle of biologically available phosphorus, phytoplankton, and zooplankton in Oneida Lake when the July–August standing crop of young yellow perch is >20–40 kg/ha (—) and <10–15 kg/ha (---).*

interactions and stressed the cascading impact of predation. To quantify interactions and assess their strength, linear regressions were fitted to mean June–October estimates of age-0 yellow perch, *D. pulex*, and phytoplankton biomass (table 8.5). In all cases, the upper trophic level represented the independent variable and the lower level the dependent variable. Predictions of *D. pulex* biomass from young yellow perch biomass and prediction of phytoplankton biomass from *D. pulex* biomass were both significant (F-test; p, <0.05). For individual regres-

Table 8.5. *Predictive equations between individual trophic levels in Oneida Lake based on June–October averages between 1975 and 1983 (only years in which Daphnia were >15% of the total biomass are included)*

Variable Dependent X Independent	Years (no.)	Relationship	r^2	Significance level (p)
D. pulex X yellow perch (μg/L) (kg/ha)	6	$Y = 303.8 \exp(-0.1X)$	0.71	<0.05
D. pulex X yellow perch (%) (kg/ha)	6	$Y = 129.8 \exp(-0.09X)$	0.87	<0.01
Secchi disc X D. pulex (m) (%)	6	$Y = 1.242 \exp(0.17X)$	0.71	<0.005
Phytoplankton X D. pulex (μg/L) (μg/L)	5	$Y = 6343.7 - 9.19(X)$	0.77	<0.05

Table 8.6. *Dependence of Daphnia, phytoplankton, Secchi disc, and total phosphorus on yellow perch biomass in Oneida Lake (only years in which Daphnia represented >15% of the total zooplankton biomass are included)*

Variable Dependent X Independent	Years (no.)	Relationship	r^2	Significance level (p)
D. pulex X yellow perch (μg/L) (kg/ha)	6	$Y = 303.8 \exp(-0.1X)$	0.71	<0.05
D. pulex X yellow perch (%) (kg/ha)	6	$Y = 129.8 \exp(-0.09X)$	0.87	<0.01
Secchi disc X yellow perch (m) (kg/ha)	6	$Y = 2.96 \exp(-0.018X)$	0.57	<0.10
Phytoplankton X yellow perch (μg/L) (kg/ha)	5	$Y = 4582.7 + 60.7(X)$	0.36	<0.25
Total phosphorus X yellow perch (μg/L) (kg/ha)	6	$Y = 52.65 \exp(-0.01X)$	0.093	—

sions, each independent variable accounted for 71 to 87% of the variance of each dependent variable, providing firm evidence that links exist at each trophic level.

The question remains, how much ecosystem structure and function is determined by planktivore abundance? In table 8.6, the dependence of *D. pulex* (expressed as a percentage of the total zooplankton biomass), total phytoplankton biomass, Secchi disc depth, and total phosphorus concentration on age-0 yellow perch biomass is examined. To assess trophic-level interactions with *Daphnia*, only years in which *D. pulex* represented at least 15% of the total zooplankton biomass were examined. As might be expected, annual fluctuations in *Daphnia* abundance are most closely linked to annual changes in yellow perch abundance, and the effects of fish predation are indirectly related to both water clarity and abundance of algae, but yel-

low perch abundance had no detectable effect on total phosphorus. Associated with the cascading effect of yellow perch predation is a "cascade" in both the percentage of the variance of each dependent variable explained by yellow perch, and the significance of each trophic-level interaction with yellow perch as the food chain length increases downward from fish. These results are consistent with those of McQueen and Post (1985) and indicate the importance of biological structure in aquatic systems; fish predation can have impacts on zooplankton and phytoplankton that are predictable, yet predictability declines with trophic distance between variables.

MANAGEMENT IMPLICATIONS

Annual fluctuations in walleye abundance in Oneida Lake are considerable, and the risk exists that yellow perch could swamp the predator population. The destabilizing effects of changes in prey abundance on predator stocks can be amplified by the impact of humans. Consequently, the combined effects of intense exploitation by humans and low recruitment can drive the walleye biomass to low levels. To achieve the dual objective of minimizing risk and maximizing yield, regulations have been developed to harvest surplus walleye stock when biomass is high and restrict harvest when predator recruitment and biomass is low.

Enhancement of water quality through manipulation of fish populations is a potentially appealing lake management strategy. Addition or removal of planktivorous fish from ponds and lake enclosures has demonstrated that top-down predation can alter size composition of the zooplankton community and indirectly modify algal abundance and species composition. Improvement in water quality through the control of planktivorous fish populations has generated the theory of trophic "biomanipulation" (Shapiro et al. 1982; Shapiro and Wright 1984). In Oneida Lake, walleye prey on young and yearling yellow perch, and the impact of predation is transmitted through several trophic levels (fig. 8.9). Young yellow perch serve as the primary link in the transfer of energy from secondary to higher trophic levels. These fish are not only prey for walleye but are predators on *Daphnia*. Because of their impact on *Daphnia*, young yellow perch can indirectly impact on algal communities through alteration of the zooplankton community and can either enhance or reduce water transparency. Despite such "consumer controlled" effects on biological structure, management of size and species structure of the fish community is difficult to control, and the goals of fishery and water quality management will often conflict. For example, growth and recruitment of walleye are in part dependent on abundance of age-0 yellow perch. However, in years when walleye production is high, age-0 yellow perch are abundant, daphnids are scarce, and water quality declines. On the other hand, efforts to reduce age-0 yellow perch abundance by increasing walleye biomass enhances the *Daphnia* population and indirectly enhances water quality, but such a management strategy simultaneously reduces walleye production. The management implication from Oneida Lake is

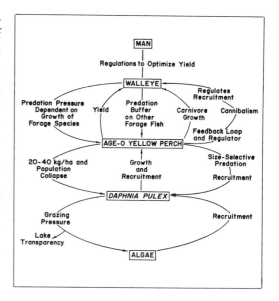

Figure 8.9. *Trophic level interactions in the Oneida Lake food chain. Arrows indicate direction of greatest impact.*

that because of potential conflicts between optimizing fish yield, controlling forage fish stocks, and desiring large-bodied daphnids, strategies for lake management must consider tradeoffs between enhancing water quality and maximizing fish production.

One clear lesson from the long-term Oneida Lake data set is that because of high annual variability in both predator recruitment and forage abundance, effective management is hampered by stochastic events. However, from experience and knowledge of the Oneida Lake system the impact of a strong year class of yellow perch is predictable, but year-class strength, which is influenced by annual fluctuations in reproductive success, must be known in advance in order to make predictions of events at other trophic levels. This is significant and of substantial value, but it limits the use of manipulation as a management strategy in structuring aquatic food chains.

ACKNOWLEDGMENT

This research was supported by research grants from Cornell University and the New York State Department of Environmental Conservation. Contribution #92 of the CUBFS.

REFERENCES

Brooks, J. L., and S. Dodson. 1965. Predation, body size, and competition of plankton. *Science* 150 : 28–35.

Carpenter, S. R., and J. K. Kitchell. 1984. Plankton community structure and limnetic primary production. *Amer. Naturalist* 124(2) : 159–72.

Chamberlain, H. D. 1975. A comparative study of the zooplankton communities of Skaneateles, Owasco, Hemlock, and Conesus lakes. Ph.D. thesis, Cornell University, Ithaca, N.Y.

Chapman, D. W. 1967. Production in fish populations. In Gerking, S. D. (eds.), *The biological basis of freshwater fish production*, pp. 3–29. New York: John Wiley and Sons.

Chevalier, J. R. 1973. Cannibalism as a factor in first-year survival of walleyes in Oneida Lake. *Trans. Amer. Fish. Soc.* 102 : 739–44.

Clady, M. D. 1976. Influence of temperature and wind on the survival of early stages of yellow perch, *Perca flavescens*. *J. Fish. Res. Board Can.* 33(9) : 1887–93.

Clady, M., and B. Hutchinson. 1976. Food of the yellow perch, *Perca flavescens*, following a decline of the burrowing mayfly, *Hexagenia limbata*. *Ohio J. Sci.* 76 : 133–38.

Clepper, H. (ed.). 1979. Predator–prey systems in fisheries management. Washington, D.C.: Sport Fishing Institute.

Dence, W. A., and D. F. Jackson. 1959. Changing chemical and biological conditions in Oneida Lake, New York. *School Sci. Math.* 59 : 317–24.

Edmondson, W. T. (ed.). 1963. Freshwater biology. New York: J. Wiley and Sons.

Forney, J. L. 1971. Development of dominant year-classes in a yellow perch population. *Trans. Amer. Fish. Soc.* 100(4) : 739–49.

———. 1974. Interactions between yellow perch abundance, walleye predation, and survival of alternate prey in Oneida Lake, New York. *Trans. Amer. Fish. Soc.* 103 : 15–24.

———. 1976. Year-class formation in the walleye (*Stizostedion vitreum vitreum*) population in Oneida Lake, New York. *J. Fish. Res. Board Can.* 33(4) : 1812–20.

———. 1977a. Evidence of inter- and intraspecific competition as factors regulating walleye (*Stizistedion vitrem vitreum*) biomass in Oneida Lake, New York. *J. Fish. Res. Board Can.* 34(10) : 1812–20.

———. 1977b. Reconstruction of yellow perch (*Perca flavescens*) cohorts from examination of walleye (*Stizostedion vitreum vitreum*) stomachs. *J. Fish. Res. Board Can.* 34(7) : 925–32.

———. 1980. Evolution of a management strategy for walleye in Oneida Lake, New York. *New York Fish and Game Journal* 27 : 105–41.

Galbraith, M. G. 1967. Size-selective predation of *Daphnia* by rainbow trout and yellow perch. *Trans. Amer. Fish. Soc.* 96 : 1–10.

Haney, J. F., and D. J. Hall. 1975. Diel vertical migration and filter feeding activities of *Daphnia*. *Arch. Hydrobiol.* 75 : 413–41.

Hansen, M. J., and D. H. Wahl. 1981. Selection of small *Daphnia pulex* by young yellow perch in Oneida Lake, New York. *Trans. Amer. Fish. Soc.* 110 : 64–71.

Hebert, P. D. N. 1978. The population biology of *Daphnia* (Crustacea: Daphnidae). *Biol. Rev.* 53 : 387–426.

Henrikson, L., H. G. Nyman, H. G. Oscarson, and J. A. E. Stenson. 1980. Trophic changes, without changes in the external nutrient loading. *Hydrobiologia* 68 : 257–63.

Hrbàček, J. 1962. Species composition and the amount of zooplankton in relation to fish stock. *Rozpr. Česk. Akad. Ved. Řapa Mat. Přír. Yed* 72 : 1–116.

Hrbàček, J., M. Dvořakova, M. Kořínek, and L. Procházková. 1961. Demonstration of the effect of the fish stock on the species composition of zooplankton and the intensity of metabolism of the Whole plankton association. *Verh. Int. Verein. Limnol.* 14 : 192–95.

Hurlbert, S. H., and M. S. Mulla. 1981. Impacts of mosquito fish (*Gambusia affinis*) predation on plankton communities. *Hydrobiologia* 83 : 125–51.

Hurlbert, S. H., J. Zedler, and D. Fairbanks. 1972. Ecosystem alternation by mosquito fish (*Gambusia*) predation. *Science* 175 : 639–41.

Jacobsen, T. V. 1966. Trends in abundance of the mayfly (*Hexagenia limbata*) and chironomids in Oneida Lake, New York. *N.Y. Fish. Game J.* 13 : 168–75.

Lampert, W. 1978. A field study on the dependence of the fecundity of *Daphnia* sp. on food concentration. *Oecology* 36 : 363–69.

LeCren, E. D. 1962. The efficiency of reproduction and recruitment in freshwater fish. In E. D. LeCren, and M. W. Holdgate (eds.), *The exploitation of natural animal populations*, pp. 283–96. Oxford: Blackwell Scientific Publications.

Levitan, C., W. C. Kerfoot, and W. R. DeMott. 1985. Ability of *Daphnia* to buffer trout lakes against periodic nutrient inputs. *Verh. Internat. Verein Limnol.* 22 : 1–7.

Lund, J. W. G., C. Kipling, and E. D. LeCren. 1958. The inverted microscope method of estimating algal numbers and the statistical basis of estimations by counting. *Hydrobiologia* 11 : 142–70.

Lynch, M., and J. Shapiro. 1981. Predation, enrichment, and phytoplankton community structure. *Limnol. Oceanogr.* 26(1) : 86–102.

McQueen, D. J., and J. R. Post. 1985. Effects of planktivorous fish on zooplankton, phytoplankton, and water chemistry. In *Lake and reservoir management: Practical applications*. Proceedings of the Fourth Annual Conference and International Symposium. McAfee, N.J.: North American Lake Management Society.

Menzel, D., and N. Corwin. 1965. The measurement of total phosphorus on the liberation of the organically bound fractions by persulfate oxidation. *Limnol. Oceanogr.* 10 : 280–82.

Mills, E. L., and J. L. Confer. 1986. Computer processing of zooplankton: Application in fisheries studies. *Fisheries* (in press).

Mills, E. L., J. L. Confer, and R. C. Ready. 1984. Prey selection by young yellow perch: The influence of capture success, visual acuity, and prey choice. *Trans. Amer. Fish Soc.* 113 : 579–87.

Mills, E. L., and J. L. Forney. 1981. Energetics, food consumption, and growth of young yellow perch in Oneida Lake, New York. *Trans. Amer. Fish. Soc.* 110 : 479–88.

———. 1983. Impact on *Daphnia pulex* of predation by young yellow perch in Oneida Lake, New York. *Trans. Amer. Fish. Soc.* 112 : 154–61.

Mills, E. L., J. L. Forney, M. D. Clady, and W. R. Schaffner. 1978. Oneida Lake. In J. A. Bloomfield (ed.), *Lakes of New York State*, vol. 2, pp. 367–451. New York: Academic Press.

Nauwerck, A. 1963. The relation between zooplankton and phytoplankton in Lake Erken. *Symb. Bot. Ups.* 17 : 163.

Nielsen, L. A. 1980. Effect of walleye (*Stizostedion vitreum vitreum*) predation on juvenile mortality and recruitment of yellow perch (*Perca flavescens*) in Oneida Lake, New York. *Can. J. Fisheries Aquat. Sci.* 37 : 11–19.

Noble, R. L. 1968. Mortality rates of pelagic fry of the yellow perch, *Perca flavescens* (Mitchill), in Oneida Lake, New York, and an analysis of the sampling problem. Ph.D. thesis, Cornell University, Ithaca, N.Y.

———. 1970. An evaluation of the Miller highspeed sampler for sampling yellow perch and walleye fry. *J. Fish. Res. Board Can.* 27(6) : 1033–44.

O'Brien, W. J., and F. de Noyelles. 1974. Relationship between nutrient concentration, phytoplankton density, and zooplankton density in nutrient enriched experimental ponds. *Hydrobiologia* 44 : 105–25.

Oglesby, R. T., and W. R. Schaffner. 1976. The response of lakes to phosphorus. In K. H. Porter (ed.), *Nitrogen and phosphorus: Agriculture,* wastes, and the environment, pp. 25–60. Ann Arbor, Mich.: Ann Arbor Science Publishers.

Shapiro, J., B. Forsberg, V. Lamarra, G. Lindmark, M. Lynch, E. Smeltzer, and G. Zoto. 1982. Experiments and experiences in biomanipulation—studies of biological ways to reduce algal abundance and eliminate blue-greens. EPA-600/3-82-096. Corvallis, Oreg.: Corvallis Environmental Research Laboratory. U.S. Environmental Protection Agency.

Shapiro, J., V. A. Lammara, and M. Lynch. 1975. Biomanipulation: An ecosystem approach to lake restoration. In P. L. Brezonik and J. L. Fox (eds.), *Proceedings of the Symposium on Water Quality Management through Biological Control.* Gainesville, Fla.: University of Florida and USEPA.

Shapiro, J., and D. I. Wright. 1984. Lake restoration by biomanipulation: Round Lake, Minnesota, the first two years. *Freshwater Biol.* 14 : 371–83.

Sprules, W. G. 1972. Effects of size-selective predation and food competition on high altitude zooplankton communities. *Ecology* 53 : 375–86.

Strickland, J. D. H., and T. R. Parsons. 1972. *A practical handbook of seawater analysis.* 2nd ed. Fish. Res. Board Can. Bull. 167.

Tarby, M. J. 1974. Characteristics of yellow perch cannibalism in Oneida Lake and the relation to first-year survival. *Trans. Amer. Fish. Soc.* 103 : 462–71.

———. 1977. Energetics and growth of walleye (*Stizostedion vitreum vitreum*) in Oneida Lake, New York. M.S. thesis, Cornell University, Ithaca, N.Y.

Threlkeld, S. T. 1979. The midsummer dynamics of two *Daphnia* species in Wintergreen Lake, Michigan. *Ecology* 60 : 165–79.

Vashro, J. E. 1975. Production of yellow perch in Oneida Lake. M.S. thesis, Cornell University, Ithaca, N.Y.

Wagner, K. J. 1983. The impact of natural phytoplankton assemblages on *Daphnia pulex* reproduction in Oneida Lake, New York. M.S. thesis, Cornell University, Ithaca, N.Y.

Wellen, T. 1959. The phytoplankton of Gorwalm—a bay of Lake Malaren. *Oikos* 10 : 241–74.

Wells, L. 1970. Effects of alewife predation on zooplankton populations in Lake Michigan. *Limnol. Oceanogr.* 15 : 556–65.

9. Piscivores, Planktivores, Fossils, and Phorbins

James F. Kitchell

Stephen R. Carpenter

Analysis of a sediment core from Lake Michigan suggests a cascade of complex interactions resulting from a history of fish perturbations. Reduction of piscivory and increased size-selective zooplanktivory are clearly recorded in the morphology of herbivorous zooplankton. Changes in the abundance and antipredator morphology of *Bosmina longirostris* correspond to historically documented food web events: collapse of the lake trout populations, increase in alewife abundance, and reduction of large zooplankton. As zooplankton community shifted toward small herbivorous copepods, grazing favored increased loading of phorbins to sediments. Chlorophyll and pheophytin concentrations did not change, but increases in a grazing indicator, pheophorbide, were recorded. A suite of recent changes in Lake Michigan's food web suggests that reversal of the previous shift is currently underway.

Some ecologists view predators as largely the result of production dynamics at lower trophic levels (Lindeman 1942; Odum 1957). Alternatively, the predation process has been viewed as a major determinant of population, community, and ecosystem dynamics (Glasser 1979; Huston 1979). Both are true, in part, as the principles of thermodynamics must apply to energy transfer up food webs, yet structures of food webs are strongly influenced by selective predators (Paine 1980; Lubchenco and Gaines 1981; Mills, et al., this volume). The interplay between selective predation and ecosystem behavior seems essential to understanding the ecological variation observed in nature. Yet information on these variations is so limited that strong arguments are increasingly advanced in support of the need for and value of long-term ecological perspectives (Likens 1983).

The interannual variation observed in lakes continues to surprise even those with the advantage of many years of site-specific study (Edmondson and Lehman 1981; Edmondson and Litt 1982). Nearly an order of magnitude of variation in primary production has been reported for one well-studied lake (Henrikson et al. 1980). At the opposite end of the food web, long-term studies of fish populations reveal similar variability. They are secondarily similar in that surprises continue to appear and the

understanding of cause and effect remains elusive (Ursin 1982, Steele and Henderson 1984).

Two basic approaches may be adopted in attempts to develop understanding and predictability that include both inherent variability and the cause–effect components of dynamics observed over long periods in lentic ecosystems. One approach can derive from a theoretical view that encompasses the full range of likely variation around some steady-state condition. The second takes advantage of the sediment archive that preserves some subset of the historical behavior of the ecosystem. The theoretical approach gains credence as the basic mechanisms of species interaction and nutrient cycling become better understood through observation and experimentation. The bulk of this volume is directed to those goals. Similarly, interpretation of the paleolimnological record is enhanced as we learn more of the mechanisms whose effects are archived in the sediments (Binford et al. 1983).

Our goal for this chapter is to combine both the theoretical and paleoecological perspectives. Toward that end, we focus on Lake Michigan, an ecosystem that has been subjected to massive food web manipulation and offers the advantage of independent observations made prior to and during those manipulations. Thus, on the one hand we can assess the adequacy of the sediment record as a representation of the observed dynamics. On the other hand, we can compare the observed responses to likely possibilities derived from a modeling approach.

We reason from previous work (Henrikson et al. 1980; DiBernardi 1981; Shapiro and Wright 1984; Carpenter et al. 1985) that much of the observed variation in system productivity can be attributed to the dynamics of food web interactions. In particular, we argue that predation effects cascade through food webs, and at a fixed level of nutrient loading, are responsible for up to a hundredfold variation in primary production (Carpenter and Kitchell 1984). Specifically, we address several questions: (1) What kinds of long-term dynamics (piscivory, planktivory, and herbivory) can be discerned from the sediment record? (2) Given that the fossil record is incomplete, what evidence of complex interactions can be unequivocally identified? (3) What are the indicators of key mechanisms operating in this and similar food webs? (4) How can the lessons derived from this analysis be employed as predictors of ecosystem behavior?

BRIEF HISTORY OF
LAKE MICHIGAN'S FOOD WEB

As a basis for expectations, we briefly review the history of predator–prey interactions in Lake Michigan. More complete reviews of fish population dynamics are available in Smith (1970), Wells and McLain (1973), Christie (1974), and Smith and Tibbles (1980). The changes in zooplankton populations are derived from Wells (1970), Evans et al. (1980), and Gitter (1982).

During the first half of this century expansion of fishery exploitation coupled with invasion of the sea lamprey (*Petromyzon marinus*) combined to cause a total collapse of the lake trout (*Salvelinus namaycush*), a species that had been the dominant piscivore in Lake Michigan. Both the fisheries and the sea lamprey switched to alternate species, resulting in the decline and, in some cases, local extinction of many other native species.

In the absence of effective piscivores, invading planktivore populations expanded. Alewife (*Alosa pseudoharengus*), in particular, flourished during the 1950s and early 1960s until a major

density-dependent mortality occurred during the winter of 1966–1967 (Hatch et al. 1981). Until quite recently, the alewife remained the dominant component of total fish biomass. As a consequence of intense interactions with alewife, nearly one-half of the 20 or so native fish species that fed heavily on zooplankton during some stage of their life history were displaced (Crowder 1980).

Lake Michigan's zooplankton changed as a result of the alewife expansion. Large-bodied copepods and cladocerans virtually disappeared from the lake during the late 1960s and were replaced by small cladocerans and copepods (Wells 1970). In a later section of this paper we describe the effects of alewife on zooplankton species composition and biomass and the likely changes in herbivory that followed.

The development of an effective lampricide and a large-scale lamprey control program allowed re-establishment of salmonid populations (Smith and Tibbles 1980). In addition to lake trout, Pacific salmon species were stocked in the lake as an attempt to develop both a biological control for the alewife nuisance and as an attempt to re-establish fisheries. The result was spectacularly successful. A sport fishery of substantial economic and social value is now active. Because there is virtually no natural reproduction by salmonids, the fisheries are entirely dependent on stocking programs, which plant 15 to 16 million juvenile salmon and trout each year (Stewart et al. 1981).

There exist two contrasting views of the future of this intensively managed, artificial predator–prey system. One approach based on traditional fisheries population models holds that the alewife population could serve as a stable forage resource for up to several times the current levels of piscivory (Eck and Brown 1985). The alternative view, based on bioenergetic and predation theory, suggests that the system is currently at or in excess of predator carrying capacity (Stewart et al. 1981; Kitchell and Crowder, in press). These authors offered several predictions of system changes to be expected as the effect of increased piscivory induces major reductions in the alewife and the consequent community changes cascade from piscivore to plankton.

As stated above, one of the goals of this chapter is to describe paleoecological and theoretical tools that can be used to evaluate the historic, present, and likely future conditions of this complex suite of food web interactions. There are reasons to be concerned about the trajectory and stability of this system because one of the main components of the food web (that of salmonid predators and their prey) is uncoupled from typical feedback and damping mechanisms due to the fact that virtually all of the predators are derived from fish hatcheries. An ecological rationale to govern stocking policy has not yet been developed. In addition, the major interactions occurring at levels above the zooplankton involve predominantly non-native species. Sea lampreys originally from the Atlantic prey on salmon and trout derived from Pacific Ocean and European habitats. Salmonid diets are primarily (75–90%) composed of alewife and smelt (*Osmerus mordax*) that invaded the Great Lakes from the' Atlantic (Hagar 1984). There is little basis for expectation of co-evolutionary stabilizing factors in these interactions now occurring to varying degrees in each of the five major lakes. That, in fact, is one of their interesting features; there is opportunity in these large systems for testing of ecological principles and development of new insights.

PALEOECOLOGICAL APPROACH

The paleoecological evidence developed in this paper is of two kinds. The abundance and morphology of *Bosmina longirostris* fossils present in a core from Lake Michigan were analyzed as a basis for evidence and inferences regarding the nature of planktivory. This organism has been the object of insightful experimentation, field studies, and paleoecological analysis (Kerfoot 1977, 1981; Sprules et al. 1984). As a clonal species, it is rapidly responsive to the changing balance of selection pressure and exhibits predator-induced changes in its external morphology (Kerfoot 1981; Stemberger and Gilbert 1984; Harwell 1984).

The second class of evidence derives from plant pigments found in core samples. Among those, pheophorbide *a* offers insights about the nature and intensity of herbivory. The suggestion of Daley and Brown (1973) that pheophorbide is a grazing indicator has now been well substantiated. Chlorophyll *a* decomposes to pheophorbide *a* as follows:

Chlorophyll a → pheophorbide a + Mg

+ phytol

(Brown et al. 1977). In a comparison of several chlorophyll degradation processes (photodegradation; mechanical, viral, and bacterial cell lysis; grazing), grazing was the only consistent source of pheophorbide *a* (Daley and Brown 1973; Daley 1973). Pheophorbide formation was directly proportional to the biomasses of zooplankton added to experimental mesocosms in lakes (Carpenter and Bergquist 1985). In three small Michigan lakes, flux of pheophorbide to sediment traps was well correlated with grazer biomass and mean size (Carpenter et al. 1986).

Biological oceanographers have been quick to make use of pheophorbide flux measurements. Of the chlorophyll *a* degraded by marine copepods, 100% is egested as pheophorbide *a* (Shuman and Lorenzen 1975). This discovery prompted the use of pheophorbide as a nonassimilated tracer for measurements of assimilation efficiency (Landry et al. 1984), and the use of pheophorbide flux in sediment traps to estimate rates of grazing by calanoid copepods (Welschmeyer et al. 1984). In the oceans, ambient pheophorbide concentrations result from a dynamic balance between photodegradation of pheophorbide and its production in the small, slowly sinking egesta of microzooplankton (Soo Hoo and Kiefer 1982a, 1982b; Stoecker 1984; Welschmeyer and Lorenzen 1985). Most of the vertical flux of pheophorbide occurs in the large, rapidly sinking fecal pellets of copepods (Welschmeyer et al. 1984).

The limited data available support the hypothesis that cladocera convert chlorophyll to pheophorbide less efficiently than calanoid copepods. The molar ratio of chlorophyll to pheophorbide is less than 0.07 in fecal pellets of marine calanoids (Shuman and Lorenzen 1975; Landry et al. 1984). By contrast, several chlorophyll degradation products are found in cladoceran feces, and only 10 to 30% of the molecules are pheophorbide (Daley 1973; S. Carpenter, personal communication). This observation is consistent with egestion of viable algae from cladocerans (Porter 1975). However, a comparative study of pheophorbide production by freshwater zooplankton under standardized conditions has not yet been performed.

MODEL OF PHEOPHORBIDE DEPOSITION

Many variables, such as grazing intensity and its vertical distribution, light penetration, and the size distribution of fecal particles, determine the deposition rate of pheophorbide. The outcomes of these complex interactions are difficult to understand intuitively and are not easily accessible to experimentation. Therefore, we developed a model of formation, degradation, sinking, and deposition of pheophorbide in a stratified water column to determine the possible effects of limnological changes in Lake Michigan on accumulation rates of pheophorbide in the sediments.

The net change in pheophorbide (dP) during a time interval, dt, is

$$dP/dt = \text{Production} - \text{Sinking} \qquad (9.1)$$

$$- \text{Photodegradation}.$$

Production of pheophorbide by grazing is EG, where E is molar fraction of ingested chlorophyll that is egested as pheophorbide and G is the rate at which chlorophyll is grazed (see table 9.1 for explanation of symbols). Sinking follows the settling model for algae developed by Smith (1982) and elaborated by Reynolds (1984, chap. 2). The loss rate from a layer of thickness, z, is $- VP/z$, where sinking velocity, V, is obtained from Stokes's Law:

$$V = 2 g r^2 (\varrho' - \varrho)/9n\phi. \qquad (9.2)$$

Photodegradation follows first-order decay kinetics, i.e. the rate is a linear function of irradiance (Soo Hoo and Kiefer 1982a, 1982b; Welschmeyer and Lorenzen 1985).

We developed a specific model from equation 9.1 for two layers, epilimnion (e) and hypolimnion (h):

$$dP_e/dt = EG_e - (V_eP_e/Z_e) - kI_eP_e \qquad (9.3)$$

$$dP_h/dt = EG_h + (V_eP_e/Z_e) \qquad (9.4)$$

$$- [V_hP_h/(Z_{max} - Z_e)] - kI_hP_h.$$

In equation 9.4, the first term in parentheses represents sinking of pheophorbide from the epilimnion to the hypolimnion. Equations 9.3

and 9.4 have a single, globally stable equilibrium point:

$$P_e^* = EG_e/[(V_e/Z_e) + k I_e] \qquad (9.5)$$

$$Ph^* = [EG_h + P_e^*(V_e/Z_e)]/\{kI_h \qquad (9.6)$$

$$+ [V_h/Z_{max} - Z_e)]\}.$$

For the purposes of paleolimnology we wish to know the fraction of grazed chlorophyll that is deposited in the sediments as pheophorbide, F:

$$F = P_h^*[V_h/(Z_{max} - Z_e)]/(G_e + G_h). \qquad (9.7)$$

To economize on parameters, let Q be the proportion of total grazing that occurs in the epilimnion:

$$Q = G_e/(G_e + G_h). \qquad (9.8)$$

Then equation 9.7 reduces to

$$F = E[(1 - Q) + QR_e]R_h, \qquad (9.9)$$

where

$$R_e = (V_e/Z_e)/[kI_e + (V_e/Z_e)] \qquad (9.10)$$

$$R_h = [V_h/(Z_{max} - Z_e)]/\{kI_h + [V_h/(Z_{max} - Z_e)]\}. \qquad (9.11)$$

R_e and R_h are the ratios of sinking losses to total losses of pheophorbide from the epilimnion and hypolimnion, respectively.

Our goal is to study the effects on F of variations in E, Q, particle size (which influences sinking rate), and light intensity. All other variables were fixed at typical values for Lake Michigan. Values of E were varied from 0 to 1. Effects of Q were compared at E values of 0.25 and to 0.75.

Sinking rates in the epilimnion and hypolimnion were calculated for mean temperatures of 15°C and 4°C, respectively, using standard tables of density and viscosity. The density of sinking particles (ϱ') was 1,100 kg/m³, comparable to values for algae collected by Reynolds (1984, table 8). Three values for the coefficient of form resistance, ϕ, were used: unity and 4, which span the reported range for moribund algae (Reynolds 1984, table 10), and a value de-

Table 9.1. *Symbols used in the phorbin dynamics model*

Symbol	Meaning (units)
E	Conversion efficiency (μmole pheophorbide produced/μmole chlorophyll grazed)
F	Deposition efficiency (μmole pheophorbide deposited/μmole chlorophyll grazed)
G	Grazing rate (μmole chlorophyll $\cdot$ m^{-2} $\cdot$ d^{-1})
g	Gravitational constant (m $\cdot$ d^{-2})
I	Irradiance (ϵm^{-2} $\cdot$ d^{-1})
k	Photodegradation constant (m^{2}d $\cdot$ E^{-1})
P	Pheophorbide density (μmole $\cdot$ m^{-2})
Q	Proportion of grazing in epilimnion (dimensionless)
R	Proportion of pheophorbide loss due to sinking (dimensionless)
r	Radius (μm)
t	Time (d)
V	Velocity (m/d)
VOL	Volume (μm^3)
Z_e	Thickness of epilimnion (m)
Z_{max}	Water column depth (m)
n	Dynamic viscosity (kg $\cdot$ m^{-1} $\cdot$ d^{-1})
ϱ	Density of water (kg $\cdot$ m^{-3})
ϱ'	Density of particles (kg $\cdot$ m^{-3})
ϕ	Coefficient of form resistance (dimensionless)

Subscripts *e* and *h* denote epilimnion and hypolimnion, respectively. Asterisks denote equilibrium values.

rived for copepod fecal pellets. To estimate ϕ for fecal pellets, we calculated Stokes's velocities in seawater at 14°C for fecal pellets equivalent in spherical volume (VOL) to those described by Small et al. (1979):

$$V = 0.00677 \, (VOL)^{0.667}. \qquad (9.12)$$

Empirically, Small et al. found

$$V = 0.0611 \, (VOL)^{0.513}. \qquad (9.13)$$

The ratio of Stokes's velocity to actual velocity is ϕ (Reynolds 1984):

$$\phi = 0.111 \, (VOL)^{0.154}. \qquad (9.14)$$

Over the range of pellet sizes studies by Small et al., ϕ varies from 0.65 for pellets 29 μm in equivalent spherical radius to 1.9 for pellets 290 μm in equivalent spherical radius.

The photodegradation constant *k* was 0.0235 m^{2}dϵ^{-1} (SE, 0.0043), based on six decay curves with 6 to 13 data points each (Carpenter et al. 1986). Surface photon flux was set at 70 ϵm^{-2}d^{-1}, typical of a sunny summer day at the latitude of southern Lake Michigan. Light extinction was calculated from Secchi depth, which coincides with approximately 19% of

surface irradiance in southern Lake Michigan (D. Scavia, personal communication). Secchi depth varies from about 5 m during periods of calcite formation to 15 m during periods of intense grazing (D. Scavia, personal communication). From Secchi depth, we calculated an extinction coefficient and the geometric mean light intensities of the epilimnion (I_e) and hypolimnion (I_h). Thickness of the epilimnion (Z_e) and water column depth (Z_{max}) were set at 10 and 100 m, respectively (D. Scavia, personal communication).

MODEL RESULTS

Model results are expressed as isopleths of the percentage of chlorophyll grazed that is deposited as pheophorbide (figs. 9.1, 9.2). On each panel, pheophorbide deposition is shown as a function of conversion efficiency of chlorophyll to pheophorbide and the equivalent spherical radius of sinking particles. We compared response surfaces for three coefficients of form resistance at two Secchi depths (fig. 9.1) and at two depth distributions of grazing activity (fig. 9.2).

In all cases, pheophorbide deposition was greatest at high conversion efficiency and large particle sizes (figs. 9.1, 9.2). At low particle

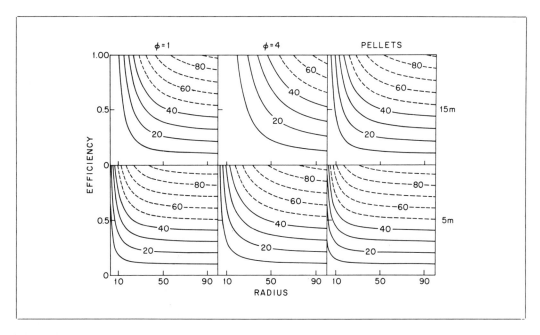

Figure 9.1. *Isopleths of percentage of grazed chlorophyll deposited as pheophorbide as a function of conversion efficiency and equivalent spherical particle radius. Columns (left to right) correspond to coefficients of form resistance of 1, 4, and values for fecal pellets. Rows correspond to Secchi depth of 15 (top) and 5 (bottom) m. All calculations assume 50% of total grazing in epilimnion.*

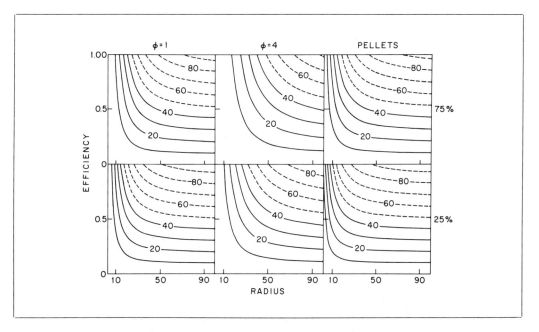

Figure 9.2. *Isopleths as in fig. 9.1, except that rows correspond to percentages of total grazing in the epilimnion of 75 (top) and 25 (bottom). All calculations assumed Secchi depth = 10 m.*

sizes, pheophorbide deposition increased as particle size increased but was relatively insensitive to conversion efficiency. At high particle sizes, pheophorbide deposition increased as conversion efficiency increased but was relatively insensitive to particle size.

For particles 2 to 30 μm, particle size, form resistance to sinking, and light intensity interacted strongly to determine pheophorbide deposition (fig. 9.1). Both pheophorbide deposition and its rate of increase with particle radius decline as light intensity and/or form resistance increase.

The depth at which grazing occurred also interacted strongly with particle size and form resistance to determine pheophorbide deposition for particles 2 to 30 μm in radius (fig. 9.2). As the proportion of grazing in the epilimnion rises, pheophorbide deposition declines but becomes more sensitive to changes in particle size or form resistance.

ZOOPLANKTON GRAZING

As a basis for comparison, biomass of Lake Michigan zooplankton in 1954 and 1966 was calculated from lengths and population densities given by Wells (1970). We prefer the computational simplicity offered by the single equation of Peters and Downing (1984) over a set of species-specific equations. We doubt that substantial additional error results from use of the Peters–Downing equation because masses commonly vary more than 10-fold for conspecific zooplankters of the same length (Downing and Rigler 1984).

To compare grazing intensities in 1954 and 1966, it was necessary to adjust for differences in size distribution because zooplankter size affects grazing rate. To make this comparison, we calculated a relative grazing ratio (RGR) using the empirical equation 5 of Peters and Downing (1984). This equation, based on measurements of ingestion rate per animal for a wide variety of zooplankton taxa, predicts ingestion rate as a function of animal mass, food concentration and particle size, experiment duration, and volume of the experimental container. We calculated RGR as SIR(1966)/SIR(1954), where standardized ingestion rate (SIR) was calculated from equation 5 of Peters and Downing by multiplying ingestion rate per animal by animals m^{-3} and animal sizes as reported by Wells (1970). All other variables

were set to the median values reported by Peters and Downing. Therefore, RGR reflects changes in grazing rate due solely to changes in zooplankton size and abundance. Other factors that influence grazing rate, e.g., changes in phytoplankton abundance and composition, do not affect RGR.

PALEOLIMNOLOGICAL METHODS

The sediment samples used in this study were from a box core taken in 1981 as part of an NOAA/Great Lakes Environmental Research Laboratory program. This core is from a deep-water site of southeastern Lake Michigan in an area of known high net sedimentation rates. Characteristics of the core (LM-HS-81) are described elsewhere (Eadie et al. 1984).

Chlorophyll degradation products were determined by the method described in detail by Carpenter and Bergquist (1985) with minor modifications. A brief description of the method will be given here. Freeze-dried sediment was sonicated in methanol-acetone (15 : 85) and centrifuged. Supernatant was filtered through 0.2-μm-porosity Acropore filters into vials, from which the solvent was then evaporated under a nitrogen stream. Dried samples were kept frozen under N_2 until they could be chromatographed. Normal-phase high-pressure liquid chromatography used a dual-pump Beckman system and two solvents: first, 15% acetone in petroleum ether and, second, a 50–50 mixture of the first solvent with methanol, petroleum ether, and acetone in the ratio 27 : 30 : 43. Absorbance peaks (660 μm) were integrated electronically. Calibration standards were purified by preparative paper chromatography of spinach and sediment extracts (Brown et al. 1977). We will not report data on chlorophyll a' or chlorophyllide a because they were minor pigments in this core. Results presented here are based on chlorophyll a and its two major degradation products in this core: pheophytin a and pheophorbide a. We refer to these three pigments collectively as phorbins.

Samples were prepared for zooplankton analyses following conventional procedures (e.g., Kitchell and Kitchell 1980) except that a known amount of *Eucalyptus* pollen was added to each sample so that fossil densities per unit of dry sediment could be calculated from ratios. Zooplankton morphometric measurements followed the procedures of Sprules et al. (1984). In addition, comparisons were made with *B. longirostris* collected during August 1979–1981 (Gitter 1982) off Grand Haven, Michigan, and the remains of *Bosmina* found in deep-water sediment traps operated in the same region of the lake during summer of 1980 (Eadie et al. 1984).

Most of the *Bosmina* carapaces and head-shields observed were separated and broken. However, mucro spines on the carapaces were usually intact and could be readily measured. Although *Bosmina* parts are rare in these sediments (which are only 2–4% organic by dry weight), at least 50 measureable mucrones were found in each sample. Paired measurements of carapace and mucro lengths were made if the carapace edges were not damaged. Fewer of the fossils met that condition, and we report data only for those samples with at least 10 undamaged carapaces. Antennules were usually broken, and few lengths could be measured. Antennule basal width is, however, highly correlated with length (Kerfoot 1977, 1981) and could be readily measured. As for mucrones, a total of 50 was measured in each sample.

Parametric statistics were used for analysis of zooplankton morphometrics, as the data appeared normally distributed and the variances were generally independent of the means.

RESULTS: ZOOPLANKTON GRAZING

Following the irruption of alewife ca. 1960, zooplankton biomass declined, and dominance shifted from large cladocerans to small copepods (fig. 9.3). The large cladocerans *Leptodora kindtii*, *Daphnia galeata*, and *D. rectrocurva* declined precipitously, whereas *B. longirostris* increased. The large copepods *Epischura lacustris*, *Limnocalanus macrurus*, and *Mesocyclops edax* declined. Small copepods, especially the calanoids *Diaptomus ashlandi*, *Diaptomus minutus*, and *Diaptomus oregonensis*, exhibited a major increase in biomass between 1954 and 1966.

Relative grazing ratios (RGRs) indicated a decrease in grazing intensity from 1954 to 1966, particularly in summer. RGR < 1 indicates lower grazing in 1966. RGRs calculated from Wells's (1970) data were as follows: 1.6 in early June, 0.92 in late June, 0.66 in mid-July, and 0.46 in August. We excluded the predaceous taxa *Leptodora kindtii* and *M. edax* and

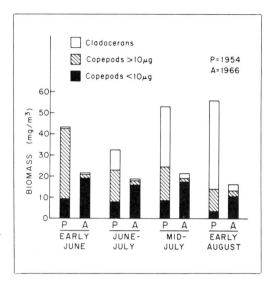

Figure 9.3. *Biomass (mg dry mass/m^{-3}) of zooplankton in three categories prior to alewife invasion (P) (1954), and during a year of abundant alewife (A) (1966). Calculated from data of Wells (1970).*

the large omnivorous calanoids *E. lacustris* and *Limnocalanus macrurus* from the RGR calculations. All of these taxa were abundant in 1954 and greatly reduced in 1966. Therefore, we overestimated RGR by a factor equal to the proportion of total grazing accounted for by these taxa. Grazing by early instars of *E. lacustris* and *Limnocalanus macrurus* may be substantial (Balcer et al. 1984), reinforcing the conclusion that grazing was less in 1966 than in 1954.

ZOOPLANKTON FOSSILS

Of the miscellaneous zooplankton remains found in core samples only those of *B. longirostris* were sufficiently abundant to allow quantitative analyses. Numerically, *Bosmina* is a dominant cladoceran of the Lake Michigan zooplankton (Wells 1970). The relationships of key variables are given as a function of core depth in figure 9.4. Mucro length was independent of carapace length but strongly related to *Bosmina* density and antennule width. Collectively, the changes in "exuberant structures" suggested a morphological response that was independent of *Bosmina* size but changed over time (sediment depth) and was strongly related to *Bosmina* density.

Although age dating for this core is compli-

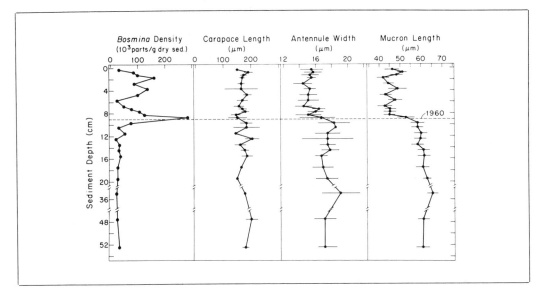

Figure 9.4. *Densities, carapace sizes, antennule width at the base and mucron lengths of* Bosmina longirostris *fossils derived from a core (LM-HS-81) taken from Lake Michigan. Means and their 95% confidence intervals are given for morphometric measurements.*

cated by as yet unknown differential bioturbation of clay particles bearing the ^{210}Pb isotope (Robbins 1982), the ^{137}Cs profile allows the conservative interpretation that any sediment below 9 cm in depth was deposited before about 1960. Accordingly, we have drawn a 1960 time line on each figure. This corresponds approximately with the exponential rise of the alewife population. Thus, the core samples deeper than 10 cm represent a pre-alewife or ancestral food web condition for Lake Michigan. Samples in the range of 8 to 10 cm represent the period when the alewife population was expanding and those at depths less than 8 cm represent conditions with alewife as the dominant zooplanktivore and primary prey resource of stocked salmonids.

The importance of mixing and diagensis are unknown for this core. Accordingly, we interpret apparent changes cautiously and discuss only those differences that are unequivocal. We choose, however, not to belabor statistical comparisons and will highlight only those features of the data that are statistically defensible. Cursory examination of samples from other Lake Michigan cores reveals the same patterns we discuss herein.

B. longirostris fossils in this core reveal major changes in density and morphology at about the time that alewife populations were expanding. At sediment depths greater than 10 cm, *Bosmina* densities are low and relatively nonvariable (fig. 9.4). Densities increase sharply, are more variable, and remain generally higher in more recent sediments. Loading rates clearly increased during and after 1960, which corresponds with the increased abundance of small zooplankton reported by Wells (1970). *Bosmina* carapace size (fig. 9.4) is variable, due in part to small sample size, but shows no apparent pattern over core depth.

Antennule widths and mucro lengths exhibit striking changes at the same depths that densities change. Mucro lengths offer the strongest evidence and serve as the basis for subsequent analysis of morphological changes. Long mucrones and low variation occurred at all depths greater than 10 cm, i.e., during the pre-alewife period. *Bosmina* morphology shifted abruptly to shorter featured forms during the late 1950s and early 1960s when alewife were expanding. Although more variable in the sediments deposited since 1960, the shorter form persisted until the most recent samples.

COMPARISON OF CORE AND ZOOPLANKTON SAMPLES

To what extent do samples from the core represent populations in the lake? Mucro lengths in surficial (0.5 cm) samples are intermediate ($\bar{x}$ = 51.0 ± 2.3 μm; N = 50) between the long-featured forms of the pre-alewife years and many of the shorter featured forms represented in discrete core cuts above 9 cm. Zooplankton samples collected in 1979 (Gitter 1982) reveal mucro lengths that are also intermediate ($\bar{x}$ = 53.6 ± 0.8 μm; N = 50) and not different from those in the 0.5-cm core sample. In addition, *Bosmina* remains in a 1980 sediment trap sample had mucro lengths ($\bar{x}$ = 54.5 ± 2.3 μm; N = 15) similar to those in both the most recent sediment sample and the 1979 zooplankton sample.

In short, there is direct correspondence between those data derived from analysis of core materials, those that derive from sediment traps, and the populations in the plankton. In answer to the original question, the population and morphological history of *B. longirostris* in Lake Michigan is effectively archived in the lake's sediments.

PHORBIN ANALYSES

Pheophorbide *a* was the most abundant chlorophyll *a* degradation product in the core (fig. 9.5). Concentrations of pheophorbide *a* increased abruptly between 9.5 and 7.5 cm downcore (ca. 1960) and remained high between 7.5 cm and 2.5 cm. In the uppermost sediments we analyzed, pheophorbide concentration returned to a low value comparable to those found below 10 cm.

In contrast, no trends are evident in the profiles of chlorophyll *a*, pheophytin *a*, or their ratio (fig. 9.5). The ratio of pheophorbide *a* to pheophytin *a* follows a profile similar to that of pheophorbide *a* concentration. These results suggest that the change in pheophorbide *a* concentration was not a diagenetic effect because no comparable change occurred in concentrations of its precursors.

DISCUSSION

Analysis of the Lake Michigan core allows an estimate of the cause–effect responses in a com-

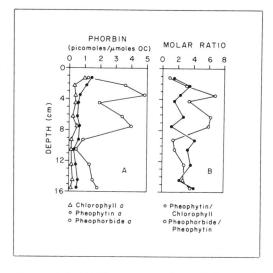

Figure 9.5. (A) Concentrations (picomole/micromole organic C) of chlorophyll a, pheophytin a, and pheophorbide a versus depth in the core. (B) Ratios of pheophytin a to chlorophyll a and pheophorbide a to pheophytin a versus depth in the core.

plex food web. Both the timing and magnitude of major perturbation to the food web are recorded. The decimation of piscivore populations in Lake Michigan and explosive increase in size-selective predation on zooplankton were clearly recorded at the herbivore level and in the sediments both as increased density of small herbivores and their effect in loading phorbins to the sediments. We outline these effects in a conceptual model of causality (fig. 9.6).

There is ample evidence from both field and laboratory studies that the abundance and morphological dynamics of *B. longirostris* are direct indications of the intensity of predation by copepods (Kerfoot 1977, 1981; Sprules et al. 1984). In Lake Michigan (fig. 9.6) long mucrones and lower *Bosmina* densities correspond with higher densities of omnivorous or predaceous copepods, including many cyclopoid species and the calanoid *Limnocalanus macrurus*. As piscivore populations declined and alewife increased, intense size-selective predation on large-bodied zooplankton reduced the densities of *Bosmina* predators and released selection pressures for antipredatory morphology. In other words, *Bosmina* mucro length is inversely

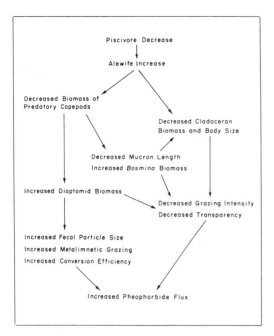

Figure 9.6. *Conceptual model of changes in Lake Michigan reflected in the sediment record. See text for further discussion.*

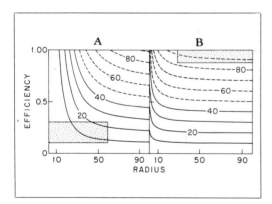

Figure 9.7. *Pheophorbide deposition isopleths* **(A)** *prior to alewife invasion and* **(B)** *during period of alewife abundance. Stippled areas denote prevalent efficiencies of chlorophyll conversion and sizes of pheophorbide-bearing particles at each time. Simulation conditions:* **(A)** *Secchi = 15 m, φ = 1, Q = 0.75;* **(B)** *Secchi = 5 m, φ for fecal pellets, Q = 0.25. M*

related to the intensity of zooplanktivory by size-selective fish and therefore directly related to the intensity of piscivory and those zooplanktivorous fish. Relative densities of *Bosmina* vary conversely. Abundant piscivores suppress size-selective zooplanktivorous fish allowing invertebrate predator populations to expand, resulting in fewer *Bosmina* and greater evidence of antipredatory morphology.

Large invertebrate predators such as *Mysis relicta* or *Chaoborus* are likely less important in the responses of *Bosmina*. The evidence for that assertion is indirect, as Lake Michigan *Mysis* population dynamics are poorly known in the past. The strongest evidence is that small *Bosmina* exhibit the most extreme development of antipredator morphology (Kerfoot 1981; Sprules et al. 1984). Thus, it seems most likely that relatively small predators (e.g., *Epischura* and *Limnocalanus*) are the primary source of selection pressure on this plastic trait.

Increased pheophorbide flux to the sediments of Lake Michigan resulted from interacting concomitant changes in the light field and a suite of grazer characteristics (fig. 9.7). Overall, grazer biomass declined and grazing intensity (as estimated by RGR) declined. Decreased grazing led to increased algal biomass and decreased transparency. At the same time, dominance of the grazing assemblage shifted from large cladocerans to small diaptomids. This caused three important changes in the nature of grazing, which altered pheophorbide flux.

1. Fecal particle size increased. The feces of cladocera are a slurry of partially digested algae and gut fluids that dissipate rapidly (Rigler 1971). Particulate matter in cladoceran feces is likely smaller on average than the food particles, with sinking characteristics comparable to those of small moribund algae. Such particles lie on the extreme left of the response surfaces in figures 9.1 and 9.2. In contrast, calanoid copepods form fecal pellets 30 to 300 μm in equivalent spherical radius that sink rapidly from the water column (Small et al. 1979). Some fecal pellets break up in the water column before reaching the sediment (Ferrante and Parker 1977). Even so, they transport pheophorbide from the photic zone to aphotic, cold water where the likelihood of preservation and deposition is high.

2. The locus of grazing moved down in the water column. The cladoceran grazers that predominated before the alewife invasion (*Daphnia galeata* and *Daphnia retrocurva*) were migratory, moving near the surface by night and returning to the metalimnion by day (Balcer et al. 1984). Because of the slow sinking of cladoceran feces, pheophorbide produced during nocturnal grazing could remain in the epilimnion several days and be extensively photodegraded. In contrast, the calanoid grazers that predominated after the alewife invasion (*Diaptomus ashlandi*, *Diaptomus minutus*, and *Diaptomus oregonensis*) were metalimnetic (Balcer et al. 1984). Therefore, pheophorbide formation occurred at greater depth, where light intensity and photodegradation rate were relatively low.

3. Conversion efficiency of chlorophyll to pheophorbide probably increased. On a molar basis, the percentage of pheophorbide *a* among chlorophyll *a* degradation products is 10 to 30% for cladoceran feces (Daley 1973; Carpenter, unpublished data) and 93 to 100% for marine calanoid feces (Shuman and Lorenzen 1975; Landry et al. 1984).

Collectively, these changes led to a large increase in the percentage of grazed chlorophyll deposited as pheophorbide (fig. 9.7). Prior to the alewife irruption, we estimate that less than 25% of the grazed chlorophyll was deposited as pheophorbide. When alewife were abundant, 82 to 97% of the grazed chlorophyll was deposited as pheophorbide. Pheophorbide deposition would be expected to double even if grazing were halved.

Although there are other possible explanations for the data we report from this core, recent events in Lake Michigan suggest that the cascading effects derived from piscivory are most parsimonious. Based on their estimate of the intensity of salmonid predation on alewife, Stewart et al. (1981) forecast that the alewife population was increasingly likely to decline and possibly collapse as salmonid stocking practices continued to increase the intensity of piscivory in the system. In terms of the past, piscivory is being re-established at increasing levels, and the primary indicator of change in food web dynamics—*Bosmina* morphology—should evidence the treatment effect (fig. 9.6).

Increased mucro length evident at the top of our core in the recent plankton (Gitter 1982) and in sediment traps suggest that Lake Michigan's food web is again changing. This is accompanied by a suite of independent observations, all of which implicate increasing predation on alewife by salmonids. Among the observations are a continuous and several-fold increase in the abundance of large and predaceous copepods in southern Lake Michigan during recent years (Evans et al. 1980; Gitter 1982). In addition, a sudden increase of large *Daphnia* recorded in midsummer 1983 at deepwater stations off Grand Haven, Michigan (Scavia et al. 1984) was accompanied by increased water clarity ascribed to intense grazing pressure in the epilimnion. Both observations suggest a recent reduction in size-selective predation on large zooplankton. In all, the changes in species composition and distribution within the water column should yield a major *reduction* in pheophorbide flux during 1983. At the next trophic level, populations of many native fishes suppressed by alewife are now recovering, and the lakewide alewife biomass in 1983 and again in 1984 was estimated at only 10 to 20% of the mean levels sustained since 1967 (Wells and Hatch 1984). As expected, salmon and trout growth rates have declined in the recent past, and their diets, although still dominated by alewife, have expanded to include alternate prey (Hagar 1984).

We do not know what the conditions of a new steady state will be for this system of food web interactions that is now either in a transient condition en route to some semblance of the pre-alewife steady state or simply evidencing short-term dynamics and substantial variability. Based on the core results, we have a reasonable idea of the conditions that persisted with relatively little variation in the ancestral state. As a testable hypothesis, we argue that the morphology of *Bosmina* will indicate the levels of piscivory that can sustain the complex predator–prey interactions represented among native species in the ancestral plankton assemblage. The test of this hypothesis is currently underway in Lake Michigan. This test will be repeated, as the salmonid stocking programs in Lakes Superior, Huron, Erie, and Ontario are not yet as developed as that for Lake Michigan. Their expansion should offer some interesting opportunities for tests of the role of higher-order predators in these ecosystems.

In answer to the last of our original questions: What can we predict about ecosystem behavior as a result of this analysis? We offer the following: Sediments deposited in Lake Michigan since about 1980 should contain lower pheophorbide concentrations, lower densities of *Bosmina* remains, and greater abundance of longer-featured *Bosmina*. In fact, we expect that conditions within the current food web should produce integrated evidence in sediment traps and/or future paleoecological studies suggesting a rapid change in system state similar to that observed about 1960 and that a new steady state will have been established that resembles that of the ancestral food web. The participating species will likely be different, as the top of Lake Michigan's food web will undoubtedly contain many non-native forms; but the mechanisms of species interactions and integrated behavior of the system will look much as they did several decades ago.

ACKNOWLEDGMENTS

This work was supported in part by a grant from the University of Wisconsin Sea Grant program (to James F. Kitchell) and a National Science Foundation grant (to Stephen R. Carpenter and James F. Kitchell). We thank Brian Eadie for his aid and cooperation in providing core samples and organic C data. Don Scavia kindly provided unpublished information in support of model development. Sharon Barta patiently and carefully measured the fossil zooplankton. We appreciate the assistance of Monica and Jim Elser in performing HPLC analyses. Mary Smith repeatedly brought order from chaos during preparation of the manuscript.

REFERENCES

Balcer, M. D., N. L. Korda, and S. I. Dodson. 1984. *Zooplankton of the Great Lakes.* University of Wisconsin Press. Madison, Wis.

Binford, M., E. Deevey, and T. Crisman. 1983. Paleolimnology: An historical perspective on lacustrine ecosystems. *Annu. Rev. Ecol. Syst.* 14 : 255–86.

Brown, S. R., R. J. Daley, and R. N. McNeely. 1977. Composition and stratigraphy of the fossil phorbin derivatives of Little Round Lake, Ontario. *Limnol. Oceanogr.* 22 : 336–48.

Carpenter, S. R., and A. M. Bergquist. 1985. Experimental tests of grazing indicators based on chlorophyll *a* degradation products. *Arch. Hydrobiol.* 102 : 303–17.

Carpenter, S. R., M. M. Elser, and J. J. Elser. 1986. Chlorophyll production, degradation, and sedimentation: Implications for paleolimnology. *Limnol. Oceanogr.* 31 : 112–24.

Carpenter, S. R., and J. F. Kitchell. 1984. Plankton community structure and limnetic primary production. *Amer. Naturalist* 124(2) : 159–72.

Carpenter, S. R., J. F. Kitchell, and J. R. Hodgson. 1985. Cascading trophic interaction and lake ecosystem productivity. *Bioscience* 35 : 635–39.

Christie, W. J. 1974. Changes in fish species composition of the Great Lakes. *Journal of the Fisheries Research Board of Canada,* 31 : 831–54.

Crowder, L. B. 1980. Alewife, rainbow smelt and native fishes in Lake Michigan: Competition or predation? *Environ. Biol. Fish.* 5 : 225–33.

Daley, R. J. 1973. Experimental characterization of lacustrine chlorophyll diagenesis. II. Bacterial, viral, and herbivore grazing effects. *Arch. Hydrobiol.* 72 : 409–39.

Daley, R. J., and S. R. Brown. 1973. Experimental characterization of lacustrine chlorophyll diagenesis. I. Physiological and environmental effects. *Arch. Hydrobiol.* 72 : 277–304.

DiBernardi, R. 1981. Biotic interactions in freshwater and effects on community structure. *Boll. Zool.* 48 : 353–71.

Downing, J. A., and F. H. Rigler. 1984. *A manual on methods for the assessment of secondary productivity in fresh waters.* Oxford: Blackwell.

Eadie, B. J., R. L. Chambers, W. S. Gardner, and G. L. Bell. 1984. Sediment trap studies in Lake Michigan: Resuspension and chemical fluxes in the southern basin. *J. Great Lakes Res.* 10(3) : 307–21.

Eck, G. W., and E. H. Brown, Jr. 1985. An estimate of Lake Michigan's capacity to support lake trout and other salmonines based on the status of prey populations in the 1970s. *Can. J. Fish. Aquat. Sci.* 42 : 449–54.

Edmondson, W. T., and J. T. Lehman. 1981. The effect of changes in the nutrient income on the condition of Lake Washington. *Limnol. Oceanogr.* 26 : 1–29.

Edmondson, W. T., and A. H. Litt. 1982. *Daphnia* in Lake Washington. *Limnol. Oceanogr.* 27 : 272–93.

Evans, M., B. E. Hawkins, and D. W. Sell. 1980. Seasonal features of zooplankton assemblages in nearshore area of southeastern Lake Michigan. *J. Great Lakes Res.* 6 : 275–89.

Ferrante, J. G., and J. I. Parker. 1977. Transport of diatom frustules by copepod fecal pellets to the sediments of Lake Michigan. *Limnol. Oceanogr.* 22 : 92–98.

Gitter, M. J. 1982. Thermal distribution and community structure of Lake Michigan zooplankton with emphasis on interactions with young-of-year fishes. M.S. thesis, University of Wisconsin, Madison, Wis.

Glasser, J. W. 1979. The role of predation in shaping and maintaining the structure of communities. *Amer. Naturalist* 113 : 31–41.

Hagar, J. M. 1984. Diets of Lake Michigan salmonids: An assessment of the dynamics of predator-prey interactions. M.S. thesis, University of Wisconsin, Madison, Wis.

Harwell, C. D. 1984. Predator-induced defense in a marine bryozoan. *Science* 224 : 1357–59.

Hatch, R. W., P. M. Haack, and E. H. Brown. 1981. Estimation of alewife biomass in Lake Michigan, 1967–78. *Trans. Amer. Fish. Soc.* 110 : 575–84.

Henrickson, L., H. G. Nyman, H. G. Oscarson, and J. A. E. Stenson. 1980. Trophic changes, without changes in the external nutrient loading. *Hydrobiology* 68 : 257–63.

Huston, M. 1979. A general hypothesis of species diversity. *Amer. Naturalist* 113 : 81–101.

Kerfoot, W. C. 1977. Competition in cladoceran communities: The cost of evolving defenses against copepod predation. *Ecology* 58 : 303–13.

———. 1981. Long-term replacement cycles in cladoceran communities: A history of predation. *Ecology* 62 : 216–33.

Kitchell, J. F., and L. B. Crowder. 1986. Predator-prey interactions in Lake Michigan: Model predictions and recent dynamics. *Environ. Biol. Fish* (in press).

Kitchell, J. A. and J. F. Kitchell. 1980. Size-selective predation, light transmission and oxygen stratification: evidence from the recent sediments of manipulated lakes. *Limnol. Oceanogr.* 25 : 389–402.

Landry. M. R., R. P. Hassett, V. Fagerness, J. Downs, and C. J. Lorenzen. 1984. Effect of food acclimation efficiency of *Calanus pacificus. Limnol. Oceanogr.* 29 : 361–64.

Likens, G. E. 1983. A priority for ecological research. *Bull. Ecol. Soc. Amer.* 64 : 234–43.

Lindeman, R. L. 1942. The trophic-dynamic aspect of ecology. *Ecology* 23 : 399–418.

Lubchenco, J., and S. D. Gaines. 1981. A unified approach to marine plant-herbivore interactions. I. Populations and communities. *Annu. Rev. Ecol. Syst.* 12 : 405–37.

Odum, H. T. 1957. Trophic structure and productivity of Silver Springs, Florida. *Ecol. Monogr.* 27 : 55–112.

Paine, R. T. 1980. Food webs: Linkage, interaction strength and community infrastructure. *J. Anim. Ecol.* 49 : 667–85.

Peters, R. H., and J. A. Downing. 1984. Empirical analysis of zooplankton filtering and feeding rates. *Limnol. Oceanogr.* 29 : 763–84.

Porter, K. G. 1975. Viable gut passage of gelatinous green algae ingested by *Daphnia. Verh. Internat. Verein. Limnol.* 19 : 2840–50.

Reynolds, C. S. 1984. *The ecology of freshwater phytoplankton.* London: Cambridge University Press.

Rigler, F. H. 1971. Methods for measuring the assimilation of food by zooplankton. In W. T. Edmondson and G. G. Winberg (eds.), *A Manual for the assessment of secondary productivity in freshwaters*, pp. 264–69. Oxford: Blackwell.

Robbins, J. A. 1982. Stratigraphic and dynamic effects of sediment reworking by Great Lake zoobenthos. *Hydrobiology* 92 : 611–22.

Scavia, D., G. L. Fahnensteil, M. S. Evans, D. Jude, and J. T. Lehman. 1984. Phosphorus loading and fisheries impact on Lake Michigan water quality. XXVII Conference, International Association for Great Lakes Research (Abstract).

Shapiro, J., and D. I. Wright. 1984. Lake restoration by biomanipulation. *Freshwater Biol.* 14 : 371–83.

Shuman, F. R., and C. J. Lorenzen. 1975. Quantitative degradation of chlorophyll by a marine herbivore. *Limnol. Oceanogr.* 20 : 580–86.

Small, L. F., S. W. Fowler, and M. Y. Unlu. 1979. Sinking rates of natural copepod fecal pellets. *Mar. Biol.* 51 : 233–41.

Smith, B. R., and J. J. Tibbles. 1980. Sea Lamprey (*Petromyzon marinus*) in Lakes Huron, Michigan, and Superior: History of invasion and control, 1936–78. *Can. J. Fisheries Aquat. Sci.*, 37 : 1780–1801.

Smith, I. R. 1982. A simple theory of algal deposition. *Freshwater Biol.* 12 : 445–49.

Smith, S. H. 1970. Species interactions of the alewife in the Great Lakes. *Trans. Amer. Fish. Soc.* 99 : 754–65.

SooHoo, J. B., and D. A. Kiefer. 1982a. Vertical distribution of phaeopigment I. A simple grazing and photooxidative scheme for small particles. *Deep-Sea Res.* 29 : 1539–52.

———. 1982b. Vertical distribution of phaeopigments II. Rates of production and kinetics of photo-oxidation. *Deep-Sea Res.* 29 : 1553–64.

Sprules, W. G., J. C. H. Carter, and C. W. Ramcharan. 1984. Phenotypic associations in the Bosminidae (Cladocera): Zoogeographic patterns. *Limnol. Oceanogr.* 29 : 161–69.

Steele, J. H., and E. W. Henderson, 1984. Modeling long-term fluctuations in fish stocks. *Science* 224 : 985–87.

Stemberger, R. S., and J. J. Gilbert. 1984. Spine development in the rotifer *Keratella cochlearis* development by cyclopoid copepods and *Aspanchna. Freshwater Biol.* 14 : 639–47.

Stewart, D. J., J. F. Kitchell, and L. B. Crowder. 1981. Forage fishes and their salmonid predators in Lake Michigan. *Trans. Amer. Fish. Soc.* 110 : 751–63.

Stoecker, D. K. 1984. Particle production by planktonic ciliates. *Limnol. Oceanogr.* 29 : 930–40.

Ursin, E. 1982. Stability and variability in the marine ecosystem. *Dana* 2 : 51–67.

Wells, L. 1970. Effects of alewife predation on zooplankton populations in Lake Michigan. *Limnol. Oceanogr.* 15 : 556–65.

Wells, L., and R. W. Hatch. 1984. Status of bloater chubs, alewives, smelt, slimy sculpins, deepwater sculpins, and yellow perch in Lake Michigan, 1984. Great Lakes Fishery Commission Lake Michigan Committee Meeting, March 19, 1985. Agenda Item 6.a.

Wells, L., and A. L. McLain. 1973. Lake Michigan: Man's effects on native fish stocks and other biota. Great Lakes Fisheries Commission Technical Report 20 : 55 p.

Welschmeyer, N. A., A. E. Copping, M. Vernet, and C. J. Lorenzen. 1984. Diel fluctuations in zooplankton grazing rate as determined from the downward vertical flux of pheopigments. *Mar. Biol.* 83 : 263–70.

Welschmeyer, N. A., and C. J. Lorenzen. 1985. Chlorophyll budgets: Zooplankton grazing and phytoplankton growth in a temperate fjord and the Central Pacific Gyres. *Limnol. Oceanogr.* 30 : 1–21.

B. Indirect Facilitation

10. Indirect Effect of Predators on Age-Structured Prey Populations: Planktivorous Fish and Zooplankton

Michael J. Vanni

Selective predators can remove or reduce the abundance of specific age (or size) classes of coexisting prey populations. If the preferred age (size) classes share resources with classes that are preyed on to a lesser extent, and prey resources are limiting, the interesting situation arises in which the predator actually can benefit the less-preferred age (size) classes of prey. This occurs when the predator reduces the abundance of certain age (size) classes, thereby providing more resources for remaining individuals in the less-preferred classes.

An example of such an interaction among fish, zooplankton, and phytoplankton is presented. An *in situ* enclosure experiment showed that fish remove the largest individuals within populations of coexisting cladocerans, thereby increasing food resources (phytoplankton) for remaining zooplankton individuals. Life table experiments on the cladocerans, using ambient phytoplankton communities as food, revealed that both survivorship and fecundity of these cladocerans could be elevated by the predator-mediated increase in phytoplankton. In some cases the increase in fitness of remaining individuals compensates for predator loss, the net result being that predation has little effect on population density.

Additional anecdotal evidence from the literature suggests that such indirect effects of predators, both vertebrate and invertebrate, may be common in planktonic food webs. The implications of these indirect effects on community structure and dynamics are discussed.

Predation in natural communities may involve complex and counterintuitive interactions among predators and prey. Predators that prey on several species but exhibit preference for one species over others may indirectly benefit a less-preferred prey species by reducing densities of the preferred species, thereby freeing resources for less-preferred prey. The classic studies of Paine (1966, 1984) with intertidal food webs illustrate this concept. In this system a top predator, the sea-star *Pisaster*, selectively preys on the mussel *Mytilus*, reducing its abundance and as a consequence making a limiting resource (in this case unoccupied substrate) more available to other less-preferred prey species. The net result is an increase in the abundance of less-preferred species, which in the absence of predation are competitively excluded by *Mytilus*. The less-preferred species are thus dependent on the presence of the predator for persistence in the community, thereby representing a case of indirect mutualism (Boucher et al. 1982) between a predator and potential prey species. Circumstances in which potential prey species and predators are indirectly mutualistic are fairly common and include examples from both aquatic and terrestrial systems (Hrbáček 1962; Brooks and Dodson 1965; Harper 1969; Porter 1976; Lynch 1979; McCauley and Briand 1979, Lynch and Shapiro 1981; Wilbur et al. 1983), although the particular mechanism responsible for the apparent indirect mutualism between predator and prey may differ among communities.

Predators may have similar indirect, mutualistic interactions, not only within a natural community but also with a single natural prey population. That is, a predator may exert both positive and negative effects on a single species. This is particularly likely if the prey species exhibits age (or size) structure and the predator selectively preys on particular age (size) classes. In such a situation the selective removal of individuals of specific classes may result in greater resource availability for the remaining individuals in classes that are less heavily preyed on. As such, the interaction between the predator and the preferred age (size) classes can be viewed as a classical predator–prey interaction; and the predator and the less-preferred, benefiting age classes may be viewed as indirect mutualists (or more properly, commensalists because it is not apparent that the predator benefits from the presence of the less-preferred age classes).

These indirect interactions among predator and prey are probably common in natural food webs. Many predators prey selectively on a particular fragment of a prey population, and it is common for age or size classes of prey to share some limiting resource. When these conditions are met, those age (size) classes that are not heavily preyed on should benefit from the presence of the predator. The net effect of a predator on a prey population will depend on the strength of the positive and negative effects.

It is difficult to uncouple the direct and indirect effects of predation on a natural prey population in an experimental fashion. Many factors simultaneously influence mortality and fecundity, and even one factor such as predation can have several effects on prey demography, making causal detection difficult. Interactions between planktivorous fish and freshwater zooplankton populations are amenable to experiments in which indirect, complex processes are isolated from other processes because of the relative ease with which manipulative experiments can be conducted and because a great deal of knowledge exists concerning fish–zooplankton interactions.

The size-selective nature of fish predation and its direct effects on zooplankton are well known (Hrbáček 1962; Brooks and Dodson 1965; Lynch 1979; O'Brien 1979; Zaret 1980). Planktivorous fish typically select the largest encountered individuals and can exert enough predation pressure to drive larger species to extinction. Only relatively small species of zooplankton (carapace length generally < 1.5 mm, often < 1.0 mm) are able to coexist with fish. Fish also prey size-selectively within a particular small species (Lynch et al. 1981; Bartell 1982; Wright and O'Brien 1984). Both within and among species, large size classes of zooplankton exert a greater grazing pressure on phytoplankton than do smaller sizes (Burns 1969; DeMott 1982; Peters and Downing 1984), and selective removal of large zooplankton can result in an increase in phytoplankton abundance (Losos and Heteša 1973; Hurlbert and Mulla 1981; Lynch and Shapiro 1981; Carpenter and Kitchell 1984). Reduction in grazing also can shift the phytoplankton community toward more edible, less resistant (to grazers) species (Porter 1976, 1977; McCauley and Briand 1979; DeMott and Kerfoot 1982; Schoenberg and Carlson 1984).

These observations taken together suggest that fish may have indirect, positive effects on the smaller zooplankton species that coexist with fish. These species are preyed on size-selectively, which can potentially reduce grazing pressure on phytoplankton and result in greater phytoplankton abundance and/or a greater percentage of high-quality phytoplankton (i.e., those without gelatinous sheaths or other protective structures). If so, individuals not preyed on can potentially benefit from the increase in food abundance/quality, assuming food is a limiting resource. Note that both negative (predatory mortality) and positive (increased fecundity and/or survivorship due to increased food availability) effects may be experienced within a single prey population. An illustration of this potential series of interactions is outlined in fig. 10.1.

I isolated the indirect effects of fish predation (i.e., those occurring through modification of the food base of zooplankton) on coexisting populations of zooplankton, using an in situ enclosure experiment combined with a series of life table experiments, and found that the positive, indirect effects do occur: survivorship and reproduction of remaining cladoceran individuals were increased as a result of a predator-mediated increase in phytoplankton abundance. Details of these experiments appear elsewhere (Vanni 1986). Here I wish to review the experimental procedure employed, highlight some results, and provide a detailed example of how the predator-mediated increase in food availability can influence prey life history patterns, review other evidence in support of such indirect effects of predation, and discuss the implications of the indirect effects for demography and community structure.

EXPERIMENTAL DESIGN

The general experimental procedure consisted of manipulating levels of a fish predator, bluegill sunfish (*Lepomis macrochirus*), within large *in situ* enclosures containing natural phytoplankton and zooplankton densities, and conducting two life table experiments using the enclosure water (with ambient phytoplankton communities) as experimental media. Details of the experimental protocol can be found in Vanni (1986).

The enclosure experiment was conducted in Dynamite Lake, Illinois, an oligomesotrophic lake with a dense population of bluegill sunfish and a zooplankton community consisting entirely of small (<1.0 mm carapace length) species (Lynch et al. 1981; Vanni 1986). Enclosures were made of thin, clear polyethylene and were similar to those used by Lynch (1979) and Weider (1984). Two treatments were employed: (1) a "control" in which fish were excluded and (2) a "fish" treatment in which fish were stocked at a density approximating that in the lake. Enclosures were filled with lake densities of plankton at the beginning of the experiment.

Approximately 2 weeks after the initiation of the enclosure experiment, the first life table experiment was begun. To do so, I isolated individuals of the Dynamite Lake cladocerans *Bosmina longirostris*, *Ceriodaphnia lacustris* and *Diaphanosoma birgei* into vials containing ≈ 15 ml of either fish enclosure water or control enclosure water. The enclosure water was filtered (44-μm mesh) to remove all zooplankton, but not phytoplankton, before individuals were added to it in the vials. Cohorts were grown on experimental media for one generation, and their offspring were used in the first life table experiment. Vials were kept in a controlled-temperature cabinet maintained at lake temperatures. New medium, taken from the enclosures and filtered to remove zooplankton, was replaced every 2 to 3 days. The first life table experiment was terminated ≈ 2 weeks after its commencement. The second life table experiment was begun by isolating offspring of the first experiment and growing them on the same experimental medium as their mothers. The second experiment also lasted ≈ 2 weeks.

The procedure employed in these experiments allowed the uncoupling of the indirect effects (i.e., those arising from fish-mediated alteration of the phytoplankton community) from other processes affecting demography in the lake (enclosures). The only difference in conditions experienced by individuals between treatments was the difference in phytoplankton availability stemming ultimately from size-

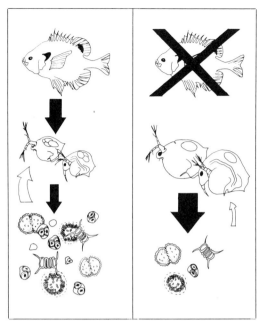

Figure 10.1. *The indirect effect of fish predation on zooplankton through modification of the phytoplankton community.* **Left panel:** *Fish are present and prey on larger size classes, leaving only small individuals. Grazing pressure by the small individuals on phytoplankton is relatively low (denoted by thickness of solid arrow pointing downward from zooplankton to phytoplankton). Phytoplankton abundance is relatively high, and therefore food is plentiful for remaining zooplankton (denoted by thickness of open arrow pointing upward from phytoplankton to zooplankton). The net result (the indirect effect on zooplankton) is relatively large clutch size, denoted by the number of eggs carried by individuals.* **Right panel:** *Fish are excluded, resulting in larger individuals of the same zooplankton species and a relatively high grazing pressure on phytoplankton (denoted by thickness of solid arrow). Phytoplankton levels are thus relatively low, and zooplankton therefore obtain less food (denoted by thickness of open arrow) than when fish are present. The net result is a relatively small clutch size.*

selective fish predation. A schematic illustration of the experimental design is provided in figure 10.2.

RESULTS

Detailed results of the experiments are given in Vanni (1986). The major trends are summarized below.

Diaphanosoma and *Ceriodaphnia* grew to greater median and maximum body lengths in

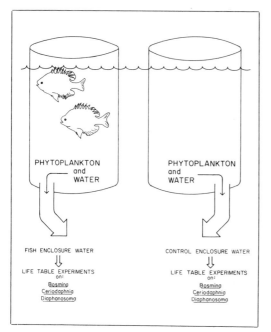

Figure 10.2. *Experimental design used to test for indirect effects of fish predation. Enclosures are set up with fish either present or absent. Water containing ambient phytoplankton communities is taken from each enclosure type and used to rear zooplankton in life table experiments. Thus, life table individuals are isolated from all predation, and individuals of different treatments are exposed to different food conditions as determined by the presence or absence of fish.*

the absence of fish than in the presence of fish. Median *Diaphanosoma* body length in the presence and absence of fish was 0.40 and 0.52 mm, respectively; and for *Ceriodaphnia*, 0.38 and 0.45 mm in the presence and absence of fish. Maximum body length (average of four dates) in the presence and absence of fish was 0.77 and 1.00 mm for *Diaphanosoma* and 0.57 and 0.70 mm for *Ceriodaphnia*. In addition, *Diaphanosoma* was slightly less abundant in the presence of fish than in their absence, whereas *Bosmina* was more abundant in the presence of fish than in their absence. *Ceriodaphnia* abundance was unaffected by fish predation.

Effects of fish predation on the cladocerans, in terms of the effects the grazers have on phytoplankton abundance, may be most properly evaluated by comparing cladoceran biomass in the two treatments. To do so, I estimated the mean individual biomass (micrograms per individual) of each species, as well as cladoceran

community biomass (mg/m^2), assuming biomass is related to size according to equations given by Peters and Downing (1984), as described in Vanni (1984). Since the effects of zooplankton on phytoplankton are likely to depend on mean individual biomass as well as community biomass (Carpenter and Kitchell 1984), biomass patterns should provide insight into how changes in zooplankton size–structure translate into changes in phytoplankton abundance. Cladoceran community biomass was only slightly reduced by fish predation, but mean individual biomass was greatly reduced (table 10.1). Thus, grazing pressure on phytoplankton was probably much less in the presence of fish than in the absence of fish (see also Vanni 1986).

Phytoplankton volume (μm^3/ml) was greater in the presence of fish than in the absence of fish. Volume of high-quality phytoplankton (those species without gelatinous sheaths or other protective coverings) tended to be higher in the presence of fish. These trends agree with the trend of smaller cladocerans in the presence of fish.

The life table experiments showed that age-specific fecundity of *Diaphanosoma* was greater when grown on fish enclosure water, relative to control enclosure water, in both experiments. In addition, survivorship of *Diaphanosoma* was greater on fish enclosure water in experiment 1, but not in experiment 2. Age-specific fecundity of *Ceriodaphnia* was higher on fish enclosure water in experiment 2, but not in experiment 1.

In all three cases in which age-specific fecundity was greater on fish enclosure water, cohort finite rate of increase (λ) was greater on fish enclosure water as well. To compare rates of increase between life table treatments, life table data were analyzed according to methods outlined by Lenski and Service (1982). This method allows one to decompose rate of increase into each life table individual's contribution to the rate. The mean of the individual contributions, $\overline{F}'$, is equal to λ when individuals from both treatments are considered as one population (Lenski and Service 1982; Service and Lenski 1982). In both experiments, *Diaphanosoma* $\overline{F}'$ was greater on fish enclosure water than on control enclosure water, whereas *Ceriodaphnia* $\overline{F}'$ was elevated in fish enclosure water only in experiment 2 (table 10.2). This indicates that fitness of individuals that

Table 10.1. *Cladoceran community biomass and individual biomass in the Dynamite Lake enclosures[a]*

Biomass	Control	Fish
Cladoceran community biomass (mg $\cdot$ m^{-2})	84.44	77.50
Mean individual biomass (μg $\cdot$ individual^{-1})		
Bosmina	0.801	0.703
Ceriodaphnia	2.130	1.478
Diaphanosoma	3.742	1.842
Mean	1.880	0.969

[a]Adapted from Vanni (1984).

Table 10.2. *F'[a] values (mean ± SE) from the life table experiments*

Subjects	Control enclosure water	Fish enclosure water	p
CERIODAPHNIA			
Experiment 1	1.397 (0.111)	1.274 (0.072)	>0.05
Experiment 2	1.126 (0.062)	1.439 (0.108)	<0.05
DIAPHANOSOMA			
Experiment 1	0.773 (0.134)	1.632 (0.116)	<0.001
Experiment 2	0.863 (0.065)	1.373 (0.097)	<0.001

[a]F' is equivalent to λ, the finite rate of increase, when individuals of both treatments are considered together as one cohort. See text and Vanni (1986) for further explanation on the calculation of F'.

manage to avoid fish predation can be substantially elevated by the fish-mediated increase in phytoplankton abundance.

Bosmina exhibited poor survivorship in the life table experiments because of handling difficulties; many individuals became trapped in the surface tension of the water and perished. Because of this, survivorship differences between treatments could not be determined. Reproduction of *Bosmina* individuals surviving the experiment was not affected by enclosure water type.

Statistical analysis of each individual's contributions to λ and the life history traits that influence those contributions (Lenski and Service 1982), revealed that age of first reproduction, number of clutches produced in a lifetime, and mean clutch size (number of offspring per clutch) were important determinants of an individual's contribution to cohort rate of increase. These life history traits were sensitive to the increase in food arising from the presence of fish. An example of how the life history traits differed between life table treatments is given for

the first *Diaphanosoma* life table experiment (fig. 10.3). Figure 10.3, which provides details on each individual's schedule of reproduction and survivorship, demonstrates that several trends are apparent. First, more individuals tended to initiate reproduction at an earlier age when grown on fish enclosure water than when grown on control enclosure water. All individuals in the fish enclosure water treatment began reproducing at age 6 days or earlier, whereas 41% of control-enclosure-water treatment individuals had not reproduced by the time they reached 6 days (fig. 10.3). Since instar durations were similar between treatments, control-enclosure-water individuals apparently, on average, needed to develop to later instars (i.e., grow to larger sizes), relative to fish-enclosure-water individuals, in order to procure enough food to produce offspring. The second trend is that average age-specific clutch size was greater in the fish-enclosure-water treatment than in the control-enclosure-water treatment (fig. 10.3). Third, individuals in the fish-enclosure-water treatment produced a

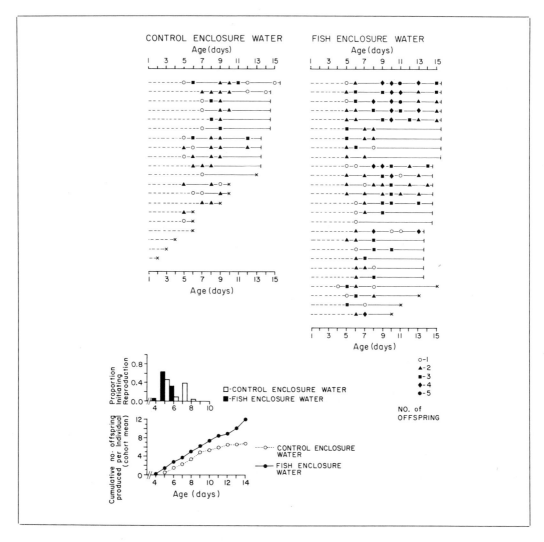

Figure 10.3. *Individual schedules of reproduction and survivorship of* Diaphanosoma *in the first life table experiment. Each horizontal line with symbols represents one individual. Dashed lines, prereproductive life; solid lines, postreproductive life. Symbols represent the num-* *ber of offspring released on that day as noted in the legend. Lines ending with a vertical bar represent individuals that survived the entire experiment; those ending with an X indicate that the individual died on that date.*

greater number of clutches than those in the control-enclosure-water treatment (fig. 10.3). This resulted from the fact that control-enclosure-water individuals "skipped" clutches more often that fish-enclosure-water individuals. That is, although instar duration was equal between treatments, control individuals more often developed through instars during which no offspring were produced (fig. 10.3). Fourth, a greater percentage of individuals in the fish-enclosure-water treatment survived

the duration of the experiment (fig. 10.3), suggesting that starvation was more prevalent among the control-enclosure-water individuals. Most mortality of control-enclosure-water individuals occurred during early instars, suggesting that juvenile survivorship is increased by fish-mediated increases in phytoplankton abundance (fig. 10.3). The younger age at first reproduction, greater clutch size, and more frequent clutch production exhibited by the fish-enclosure-water treatment individuals resulted in a greater cohort reproductive output (fig. 10.3) and as a consequence, a higher cohort rate of increase (λ).

It could be argued that the positive effects of the fish-mediated increases in phytoplankton abundance may be overestimated by these life table experiments because individuals living in the presence of fish will not survive as long as control-enclosure individuals because they are preyed on and therefore will not be able to reap the benefits of increased food availability. To test directly if this is so requires knowledge of the average age to which individuals live in the presence and absence of fish, which, unfortunately, could not be determined by the enclosure experiments. However, based on size-frequency distributions of *Diaphanosoma*, it is clear that virtually all individuals >0.75 mm are eliminated by fish predation, whereas individuals often grow to 1.00 mm in the absence of fish (Vanni 1986). Although I have not directly linked age to size in the Dynamite Lake cladocerans, Polishchuk and Ghilarov (1981) have done so for *Diaphanosoma brachyurum* and found that individuals reach a carapace length of 0.75 mm in ≈ 11 days. Thus, one can assume that *Diaphanosoma* individuals will not live beyond 11 days in the presence of fish, and recalculate intrinsic rates of increase using life table data only through the first 11 days of life. I did such a calculation for the first *Diaphanosoma* life table and found that truncating life span at 11 days lowered r, the intrinsic rate of increase, from 0.275 to 0.268, a reduction of only 2.5%. Even after elimination of these age 11+ individuals, r was $\approx 50\%$ greater on fish enclosure water than on control enclosure water. This is because the first few clutches contribute to nearly 100% of r (Vanni 1986).

The other life table experiments (not presented in detail here) demonstrated that it was not necessary for all of these life history traits to be affected in order for a significant increase in λ to result. For example, in the second life table experiment, neither *Diaphanosoma* age at first reproduction nor survivorship was affected by the greater food availability in the fish enclosure treatment, yet because mean clutch size and the number of clutches produced per individual was greater in the fish-enclosure-water treatment, λ was also greater (Vanni 1986). This is in agreement with the observation that mean clutch size and number of clutches produced were the greatest deter-minants of an individual's contribution to cohort rate of increase (Vanni 1986).

DISCUSSION

The results of the Dynamite Lake experiments demonstrate that individuals that manage to avoid being preyed on by fish actually can benefit from the presence of fish because of the fish-mediated increase in phytoplankton abundance. Additional, mostly anecdotal, evidence exists in the literature in support of such indirect, beneficial effects of predation on zooplankton (table 10.3). None of the previous studies experimentally isolated the indirect pathway from other processes affecting demography, except for Neill's (1975) work in laboratory microcosms. In addition, again except for Neill (1975), none of the investigators was specifically testing for the occurrence of indirect effects; yet the effects suggested by the results, in a variety of predator–prey systems, indicate that this type of indirect effect may be common in planktonic food webs.

The results of those previous studies also suggest that the major response of zooplankton to predator-mediated increase in food abundance is increased reproductive success, rather than increased survival of remaining individuals (table 10.3). However, this apparent trend may have arisen because of the difficulty in detecting causal factors that influence survivorship in populations in lakes or enclosures. Without conducting a life table experiment, it is impossible to separate the factors causing mortality patterns in such populations. Survivorship of some individuals may be increased by elevated food availability and that of others decreased by predation when fish are present, but standard sampling techniques will measure only net mortality patterns. Since, of those studies listed in table 10.3, only Neill (1975) attempted a life table experiment along with manipulations of predators, the apparent lack of response in survivorship may simply be due to the fact that previous experiments were not designed to detect such a response. The Dynamite Lake life table experiment presented in figure 10.3 demonstrates that survivorship of individuals avoiding predation can in fact increase as a result of a predator-mediated increase in food. Nevertheless, considering the Dynamite Lake experiments as a whole, fecundity was affected much more by the increases

Table 10.3. *Evidence of indirect effects (i.e., predator-mediated increase in survivorship and/or reproduction in the literature)*

Predator	Prey	Prey response[a]	Setting
Mosquitofish (Gambusia affinis)	Several crustaceans	Increased juvenile survivorship; earlier age at 1st reproduction	Experimental laboratory microcosms (Neill [1975])
Bluegill (Lepomis macrochirus)	Ceriodaphnia reticulata, Bosmina longirostris	Greater size-specific clutch size	In situ enclosures; Pleasant Pond, MN (Lynch [1979])
Rainbow trout (Salmo gairdneri)	Daphnia rosea	Greater number of eggs per adult female	In situ enclosures; Lake Mitchell, VT (DeMott and Kerfoot [1982])
Carp (Cyprinus carpio) and Bream (Abramis brama)	Several cladocerans and copepods	Greater number of eggs per female	Experimentally increased stock in Lake Warniak, Poland (Grygierek [1967])
Sardine (Mirogrex terraesanctae)	Mesocyclops leuckarti	Greater number of eggs per female	Lake Kinneret, Israel (Gophen and Landau [1977])
Chaoborus trivattatus and C. americanus	Daphnia rosea, Diaphanosoma brachyurum, Bosmina longirostris	Increased % of ovigerous females; greater number of eggs per female	In situ enclosures; Gwendoline Lake, British Columbia (Neill [1981])
Chaoborus americanus	Daphnia laevis, Daphnia pulex	Increased birth rate	Laboratory experiments (Cooper and Smith [1982])
Notonecta hoffmani	Daphnia pulex	Greater number of eggs per adult female	Laboratory experiments (Murdoch and Scott [1984])

[a]Response of the prey to increased levels of predation.

in food availability than was survivorship (Vanni 1986).

Although it is clear that the proposed indirect effects of predation do occur, some predictions regarding the generality of occurrence of such effects can be made. First, the indirect effects may be stronger in eutrophic lakes than in oligotrophic lakes. Phytoplankton abundance in eutrophic lakes is more likely to be responsive to fluctuations in zooplankton grazing pressure, because nutrients are not limiting the production of most phytoplankton species; whereas in oligotrophic lakes phytoplankton is constrained by nutrient levels and grazing pressure (Lynch and Shapiro 1981). Thus, any given reduction in mean zooplankton size (and therefore grazing pressure) is likely to produce a greater increase in phytoplankton abundance in eutrophic lakes than in oligotrophic lakes. As a result, the response of remaining zooplankton (i.e., increased survivorship and/or reproduction) should be greater in eutrophic lakes.

Second, the strength of the indirect effects of fish predation should increase with increasing zooplankton size. This results from the fact that zooplankton grazing rate is an exponential function of body length. Thus, the larger the size of the prey species, the more fish predation on larger size classes will reduce grazing pressure on phytoplankton. Consider, for example, the case of *Diaphanosoma*, the largest species in Dynamite Lake. Maximum and mean carapace length were reduced 0.23 mm and 0.12 mm, respectively, by fish predation. This reduction, along with the reduction in *Ceriodaphnia* body size, resulted in increased phytoplankton abundance. However, if the prey is a larger species than can coexist with fish—for example, *Daphnia pulex*—reduction in body length, and therefore grazing pressure, may be even greater. As a result, phytoplankton may increase more as a result of fish predation than when only very small species (such as those in Dynamite Lake) are present. Thus, the potential for increased survivorship and/or reproduction of remaining individuals is greater if relatively larger species are prey. Of course, the greatest increase in phytoplankton will occur when the very largest species (e.g., large *Daphnia* species) are preyed on by fish, but these species often are reduced to extinction by fish. Obviously, the predator-mediated increase in phytoplankton will be of little sig-

nificance to the species being preyed on if no individuals can survive in the presence of fish. Therefore, the indirect effects should be most pronounced in species such as intermediate-size *Daphnia* (e.g., *D. galeata*, *D. rosea*), which can coexist with fish but probably incur a relatively great reduction in mean body size.

The Dynamite Lake cladocerans that avoid predation clearly can derive some benefits from the actions of planktivorous fish. Further, the population densities of most species in the community are not greatly influenced by fish predation (Vanni 1984, 1986). This suggests that the positive and negative effects of predation cancel out. That is, in terms of factors controlling population growth, predatory mortality merely substitutes for the reduction in survivorship and/or reproduction that occurs when age class interactions are strong (i.e., when predators are absent). Other recent studies have shown that population densities of some species are affected little by the presence of coexisting predators (Neill 1981; DeMott and Kerfoot 1982; Murdoch and Scott 1984), and some workers have suggested that the increase in fecundity in the presence of predators may balance predatory mortality (Grygierek 1967; Neill 1981; Murdoch and Scott 1984). Presumably, the positive and negative effects of predation are precisely balanced at some level of predation. At higher predation levels, prey populations may not be able to compensate for predatory losses and will suffer a net reduction in abundance.

In general, age-specific predation coupled with strong interactions among age classes, such as sometimes occurs in planktonic food webs, may increase or decrease the stability of prey populations. Predatory removal of adults may reduce competitive interactions between adult and juvenile classes. Relaxing this predation could then result in stronger adult–juvenile interactions within the prey population, potentially leading to cycles in prey population density corresponding to oscillations in food availability. If interactions among age classes are very strong, oscillations of prey abundance may be of greater amplitude and eventually lead to extinction of the population. Thus, by decreasing adult–juvenile competition, predation may increase the stability of age-structured prey populations (Tschumy 1982).

The interactions among age (size) classes in Dynamite Lake populations are apparently not strong enough to cause such dramatic oscillations; population densities were fairly similar in the presence and absence of fish (Vanni 1984, 1986). Hastings (1983) showed that age-specific predation can, in theory, stabilize or destabilize prey populations, depending on the degree of age specificity of predation and the length of juvenile period. Charlesworth (1980) contends that age-structured populations subject to density-dependent growth are more likely to show cyclical fluctuations in density if fecundity is high, as opposed to when fecundity is low. On the whole, fecundity of Dynamite Lake cladocerans was apparently sufficiently low to prevent oscillations. Populations in eutrophic lakes, where fecundity is likely to be greater, may be more likely to exhibit cyclic behavior; note that this boom-and-bust behavior is common in highly enriched culture vessels when populations are not regularly cropped. Understanding the indirect effects of predation under various levels of predation and trophic conditions will aid in elucidating the mechanisms by which predation and food limitation can control the dynamics of age-structured populations.

The predator-mediated increase in phytoplankton may have substantial effects on the size structure of zooplankton populations (Vanni 1986). Because fecundity is increased in the presence of fish, populations exposed to fish predation may have proportionately more newborns than those not experiencing fish predation, resulting in an age (size) structure more skewed toward younger (smaller) individuals. It is common for zooplankton size-frequency distributions to be skewed toward smaller individuals in the presence of fish (Wells 1970; Warshaw 1972; Lynch 1979; Gliwicz et al. 1981), but usually this is interpreted as being due solely to the removal of large individuals by fish. Although the absence of larger size classes in the presence of fish is no doubt caused by direct predatory actions, the preponderance of smaller size classes is likely partly the result of the indirect effects of fish (i.e., increased production of offspring). Indeed, simulations of the dynamics of prey age structure under the Dynamite Lake control-enclosure-water and fish-enclosure-water conditions, using the life table data in projection matrices, showed that the indirect effects of fish (i.e., elevated phytoplankton abundance and resultant increase in cladoceran fecundity)

by themselves can result in populations dominated by younger (smaller) individuals (Vanni 1986). This demonstrates that fish predation may indirectly alter cladoceran age (size) structure in a manner that parallels the direct action of fish predation and therefore that care must be taken in interpreting trends in the size structure of natural populations.

The indirect effects of predation may have effects beyond those on prey population dynamics. The increase in food availability resulting from fish predation may affect the genetic structure of prey populations. For example, certain genotypes, e.g., those that mature at a smaller size, may be less vulnerable than others to fish predation. Such genotypes will be favored in the presence of fish because they will remain to reap the benefits of increased food availability. Genetic variation in vertical migration behavior also exists in zooplankton populations (Weider 1984); genotypes that avoid predation by vertically migrating may be selectively favored by the fish-mediated increase in phytoplankton.

Indirect effects of predation in natural communities may extend to species other than the prey. Often, small invertebrate predators (of zooplankton), such as cyclopoid copepods and the rotifer Asplanchna, are present in lakes containing fish. These small predators apparently cannot drive their prey (small zooplankton species) to extinction (Lynch 1979) but rather are probably limited to some extent by prey availability. Thus, an increase in herbivore offspring production and juvenile survivorship, resulting from a fish-mediated increase in phytoplankton, may provide more food for the invertebrate predators inhabiting the community, potentially increasing the survivorship and/or fecundity of the invertebrate predators. Further experiments, which isolate the indirect effects of predation on herbivores as well as on secondary (invertebrate) predators, will be needed to determine the extensiveness of such effects.

Throughout this discussion I have implicitly and in some cases explicitly assumed that the greater phytoplankton availability in the presence of fish is the result of reduced grazing pressure by zooplankton. However, the possibility exists that the increase in phytoplankton is the result of increased nutrient regeneration by fish and/or by smaller zooplankton, which excrete nutrients at a greater rate (per unit biomass) than large zooplankton (Peters 1975;

Bartell 1981). Although I argue elsewhere that it is more likely that reduced grazing, rather than increased nutrient regeneration, caused the increased phytoplankton abundance in the Dynamite Lake experiments (Vanni 1986), nutrient regeneration may be important in this regard in other communities, especially in very dilute lakes. Note that the particular mechanism causing the increase in phytoplankton in the presence of predators does not alter any conclusions as to the significance of the predator-mediated increase in food availability to the zooplankton populations; whatever the mechanism, this indirect effect of predation clearly can greatly influence prey population dynamics.

REFERENCES

Bartell, S. M. 1981. Potential impact of size-selective plantivory on P release by zooplankton. *Hydrobiologia* 80 : 139–46.

———. 1982. Influence of prey abundance on size-selective predation by bluegills. *Trans. Amer. Fish. Soc.* 111 : 453–61.

Boucher, D. H., S. James, and K. H. Keeler. 1982. The ecology of mutualism. *Annu. Rev. Ecol. Syst.* 13 : 315–47.

Brooks, J. L., and S. I. Dodson. 1965. Predation, body size and composition of the plankton. *Science* 150 : 28–35.

Burns, C. W. 1969. Relationship between filtering rate, temperature, and body size in four species of *Daphnia*. *Limnol. Oceanogr.* 14 : 693–700.

Carpenter, S. R., and J. F. Kitchell. 1984. Plankton community structure and limnetic primary production. *Amer. Naturalist* 124 : 159–72.

Charlesworth, B. 1980. Evolution in age-structured populations. Cambridge: Cambridge University Press.

Cooper, S. D., D. W. Smith. 1982. Competition, predation, and the relative abundances of two *Daphnia* species. *J. Plankton Res.* 4 : 859–79.

DeMott, W. R. 1982. Feeding selectivities and relative ingestion rates of *Daphnia* and *Bosmina*. *Limnol. Oceanogr.* 27 : 518–27.

DeMott, W. R., and W. C. Kerfoot. 1982. Competition among cladocerans: Nature of the interaction between *Bosmina* and *Daphnia*. *Ecology* 63 : 1949–66.

Gliwicz, Z. M., A. Ghilarov, and J. Pijanowska. 1981. Food and predation as major factors limiting two natural populations of *Daphnia cucullata* Sars. *Hydrobiologia* 80 : 205–18.

Gophen, M., and R. Landau. 1977. Trophic interactions between zooplankton and sardine *Mirogrex terraesanctae* population in Lake Kinneret, Israel.

Oikos 29 : 166–71.

Grygierek, E. 1967. Formation of fish pond biocenosis exemplified by planktonic crustaceans. *Ekologia Polska (A)* 15 : 155–81.

Harper, J. L. 1969. The role of predation in vegetational diversity. *Brookhaven Symposium on Biology* 22 : 48–62.

Hastings, A. 1983. Age-dependent predation is not a simple process. 1. Continuous time models. *Theoretical Population Biology* 23 : 347–62.

Hrbáček, J. 1962. Species composition and the amount of zooplankton in relation to the fish stock. *Rozpravy Ceskoslovenske Akademie Ved Rada Matematickych a Priorodnich Ved* 72 : 1–116.

Hurlbert, S. H., and M. S. Mulla. 1981. Impacts of mosquito fish (*Gambusia affinis*) predation on plankton communities. *Hydrobiologia* 83 : 125–51.

Lenski, R. E., and P. M. Service. 1982. The statistical analysis of population growth rates calculated from schedules of survivorship and fecundity. *Ecology* 63 : 655–62.

Losos, B., and J. Heteša. 1973. The effect of mineral fertilization and of carp fry on the composition and dynamics of plankton. *Hydrobiological Studies* 3 : 173–217.

Lynch, M. 1979. Predation, competition, and zooplankton community structure: An experimental study. *Limnol. Oceanogr.* 24 : 253–72.

Lynch, M., B. Monson, M. Sandheinrich, and L. Weider. 1981. Size-specific mortality rates in zooplankton populations. *Verhandlungen Internationale Vereinigung für theoretische und Angewandte Limnologie* 21 : 363–68.

Lynch, M., and J. Shapiro. 1981. Predation, enrichment, and phytoplankton community structure. *Limnol. Oceanogr.* 26 : 86–102.

McCauley, E., and F. Briand. 1979. Zooplankton grazing and phytoplankton species richness: Field tests of the predation hypothesis. *Limnol. Oceanogr.* 24 : 243–52.

Murdoch, W. W., and M. A. Scott. 1984. Stability and extinction of laboratory populations of zooplankton preyed upon by the backswimmer *Notonecta*. *Ecology* 65 : 1231–48.

Neill, W. E. 1975. Experimental studies of microcrustacean competition, community composition and efficiency of resource utilization. *Ecology* 56 : 809–26.

Neill, W. E. 1981. Impact of *Chaoborus* predation upon the structure and dynamics of a crustacean zooplankton community. *Oecologia* 48 : 164–77.

O'Brien, W. J. 1979. The predator-prey interaction of planktivorous fish and zooplankton. *American Scientist* 67 : 572–81.

Paine, R. T. 1966. Food web complexity and species diversity. *Amer. Naturalist* 100 : 65–75.

———. 1984. Ecological determinism in the competition for space. *Ecology* 65 : 1339–48.

Peters, R. H. 1975. Phosphorus regeneration by

natural populations of limnetic zooplankton. *Verhandlungen Internationale Vereinigung für Theoretische und Angewandte Limnologie* 19 : 273–79.

Peters, R. H., and J. A. Downing. 1984. Empirical analysis of zooplankton filtering and feeding rates. *Limnol. Oceanogr.* 29 : 763–84.

Polishchuk, L. V., and A. M. Ghilarov. 1981. Comparison of two approaches used to calculate zooplankton mortality. *Limnol. Oceanogr.* 26 : 1162–67.

Porter, K. G. 1976. Enhancement of algal growth and productivity by grazing zooplankton. *Science* 192 : 1332–34.

———. 1977. The plant-animal interface in freshwater ecosystems. *American Scientist* 65 : 159–70.

Schoenberg, S. A., and R. E. Carlson. 1984. Direct and indirect effects of zooplankton grazing on phytoplankton in a hypereutrophic lake. *Oikos* 42 : 291–302.

Service, P. M., and R. E. Lenski. 1982. Aphid genotypes, plant phenotypes, and genetic diversity: a demographic analysis of experimental data. *Evolution* 36 : 1276–82.

Tschumy, W. D. 1982. Competition between juveniles and adults in age-structured populations. *Theoretical Population Biology* 21 : 255–68.

Vanni, M. J. 1984. Effects of nutrients and top predators on the structure and dynamics of a freshwater plankton community: Experimental studies. Ph.D. diss. University of Illinois at Urbana–Champaign, Urbana, Ill.

———. 1986. Fish predation and zooplankton demography: Indirect effects. *Ecology.* 67 : 337–54.

Warshaw, S. J. 1972. Effects of alewives (*Alosa pseudoharengus*) on the zooplankton of Lake Wononskopomuc, Connecticut. *Limnol. Oceanogr.* 17 : 816–25.

Weider, L. J. 1984. Spatial heterogeneity of *Daphnia* genotypes: Vertical migration and habitat partitioning. *Limnol. Oceanogr.* 29 : 225–35.

Wells, L. 1970. Effects of alewife predation on zooplankton populations in Lake Michigan. *Limnol. Oceanogr.* 15 : 556–65.

Wilbur, H. M., P. J. Morin, and R. N. Harris. 1983. Salamander predation and the structure of experimental communities: Anuran responses. *Ecology* 64 : 1423–29.

Wright, D. I., and W. J. O'Brien. 1984. The development and field test of a tactical model of the planktivorous feeding of white crappie (*Poxomis annularis*). *Ecol. Monogr.* 54 : 65–98.

Zaret, T. M. 1980. Predation and freshwater communities. New Haven, Conn.: Yale University Press.

11. Experimental Evaluation of Trophic-Cascade and Nutrient-Mediated Effects of Planktivorous Fish on Plankton Community Structure

Stephen T. Threlkeld

The effects on phytoplankton and zooplankton community structure of three fish species—a particulate feeding zooplanktivore, *Menidia beryllina*, an omnivorous pump filter feeder, *Dorosoma cepedianum*, and a facultative omnivore, *Tilapia aurea*—and various mechanical and chemical analogs of planktivorous fish were measured during two seasons in a set of 18 large outdoor experimental tanks. Nutrient regeneration rates of *Dorosoma* and *Menidia* measured in the laboratory were used as a basis for simulating the nutrient regeneration effects of planktivores on plankton communities. A plankton net filter was used separately and in combination with nutrient additions to simulate the combined predatory and nutrient regeneration effects of planktivores. The effects of accidental fish mortality associated with experimental manipulations of planktivorous fish were evaluated by introducing dead fish into some tanks.

The three fish species differed little in their overall impact on zooplankton composition, even though their feeding behaviors and efficiencies have been reported as different. Artificial removal of zooplankton by plankton netting resulted in phytoplankton and zooplankton communities similar to those in the three fish treatments. Filter and fish manipulations did not involve accidental fish mortality and altered zooplankton community structure without appreciably affecting phytoplankton community structure. Dead fish and nutrient regeneration treatments enhanced phytoplankton growth and increased turbidity. In combination with artificial removal of zooplankton by plankton netting, the dead fish and nutrient regeneration treatments were most similar to previously reported results of fish effects on zooplankton and phytoplankton.

In an experimental setting, nutrient-mediated effects of planktivore introductions may exceed trophic-cascade effects associated with reduced grazing pressure that results from zooplanktivory. Experimental protocols used to examine fish-zooplankton-phytoplankton interactions need to be examined closely for the effects of nuisance variables such as fish mortality.

It is well known that planktivorous fish can dramatically alter the species composition and size structure of zooplankton communities (Hrbáček 1962, Hillbricht-Ilkowska 1964; Hrbáček and Novotná-Dvořáková 1965; Brooks and Dodson 1965; Novotná and Kořínek 1966; Hall et al. 1970, 1976; Hurlbert et al. 1972; Hurlbert and Mulla 1981; Losos and Heteša 1973; Grygierek 1973; Lynch 1979; Lynch and Shapiro 1981; Drenner et al. 1982, 1984a, 1984b). As first demonstrated by Hrbáček (1962), a common effect of an increase in planktivorous fish is an increase in phytoplankton biomass or a shift in the size structure of the algal community to smaller cells (many of the same references as above). This chapter will focus on the reasons for this enhancement of phytoplankton by planktivorous fish.

Two major hypotheses attempt to account for the enhancement of algae by planktivorous fish. The first hypothesis is that phytoplankton are enhanced when zooplankton grazing pressure is reduced by planktivorous fish (a trophic cascade). This hypothesis relies on two critical features: (1) that grazing by zooplankton can control the biomass and size structure of the algal community, and (2) that when zooplankton grazing pressure is reduced, the phytoplankton are able to respond with increased growth (i.e., there are sufficient nutrients that a measurable difference exists between grazer-controlled and nutrient-limited algal communities). It has been argued that cascading effects are seen only in lakes or experimental systems with high nutrient loading where phytoplankton can have a large and rapid response (Dodson et al. 1976; Bürgi et al. 1979; Leah et al. 1980; Reynolds 1984; but see Henrikson et al. 1980).

The second hypothesis is that planktivorous fish enhance phytoplankton by enhancing nutrient cycling. This hypothesis subsumes a number of specific mechanisms that have been shown to exist and be potentially important in some systems. These include (1) a shift in the size structure or vertical distribution of zooplankton that alters biomass-specific excretion rates or nutrient availability to phytoplankton (Bartell 1981; Bartell and Kitchell 1978; Carpenter and Kitchell 1984; Kitchell et al. 1979; Wright and Shapiro 1984; Shapiro and Wright 1984); (2) excretion or defecation by the fish (Bray et al. 1981; Meyer et al. 1983; Robison and Bailey 1981); (3) nutrient release from fish that have died after being introduced as part of the experimental manipulation (Durbin et al. 1979; Drenner et al. 1985); and (4) release of nutrients from the sediments or benthic organisms to the overlying water column by fish that include benthic habitats in their foraging or other activities (Lamarra 1975). With respect to this last mechanism, many of the fish species that have been studied as planktivores also are known to have major effects on benthic community structure (Andersson et al. 1978; Ball and Hayne 1952; Crowder and Cooper 1982; Hall et al. 1970; Hayne and Ball 1956; Lellak 1966; Post and Cucin 1984), either directly through predation or physical disturbance of the habitat or indirectly through predator-mediated effects on the quality and quantity of planktonic material sedimenting to the bottom (Lellak 1966; Bürgi et al. 1979).

Previous consideration of these two hypotheses has depended primarily on budgetary approaches, seeking to answer questions related to amounts and rates of nutrient release or phytoplankton loss (e.g., Nakashima and Leggett 1980; review by Lehman 1984). Experimental or observational approaches have rarely focused on the dichotomy of processes implied by these hypotheses or sought data that clearly distinguished between them. This chapter presents results of experiments in which I have attempted to distinguish between these hypotheses, and reviews the responses of phytoplankton community structure to experimental manipulations of planktivorous fish in this context.

METHODS AND MATERIALS

Two experiments were conducted, one in May–June and one in October–November 1984, in a set of 18 large insulated fiberglass tanks located at the University of Oklahoma Biological Station. The tanks are 2.3 m (diameter) • 2 m and hold approximately 7,000 L. The tanks are arranged in three rows of six tanks each and are conveniently sampled from elevated walkways between the three rows (fig. 11.1). Water is mixed within each tank by an airlift mixer; air is delivered through a pvc manifold from a 1-hp Rotron blower and injected at 75-cm depth into a 5-cm diameter pvc airlift suspended from a surface float. Water is entrained from a depth of 120 cm, or approximately 50 cm above the bottom of each tank. An average of 85 L/min passes through the airlift, ensuring homogeneity of the tank environments. Vertical heterogeneities in physical or chemical characteristics in 12 experiments conducted since airlift installation in March 1983 have been minimal (usually less than 1% of the mean). The airlifts seem to offer no impediment to the development of diverse plankton communities or high densities under selected treatment conditions.

Prior to each experiment, the tanks were drained and the walls and bottoms scrubbed clean of any visible algae, and all particulate material was removed by vacuum. Each experiment was initiated by filling the tanks with unfiltered Lake Texoma water, delivered to the tanks by centrifugal pump through a 5-cm pvc pipe with intake 65 m offshore in Lake Texoma (at the Biological Station on the north shore of the Red River arm of the lake). Tank filling took approximately 22 hours to complete on each occasion, and homogeneity of starting conditions was attempted by initially filling each tank to about one-half volume, followed by final filling of all tanks. Each tank was then innoculated with 1/18 of the contents of 10 composited vertical net hauls of an 80-μm mesh plankton net from an offshore station. The tanks were sampled after filling, and 1 to 2 times again (during 5–10 days) before treatments were initiated. After treatments were initiated, tanks were sampled at 2 to 6 day intervals.

Fish used in the experiments were obtained as follows: *Dorosoma cepedianum* (Clupeidae) and *Menidia beryllina* (Atherinidae) were seined from Lake Texoma and held for 3 to 7 days prior to initiation of experiments. *Tilapia aurea* (Cichlidae) were obtained from Dr. William Shelton at the University of Oklahoma. It was my intention to provide a broad basis for comparing fish effects on plankton community development by using three species known to differ in their feeding behavior and impacts (Drenner and McComas 1980; Drenner et al. 1982, 1984a, 1984b). *Menidia* directs its feeding movements toward individual prey items and with its coarse gillrakers selectively removes zooplankton. *Dorosoma* and *Tilapia* feed by pump filter-feeding and with their finely spaced gillrakers (and microbranchial spines in the case of *Tilapia*), remove nonevasive zooplankton prey and large algae. In order to distinguish more fully between nutrient regeneration and predation effects, artificial planktivores also were devised as treatments in these experiments. To mimic the predation impacts of the planktivores, the outflow of the airlift mixer could be fitted with a 30-cm-diameter plankton net of 145-μm mesh that removed zooplankton entrained in the airlift flow but did not impede the recirculation of phytoplankton in the tanks. Zooplankton caught in the net were removed daily and not returned to the tanks. To simulate nutrient regeneration, inorganic nutrients were added as Na_2PO_4 and NH_4NO_3 in proportion to biomass-specific nutrient regeneration rates estimated for *Menidia* and *Dorosoma* by Korstad and Threlkeld (unpublished data). In the October–November experiment, dead fish (freshly killed *Dorosoma* and *Menidia*) were added to some tanks as an additional treatment factor. Controls lacked fish, nutrient or filter manipulations. Exact treatment specifications, protocol and amount of replication are given in table 11.1. Figure 11.1 shows the spatial distribution of treatments to the tanks, which were assigned randomly. ANOVA of pretreatment measurements of algal chlorophyll fluorescence, turbidity, and Secchi depths did not reveal any variation likely to be manifest as treatment effects due to tank location or pretreatment conditions.

A Hydrolab 4041 water-quality monitoring unit was used to measure surface and bottom

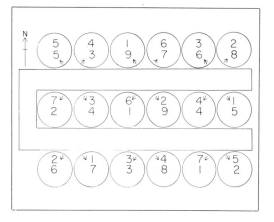

Figure 11.1. *Outline of the experimental tanks, showing location and direction of outflow from airlift mixers (arrows) and sampling walkway. Assignment of treatments to the tanks is indicated by the numbers within the circles (top digit is treatment for the May–June experiment; the lower number is for the October–November experiment). Refer to table 11.1 for treatment numbers and descriptions and to the text for rationale.*

temperature, dissolved oxygen, conductivity and pH. Water samples were filtered through Gelman HA (0.45 μm) glass-fiber filters and analyzed for nitrate and soluble reactive phosphorus, using cadmium reduction and molybdate methods, respectively. Alkalinity was determined by potentiometric methods. Phytoplankton were sampled by an integrated 1.5-cm-diameter tube sampler and preserved in Lugol's; algal samples were examined with an Omnicon 5000 image analysis system linked to a Wild Inverted microscope. Algal cells greater than 2 μm in their longest dimension were counted and distinguished from blue-green algal filaments, which were counted separately. Fresh water samples were examined for algal chlorophyll fluorescence and DCMU-enhanced fluorescence, following the methods of Vincent (1981). Secchi depth and turbidity (as NTU) also were measured. Zooplankton were sampled by a vertical net haul of an 80-μm-mesh Wisconsin net from the tank bottom to the surface. Fish mortality was monitored daily, and any dead fish not intentionally added to the tanks were removed and weighed. Table 11.2 gives a schedule of samples and measurements taken in each experiment.

Analysis of the resulting data sets was done primarily by split-plot ANOVA for factors with repeated measures (Winer 1971, Gill

Table 11.1. *Treatment specifications and protocols in*
May–June (MJ) and October–November (ON) tank experiments

1. Control: no manipulations; replication—MJ: 3 tanks; ON: 2 tanks.
2. *Menidia beryllina*: 75/tank = 350 g wet weight; 60–75 mm std. length; replication—MJ: 3 tanks; ON: 2 tanks.
3. *Dorosoma cepedianum*: 3/tank = 350 g wet weight; 135–190 mm std. length; replication—MJ: 3 tanks; ON: 2 tanks.
4. *Tilapia aurea*: 15/tank = 350 g wet weight; 80–100 mm std. length; replication—MJ: 3 tanks; ON: 2 tanks.
5. Filter: 145-μm mesh filter placed over outflow of airlift mixer for at least 36 h of each 4-day period between sampling times; removed and cleaned after 12–24 h on an airlift; replication—MJ: 2 tanks; ON: 2 tanks.
6. Nutrients: 0.95 g $NH_4\,NO_3$ and 20.5 g $Na_2\,PO_4$ per tank per day, dissolved in tank water and dispersed at the surface by the airlift mixers; replication—MJ: 2 tanks; ON: 2 tanks.
7. Nutrients and filter: combination of manipulations 5 and 6; replication—MJ: 2 tanks; ON: 2 tanks.
8. Dead fish: 350 g freshly killed *Dorosoma cepedianum* and *Menidia beryllina* added at the beginning of the experiment; replication–ON: 2 tanks; none in MJ.
9. Dead fish and filter: combination of manipulations 5 and 8; replication—ON: 2 tanks; none in MJ.

Table 11.2. *Schedule of experiments conducted and measurements taken*

Date	Temp. range (°C)	Samples/Measurements
EXPERIMENT I: 12 MAY–16 JUNE 1984		
11–12 May 1984		Tank filling, Zoop. spike
14 May	24.5–25.2	H,Fl
21 May	22.7–23.2	H,Fl,Zoop
24 May	24.5–25.0	H,Fl,Wc,Zoop
		Treatments initiated after sampling
26 May	25.4–26.0	H,Fl,Wc,Zoop
29 May	22.0–22.6	H,Fl,Zoop,Algae
3 June	23.8–24.4	H,Fl,Wc,Zoop
7 June	24.0–24.5	H,Fl,Zoop,Algae
12 June	24.6–25.0	H,Fl,Zoop
16 June	26.5–27.0	H,Fl,Wc,Zoop,Algae
16 June–11 July		Changeover period
11–12 July		Tanks drained, Fish recovered
EXPERIMENT II: 25 OCTOBER–28 NOVEMBER 1984		
25–26 Oct		Tank filling, Zoop spike
31 Oct	17.0–17.3	H,Fl,Wc,Zoop
1 Nov		Treatments initiated
5 Nov	18.8–19.1	H,Fl,Wc,Zoop,Algae
10 Nov	15.8–16.5	H,Fl,Wc,Zoop
15 Nov	14.6–15.3	H,Fl,Zoop,Algae
20 Nov	7.2–8.3	H,Fl,Wc,Zoop
25 Nov	7.8–8.5	H,Fl,Zoop
28 Nov	7.5–7.9	H,Fl,Wc,Zoop,Algae
28 Nov–28 Dec		Changeover period
29 Dec		Tank draining, Fish recovered

H, Hydrolab; Fl, fluorescence, Secchi, and turbidity; Wc, phosphate, nitrate, and alkalinity; Zoop, zooplankton.

1978), modified in the May–June experiments for unequal replication (see tables 11.1 and 11.3). A posteriori tests of differences between treatment means and the control were done by Dunnett's test. Given the replication involved in these experiments (two or three per treatment), a deviation of $\pm 2 (MS_e)^{-1/2}$ could be detected as significantly different at $\alpha = 0.05$ with a probability of 0.75 (experimental power); smaller deviations ($< 0.5 (MS_e)^{-1/2}$ would only be accurately assessed as different at $\alpha = 0.05$ with a probability of 0.1 to 0.2. Multiple regression analysis also was done on the pooled results of the two experiments (36 tank-specific means for each variable measured).

Changeover periods of 25 to 30 days followed each of the experimental periods reported here, during which additional treatments were imposed on the tank phytoplankton communities (Threlkeld and Soballe, unpublished data). During this time daily records of any fish mortality were continued. Recovery of all remaining fish occurred at the end of these periods. Accidental fish mortality during the experiment could be apportioned to the following categories: (1) removed from the tanks as death occurred (dead but without significant decomposition or contribution to the soluble nutrient pools); (2) recovered at the end of the experiment as living or recently dead (from condition of carcass, a qualitative assessment of time since death could be made and the fish biomass categorized as primarily living or dead during the experimental period or changeover period); and (3) lost and unaccounted for (these fish were assumed to have died and decomposed in the

tanks throughout the course of the experiments).

There was minimal accidental fish mortality and lost biomass during the two experiments. After the May–June experiment, all *Dorosoma* and *Tilapia* were recovered alive on 11 July. *Menidia* density declined by an average of 12%. Average *Menidia* biomass during this experiment was thus 3% less than for *Dorosoma* or *Tilapia* because missing fish biomass is assumed to have decomposed continuously during the experimental period and changeover period. In the October–November experiment, there was no lost biomass of *Dorosoma* and *Tilapia*, even though all *Tilapia* died before 28 December when the tanks were drained. However, the integrity of the carcasses recovered at this time suggested that the *Tilapia* had died during the changeover period (28 Nov–28 Dec), probably as a result of decreasing water temperatures (Shelton, personal communication). *Menidia* losses averaged 3% in this experiment.

RESULTS

In tables 11.4 and 11.5, treatment means and mean square errors are shown along with probabilities derived from split-plot ANOVAS of each variable in the two experiments (see table 11.3 for design, degrees of freedom and method of calculation of F ratios). Figure 11.2 shows the data from these experiments after standard transformation of each treatment mean $[(\text{treatment mean}-\text{control mean})/(MS_e)^{-1/2}]$. This transformation permits direct comparison of

Table 11.3. *Summary of analysis of variance with repeated measures on treatment factors in May–June (MJ) and October–November (ON) experiments*

Source of variation	df (MJ)	df (ON)	F-ratio
BETWEEN TANKS			
A (treatments)	6	8	A/E (between)
Error (between treatments)	11	9	
WITHIN TANKS			
B (time)[a]	B-1	B-1	B/E (within)
AB*	6(B-1)	8(B-1)	AB/E (within)
Error (within treatments)	11(B-1)	9(B-1)	

[a]Degrees of freedom on time and treatment–time interactions vary according to frequency of sampling for particular parameters; one date immediately before and all dates after initiation of treatments are included (see table 11.2 for schedule).

Table 11.4. *Probabilities, $\sqrt{MS_e}$, and treatment means for May–June 1984 experiment*

Parameter	Split-plot ANOVA Prob.	$\sqrt{MS_e}$	Control	Menidia	Dorosoma	Tilapia	Filter	Nutrients	Nutrients & Filter
Fluorescence	<.0001	45.2	58.3	51.2	54.8	51.4	46.4	249.9	87.9
Blue-green filaments (x 10^4/ml)	.038	3.02	3.68	3.65	1.17	1.24	1.55	1.44	0.81
Algal particles >2 μm (x 10^5/ml)	<.0001	2.52	7.24	1.79	3.82	4.52	1.95	30.45	4.36
Fl/DCMU-enhanced fluorescence	.0014	0.07	0.46	0.53	0.50	0.47	0.56	0.41	0.46
Secchi depth (cm)	.0053	38	122	131	121	141	143	80	108
Turbidity (NTU)	.0007	1.1	2.1	2.1	2.1	1.5	1.5	3.9	2.8
Phosphate (mg/L)	<.0001	.033	.002	.022	.013	.002	.008	.220	.031
Nitrate (mg/L)	.213	.004	.008	.011	.008	.006	.007	.008	.008
Alkalinity (meq/L)	<.0001	11	104	100	110	108	101	132	138
Keratella quadrata (ind/L)	.0086	15.7	2.5	24.7	2.1	3.9	2.0	2.2	1.2
Brachionus (ind/L)	.95	28.7	1.1	35.2	8.3	1.6	5.1	1.3	15.0
Bosmina (ind/L)	.069	40.4	49.4	7.0	11.3	9.2	6.2	17.2	4.1
Ceriodaphnia (ind/L)	<.0001	17.4	105.1	5.6	27.2	9.1	25.0	159.2	14.3
Daphnia parvula (ind/L)	.294	8.0	9.6	5.6	8.3	5.9	11.0	7.4	5.0
Diaptomus copepodids (ind/L)	.055	4.5	3.1	2.8	5.8	5.9	1.7	1.9	1.7
Eurytemora copepodids (ind/L)	.99	9.6	10.3	9.7	11.3	11.6	9.7	10.1	9.9
Cyclopoid copepodids (ind/L)	<.0001	11.1	2.3	3.6	3.7	1.8	25.1	13.9	23.8
Nauplii (ind/L)	.041	105	135	173	68	113	164	152	209

Power of design (1-B) with $\alpha = 0.05$ to detect differences of $0.5\sqrt{MS_e}$, $1.0\sqrt{MS_e}$, $1.5\sqrt{MS_e}$, and $2\sqrt{MS_e}$ are 0.1, 0.25, 0.50, and 0.75, respectively. ind, individuals.

Table 11.5. Probabilities, $\sqrt{MS_e}$, and treatment means for October–November 1984 experiment

Parameter	Split-plot ANOVA Prob.	$\sqrt{MS_e}$	Control	Menidia	Dorosoma	Tilapia	Filter	Nutrients	Nutrients & Filter	Dead Fish	Dead Fish & Filter
Fluorescence	.009	66.0	29.8	42.7	40.8	44.7	25.4	124.7	81.8	141.8	75.5
Blue-green filaments (x 10^4/ml)	.015	3.40	0.24	4.98	5.79	2.1	3.80	3.73	10.70	1.19	8.92
Algal particles >2 μm (x 10^5/ml)	.002	8.12	3.9	5.3	6.6	6.1	3.2	16.1	10.0	23.8	13.4
Fl/DCMU-enhanced fluorescence	.0004	0.035	0.51	0.48	0.46	0.48	0.54	0.44	0.44	0.46	0.44
Secchi depth (cm)	<.0001	14	151	119	128	135	142	84	70	80	88
Turbidity (NTU)	.001	1.0	1.8	2.6	2.2	1.9	2.1	3.1	4.6	3.2	3.2
Phosphate (mg/L)	<.0001	.086	0	0	0	0	0	1.05	1.07	0	0
Nitrate (mg/L)	<.0001	.0008	.001	.001	.001	.001	0	.012	.004	.002	.001
Alkalinity (meq/L)	.0003	4	95	96	95	94	96	108	106	91	94
Keratella quadrata (ind/L)	.021	2.7	0.1	0.4	0.4	0.1	0.4	0	4.7	0.3	1.0
Brachionus (ind/L)	.66	1.4	0.5	0.1	0.9	0.2	0.5	0.1	0.4	0.1	0.9
Bosmina (ind/L)	.17	7.6	8.6	8.7	3.6	9.2	4.5	7.9	2.4	8.7	3.1
Ceriodaphnia (ind/L)	.017	2.5	3.0	0.2	0.4	0.9	0.4	3.6	0.4	3.2	0.5
Daphnia parvula (ind/L)	<.0001	4.2	6.0	0.5	0.7	1.4	0.4	22.9	0.4	21.7	0.7
Diaptomus copepodids (ind/L)	.06	2.2	0.7	0.1	1.2	2.8	1.1	1.7	1.0	3.2	0.6
Eurytemora copepodids (ind/L)	<.0001	9.1	29.7	10.5	6.4	43.7	12.5	94.9	7.0	93.9	7.7
Cyclopoid copepodids (ind/L)	.55	12.8	6.5	1.5	2.3	3.1	9.5	6.3	4.5	11.2	4.0
Nauplii (ind/L)	.14	101	56	32	33	59	74	127	95	131	123

ind, individuals.

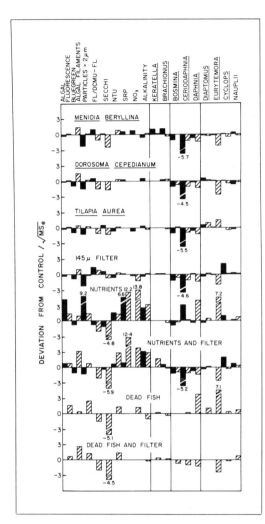

Figure 11.2. *Standard deviation from control mean values for all parameters measured for the two experiments, by treatment. Solid bars, deviations in May–June experiment; hatched bars, October–November experiment. The critical values for Dunnett's test of difference between a control and a treatment mean at p = 0.05, under the circumstances of two experiments, are ±2.8 and 3.09 (MS$_e$)$^{-1/2}$ (MJ and ON experiments, respectively). Vertical lines separate water quality and nutrient parameters from rotifer, cladoceran, and copepod parameters.*

each treatment mean with the control and is expressed in proportion to each parameter's measured variance $((MS_e)^{-1/2})$ under the experimental conditions.

The May–June experiment included three fish species, nutrient and filter additions (singly and in combination), and a control. With some exceptions (nitrate, *Brachionus*, *Daphnia*

parvula, and *Eurytemora* copepodids), most parameters showed significant experimental treatment effects. Water quality parameters (algal fluorescence, algal cell and blue-green filament counts, Secchi depth, turbidity, and nutrients) were influenced most by the nutrient additions, whereas zooplankton parameters were more variable (higher mean square errors) and only occasionally (e.g., *Ceriodaphnia*) showed clear responses to nutrient manipulations. In contrast, zooplankton population densities often showed negative responses to the fish and filter manipulations. The fish species had similar effects on phytoplankton and zooplankton, and these were also similar to the effects of the filter manipulations. Only in the case of *Menidia* were rotifers slightly enhanced. Drenner et al. (1982, 1984a, 1984b, 1986) showed that *Dorosoma* and *Tilapia* can filter and suppress rotifer populations. Fish and filter treatments differed somewhat in their effect on *Cyclops*. With respect to phytoplankton and water quality parameters, fish and filter manipulations were not significantly different from controls (at $p < 0.05$; see fig. 11.2), in direct contrast to the usual observation that when fish suppress zooplankton, phytoplankton are enhanced.

The results of the October–November experiment were similar (table 11.5, fig. 11.2), showing a general pattern of independence of phytoplankton and water quality parameters from the fish and filter manipulations, in spite of strong impacts of these treatments on zooplankton. Conversely, phytoplankton and water quality parameters were greatly affected by nutrient and dead fish treatments; zooplankton were either enhanced or not affected.

Table 11.6 shows the partial r^2 values which resulted from multiple regression of each response variable (data for both experiments pooled in this analysis) on live fish, dead fish, nutrient and filter effects. Fish and filter effects were similar in magnitude and direction (suppression of zooplankton), and distinct from nutrient and dead fish effects that typically enhanced phytoplankton, zooplankton, or nutrient parameters.

DISCUSSION

The results of these two tank experiments are in sharp contrast to many bag, pond, and lake

Table 11.6. *Results of multiple regression analysis of all response variables on fish, filter, nutrient and dead fish factors in the 2 experiments (36 tanks—specific means; df = 30)*

Parameter	Explained variation (%)	Live fish	Filter	Nutrients	Dead fish	Experiment
Fluorescence	72		−.35	+.58	+.38	−.16
Algal particles >2 μm	77	−.11	−.46	+.50	+.55	
Blue-green algal filaments	53				+.19	+.28
Fl/DCMU-fluorescence ratio	56			−.48	−.23	
Secchi depth	77			−.67	−.55	
Turbidity	69			+.58	+.27	
Phosphate	75			+.64		+.23
Nitrate	60			+.24		−.44
Alkalinity	81			+.66		−.60
Keratella quadrata	21					−.11
Brachionus	26					−.17
Bosmina	41	−.28	−.22		+.07	+.12
Ceriodaphnia	62	−.37	−.33			−.40
Daphnia parvula	48	−.22	−.37		+.11	
Diaptomus copepodids	46					−.32
Eurytemora copepodids	55		−.37		+.13	+.19
Cyclopoid copepodids	43		+.13			−.13
Nauplii	55			+.11	+.14	−.47

Partial r^2 values (where $p < 0.05$) are given and preceded by sign of regression coefficient.

studies of planktivorous fish impacts, and superficially also to recent studies by Drenner et al. (1986) conducted in the same experimental tanks. The major differences between this study and that of Drenner et al. (1986) are as follows: (1) the different fish species (or dead fish) were stocked at equal biomasses in the current study [in Drenner et al. (1986), *Dorosoma* biomass exceeded *Menidia* biomass by a factor of 5–20]; (2) the biomass of fish added in the present experiments was low and roughly comparable to the treatment of *Menidia* without *Dorosoma* in the experiments of Drenner et al. (1986); (3) fish mortality in the current study was minimal, and there was little fish decomposition, and (4) the cross-classified design and threefold replication of treatments in Drenner et al. (1986) were sacrificed for additional treatments intended to directly distinguish between nutrient-mediated and trophic-cascade effects of planktivorous fish. It is noteworthy that Drenner et al. (1986) found consistent enhancement of phytoplankton by *Dorosoma* but not *Menidia*; fish mortality and biomass (*Dorosoma* > *Menidia*) were both important factors in the enhancement of phytoplankton by *Dorosoma cepedianum*. But in the present study, where fish mortality and decomposition were not ma-

jor problems and the biomasses of the three fish were equal and low, there was no significant enhancement of phytoplankton or differences among fish treatments. However, fish densities were high enough to result in significant changes in the zooplankton community, suggesting a less than perfect relationship between zooplankton suppression and phytoplankton enhancement.

Although the possibility of direct predatory impact of planktivorous fish on zooplankton and large rigid phytoplankton is easily appreciated, the role of fish in nutrient-mediated effects on plankton is less clear. In this context, it may be helpful to reassess how some major features of experiments to date influence how we interpret phytoplankton responses to planktivorous fish.

Fish as Nutrients

Experiments in which fish are added to bags, tanks, or ponds usually involve some fish mortality and decomposition. The role of the fish introductions as nutrient additions has not been considered in the analysis or interpretation of these experiments, even though in some cases that information has been available

(Hall et al. 1970; Durbin et al. 1979; Lynch 1979; Drenner 1984a). Only in the recent tank experiments of Drenner et al. (1986) have multiple regression analyses been performed to attempt to understand the relative importance of this nutrient mediated effect. In Durbin et al. (1979) and the present study, dead fish were for the first time an intentional part of the experiment, even though phytoplankton enhancement by dead fish was confounded with nutrient regeneration by live fish in Durbin et al. (1979). It is problematic that the experimental control for a planktivorous fish is an absence of fish, even though it is usually only the predatory impacts of the fish that are being addressed with the resulting data. All other effects of planktivorous fish (from turbulence associated with swimming to nutrient regeneration to the production of semiochemicals) are confounded in these designs with the trophic effects of primary interest.

Similarly, in experiments in which fish are removed from natural ponds or lakes by poisoning, the effects on phytoplankton of loss of nutrients associated with fish biomass are unknown. The fate of nutrients in the fish carcasses is unknown, with basin morphometry, mixing regime, and external nutrient loading all probably being of major importance in determining the effects of the dead fish on future nutrient availability. The experiments reported here demonstrate clear short-term responses to nutrient and dead fish additions, whereas most post-rotenone fish pond or lake experiments begin the following spring after rotenone effects have dissipated. The time course of nutrient availability from decomposing fish carcasses to the phytoplankton appears highly variable (Anderson 1970; Henrikson et al. 1980) although clearly relevant to interpreting fish removal or mortality effects.

Choosing the Right Fish Density

Most experiments examining planktivore effects use one or, in a few cases, two fish densities in addition to the fishless control. In most cases, the high densities chosen ensure dramatic food web effects. The statistical effect of this design feature is that any effects that would have occurred at low or intermediate fish densities are pooled with effects resulting from the highest experimental fish density. This is especially true where only two extreme densities are used (fishless and a high fish density). Thus, even though low densities of fish may be sufficient to eliminate large cladoceran zooplankton, it may require much higher densities of fish (and associated nutrient regeneration or fish mortality or weight loss) to stimulate measurable enhancement of phytoplankton. Multiple density experiments provide another approach for examination of planktivorous fish effects on plankton communities. In the few cases where three or more densities of fish have been included in experimental designs (Hillbricht-Ilkowska 1964; Januszko 1978; Neill 1975), no attempts have been made to examine the data on plankton responses for evidence of nonlinearities in food web responses (e.g., if phytoplankton enhancement is independent of zooplankton suppression at low fish densities [Threlkeld and Choinski, in press]).

Recent empirical studies (e.g., Hanson and Leggett 1982; Jones and Hoyer 1982; but see Sprules and Knoechel 1984) have shown a positive relationship between fish biomass or production and zooplankton, benthos and phytoplankton biomass and nutrient loading. The extent to which planktivore changes merely reflect changes in system trophy, or in themselves are responsible for the observed changes, is not yet clear. In light of these strong positive relationships, the effect of adding or removing fish biomass from a plankton assemblage that has developed at a certain level of productivity may be especially important.

Comparing Different Planktivores

The impacts of different kinds of planktivores are rarely compared under similar conditions. However, it is conceivable that species-specific differences in predatory behavior or feeding rates will provide additional information on how food web compartments are linked. Drenner et al. (1986) stocked Dorosoma at 5 to 20 times the biomass of Menidia because of previous studies (Drenner and McComas 1980) that showed Menidia was more efficient, on a biomass-specific basis, at consuming zooplankton. Drenner et al. (1986) also showed that phytoplankton were generally enhanced by Dorosoma, but not Menidia, in those experiments. Their experiment, through a comparative approach, thus provided an intriguing confirmation of the idea that the enhancement of phytoplankton is proportional to fish biomass, and not zooplankton suppression.

Artificial Planktivores

Few investigators have attempted to experimentally mimic trophic or other direct effects of fish, although such attempts have often been very instructive. Neill (1975) maintained nutrient regeneration effects associated with fish treatments by adding equivalent amounts of fish-influenced water to the fishless controls. And Lamarra (1975) attempted to evaluate the direct sediment disturbance activities of carp on plankton by stirring sediments in fishless containers with a paddle. Application of excessive nutrient loads to experimental vessels might provide another means of eliminating confounding effects of nutrient regeneration or accidental fish mortality (see Lehman 1980 for a parallel development regarding zooplankton nutrient regeneration effects). Other manipulations, involving removal of zooplankton with plankton netting (Bürgi et al. 1979) or allowing or eliminating access by fish to the benthic substrate are additional means for distinguishing fish trophic effects associated with particular activities or habitats from the Gestalt of planktivory. All too often, the diversity of available experimental approaches are not brought to bear on a given community (or set of responses) to distinguish even the most basic of trophic and nutrient-mediated effects.

CONCLUSIONS

The present experiments and consideration of experimental design and interpretation suggest some reason for restraint in the application of the trophic-cascade concept to aquatic communities. Although experimental manipulations of planktivorous fish often demonstrate inverse relationships between fish and zooplankton biomass and zooplankton and phytoplankton biomass, they do not lend exclusive support to the hypothesis of a trophic cascade. Changes in nutrient loading associated with manipulations of planktivorous fish (nutrient regeneration, accidental fish mortality) may be fundamental to phytoplankton responses. It is not yet clear to what extent our perception of lacustrine food web interactions is biased by

experimental results in which these critical features of experimental design are left unreconciled. Certainly, the issues of nutrient release by fish (living or dead) and the strength of interaction between phytoplankton and zooplankton are not resolved, although many experimental approaches are available and not yet fully exploited.

ACKNOWLEDGMENTS

I gratefully acknowledge the helpful comments and interaction of Dr. Ray Drenner, whose research provided much of the foundation of the experiments reported here. Drs. Riccardo deBernardi, Nan Duncan, and Gary Sprules contributed helpful comments about this work during its gestation. Elizabeth Choinski and Pat Cantrell provided assistance in the tank experiments. The tank experiments were supported by NSF grants BSR 8206850, BSR 8206865, and BSR 8315127.

REFERENCES

Anderson, R. S. 1970. Effects of rotenone on zooplankton communities and a study of their recovery patterns in two mountain lakes in Alberta. *J. Fish. Res. Board Can.* 27 : 1335–56.

Andersson, G., H. Berggren, G. Cronberg, and C. Gelin. 1978. Effects of planktivorous and benthivorous fish on organisms and water chemistry in eutrophic lakes. *Hydrobiologia* 59(1) : 9–15.

Ball, R. C., and D. W. Hayne. 1952. Effects of the removal of the fish population on the fish-food organisms of a lake. *Ecology* 33 : 41–48.

Bartell, S. M. 1981. Potential impact of size-selective planktivory on phosphorus release by zooplankton. *Hydrobiologia* 80 : 139–45.

Bartell, S. M., and J. F. Kitchell. 1978. Seasonal impact of planktivory on phosphorous release by Lake Wingra zooplankton. *Verh. Internat. Verein. Limnol.* 20 : 466–74.

Bray, R. N., A. C. Miller, and G. G. Geesey. 1981. The fish connection: A trophic link between planktonic and rocky reef communities? *Science* 214 : 204–205.

Brooks, J. L., and S. I. Dodson. 1965. Predation, body size, and composition of plankton. *Science* 150 : 28–35.

Bürgi, H. R., H. Bührer, J. Bloesch, and E. Szabo. 1979. Der Einfluss experimentell variierter Zooplanktondichte auf die' Produktion und Sedimentation im hocheutrophen See. [The influence of experimentally varied zooplankton density on production and sedimentation in a highly eutrophic lake.] *Schweiz. Z. Hydrol.* 41 : 38–63.

Carpenter, S. R., and J. F. Kitchell. 1984. Plankton community structure and limnetic primary production. *Amer. Naturalist* 124 : 159–72.

Crowder, L. B., and W. E. Cooper. 1982. Habitat structural complexity and the interaction between bluegills and their prey. *Ecology* 63 : 1802–13.

Dodson, S. I., C. Edwards, F. Wiman, and J. C. Normandin. 1976. Zooplankton: Specific distribution and food abundance. *Limnol. Oceanogr.* 21(2) : 309–13.

Drenner, R. W., F. deNoyelles, Jr., and D. Kettle. 1982. Selective impact of filter-feeding gizzard shad on zooplankton community structure. *Limnol. Oceanogr.* 27(5) : 965–68.

Drenner, R. W., and S. R. McComas. 1980. The roles of zooplankter escape ability and fish size selectivity in the selective feeding and impact of planktivorous fish. In W. C. Kerfoot (ed.), *Evolution and ecology of zooplankton communities. Amer. Soc. Limnol. Oceanogr. Spec. Symp.* 3 : 587–593.

Drenner, R. W., J. R. Mummert, F. deNoyelles, Jr., and D. Kettle. 1984a. Selective particle ingestion by a filter-feeding fish and its impact on phytoplankton community structure. *Limnol. Oceanogr.* 29(5) : 941–48.

Drenner, R. W., S. B. Taylor, X. Lazzaro, and D. Kettle. 1984b. Particle-grazing and plankton community impact of an omnivorous cichlid. *Trans. Amer. Fish. Soc.* 113 : 397–402.

Drenner, R. W., S. T. Threlkeld, and M. D. McCracken. 1986. Experimental analysis of the direct and indirect effects of filter feeding clupeid on plankton community structure. *Can. J. Fisheries Aquat. Sci.* 43 : 1935–45.

Durbin, A. G., S. W. Nixon, and C. A. Oviatt. 1979. Effects of the spawning migration of the alewife, *Alosa pseudoharengus*, on freshwater ecosystems. *Ecology* 60(1) : 8–17.

Gill, J. L. 1978. *Design and analysis of experiments in the animal and medical sciences.* Ames, Iowa: Iowa State University Press.

Grygierek, E. 1973. The influence of phytophagous fish on pond zooplankton. *Aquaculture* 2 : 197–208.

Hall, D. J., W. E. Cooper, and E. E. Werner. 1970. An experimental approach to the production dynamics and structure of freshwater animal communities. *Limnol. Oceanogr.* 15(6) : 839–928.

Hall, D. J., S. T. Threlkeld, C. W. Burns, and P. H. Crowley. 1976. The size-efficiency hypothesis and the size structure of zooplankton communities. *Annu. Rev. Ecol. Syst.* 7 : 177–208.

Hanson, J. M., and W. C. Leggett. 1982. Empirical prediction of fish biomass and yield. *Can. J. Fisheries Aquat. Sci.* 39 : 257–63.

Hayne, D. W., and R. C. Ball. 1956. Benthic productivity as influenced by fish predation. *Limnol.*

Oceanogr. 1 : 162–75.

Henrikson, L., H. G. Nyman, H. G. Oscarson, and J. A. E. Stenson. 1980. Trophic changes, without changes in the external nutrient loading. *Hydrobiologia* 68(3) : 257–63.

Hillbricht-Ilkowska, A. 1964. The influence of the fish population on the biocenosis of a pond, using Rotifera fauna as an illustration. *Ekologia Polska (A)* 12(28) : 453–503.

Hrbáček, J. 1962. Species composition and the amount of zooplankton in relation to the fish stock. *Rozpr. Cesk. Akad. Ved, Rada Mat. Prir. Ved* 72(10) : 1–116.

Hrbáček, J., and M. Novotná-Dvořáková. 1965. Plankton of four backwaters related to their size and fish stock. *Rozpr. Cesk. Akad. Ved, Rada Mat. Prir. ved* 75(13) : 1–65.

Hurlbert, S. H., and M. S. Mulla. 1981. Impacts of mosquitofish (*Gambusia affinis*) predation on plankton communities. *Hydrobiologia* 83 : 125–51.

Hurlbert, S. H., J. Zelder, and D. Fairbanks. 1972. Ecosystem alteration by mosquitofish (*Gambusia affinis*) predation. *Science* 175 : 639–41.

Januszko, M. 1978. The influence of silver carp (*Hypophthalmichthys molitrix* Val.) on eutrophication of the environment of carp ponds. Part III. Phytoplankton. *Rocz. Nauk Roln., Ser. H* 99(2) : 55–79.

Jones, J. R., and M. V. Hoyer. 1982. Sportfish harvest predicted by summer chlorophyll-*a* concentration in midwestern lakes and reservoirs. *Trans. Amer. Fish. Soc.* 111 : 176–79.

Kitchell, J. R., R. V. O'Neill, D. Webb, G. W. Gallepp, S. M. Bartell, J. F. Koonce, and B. S. Ausmus. 1979. Consumer regulation of nutrient cycling. *Bioscience* 29 : 28–34.

Lamarra, V. A., Jr. 1975 Digestive activities of carp as a major contributor to the nutrient loading of lakes. *Verh. Internat. Verein. Limnol.* 19 : 2461–68.

Leah, R. T., B. Moss, and D. E. Forrest. 1980. The role of predation in causing major changes in the limnology of a hyper-eutrophic lake. *Int. Revue ges. Hydrobiol.* 65(2) : 223–47.

Lehman, J. T. 1980. Release and cycling of nutrients between planktonic algae and herbivores. *Limnol. Oceanogr.* 25 : 620–32.

———. 1984. Grazing, nutrient release, and their impacts on the structure of phytoplankton communities. In D. G. Meyers and J. R. Strickler (eds.), *Trophic interactions within aquatic ecosystems.* AAAS Selected Symp. 85 : 49–72. Boulder, Colo. Westview Press.

Lellak, J. 1966. Influence of the removal of the fish population on the bottom animals of the five Elbe backwaters. *Hydrobiol. Studies* 1 : 323–80.

Losos, B., and J. Heteša. 1973. The effect of mineral fertilization and of carp fry on the composition and dynamics of plankton. *Hydrobiol. Studies* 3 : 173–217.

Lynch, M. 1979. Predation, competition, and zooplankton community structure: an experimental study. *Limnol. Oceanogr.* 24 : 253–72.

Lynch, M., and J. Shapiro. 1981. Predation, enrichment, and phytoplankton community structure. *Limnol. Oceanogr.* 26 : 86–102.

Meyer, J. L., E. T. Schultz, and G. S. Helfman. 1983. Fish schools: An asset to corals. *Science* 220 : 1047–49.

Nakashima, B. S., and W. C. Leggett. 1980. The role of fishes in the regulation of phosphorus availability in lakes. *Can. J. Fisheries Aquat. Sci.* 37 : 1540–49.

Neill, W. E. 1975. Experimental studies of microcrustacean competition, community composition and efficiency of resource utilization. *Ecology* 56(4) : 809–26.

Novotná, M., and V. Kořínek. 1966. Effect of the fishstock on the quantity and species composition of the plankton of two backwaters. *Hydrobiol. Studies* 1 : 297–322.

Post, J. R., and D. Cucin. 1984. Changes in the benthic community of a small Precambrian lake following the introduction of yellow perch, *Perca flavescens. Can. J. Fisheries Aquat. Sci.* 41 : 1496–1501.

Reynolds, C. S. 1984. The ecology of freshwater phytoplankton. Cambridge: Cambridge University Press.

Robison, B. H., and T. G. Bailey. 1981. Sinking rates and dissolution of midwater fish fecal matter. *Mar. Biol.* 65 : 135–42.

Shapiro, J., and D. I. Wright. 1984. Lake restoration by biomanipulation: Round Lake, Minnesota, the first two years. *Freshwater Biol.* 14 : 371–83.

Sprules, W. G. and R. Knoechel. 1984. Lake ecosystem dynamics based on functional representations of trophic components. In D. G. Meyers and J. R. Strickler (eds.), *Trophic interactions within aquatic ecosystems.* AAAS Selected Symp. 85 : 383–403. Boulder, Colo.: Westview Press.

Threlkeld, S. T., and E. M. Choinski, in press. Rotifers, cladocerans and planktivorous fish: what are the major interactions? *Hydrobiologia.*

Vincent, W. F. 1981. Photosynthetic capacity measured by DCMU-induced chlorophyll fluorescence in an oligotrophic lake. *Freshwater Biol.* 11 : 61–78.

Winer, B. J. 1971. Statistical principles in experimental design. 2nd ed. New York: McGraw-Hill.

Wright, D. I., and J. Shapiro. 1984. Nutrient reduction by biomanipulation: An unexpected phenomenon and its possible cause. *Verh. Internat. Verein. Limnol.* 22 : 518–24.

12. Salamander Predation, Prey Facilitation, and Seasonal Succession in Microcrustacean Communities

Peter Jay Morin

I experimentally manipulated salamander densities in artificial ponds to measure the effects of density-dependent predation by *Notophthalmus viridescens* and *Ambystoma tigrinum* on microcrustacean community development. Separate multivariate analyses of early and late phases of seasonal succession demonstrated strong density-, time-, and species-dependent effects of salamanders on overall patterns of species composition.

Salamanders slowed the seasonal declines in microcrustacean diversity and species richness that occurred in ponds without vertebrate predators. Salamanders also significantly altered patterns of seasonal succession and clearly facilitated some prey species. *Ambystoma* enhanced the abundance of *Chydorus sphaericus* and ostracods, and inhibited large-bodied *Daphnia laevis*. Intermediate densities of *Notophthalmus* facilitated recruitment of *Mesocyclops edax*. *Bosmina longirostris* persisted in ponds containing *Notophthalmus* throughout the experiment but declined after *Ambystoma* metamorphosed in other ponds. Species-specific differences in patterns of salamander reproduction and larval development apparently contributed to patterns of seasonal succession, although seasonal succession was not strictly driven by vertebrate predation. Resulting community patterns indicate that the indirect facilitation of prey arose from predator-mediated competition rather than from predation on invertebrate predators.

Studies of interactions between salamanders and microcrustaceans have contributed much to the development of aquatic community theory (Dodson 1970, 1974; Sprules 1972; Giguere 1979; Williams 1980; Zaret 1980 and references therein). Many of the community-level consequences of these interactions have been inferred from correlative studies and lack experimental confirmation. Other details of theoretical import, such as relations between predation intensity gradients and microcrustacean community structure (Zaret 1980), remain virtually unexplored by community-level experiments (for exceptions see Neill 1975, 1981; Kerfoot and DeMott 1980).

Although vertebrate predation is the cornerstone of recent aquatic community theory, it is only one of a suite of potentially interacting factors that may structure zooplankton communities. Competition (Allan 1973, Neill 1975, Lynch 1978, DeMott 1983), invertebrate predation (Dodson 1974), indirect interactions (Dodson 1970, Giguere 1979, Vandermeer 1980), variation in life history strategies (Allan

and Goulden 1980), and stochastic events may all contribute to variation in zooplankton communities assembled from a similar species pool. One important recurring theme, the indirect facilitation of certain small microcrustacean species by vertebrate predators, has alternately been explained as a consequence of vertebrate predation on either large invertebrate predators (Dodson 1974) or large competitors (Brooks and Dodson 1965) that would otherwise limit small species (see Kerfoot and DeMott 1984). This paper describes an experimental test of alternate hypotheses invoking indirect effects, and explores other effects of vertebrate predators on aquatic community organization.

Compelling evidence indicates that fish and salamanders can alter zooplankton species composition (e.g., Hrbáček et al. 1961; Hall et al. 1970; Sprules 1972; Giguere 1979; Zaret 1980), yet previous studies yield little insight about how spatial or temporal differences in the species or density of vertebrate predators influence community patterns. For instance, comparative studies suggest important seasonal variation in the effects of predators on zooplankton population dynamics (Kerfoot 1975). There are few experimental studies of such effects (but see Kerfoot and Peterson 1979). Here I address this problem by describing patterns of microcrustacean community development in comparable experimental communities. Experimental manipulations involved different abundances of two vertebrate predators. Such experiments can determine whether communities respond differently to predators with different seasonal patterns of reproduction, development, metamorphosis, or mortality. The experiment's temporal scale encompassed recruitment by planktivorous larvae of one predator species, and metamorphosis by another, over a period comparable to the annual cycle of community development in natural temporary ponds.

Evaluations of synthetic community theories require measurements of the impacts of predation, competition, and other factors (e.g., Zaret 1980). The contribution of vertebrate predation to overall community structure, relative to other factors, remains unmeasured in most systems. As a starting point, I suggest that predation's contribution to community organization should be weighed experimentally against the aggregate contribution of other potentially interacting factors. This can be done by estimating the fraction of total intercommunity variation in species composition that is directly attributable to experimental manipulations of predator abundance. This measurement was another major goal of the experiments outlined below.

MATERIALS AND METHODS
Ponds, Predators, and Prey

I conducted experiments in artificial ponds that were designed to mimic key features of natural temporary pond food webs (see Morin 1983a, 1983b for details of pond construction). The natural ponds are small rain-filled depressions scattered through *Pinus palustris—Aristida stricta* savannas of the Sandhills Wildlife Management Area in south-central North Carolina. These ponds may fill and dry several times during 1 year but occasionally remain full for longer intervals, depending on rainfall. Complete episodes of community development typically occur over a few months. In late spring, ponds

contain an average of five or six microcrustacean species (means ± SE for ponds sampled in 1979 and 1980 were, respectively, 5.26 ± 3.15 species per pond, N = 15 ponds, and 5.92 ± 2.62 species per pond, N = 13 ponds). The combined species pool for all sampled ponds, 22 species, is much larger than the average species richness within ponds. This assemblage provided a diverse source of potentially interacting species for experimental studies. Although many of these species are planktonic, others are littoral or benthic.

Salamanders are the only vertebrate planktivores in the natural ponds. The frequent and unpredictable drying of ponds apparently excludes fish. The numerically dominant salamanders are adult and larval broken-striped newts, *Notophthalmus viridescens dorsalis*, and larval tiger salamanders, *Ambystoma tigrinum*. Both species are facultative size-selective planktivores (Dodson 1970, Dodson and Dodson 1971, Mellors 1975, Attar and Maly 1980, Brophy 1980).

Notophthalmus and *Ambystoma* differ importantly in life history traits that influence seasonal abundance, numerical responses, and developmental responses (*sensu* Murdoch 1971). Adult *Notophthalmus* reside in ponds throughout the year and endure periods of drought by burrowing into moist debris in pond basins. Consequently, adult newts are present throughout the entire period of pond community development. Adult growth, larval growth, and the per capita production of planktivorous larvae are density-dependent (Morin 1983b, Morin et al. 1983). *Notophthalmus* can respond to low conspecific density or high prey abundance with increased growth and extended periods of oviposition. Planktivorous larvae hatch in May and eventually metamorphose into terrestrial efts in late summer or early fall. Efts return to ponds after several years of terrestrial growth and metamorphose into aquatic adults in late winter and early spring. Adult densities appear to fluctuate little during annual episodes of community development. Larval densities increase rapidly when eggs hatch in May and June and then gradually decline with mortality and metamorphosis in late summer and autumn.

Ambystoma abundance is strongly seasonal. Larvae hatch in March from eggs deposited by terrestrial adults during winter rains. *Ambystoma* begin feeding up to 2 months before the first *Notophthalmus* larvae hatch. In some ponds, *Ambystoma* coexist with *Notophthalmus* until the former metamorphose in June and July. *Ambystoma* larvae can respond to variation in prey abundance only by density-dependent growth (Morin 1983b). Since they do not reproduce, they lack the potentially rapid numerical response of *Notophthalmus*.

Species-specific body size differences also may influence the per capita impact of salamander predation. *Ambystoma* metamorphosed at mean masses of 6,053 mg and 8,437 mg in two of my experimental treatments, and can attain much larger sizes at low densities. Much smaller adult newts rarely exceeded 3,000 mg. Larval newts metamorphosed at mean masses of 271 to 538 mg, depending on salamander densities (see Morin 1983b for further details on interactions among salamanders).

Experimental Protocol

I manipulated salamander densities in 22 initially identical artificial pond communities. Artificial ponds offered two important advantages over natural ponds. First, experimental communities could be constructed with initially similar species compositions and precisely known numbers of predators, eliminating historical differences as a source of variation. Such historical variation may be extremely important in natural communities, but it can also obscure patterns caused by biological interactions (Colwell and Winkler 1984). Second, I could stock ponds with nearly the entire species pool of microcrustaceans found in natural ponds, to assay whether the small subset of coexisting species in natural ponds was a simple consequence of limited colonization or active postcolonization exclusion.

The artificial ponds (hereafter called tanks) were cylindrical steel cattle-watering tanks (1.52 m in diameter, 0.61 m in height) with interiors painted with inert white epoxy paint. The tanks were within the size range of natural temporary ponds in the Sandhills and were arranged in a rectangular array in the Duke University Botany Experimental Plot. A screen lid covered each tank and excluded planktivorous insects.

I created tank communities during the first 2 weeks of March 1980 by adding 1,000 liters of water (from the Durham, NC, municipal water supply), 550 g of grassy plant litter, 50 g of

trout chow, and 50 rooted stems of *Myriophyllum pinnatum* to each tank (see Morin 1983a for further details). The date of tank establishment corresponded approximately to the seasonal creation of natural ponds by spring rains. These replicated starting conditions generated detritus-based food webs similar to natural temporary pond communities (see Wiggins et al. 1980). Trout chow fueled a rapid bloom of bacteria and phytoplankton, which supported the subsequent growth of microcrustacean populations.

I inoculated the tanks with organisms collected by multiple horizontal tows of an 88-μm mesh plankton net across eight natural ponds. The inoculum contained most of the microcrustacean species found in natural temporary ponds. I pooled collected organisms in a 20-L carboy of pond water, and added a 650-ml inoculum from the well-mixed carboy to each tank on 14, 20, and 22 March 1980. By 11 April 1980, the first abundant microcrustaceans appeared (dense populations of *Simocephalus serrulatus* and *B. longirostris*).

I randomly assigned the following numbers of replicates (N) of each of 6 manipulations to the tanks: no salamanders (N = 4); 2, 4, or 8 adult *Notophthalmus* (N = 4), 4 *Ambystoma* larvae (N = 3), and 4 *Ambystoma* larvae plus 4 *Notophthalmus* (N = 3). These manipulations generated (1) a *Notophthalmus* density gradient to test community responses to a graded perturbation, and (2) communities containing equal densities of two predator species in syntopic and allotopic settings, to examine species-specific differences in the impact of the predators. Salamander densities fell within the range observed in natural ponds (Morin 1983a, 1983b).

I added *Notophthalmus* (sex ratio of 1 : 1) to designated tanks on 22 March 1980, and added *Ambystoma* larvae on 27 March 1980, when the latter had grown large enough to escape predation by *Notophthalmus*. Other prey added included 1,200 hatchling tadpoles of six anuran species, corixids (*Sigara* and *Hespercorixa* spp.), and the amphipod *Synurella chamberlaini*. Corixids and amphipods never became abundant in the tanks. Consequently, I consider macroinvertebrates part of the unmanipulated background variation in community composition and discount any effects of their predation on observed differences in microcrustacean abundance.

I sampled microcrustaceans at regular intervals, beginning on 11 April 1980, when the first species became abundant. Sampling continued at approximately 5-day intervals from 11 April through 23 September. Each tank sample consisted of five pooled subsamples, which together corresponded to 6.4 L water. Subsampling involved dropping an aluminum tube (6.35 cm in diameter, 60 cm long) vertically through the water column, stoppering the tube's bottom, and pouring the tube's contents into an 88-μm mesh Nitex filter cone attached to a 20-ml screw-topped glass vial. Pooled samples consisted of four subsamples from equidistant points on each tank's circumference and a fifth subsample from each tank's center. This protocol yielded directly comparable samples. Such samples undoubtedly biased estimates of total population size within tanks. The point is that relative differences among tanks were assayed with the same sampling method, and it is such relative differences that I focus on here.

I washed sampled organisms from the Nitex filter into an attached 20-ml vial, and preserved them in formalin and sucrose. The data analyzed here are from complete counts of samples from each tank on the following representative dates: 11 April, 26 April, 26 May, 20 June, 20 July, and 24 August, 1980. I ignored nauplii or copepodites of uncertain identity and lumped all Ostracoda into a single operational taxon.

STATISTICAL ANALYSIS

The rigorous analysis of temporal changes in multispecies assemblages presents special statistical problems. Measures of the same community repeated over time (e.g., time series data) are not statistically independent. Such data violate assumptions of independence inherent in most univariate parametric statistical techniques. Multiple univariate analyses repeated for each of a large number of species also pose problems, primarily because of the risk of type 1 error with multiple tests. Problems involve (1) appropriate analytical methods for repeated nonindependent measures made in the same communities over time, (2) the need for concise description and analysis of often complex multispecies patterns, typically requiring data reduction from a large number of correlated species abundances to a smaller number of derived uncorrelated variables, and (3) identification of objective criteria for the selection of illustrative

species for further detailed analysis. These problems have no simple simultaneous solution, but I have attempted to minimize sources of bias by using the conservative analysis outlined below.

I used vectors of microcrustacean species diversity (H′) and species richness to describe broad temporal patterns of community structure. For example, elements of the vector $H'_j = (H'_{1,j} \ldots , H'_{1,j})$ represent the values of H' in tank j on sequential sampling dates denoted by the subscript i. A MANOVA of such vectors tested whether predator treatments significantly affected microcrustacean diversity and richness, using an analysis appropriate for potentially correlated response variables obtained by repeated measurement of the same communities. See Morrison (1976) for further justification of this approach.

Principal component analyses (PCA) concisely described multivariate patterns of community structure by representing variation in large numbers of correlated species with a smaller number of community scores. To minimize the risk of type 1 errors associated with multiple tests, I restricted my analysis to two sets of samples collected early (April 26) and late (July 20) in community development. Vectors of the angularly transformed relative abundances of each species in each tank represented species composition on a particular sampling date. Relative abundance (defined as the proportion of the total number of all individuals in a sample represented by each species) focused the analysis on differences in dominance standardized against differences in actual abundance. The separation of dominance from total abundance is desirable when comparing communities that may differ in carrying capacity or temporal dynamics.

A PCA of the variance–covariance matrix derived from these vectors extracted principal components, which were independent linear combinations of the original relative abundances. I focused only on the three components that together described the maximal amount of variation in the original data among tanks on a given sampling date. PCA scores for each tank summarized species composition by the position of each tank sample in a space defined by the first three principal components. The components and sample scores were interpreted by evaluating correlations between PCA scores and the angularly transformed relative abundances of each species. This approach has successfully described complex patterns in zooplankton communities (Cassie 1963; Sprules 1977) and other multispecies assemblages (Pielou 1984).

This protocol offered significant analytical advantages. A graphical summary permitted visualization of complex multispecies responses to experimental treatments in three or fewer dimensions. Three PCA scores also replaced the much larger number of original variables in statistical analyses of effects of experimental treatments on community structure. Such substitution is desirable when communities contain large numbers of species because the power of multivariate tests declines with the number of variables (species) analyzed. PCA also measured the fraction of total variance in relative community composition described by community scores along each component. To the extent that experimental treatments influenced PCA scores, the fraction of total intercommunity variation described by the affected component scores measured the maximum contribution of the experimental manipulations to community patterns.

A MANOVA of three PCA scores tested whether predator treatments affected microcrustacean species composition. ANOVAs of tank scores for each principal component identified the components and associated species responding to predators. Differences among PCA score treatment means measured the magnitude of species composition differences caused by predators. The null hypothesis in all tests was that predators did not affect PCA scores (and species composition). The analysis also provided an objective criterion for selecting individual species for analyses of absolute abundance (defined here as the number of individuals per 6.4 L sample). I analyzed only the absolute abundances of those species that were affected by predators in the preceding analysis of relative abundances.

Analysis of absolute abundance vectors showed whether changes in the relative abundance of a given species simply reflected changes in its own absolute abundance or also reflected changes in the abundance of other microcrustaceans. Such latter "passive" changes in relative abundance can arise because the relative abundances of all species in a sample must sum to 1. The procedure was analogous to that previously outlined for the analysis of temporal patterns of diversity and species richness. Vectors of the absolute abundance of a species on successive sampling dates represented seasonal trends in abundance. Separate MANOVAs for each species tested whether predators affected absolute abundance.

RESULTS
Community-wide Patterns
On April 11, tanks had twice the microcrustacean species richness of natural ponds (table

Table 12.1. *Mean values of microcrustacean species richness in six salamander predation treatments on six successive sampling dates*

Treatment	Sampling date:					
	4/11	4/26	5/26	6/20	7/20	8/24
Control	11.3	9.8	8.0	7.8	8.3	6.3
	(7–13)	(7–13)	(8–8)	(6–9)	(8–9)	(5–8)
2 newts	10.5	10.8	7.8	9.8	6.5	6.3
	(9–11)	(9–14)	(5–10)	(9–11)	(3–8)	(6–7)
4 newts	10.3	12.5	8.8	9.5	9.0	9.0
	(9–11)	(12–13)	(8–9)	(8–11)	(6–12)	(7–10)
8 newts	10.5	12.8	9.0	10.0	9.3	9.8
	(10–11)	(12–13)	(8–10)	(7–13)	(6–12)	(8–11)
4 *Ambystoma*	11.0	12.0	12.0	11.0	10.0	8.0
	(10–12)	(12–12)	(12–12)	(7–14)	(6–13)	(5–13)
4 *Ambystoma* and 4 newts	10.0	10.0	12.3	12.3	7.6	6.0
	(7–11)	(7–13)	(8–15)	(11–14)	(5–11)	(4–7)

Means are based on a sample size of 3 or 4 replicates; ranges in parentheses. Potential temporal changes in predation intensity include the hatching of larval newts during May and June and metamorphosis of *Ambystoma* in June.

12.1). Species richness declined over time in control communities without salamanders, suggesting that tanks were initially supersaturated with microcrustacean species. Certain predator treatments, particularly the presence of *Ambystoma*, slowed or halted species attrition (Wilks' Lambda, 0.02751; $p < 0.01$). Once the *Ambystoma* metamorphosed, this enhanced microcrustacean species richness eventually decayed.

Microcrustacean species diversity generally mirrored patterns of species richness (fig. 12.1). Salamanders significantly affected temporal patterns of diversity (Wilks' Lambda, 0.02936; $p < 0.01$) and generally enhanced microcrustacean diversity. The most striking differences were between controls and tanks containing either *Ambystoma* or high densities of *Notophthalmus*.

A PCA of 26 April samples summarized the multispecies changes in relative abundance that contributed to patterns of species diversity (table 12.2, fig. 12.2). Three principal components described 86% of the total variation in relative abundance of 16 species. Principal components 1, 2, and 3 (abbreviated PC1, PC2, and PC3) corresponded respectively to 55, 24, and 8% of the total community variance.

Variance analysis indicated that salamanders affected only PC1 scores (Wilks' Lambda, 0.1537; $p < 0.01$; PC1: $F_{5,16} = 7.14$; $p < 0.01$). Consequently, patterns described by PC2 and PC3 are not considered further in detail. PC1 scores were positively correlated with relative abundances of two small species (i.e., adult body length less than 1 mm), *B. longirostris* and *Chydorus sphaericus*, and negatively correlated with three large species (i.e., adult body length greater than 1.5 mm), *Daphnia laevis*, *Simocephalus serrulatus*, and *Diaptomus spatulocrenatus*. Large negative PC1 scores for tank communities indicated high relative abundances of the three large species; large positive PC1 scores indicated high relative abundances of the two small species. Mean PC1 scores for the predation treatments described an inverse gradient in the relative abundances of large and small species, ranging from control communities dominated by large species to *Ambystoma* communities dominated by small species. PC1 represented the majority of community variation, but the substantial residual variation associated with other principal components remained unexplained by predator treatments.

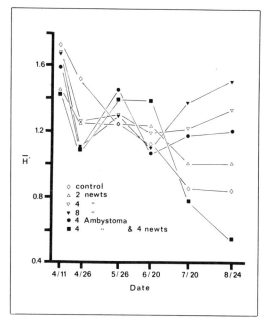

Figure 12.1. *Patterns of mean microcrustacean diversity (H') for replicates of each predation treatment over time. Maximal differences among treatments occurred late in community development.*

Variability in community structure among replicates depended on the different treatments. Small standard deviations of PC1 scores indicated very consistent patterns in controls and tanks containing *Ambystoma* (fig. 12.2). In contrast, PC1 scores for tanks containing moderate densities of *Notophthalmus* varied greatly. Consequently, communities containing moderate densities of newts were less deterministically structured than either salamander-free controls or *Ambystoma* tanks.

Salamander recruitment, growth, and metamorphosis affected predation intensities between the April and July sampling dates. *Ambystoma* completed metamorphosis in five of six tanks by 5 July, greatly reducing predation intensity. *Notophthalmus* larvae began hatching on 17 May and continued to increase in abundance and size through early July, doubtless increasing predation intensity. These changes should be kept in mind when interpreting effects of the original treatments assayed in July.

The simple predator-generated gradient in community composition seen in April was replaced by more complex patterns in late July. Three principal components described 85% of

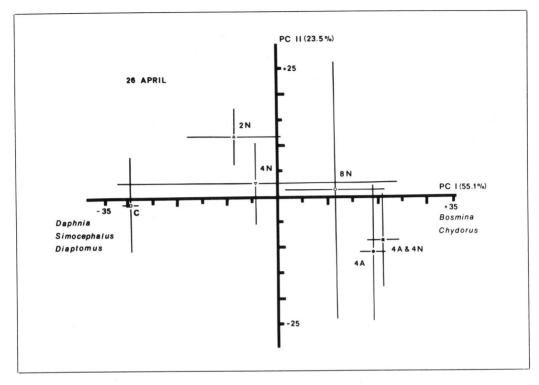

Figure 12.2. *Mean values of PCA scores for tank communities on April 26. Error bars indicate 1 SD about the mean score of each treatment with respect to each principal component. Note the community composition gradient running along PC1 from predator-free controls through tanks containing increasing densities of small* Notophthalmus *and large* Ambystoma.

the total variation in microcrustacean relative abundance on 20 July (table 12.2, fig. 12.3). Predators affected tank scores for PC1 and PC2 (Wilks' Lambda, 0.1422; $p < 0.01$; PC1: $F_{5, 16} = 3.47$; $p < 0.05$; PC2: $F_{5, 16} = 5.74$; $p < 0.01$), which respectively described 41% and 33% of total variance. PC1 scores were positively correlated with high relative abundances of Ostracoda, and negatively correlated with high relative abundances of *Diaphanosoma* and *Diaptomus*. PC2 scores were positively correlated with high relative abundances of *Diaphanosoma* and negatively correlated with *Diaptomus*, *Bosmina*, and *Mesocyclops*.

Late July communities differed from April communities in three important respects. First, several new species became abundant and replaced early dominants. For example, even in predator-free controls, *Daphnia laevis* gave way to *Diaphanosoma brachyurum*. Predators influenced the recruitment of new dominants in other tanks. Ostracoda predominated primarily in tanks that previously contained *Amby-*

stoma or still contained high densities of *Notophthalmus*. These new dominant species were either present but rare in early samples, or hatched from resting eggs. Second, the appearance of some species was highly treatment-dependent. For example, *Mesocyclops edax* attained high relative abundances only in tanks that initially contained 4 *Notophthalmus* (as inferred from PC2 scores; see table 12.2 and fig. 12.3). Third, the initial large differences in community structure between controls and tanks containing *Ambystoma* decayed. By 20 July controls and tanks that previously contained *Ambystoma* converged in species composition, as shown by their similar scores for PC1 and PC2 in figure 12.3. In April the same tank communities defined opposite ends of a community structure gradient.

Responses of Individual Species

Salamanders significantly enhanced the absolute abundance of three taxa (*Chydorus sphaeri-*

Table 12.2. *Product moment correlations between principal component scores and angularly transformed relative abundances of microcrustaceans in tank communities on April 26 and July 20*

Species	April 26			July 20		
	PC1	PC2	PC3	PC1	PC2	PC3
Daphnia laevis	−.94	.09	−.29	−.13	−.14	.64
Simocephalus serrulatus	−.77	.13	.35	−.25	.19	−.33
Diaptomus spatulocrenatus	−.72	−.15	−.23	−.51	−.61	.54
Diaphanosoma brachyurum	−.76	−.02	−.21	−.63	.76	−.11
Bosmina longirostris	.77	.13	.35	−.20	−.85	−.30
Chydorus sphaericus	.68	−.69	−.16	.20	.06	−.49
Ostracoda	—[a]	—	—	.96	.25	.08
Mesocyclops edax	.15	.08	.26	.14	−.56	−.62
Alona guttata	.31	−.69	.52	−.10	.42	−.10
Scapholeberis kingi	−.20	−.46	.70	−.02	.16	−.39
Simocephalus exspinosus	−.32	.04	−.07	—	—	—
Ceriodaphnia pulchella	−.17	−.12	.53	.13	.36	−.12
Cyclops vernalis	.23	−.34	.05	−.21	.47	−.13
Macrocyclops albidus	−.14	.11	−.05	.52	.01	.08
Macrocyclops fuscus	−.23	−.34	.37	.08	−.01	−.10
Moina micrura	.22	.01	−.08	—	—	—
Tropocyclops prasinus	.09	−.39	.14	.03	.25	.53
% Total variance	55.1%	23.5%	8.2%	40.6%	33.0%	11.8%

[a]Dashes indicate species absent on particular sampling dates.

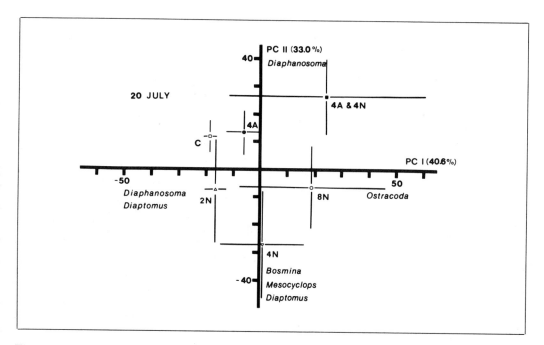

Figure 12.3. *Mean values of PCA scores for tank communities on July 20. Format as in fig. 12.2. Note the absence of the gradient described in fig. 12.2.*

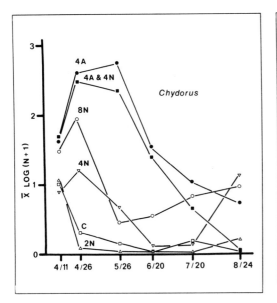

Figure 12.4. Mean values of Chydorus abundance (log_{10}(no. + 1) per 6.4 L sample) in each treatment during community development. C, salamander-free controls; 2N, 2 newts per tank; 4N, 4 newts per tank; 8N, 8 newts per tank; 4A, Ambystoma per tank; 4A & 4N, 4 Ambystoma and 4 newts per tank.

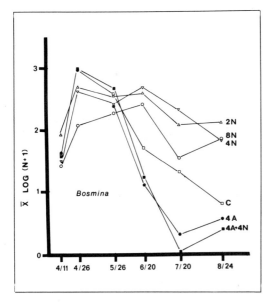

Figure 12.5. Mean abundances of Bosmina abundance over time. Format as in fig. 12.4.

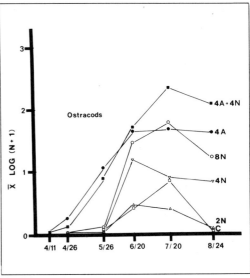

Figure 12.6. Mean abundances of Ostracoda over time. Format as in fig. 12.4.

cus, *B. longirostris,* and Ostracoda; figs. 12.4–12.6). *Chydorus* was an early successional species that became most abundant in tanks containing either *Ambystoma* or high densities of newts (Wilks' Lambda, 0.00694; $p < 0.01$). At high densities *Chydorus* also moved from the tank walls and macrophytes into the water column. *Chydorus* failed to increase in tanks without salamanders, despite the initial presence of a few individuals in samples collected from these tanks in early April. *B. longirostris* was abundant in all treatments early in community development, but it declined in *Ambystoma* tanks after the *Ambystoma* metamorphosed (Wilks' Lambda, 0.03354; $p < 0.05$). Ostracoda appeared only in late successional communities that had contained *Ambystoma* or high densities of newts.

Ostracoda possessed a novel defense against *Ambystoma* predation: viable gut passage. I noticed this phenomenon during a weekly capture–measurement–release survey designed to monitor *Ambystoma* growth. *Ambystoma* were returned to the laboratory for weighing in individual polyethylene containers filled with pond water. Within 30 minutes after placing a salamander in a container, the water swarmed with living Ostracoda defecated by the salamander. Salamander feces contained the chiti-

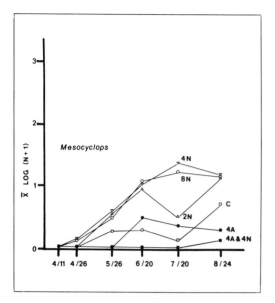

Figure 12.7. *Mean abundances of* Mesocyclops *over time. Format as in fig. 12.4.*

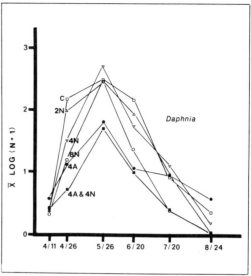

Figure 12.8. *Mean abundances of Daphnia over time. Format as in fig. 12.4.*

nous remains of a variety of microcrustaceans, including some dead Ostracoda.

Somewhat equivocal statistical evidence suggests that *Notophthalmus* may have facilitated a fourth species, the predatory cyclopoid copepod *Mesocyclops edax*. An ANOVA indicated that predators influenced *Mesocyclops* abundance on 20 July ($F_{5, 16} = 3.81$; $p = 0.018$), especially in tanks originally containing 4 *Notophthalmus* (fig. 12.7). However, the conservative MANOVA for the entire sequence of sampling dates was not significant (Wilks' Lambda, 0.1400; $p > 0.05$). This admittedly weak statistical inference should be weighed against the observation that *Mesocyclops* became abundant only in tanks containing newts.

Daphnia laevis provided the best case for a direct negative effect of predation on microcrustacean abundance (fig. 12.8). Separate ANOVAs indicated negative effects of *Ambystoma* on *Daphnia* on 26 April ($F_{5, 16} = 6.04$; $p = 0.0025$) and 26 May ($F_{5, 16} = 8.2$; $p = 0.0005$), whereas the more conservative MANOVA for predator effects on all dates was not significant (Wilks' Lambda, 0.0624; $p > 0.05$).

Salamanders did not significantly affect the absolute abundances of three numerically important species identified in the PCA. These species included large early successional *Simocephalus serrulatus*, and late successional *Diaptomus spatulocrenatus* and *Diaphanosoma brachyurum*. Reductions in the relative abundance of these species in tanks with salamanders must consequently reflect increases in the abundance of other species, rather than decreases in their own abundance.

DISCUSSION
Pathways of Facilitation

Perhaps the most striking result of this experiment was that salamanders enhanced the abundance of several microcrustacean species. Such facilitation, defined here as a positive effect of one species on the abundance of another, might occur when direct negative effects of predators on facilitated species are outweighed by indirect positive effects mediated through the food web (Dodson 1970; Levine 1976; Vandermeer 1980). The relative plausibility of different pathways of indirect facilitation can be assessed by considering the different correlated patterns of species abundance required in each scenario.

Two distinct pathways of interspecific interactions may indirectly facilitate prey abundance. Salamanders might facilitate small microcrustaceans by removing the large invertebrate predators that prey selectively on small species [hereafter termed the invertebrate predation hypothesis (e.g., Dodson 1974)]. In this case, reduced predation pressure from invertebrate predators permits small microcrustaceans to increase. Alternatively, predators may remove large species that competitively exclude smaller species [hereafter termed the predator-mediated competition hypothesis (Brooks and Dodson 1965). Here, reduced competition from larger herbivores permits smaller microcrustaceans to increase. In these scenarios specific mechanisms differ but the gross results are the same. The invertebrate predation hypothesis requires an inverse relation between abundances of large invertebrate predators and small microcrustaceans, whereas the predator-mediated competition hypothesis requires an inverse relation between abundances of large and small herbivorous microcrustaceans. Also, abundances of invertebrate predators or large competitors must be negatively correlated with salamander abundance.

Tanks without salamanders uniformly lacked abundant invertebrate predators, ruling out invertebrate predation as the cause of small microcrustacean rarity. All tanks lacked the typical invertebrate planktivores, such as *Chaoborus* or large predatory calanoid copepods, that figure prominently in other situations where the invertebrate predation hypothesis seems tenable. One large calanoid, *Diaptomus spatulocrenatus*, became abundant in most tanks, but my observations suggest that it is herbivorous. One other invertebrate predator, the cyclopoid copepod *Mesocyclops edax*, became abundant only in tanks containing moderate densities of *Notophthalmus*. This pattern suggested that vertebrate predators may have facilitated *Mesocyclops* by increasing the abundance of its smaller prey (see Kerfoot and DeMott 1984). In summary, the invertebrate predation hypothesis is inconsistent with patterns of facilitation in tank communities, although it certainly remains plausible for other systems (Dodson 1974, Zaret 1980).

Predator-mediated competition enjoys greater empirical support as a mechanism for observed patterns of indirect facilitation. Salamanders reduced abundances of *Daphnia laevis*, a large cladoceran that was common early in community development. Laboratory studies suggest that large *Daphnia* can outcompete small forms such as *Bosmina* (Goulden et al. 1982). However, salamanders failed to significantly reduce absolute abundances of other large potential competitors, including *Simocephalus*, *Diaptomus*, and *Diaphanosoma*. One possible reason for this failure is that both *Diaphanosoma* and *Diaptomus* have effective behavioral mechanisms for evading vertebrate predators (Drenner and McComas 1980).

Tank communities provided a situation where cascading effects of salamander predation might be expected, because salamanders reduced abundances of anuran tadpoles (Morin 1983a), which potentially competed with microcrustaceans for food. Tadpoles and microcrustaceans both feed on phytoplankton, and filter-feeding tadpoles can harvest food particles as small as bacteria (Wassersug 1972). Correlative studies also suggest that tadpoles can reduce the standing crop of phytoplankton in natural ponds (Seale 1980). Microcrustacean species composition was correlated with differences in anuran species composition, which were in turn caused by salamanders. For example, *Ambystoma* simultaneously facilitated *Chydorus* and extirpated tadpoles (see Morin 1983a). Unfortunately, such correlations cannot establish a causal relation between anuran and chydorid abundance, and direct experimental tests have failed to demonstrate competition between tadpoles and microcrustaceans (Morin et al. 1983). Specifically, manipulations of tadpole abundance in artificial ponds had no effect on microcrustacean species composition (Morin et al. 1983). Given the absence of documented interactions between tadpoles and microcrustaceans in other experiments, there is no need to invoke interphyletic competition between these two groups of phytoplankton consumers. Rather, I suggest that correlated changes in anuran and microcrustacean abundance simply resulted from generalist vertebrate predators feeding on two functionally independent subguilds.

The viable passage of Ostracoda through vertebrate guts shows that special properties of some prey may moderate predation and promote persistance with salamanders. Vinyard (1979) has shown that approximately 20% of the ostracods eaten by small centrarchid fish pass viably through the gut. Such resistance is directly analogous to the ability of some algae to survive ingestion by cladocerans (Porter 1973). However, such partial refuges do not explain why ostracods fail to become abundant in the absence of predators.

Other studies have shown that fish can significantly facilitate ostracod abundance in natural ponds. Comparisons of ostracod abundance inside and outside exclosures in a natural pond indicated that fish exclusion decreased ostracod abundance (Morin 1984). Low ostracod abundance was correlated with high abundances of large littoral cladocerans (*Simocephalus serrulatus*) and macroinvertebrate predators (mostly Odonata). This pattern suggests that the facilitation of Ostracoda by vertebrate predators may be a general phenomenon.

Sources of Seasonal Patterns

Microcrustacean assemblages exhibited seasonal succession with or without salamanders, but salamanders altered the successional patterns. In tanks without salamanders, the early dominance of large cladocera (*Simocephalus* and *Daphnia laevis*) gradually declined as smaller cladocera (*Diaphanosoma brachyurum*) or copepods became dominant in mid summer. This pattern loosely mirrored similar seasonal changes in other assemblages (e.g., Kerfoot 1974; Lynch 1978), where large early dominants are succeeded by smaller species or morphs. Seasonal succession in predator-free tanks was apparently driven by demographic differences among species, or by interactions between microcrustacean populations and their resources, rather than by seasonal shifts in vertebrate predation. DeMott (1983) has suggested that other seasonal shifts in zooplankton dominance are best ascribed to seasonal changes in the ability of species to exploit temporally fluctuating resources.

Other aspects of seasonal succession were correlated with temporal changes in predator abundance. One important result was the facilitation of different early or late successional species by predators (e.g., *Chydorus* versus *Daphnia* in spring; Ostracoda versus an array other species in midsummer). Certain predator effects appeared quite persistent despite temporal fluctuations in salamander abundance. For example, ostracods remained abundant long after *Ambystoma* metamorphosed. Other effects of the same predators were transient. *Bosmina* declined rapidly after *Ambystoma* metamorphosed, while remaining abundant in other tanks containing adult newts and abundant newt larvae. Such changes contributed to the apparent convergence of predator-free controls and tanks that previously contained *Ambystoma* in the 20 July PCA of microcrustacean species composition. *Diaptomus spatulocrenatus* and *Diaphanosoma brachyurum* predominated in controls and *Ambystoma* communities. It is unclear whether this similarity simply reflected a correlation between predator resistance and late successional status, or whether community dynamics rebounded to predator-free patterns following the departure of metamorphosed salamanders.

Community Dynamics along a Predation Gradient

Microcrustacean assemblages responded in a continuous fashion to the predation gradient. This response gradient does not support the existence of thresholds or breakpoints in community responses to an approximately natural range of perturbations. Rather, the community structure continuum suggested that predators influenced microcrustacean assemblages without creating discrete alternate community states (see Sutherland 1981; Connell and Sousa 1983; Peterson 1984).

Predators also generated gradients of species diversity and species richness. Predators clearly enhanced prey species richness relative to the seasonal decline in the controls, but this result was only apparent after several months of community development and species attrition. Such effects might not have appeared if communities weren't initially supersaturated with species relative to the number typically found in natural ponds.

Perturbations differed in the predictability of their effects on community structure. For example, low variance in the position of communities containing *Ambystoma* along PC1 for April samples (see fig. 12.2) indicated a highly deterministic impact of *Ambystoma* on community structure. Similarly, control communities without salamanders also exhibited little variation in structure. These deterministic patterns contrasted with the much more variable patterns of communities containing equal densities of adult *Notophthalmus*. Greater within-treatment variation suggested a weaker or more stochastic impact of newt predation on community development, compared with that of *Ambystoma*. Equal initial densities of two different salamander species clearly had drastically different effects on community composition and the variability of that composition.

One reason for the variable impact of *Notophthalmus* may have been the extremely variable production of larvae by adult newts. Newts produced from 0 to 131 planktivorous larvae per tank. The only significant relation between various salamander treatments and newt recruitment was a negative effect of *Ambystoma* on the abundance of larval newts (Morin 1983b). However, subsequent experiments have demonstrated that larval recruitment is also negatively affected by adult density (Morin et al. 1983).

Finally, salamander predation accounted for substantial fractions of community variance during early and late phases of community development. A simple pattern related to prey body size summarized predator-generated differences among communities in late April. Causes of the gradient included inhibition of a large species (*Daphnia laevis*) and facilitation of a small species (*Chydorus sphaericus*). Subsequent patterns in late July indicated a breakdown of the simple early gradient in structure and its replacement by a more complex pattern delineated by community scores with respect to two independent principal components. In both early and late stages of community development, the principal component scores affected by predator treatments also corresponded to greater than 50% of total community variance. This indicated a major contribution of predation to community structure. However, the substantial residual variation within predator treatments indicated that vertebrate predation at best offered an incomplete explanation for temporal and spatial variation in the structure of these aquatic communities. The identity of other important factors, the relative contribution of those factors to community patterns, and the mechanisms of complex interspecific interactions demand further experimental study.

ACKNOWLEDGMENTS

This research was supported by National Science Foundation Grant DEB 7911539 to H. M. Wilbur. The Zoology Department of Duke University and the Department of Biological Sciences of Rutgers University provided computer time.

REFERENCES

Allan, J. D. 1973. Competition and the relative abundances of two cladocerans. *Ecology* 54 : 484–98.

Allan, J. D., and C. E. Goulden. 1980. Some aspects of reproductive variation among freshwater zooplankton. In W. C. Kerfoot (ed.), *Evolution and ecology of zooplankton communities*, pp. 388–410. Hanover, N.H.: University Press of New England.

Attar, E. N., and E. J. Maly. 1980. A laboratory study of preferential predation by the newt *Notophthalmus v. viridescens. Can. J. Zool.* 58 : 1712–17.

Brooks, J. L., and S. I. Dodson. 1965. Predation, body size and composition of plankton. *Science* 150 : 28–35.

Brophy, T. E. 1980. Food habits of sympatric larval *Ambystoma tigrinum* and *Notophthalmus viridescens. J. Herpetol.* 14 : 1–6.

Cassie, R. M. 1963. Multivariate analysis in the interpretation of numerical plankton data. *N. Z. J. Sci.* 6 : 36–59.

Colwell, R. K., and D. W. Winkler. 1984. A null model for null models in biogeography. In D. R. Strong, D. Siberloff, L. G. Abele, and A. B. Thistle (eds.), Ecological communities: Conceptual issues and the evidences, pp. 344–59. Princeton, N.J.: Princeton University Press.

Connell, J. H., and W. P. Sousa. 1983. On the evidence needed to judge ecological stability or persistence. *Amer. Naturalist* 121 : 789–824.

Demott, W. R. 1983. Seasonal succession in a natural *Daphnia* assemblage. *Ecol. Monogr.* 53 : 321–40.

Dodson, S. I. 1970. Complementary feeding niches sustained by size-selective predation. *Limnol. Oceanogr.* 15 : 131–37.

———. 1974. Zooplankton competition and predation: an experimental test of the size-efficiency hypothesis. *Ecology* 55 : 605–13.

Dodson, S. I., and V. E. Dodson. 1971. The diet of *Ambystoma tigrinum* larvae from western Colorado. *Copeia* 1971 : 614–24.

Drenner, R. W., and S. R. McComas. 1980. The roles of zooplankter escape ability and fish size selectivity in the selective feeding and impact of herbivorous fish. In W. C. Kerfoot (ed.), *Evolution and ecology of zooplankton communities*, pp. 587–93. Hanover, N.H.: University Press of New England.

Giguere, L. 1979. An experimental test of Dodson's hypothesis that *Ambystoma* (a salamander) and *Chaoborus* (a phantom midge) have complementary feeding niches. *Can. J. Zool.* 57 : 1091–97.

Goulden, C. E., L. L. Henry, and A. J. Tessier. 1982. Body size, energy reserves, and competitive ability in three species of cladocera. *Ecology* 63 : 1780–89.

Hall, D. J., W. E. Cooper, and E. E. Werner. 1970. An experimental approach to the dynamics and structure of freshwater animal communities. *Limnol. Oceanogr.* 15 : 839–928.

Hrbácek, J., M. Dvoráková, V. Korínek, and L. Procházková. 1961. Demonstration of the effect of fish stock on the species composition of zooplankton and the intensity of metabolism of the whole plankton association. *Int. Ver. Theor. Angew. Limnol. Verh.* 14 : 192–95.

Kerfoot, W. C. 1974. Egg-size cycle of a cladoceran. *Ecology* 55 : 1259–70.

———. 1975. The divergence of adjacent populations. *Ecology* 56 : 1298–1313.

Kerfoot, W. C., and W. R. Demott. 1980. Foundations for evaluating community interactions: the use of enclosures to investigate coexistence of *Daphnia* and *Bosmina*. In W. C. Kerfoot (ed.), *Evolution and ecology of zooplankton communities*, pp. 725–41. Hanover, N.H.: University Press of New England.

———. 1984. Food web dynamics: Dependent chains and vaulting. In D. G. Myers and J. R. Strickler (eds.), *Trophic interactions within aquatic ecosystems*, pp. 347–82. AAAS Selected Symposium 85.

Kerfoot, W. C., and C. Peterson. 1979. Ecological interactions and evolutionary arguments: Investigations with predatory copepods and *Bosmina*. *Fortschrifte Zoologie* 25(2/3) : 159–96.

Levine, S. H. 1976. Competitive interaction in ecosystems. *Amer. Naturalist* 110 : 903–10.

Lynch, M. 1978. Complex interactions between natural coexploiters—*Daphnia* and *Ceriodaphnia*. *Ecology* 59 : 552–64.

Mellors, W. K. 1975. Selective predation of ephippial *Daphnia* and the resistance of ephippial eggs to digestion. *Ecology* 56 : 974–80.

Morin, P. 1983a. Competitive and predatory interactions in natural and experimental populations of *Notophthalmus viridescens dorsalis* and *Ambystoma tigrinum*. *Copeia* 1983 : 628–39.

———. 1983b. Predation, competition, and the composition of larval anuran guilds. *Ecol. Monogr.* 53 : 119–38.

———. 1984. The impact of fish exclusion on the abundance and species composition of larval odonates: results of short-term experiments in a North Carolina farm pond. *Ecology* 65 : 53–60.

Morin, P., H. M. Wilbur, and R. N. Harris. 1983. Salamander predation and the structure of experimental communities: responses of *Notophthalmus* and microcrustacea. *Ecology* 64 : 1430–36.

Morrison, D. F. 1976. Multivariate statistical methods. New York: McGraw Hill.

Murdoch, W. W. 1971. The developmental response of predators to changes in prey density. *Ecology* 52 : 132–37.

Neill, W. E. 1975. Experimental studies of microcrustacean community composition and efficiency of resource utilization. *Ecology* 56 : 809–26.

———. 1981. Impact of *Chaoborus* predation upon the structure and dynamics of a crustacean zooplankton community. *Oecologia* 48 : 164–77.

Peterson, C. H. 1984. Does a rigorous criterion for environmental identity preclude the existence of multiple stable points? *Amer. Naturalist* 124 : 127–33.

Pielou, E. C. 1984. The interpretation of ecological data. New York: J. Wiley.

Porter, K. G. 1973. Selective grazing and differential digestion of algae by zooplankton. *Nature* 244 : 179–80.

Seale, D. B. 1980. Influence of amphibian larvae on primary production, nutrient flux, and competition in a pond ecosystem. *Ecology* 61 : 1531–50.

Sprules, W. G. 1972. Effects of size-selective predation and food competition on high altitude zooplankton communities. *Ecology* 53 : 375–86.

———. 1977. Crustacean zooplankton communities as indicators of limnological conditions: an approach using principal component analysis. *J. Fish. Res. Bd. Can.* 34 : 962–75.

Sutherland, J. P. 1981. The fouling community at Beaufort, North Carolina: A study in stability. *Amer. Naturalist* 118 : 499–519.

Vandermeer, J. H. 1980. Indirect mutualism: Variations on a theme by Stephen Levine. *Amer. Naturalist* 116 : 441–48.

Vinyard, G. 1979. An ostracod (*Cypriodopsis vidua*) can reduce predation from fish by resisting digestion. *Amer. Midlands Naturalist* 102 : 188–90.

Wassersug, R. J. 1972. The mechanism of ultraplanktonic entrapment in anuran larvae. *J. Morphol.* 137 : 279–88.

Wiggins, G. B., R. J. Mackay, and I. M. Smith. 1980. Evolutionary and ecological strategies of animals in annual temporary pools. *Archive für Hydrobiologie* 58 : 97–206.

Williams, E. H. 1980. Disjunct distributions of two aquatic predators. *Limnol. Oceanogr.* 25 : 999–1006.

Zaret, T. M. 1980. Predation and freshwater communities. New Haven, Conn.: Yale University Press.

13. Indirect Mutualism: Complementary Effects of Grazing and Predation in a Rocky Intertidal Community

Michael L. Dungan

In this chapter I suggest that one of the important consequences of space effects in rocky intertidal communities is the generation of indirect interactions between algal grazers and the predators of sessile animals. Sessile organisms mediate the effects of these two groups of consumers. An experiment involving the simultaneous manipulation of limpet (*Collisella strongiana*) and predatory gastropod (*Acanthina angelica*) densities in the Gulf of California was conducted to test the hypothesis that these two species interact indirectly as mutualists. Results supported the hypothesis that algal grazing by *Collisella* freed space on the rock surface and increased the abundance of the barnacle *Chthamalus anisopoma*, the major prey of *Acanthina*. Predation on *Chthamalus* by *Acanthina* increased the abundance of both food and space for *Collisella*; the removal of *Acanthina* led to increased cover of the surface by *Chthamalus* and the near-elimination of *Collisella*. The experiment revealed strong, interdependent influences of grazing and predation. A variety of indirect interactions between grazers and predators can be hypothesized for other shores, including additional likely cases of indirect mutualism.

Trophic relationships provide a starting point for attempts to understand the organization of ecological communities, including those occurring on marine rocky intertidal shores (Paine 1966b, 1980; Menge and Sutherland 1976; Lubchenco 1978; Jara and Moreno 1984). However, many of the important interactions in rocky intertidal communities have distinctly nontrophic components. Examples of nontrophic effects that can influence the functional roles and relationships of species in these systems include competition for space between sessile organisms via preemption or interference, the provision of shelter or secondary substratum by established residents of the rock surface, and the bulldozing of sessile organisms by grazers. These nontrophic effects allow the consequences of particular interactions to be propogated horizontally or diagonally with respect to trophic levels (cf. Paine 1980). In this chapter I suggest that one of the important consequences of nontrophic effects in rocky intertidal communities is the generation of indirect interactions between grazers and predators, and an attendant interdependence of grazer–plant and predator–prey relationships. Existing evidence provides considerable, but largely indirect, support for this idea. A more direct test, involving simultaneous manipulations of grazer and predator densities in the rocky intertidal of the northern Gulf of California, is described below.

In this chapter the focus is on mid-intertidal communities typified by barnacles, mussels, attached plants, and associated mobile consumers; numerous examples are described by Stephenson and Stephenson (1972). A wealth of experimental evidence indicates the general importance of algal grazing and predation on sessile animals in these communities (Connell 1972, 1974; Paine 1977; Underwood 1979, 1985; Branch 1976, 1984; Lubchenco and Gaines 1981; Gaines and Lubchenco 1982; Hawkins and Hartnoll 1983). Either or both of these processes may ameliorate the effects of competition among sessile species in a particular locality (Jones 1948; Connell 1961; Dayton 1971; Paine 1971, 1974, 1984; Menge 1976; Lubchenco and Menge 1978; Menge and Lubchenco 1981; Underwood et al. 1983; Jara and Moreno 1984; Lubchenco et al. 1984); and Dayton (1975), Menge and Sutherland (1976),

Lubchenco (1979), Branch (1984), and Paine (1984), among others, have stressed the analogous roles of grazers and predators in these systems (but see Underwood and Denley 1984).

Although direct, trophic interactions between grazers and their predators (e.g., Paine 1969a; Garrity and Levings 1981; Fawcett 1984; Lubchenco et al. 1984; Mercurio et al. 1985) should not be downplayed, a case can be made for the prevalence of more circuitous links between grazers and predators. Numerous studies have shown or implied that predation on barnacles or mussels was preventing the exclusion of algae and/or grazers from the rock surface (Paine 1969a, 1971, 1974; Dayton 1971; Menge 1976; Lubchenco and Menge 1978; Fairweather et al. 1984; Lubchenco et al. 1984). More subtle effects of barnacle or mussel predators on grazer–plant interactions may be widespread, however. Menge (1976) and Hawkins and Hartnoll (1983) noted the tendency of fucoid algae to become established more readily on or among barnacles than on adjacent bare rock. Grazers also may be more concentrated in the vicinity of sessile animals than on larger open areas (Paine and Levin 1981; Hawkins and Hartnoll 1983; Paine and Suchanek 1983; Sousa 1984), apparently in response to the provision of shelter (Lewis and Bowman 1975; Choat 1977).

Analogous considerations suggest effects of grazers and algae on relationships between carnivores and sessile prey. Algal grazing, especially by limpets, appears in many cases to increase indirectly the abundance of barnacles by freeing space on the rock surface and thereby enhancing settlement, by preventing algae from overgrowing barnacles, or by eliminating frondose algae which sweep the surface and impede recruitment (Jones 1948; Luckens 1974; Sousa 1979; Petraitis 1983; Hawkins and Hartnoll 1983; Underwood et al. 1983; Bertness et al. 1983; Jara and Moreno 1984). Conversely, mussel recruitment may be enhanced by the presence of filamentous algae, and consequently reduced where grazers limit algal abundance (Paine 1969a; Dayton 1971; Sousa 1984). The presence of an algal canopy was shown to increase the feeding of predatory snails on mussels by Menge (1978), and Connell (1961) suggested that similar mechanisms might contribute to the reduction of barnacle abundance under a fucoid canopy in the British Isles.

An additional consideration is that grazers frequently dislodge or crush small barnacles or mussels (Connell 1961; Stimson 1970; Dayton 1971; Menge 1976; Denley and Underwood 1979; Garrity and Levings 1981; Petraitis 1983). Relatively small, flat barnacles may be less susceptible to these "bulldozing" effects (Dayton 1971; Paine 1981). Underwood (1985) discusses the interplay of positive (via algal grazing) and negative (bulldozing) effects of limpets on barnacles and suggests that the net effect of limpets on barnacle abundance depends on both limpet density and algal productivity.

Thus, indirect interactions between grazers and predators, mediated by the effects of these consumers on, and their responses to, sessile organisms, are probably widespread. Hypotheses about these interactions in particular settings are readily constructed (see Discussion) but largely untested.

In the following experiment, I examined the interacting effects of algal (*Ralfsia* sp., probably *R. pacifica*) grazing by limpets (*Collisella strongiana*), and predation on barnacles (*Chthamalus anisopoma*) by the neogastropod *Acanthina angelica* in the lower rocky intertidal zone of the northern Gulf of California. Initial observations suggested (1) the local elimination of macroscopic algae by limpet grazing; (2) the settlement of *Chthamalus* on grazed or otherwise bare areas, but not on *Ralfsia*; (3) the exclusion of both limpets and *Ralfsia* from areas with high *Chthamalus* cover; and (4) the limitation of *Chthamalus* abundance partly by *Acanthina* predation. These observations led me to hypothesize an indirect mutualistic interaction between *Acanthina* and *Collisella*, in which limpet grazing should indirectly increase the abundance of prey for *Acanthina* by clearing algae from the surface and thereby enhancing the settlement of *Chthamalus*; whereas *Acanthina* predation on *Chthamalus* should increase the availability of space and/or food for *Collisella*. Implicit in this hypothesis were strong, interdependent influences of both grazing and predation on local community structure.

Results from a series of experiments designed to tease apart the interactions involving *Collisella*, *Ralfsia*, and *Chthamalus*, coupled with long-term observations, are described in Dungan (1986). These results generally support the above hypothesis, although the experiments did not directly assess the role of *Acanthina*. A preliminary experiment, conducted between

February and August 1979 revealed increases in *Chthamalus* and decreases in algal cover where *Acanthina* and another predatory gastropod, *Morula ferruginosa*, were removed, as compared with an adjacent control site (Dungan 1984). *Morula* was much less abundant than *Acanthina* during this experiment, and the effects associated with predator removals were attributed to *Acanthina*. Limpet densities were manipulated in this experiment, but densities were relatively low and the effects of limpets on barnacle and algal abundances not conclusive (Dungan 1984). The following experiment, which spanned 2½ years, was intended to provide a more definitive test for the effects of *Acanthina* and *Collisella* against a background of seasonal variation in physical and possibly biological factors.

MATERIALS AND METHODS

This work and related studies (Dungan 1984, 1985, 1986) were conducted along the southwest shore of Pelican Point, also known as Roca del Toro (31°20′N, 113°40′W), a granitic promontory 10 km northwest of Puerto Peñasco, Sonora, Mexico, in the northern Gulf of California. The climate on the shore reflects the influence of the surrounding Sonoran desert: air temperatures range around 40°C annually, and onshore sea surface temperatures vary between 9 and 15°C during winter and 30 and 32°C during summer. The tidal range is 7 m and wave action is generally slight. For additional background, see Thomson and Lehner (1976), Thomson et al. (1979), Brusca (1980), Littler and Littler (1981, 1984), Maluf (1983), and McCourt (1984).

The experiment was done within the "*Chthamalus* zone," at 0.3 to 0.9 m above mean low water, on a 3 × 10 m outcrop situated in a small bay and further protected from wave action by seaward projections of bedrock. At this elevation at Pelican Point (see Dungan 1984, 1985, 1986), the barnacle *Chthamalus anisopoma*, the limpet *Collisella strongiana*, and a brown algal crust resembling *Ralfsia pacifica* (cf. Dawson 1966; Abbott and Hollenberg 1976; Littler and Littler 1984) are the commonest occupants of the rock surface. Other less common or ephemeral species are mentioned below where appropriate. In addition to *Acanthina angelica*, other mobile consumers, including the hermit crab *Clibanarius digueti*, the

brachyuran crab *Eriphia squamata*, the herbivorous snail *Nerita funiculata*, and the predatory snail *Morula ferruginosa* also occurred on or adjacent to the study site. The predatory starfish *Heliaster kubiniji* (see Paine 1966a) disappeared from many parts of the Gulf in 1978 (Dungan et al. 1982) and was not encountered during this study. No attempt was made to control physical and biological factors other than the densities of *Acanthina* and *Collisella*, and the results do not rule out the possible importance of factors not directly assessed.

Acanthina angelica (F. Thaididae) in the study site were typically 25 to 35 mm long. During these and other studies (Dungan 1984, 1986; see also Turk 1979), the snails appeared to move downward into the *Chthamalus* zone during fall, where they fed almost exclusively (474 of 479 of my observations of feeding by individual snails) on *Chthamalus*, until late spring or early summer when they moved back upward in the intertidal. Higher on the shore, *Acanthina* feeds mainly on a second barnacle species, *Tetraclita stalactifera* (Paine 1966b; Yensen 1979; Turk 1979; Malusa 1985; Dungan, unpublished data). The formation of breeding aggregations and the deposition of egg capsules also took place during fall and winter. Part of the population remains feeding and breeding in the upper intertidal through the fall and winter. Feeding is concentrated during evening low tides, usually within 1 m of cracks and crevices where the snails remain at other times.

Collisella strongiana is a small (to 15-mm length) acmaeid limpet that occurs primarily in the *Chthamalus* zone (Dungan 1984, 1986). Recruitment between late summer and spring is usually followed by considerable attrition, possibly due in part to crowding from *Chthamalus* (Dungan 1986). These limpets are invariably attached to bare rock at low tide, and they usually return to the same spot after foraging (Yensen 1973; Dungan 1984). This species is known to consume diatoms, green and bluegreen algae, and *Ralfsia* (Dungan 1984).

Chthamalus anisopoma is a small (to 7 mm diameter) acorn barnacle which reproduces year-round, with highest recruitment densities during summer (Malusa 1983; Dungan 1984, 1986). Lively (1984) provides evidence that the morphology of *Chthamalus* varies in direct response to the presence of *Acanthina*.

Nothing is known of the life cycle of *Ralfsia* in the Gulf (cf. Lubchenco and Cubit 1980;

Littler and Littler 1984). In my studies, the crusts appeared to become established at any time of year and persisted for several years in some places. *Chthamalus* and *Ralfsia* have similar vertical distributions at Pelican Point, with *Chthamalus* more common and *Ralfsia* less so with increasing exposure to wave action (Dungan 1984, 1985, 1986).

To assess the effects of grazing by *Collisella* and predation by *Acanthina* on local community structure, four treatments were incorporated into a split-plot (factorial) design. The treatments were (1) control—no manipulations other than those involved in setting up the experiment (see below); (2) limpets removed (since 97% of the limpets found were *Collisella strongiana* and most of the remainder were juveniles of other species less than 5 mm long, I attribute effects of this treatment to the reduction of *C. strongiana*); (3) *Acanthina* removed; (4) *Acanthina* and limpets removed. One-half of the study site (as defined by a relatively deep crevice) was chosen on the basis of a coin toss as the "removal side" and had *Acanthina* removed repeatedly (see below); the other side ("control side") served as a control with respect to *Acanthina*. Within each side, 12 10- × 10-cm plots were haphazardly located, except for the restriction that none were situated so as to span crevices or changes in the orientation of the surface. The limpet removal treatment was randomly assigned to six plots on each side, leaving the other six on each side as controls with respect to limpets.

Manual removals, as opposed to exclosures, were employed in order to minimize potential artifacts of the treatments (Underwood and Denley 1984). Removal treatments were maintained on each visit to the study site, usually at 10- to 14-day intervals in conjunction with maximum tidal ranges ("spring tides"). Limpets found on removal plots were picked off with a pocket knife after being counted. *Acanthina* found on the removal side were censused (see below), then tossed 5 to 10 m into adjacent areas. A small percentage of the removed individuals were found in or next to the crevice between the control and removal sides; these were assumed to have come from the control side and were tossed back in that direction, several meters away from any experimental plots.

Since recolonization followed the removals (see below) and undoubtedly lessened their effects (Underwood 1980), I view the experiment

as providing conservative tests for the effects of the removed species. The experimental plots were completely accessible to the mobile consumers mentioned earlier, plus fishes, shorebirds, and occasional fishermen, and fully exposed to the physical environment; I believe that these additional influences could only have tended to introduce random variation into the results.

The experiment was initiated in late October 1980. The 24 plots were first scraped repeatedly with a wire brush to remove most living material, including *Chthamalus*, *Collisella*, and *Ralfsia*. I then applied a thick coating of Easy-Off Oven Cleaner® (4% sodium hydroxide) and left plots in the sun for 1 to 2 hours, after which they were rinsed by the rising tide. The oven cleaner effectively sterilized each plot. Opposite corners of each plot were marked for future reference by excavating a shallow depression just outside of the corners and cementing in short lengths of plastic screw anchors with an underwater epoxy.

Plots were censused at 2- to 8-week intervals in the following manner. A clear plexiglas plate marked with a uniform grid of 100 dots within a 10- × 10-cm square was laid on the surface, and limpets within the area under the square were counted. The abundances of sessile organisms were measured as percent cover, determined by counting the number of dots directly over particular species (Connell 1970). Algal abundance on limpet or barnacle shells appeared very low in comparison to that on the rock surface and was not quantified; the same was true for barnacle abundance on limpet shells. With this methodology, a small barnacle covered by *Ralfsia* would have been scored as *Ralfsia*, whereas *Chthamalus* growing on *Ralfsia* would have been scored as *Chthamalus*. These sorts of overgrowth interactions were not apparent, however. *Acanthina*'s abundance on the two sides of the study site was estimated from 5 to 11 50- × 50-cm quadrat samples taken in the vicinity of the experimental plots, with samples taken immediately prior to the first removal of snails from the removal side during a particular spring tide sequence. The resulting data were converted to an average density per .25 m^2. Manipulations continued until September 1982, with plots censused several times subsequently and for the last time in May 1983.

In the design of the experiment, the effects of

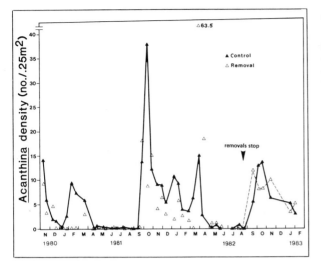

Figure 13.1. *Average densities of* Acanthina *in the vicinity of the experimental plots during the experiment, based on 5–11 0.25 m² quadrat samples.*

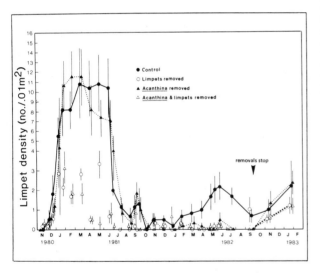

Figure 13.2. *Limpet densities on the experimental plots. The scale along the y-axis has been modified to accommodate differences in density over the course of the experiment. Data are means ±1 SE, with N = 6.*

Acanthina were confounded with any other differences between the two sides of the study site; Federer (1975) and Hurlbert (1984) discuss this sort of problem. The question is, are differences arising during the experiment due to my manipulations or to some (unknown in this case) difference between the two sides of the site? To answer the question I did a number of comparisons between the two sides of the site, involving both experimental plots and adja-

cent undisturbed areas, between 1979 and 1983. These are discussed more fully in Dungan (1984); they provide no evidence that the results obtained were biased in favor of the hypothesized effects of *Acanthina*. In this chapter I emphasize what I feel is the best test for this sort of bias, the behavior of the experimental plots after the manipulations cease. Specifically, if differences between removal and control sides arising during the experiment were due to experimentally imposed differences in the density of *Acanthina*, then the differences should disappear after the removals are stopped.

Analysis of variance (ANOVA) is used to measure the effects of *Acanthina* (confounded with site effects as noted above), limpets, and the limpet x *Acanthina* interaction on sessile organisms. The error term in the ANOVA is appropriate for testing the significance of the limpet and interaction effects because of the random assignment and replication of the limpet treatments within the two sides of the study site; it is not used to test the significance of the *Acanthina* effect because of the nonindependence of replicates with respect to the *Acanthina* treatments, termed "pseudoreplication" by Hurlbert (1984; see also Federer 1975). Pairwise t-tests or the nonparametric Mann–Whitney U-test are used to test the significance of differences in barnacle, algal, and limpet abundances on control versus removal sides of the study site, with the qualification that *Acanthina* and "side" effects are confounded.

Between 1979 and 1983, samples at random points along horizontal transects were used to assess natural variations in community structure on undisturbed portions of the study site. These data are presented in Dungan (1984) and will be referred to incidentally here.

RESULTS

Acanthina densities around the experimental plots, immediately prior to the removals on the removal side, are shown in figure 13.1. Despite removing *Acanthina* on 36 separate spring tide series (for 1–4 days after each census, densities on the removal side were maintained near zero), there were several times when densities on the removal side exceeded those on the control side. Particularly between fall 1981 and spring 1982, there were obvious effects (dead barnacles) of *Acanthina* on the removal side. As noted earlier, this introduces a conservative bias in the assessment of *Acanthina* effects. Seasonal variations in *Acanthina*'s abundance were pronounced, with the snails nearly absent during summer months.

Limpet densities are shown in figure 13.2; included are a small number (3% of the total, mostly juveniles <5 mm long) of *Collisella acutapex* and *C. turveri*. Seasonal variations in limpet abundance mirrored those of *Acanthina* and reflected recruitment by *Collisella strongiana* during winter and spring. Densities on the experimental plots early in the experiment were substantially higher than those found contemporaneously in undisturbed areas where 50 to 90% of the surface was covered with *Chthamalus* (Dungan 1984). I attribute the elevation of limpet densities to the removal of *Chthamalus* that accompanied the initial clearing of plots; this interpretation is reinforced by experimental results obtained at a site 1 km to the west (Dungan 1984, 1986), where the removal of *Chthamalus* between December 1980 and April 1981 led to significant increases in the density of *Collisella*, relative to undisturbed controls. Limpet removals were generally effective in reducing densities well below those seen on "limpet control" plots. Possible effects of *Acanthina* on *Collisella* are discussed later.

Percent cover of algae is shown in figure 13.3. All species are combined; *Ralfsia* accounted for at least 90% of total algal cover except during some of the winter censuses when the green alga *Ulva rigida* bloomed. Data for *Chthamalus* are presented in figure 13.4. The only other sessile animal averaging more than 1% cover at any time during the experiment was the small (<6 mm length) mussel *Brachiodontes semilaevis*, which averaged about 1% cover throughout summer 1981. Figures 13.3 and 13.4 are nearly mirror images after summer 1981, when most of the available surface was occupied by either *Chthamalus* or *Ralfsia*. In general, the removal of limpets was associated with increased algal cover and decreased *Chthamalus* cover, whereas the removal of *Acanthina* was associated with increased *Chthamalus* and decreased algal cover. Limpet densities were also markedly higher on the control plots than on the *Acanthina* removal plots during spring and summer 1982 (fig. 13.2).

Note that the differences between *Acanthina* removal and control plots and between *Acanthina* plus limpet removal and limpet removal plots virtually disappeared shortly after the removal treatments stopped. Not shown in the figures are data for May 1983; community composition remained essentially unchanged

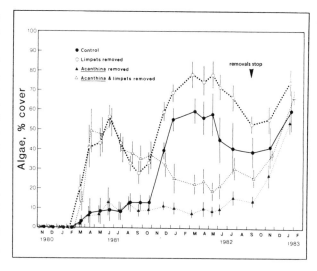

Figure 13.3. *Percent cover of algae on the experimental plots, presented as in fig. 13.2.*

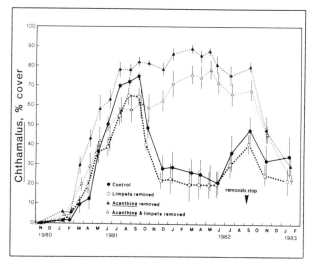

Figure 13.4. *Percent cover of* Chthamalus *on the experimental plots, presented as in figs. 13.2 and 13.3.*

between February 1983 (figs. 13.2–13.4) and May 1983 (Dungan 1984). This is strong evidence that differences associated with the *Acanthina* treatments are in fact the result of manipulation of *Acanthina* densities and not other differences between sides of the study site. Accordingly, from here on, I refer to these differences as effects of *Acanthina*.

Results of ANOVAs for *Chthamalus* and algal cover, respectively, computed date by date from March 1981 onward, are displayed graphically in figure 13.5. The figure provides a measure of the importance of grazing (limpet effect), preda-

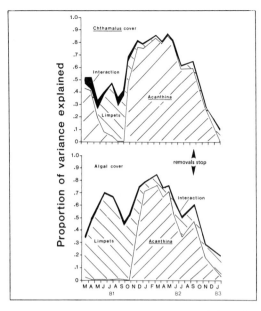

Figure 13.5. *Percent of variance in* Chthamalus *cover and algal cover, respectively, accounted for by* Acanthina, *limpets, and the interaction effect, based on 2-way ANOVAs for each sample date from March 1981 onward.*

tion (*Acanthina* effect), and the interaction effect over the course of the experiment.

In the following I attempt to summarize the major changes in local community structure, and their causes, as they occurred during the experiment. During the first 6 months after plots were cleared, densities of *Acanthina* and *Collisella* were relatively high (figs. 13.1, 13.2). The establishment of *Ralfsia* was enhanced where limpets were removed (fig. 13.3). From March through May 1981 (prior to this time, algal cover had not increased above 1%) and until the end of the experiment, F-tests on the limpet effect in the ANOVAs for algal cover on each sample date gave $p < 0.05$; the interaction effect was not significant at the 0.05 level on any date. The effect of limpets on *Chthamalus* cover was not significant (F-tests on limpet effect in ANOVAs for *Chthamalus* cover; $p > 0.05$) through May 1981; as in the ANOVA for algal cover, the interaction effect was not significant at any time.

Because the interaction effect in the ANOVAs for *Chthamalus* and algal cover was not significant at any time, tests of the effect of *Acanthina* on *Chthamalus* and algae are made on data pooled across limpet treatments, i.e., the comparisons involve control plus limpet re-

moval plots versus *Acanthina* removal plus *Acanthina* and limpet removal plots. *Chthamalus* cover, like algal cover, began to increase substantially in March 1981 (fig. 13.4). From March through May 1981, *Chthamalus* increased to a greater extent where *Acanthina* was removed (t-tests; p, <0.05); *Acanthina* had no effect on algal cover at this time (t-tests; p, >0.05). Limpet densities were not significantly different on control versus *Acanthina* removal plots at any time prior to spring, 1982 (t-tests; p, >0.05).

The first 6 months of the experiment thus reflected the effects of the two consumers on the establishment of their prey. As figure 13.5 indicates, effects of limpets on barnacles and of *Acanthina* on algae were not expressed initially. In passing, it is noteworthy that no distinct early successional stages or species were evident; the same species that were present on undisturbed areas (Dungan 1984) colonized the cleared plots.

From late spring through summer 1981, with *Acanthina* largely absent and limpet densities declining sharply (figs. 13.1, 13.2), *Chthamalus* cover increased to a much greater extent where limpets had not been removed (fig. 13.4), owing to increased settlement on grazed areas. The limpet effect in the ANOVAs for *Chthamalus* cover was significant from June through December 1981. Limpets accounted for a substantial proportion of the variance in *Chthamalus* cover during summer 1981, whereas *Acanthina*'s contribution dropped to almost zero (fig. 13.5). Algal cover declined on the limpet removal plots, although remaining well above that on plots where limpets had not been removed (fig. 13.3). Reasons for this decline were not clear; it did not appear to result from the settlement of *Chthamalus* on *Ralfsia* but may have been influenced by increased temperatures and desiccation during the summer or by other herbivores present on the study site (e.g., *Nerita funiculata*).

The increase in *Chthamalus* cover where limpets had not been removed and where algal cover was reduced, relative to the limpet removal plots, supports one-half of the hypothesis stated earlier: limpet grazing does appear to increase the abundance of prey for *Acanthina*. As of September 1981, limpet grazing was clearly of major importance in determining the extent to which *Ralfsia* became established and apparently preempted space for *Chthamalus* settlement; predation by *Acanthina* had had only temporary effects (fig. 13.5).

From fall 1981 through spring 1982, *Acanthina* was present in large numbers (fig. 13.1) and had dramatic effects on the allocation of space to *Chthamalus* and *Ralfsia* (figs. 13.3–13.5). The consumption of *Chthamalus* (vacant *Chthamalus* tests persisted for less than 4 weeks) freed space on the rock surface which was then occupied by *Ralfsia*. *Acanthina* significantly decreased *Chthamalus* cover and increased algal cover from fall 1981 until fall 1982 (t-tests; p, <0.05), and most of the variance in *Chthamalus* and algal covers was attributable to *Acanthina* (fig. 13.5).

The effect of limpets on algal cover was much reduced (although it remained significant, as noted earlier), and the previously evident effects of limpets on *Chthamalus* were, in effect, swamped by *Acanthina* predation (fig. 13.5).

Acanthina predation on *Chthamalus* also led to significant increases in limpet densities during much of 1982 (fig. 13.2; Mann–Whitney U-test based on average limpet density for each plot from January through July 1982: U = 31; p, <0.025). Samples in undisturbed portions of the study site showed limpets virtually absent from the *Acanthina* removal side of the site in 1982 until fall, when the removals were halted (Dungan 1984). This supports the second half of the hypothesis: *Collisella* benefits from *Acanthina* predation on *Chthamalus*. The fact that *Collisella* rapidly colonized the experimental plots at the start of the experiment—before any macroscopic algae became established (figs. 13.2, 13.3)—coupled with experimental data (Dungan 1986) showing an analogous, although more dramatic response by *Collisella* to the provision of barnacle-free space, suggests that the limpets responded primarily to the availability of space.

As mentioned earlier, when removals halted, limpet, algal, and barnacle abundances on the two sides of the study site converged (figs. 13.2–13.4), and *Acanthina*'s contribution to the variance (fig. 13.5) dropped to almost zero. It is perhaps significant that *Acanthina* densities on the two sides were similar after the removals stopped (fig. 13.1). Given substantially higher *Chthamalus* densities on the (former) removal side, the implication is that snails on the removal side probably had higher feeding rates than those on the control side; this was not measured, however.

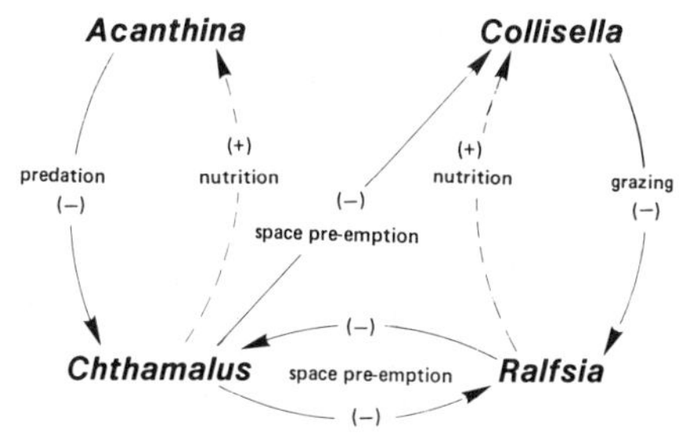

Figure 13.6. *Interactions involving* Acanthina, Chthamalus, Ralfsia, *and* Collisella, *based on this study and Dungan (1984, 1985). Direct effects and their impacts on abundance substantiated by experiments are indicated with solid lines. Dashed lines indicate effects likely but not yet verified.*

Finally, it is clear that *Acanthina* predation was a major cause of seasonal variations in community structure after August 1981 (figs. 13.2–13.4).

DISCUSSION

Figure 13.6 summarizes what seem to be the salient features of species interactions in this system. Direct effects of species on each other as suggested by the results of the above experiment and others (Dungan 1984, 1986) are indicated with solid lines; the dashed lines indicate effects that are likely but for which there is no direct experimental evidence. These direct effects form pathways that allow species to interact indirectly. There are no trophic links between the grazer–plant (*Collisella–Ralfsia*) and predator–prey (*Acanthina-Chthamalus*) components of this system. Instead, linkage (Paine 1980) is provided by space preemption.

Acanthina and *Collisella* exert complementary effects on local community structure and each might be viewed as a keystone species (Paine 1969b). Since space preemption seems to be the rule in this system, it is conceivable that either *Ralfsia* or *Chthamalus* could monopolize space in the absence of their respective consumers. In Paine's (1980) terminology, both the *Acanthina–Chthamalus* and *Collisella–Ralfsia* subsystems are probably "modules," dependent on another species for the removal of a potentially dominant competitor. Grazing and predation are equally important, and interdependent, in this community. Grazing drives local community structure in favor of *Chthamalus* at the expense of *Ralfsia* and space on the surface for the limpets; predation on *Chthamalus* by *Acanthina* reverses this tendency.

These results provide clear evidence of the sort of "indirect mutualism" envisioned in the models of Levine (1976) and Vandermeer (1980) in that the interaction is between the consumers of competing prey. Dodson (1970) provided empirical evidence of this type of relationship between size-selective predators in alpine ponds and termed it "complementary feeding niches," an apt description of the *Collisella–Acanthina* relationship in the Gulf. The *Collisella–Acanthina* interaction differs from these other situations in that the prey of one consumer competes for space with the other consumer; note that this tends to increase the benefits that *Collisella* derives from *Acanthina*.

Considerable discussion has revolved around the differences between obligate and facultative mutualism (Boucher et al. 1982; Addicott 1984). The *Acanthina–Collisella* relationship may be obligate within some habitats, like the one where the above experiment was conducted, in the sense that the eventual exclusion of one consumer seems likely to follow the elimination of the other consumer. However, it is premature to suppose that the same sorts of interactions prevail over larger areas, or that in the long run other consumers would not compensate for the loss of *Acanthina* or *Collisella*.

A situation that parallels this one in several respects, although it involves grazers and algae only, was described by Dethier and Duggins (1984). In their study, grazing by the chiton *Katharina tunicata* reduced macroalgal cover, increasing the availability of both space on the rock surface and microalgal food for limpets. They argued that limpet grazing or microalgae were unlikely to affect *Katharina*, directly or indirectly, and that therefore the interaction between *Katharina* and limpets was one of indirect commensalism.

Dethier and Duggins (1984) used their results to illustrate the dependence of indirect interactions between the consumers of competing prey on the symmetry of the competitive interaction between the prey. Asymmetry in the prey competitive interaction in systems like these will lead to an asymmetry in the benefits that the consumers derive from each other. Strictly speaking, commensalism would seem to be an extreme case in which the interaction between prey species is completely one-sided.

As implied in the opening section of this chapter, a large number of considerations enter into grazer–predator interactions on rocky intertidal shores. All of the direct pairwise interactions linking grazers, plants, sessile animals, and predators can influence the relationships between grazers and predators. For example, there is no evidence above or in Dungan (1984, 1986) that *Collisella* bulldozes small *Chthamalus* off the surface as limpets in other localities have been shown to do (e.g., Connell 1961, Dayton 1971, Denley and Underwood 1979). The consequences of barnacle bulldozing by limpets in a system such as that analyzed above would be a reduction in the benefits that the predator receives from the grazer. Indeed, the nature of the relationship could switch to exploitation if, as in Dayton (1971), the limpets severely reduce barnacle abundance while still benefiting from predation on barnacles that escape being dislodged when small (see also Branch 1976).

Where predators consume both grazers and sessile animals (e.g., Paine 1969a, 1974; Underwood et al. 1983), the grazer–predator relationship is complicated by the interplay of direct and indirect interactions. Grazers may benefit indirectly from the removal of sessile animals by predators but pay a direct cost through being eaten. Obviously, the consumption of grazers by predators adds direct benefits to the predator side of the equation, but the reduction of grazer abundance may feed back to affect the predator indirectly as well. The interaction between the predatory starfish *Pisaster ochraceus* and limpets in the Pacific Northwest (Dayton 1971; Paine 1974) provides an illustration in which the predator benefits both directly and indirectly by consuming grazers, the indirect benefits deriving from increased abundance of filamentous algae where grazers are removed (eaten), leading to enhanced recruitment of the predator's primary prey, the mussel *Mytilus californianus*. Based on Paine (1974), limpets derive a net benefit from *Pisaster*'s presence: they are excluded from the rock surface by *Mytilus* when *Pisaster* is removed. In effect, *Pisaster* lessens the extent to which it is exploited by limpets by exploiting them in turn.

In New Zealand (Luckens 1974, 1975) and the British Isles (e.g., Jones 1948; Connell 1961; Southward and Southward 1978; Hawkins and Hartnoll 1983), indirect mutualism between limpets and thaidid gastropods can be hypothesized on the basis of experiments and/or observations suggesting that (1) barnacles and algae compete for space, (2) limpets can be excluded by high barnacle densities, (3) limpet grazing limits algal abundance and thereby enhances the settlement or survival of barnacles, and (4) thaid predation limits barnacle abundance.

Studies by Menge (1976) and Lubchenco and Menge (1978) in New England suggest the elimination of algae, and possibly herbivores, where *Thais lapillus* is removed, allowing barnacles or mussels to monopolize space. Petraitis (1983), Bertness (1984), and Bertness et al. (1983) provide evidence that grazing by *Littorina littorea* in the same region may enhance barnacle settlement and even lead to the replacement of salt marsh habitats by rocky ones (Bertness 1984). Thus, mutualism between *Thais* and *Littorina* in New England can be hypothesized.

Finally, studies by Underwood and co-workers in New South Wales, Australia (e.g., Underwood et al. 1983) indicate a variety of direct and indirect interactions between grazing and predatory gastropods. In their system, the predatory *Morula marinalba* feeds on both sessile animals and limpets, whereas limpets may have beneficial effects on barnacles by removing algae but detrimental effects via bulldozing. Additionally, the barnacle *Tesseropora rosea* appears to compete with one species of limpet (*Cellana tramoserica*) but provide shelter for another (*Patelloida latistrigata*). The complexity of grazer–predator interactions in this system precludes a simple characterization.

Experimental manipulations of grazer and/or predator densities have become commonplace in rocky intertidal ecology. However, the responses of these consumers to consumer-induced variations in the abundance of sessile organisms have rarely been measured (but see Paine 1974; Dethier and Duggins 1984). A worthwhile goal for future studies is to expand the list of dependent variables in studies of grazer or predator effects so as to include grazer effects on predators and vice versa. To do this requires that experimental sites be large enough and monitored for long enough to allow behavioral, numerical, growth, or reproductive responses of consumers to be measured. Obviously, it also requires that experimental manipulations of one consumer's abundance do not involve the exclusion of other consumers. Meeting these requirements will no doubt make replication (and statistical power) more difficult to achieve. In any case, a broad understanding of the organization of rocky intertidal communities must eventually include the nontrophic relationships between grazers and predators.

Many of the contributions to this volume illustrate the propagation of predator effects via trophic pathways in aquatic foodwebs. When the interactions are "substrate-bound," as in rocky intertidal communities and perhaps in freshwater benthic systems as well (e.g., McAuliffe 1984; Hart 1985), the stage is set for a greater variety of pathways through which predator effects can be expressed and for greater complexity in the relationships between predators and co-occurring species. The preponderance of nontrophic effects and pathways in marine rocky intertidal communities suggests a fundamental difference in the way benthic and nonbenthic aquatic communities are organized.

ACKNOWLEDGMENTS

Support was provided by a grant from the University of Arizona Graduate Student Program Development Fund. The hospitality of Dutch and Paula Vreeland is deeply appreciated. Helpful comments on earlier versions of this paper were made by Jim Brown, Joe Connell, Sally Levings, Bruce Menge, Rick McCourt, Jim Munger, John Sutherland, and Don Thomson. Geri Ige was a constant inspiration. Megan Dethier kindly identified *Ralfsia* crusts for me. This paper is drawn from a Ph.D. dissertation submitted to the University of Arizona.

REFERENCES

Abbott, I. A., and G. J. Hollenberg. 1976. *Marine algae of California.* Stanford, Calif.: Stanford University Press.

Addicott, J. F. 1984. Mutualistic interactions in population and community processes. In P. W. Price, C. N. Slobodchikoff, and W. S. Gaud (eds.), *A new ecology: Novel approaches to interactive systems*, pp. 437–46. New York: Wiley Interscience.

Bertness, M. D. 1984. Habitat and community modification by an introduced herbivorous snail. *Ecology* 65 : 370–81.

Bertness, M. D., P. O. Yund, and A. F. Brown. 1983. Snail grazing and the abundance of algal crusts on a sheltered New England rocky beach. *J. Exp. Mar. Biol. Ecol.* 71 : 147–64.

Boucher, D. H., S. James, and K. H. Keeler. 1982. The ecology of mutualism. *Annu. Rev. Ecol. Syst.* 13 : 315–48.

Branch, G. M. 1976. Interspecific competition experienced by South African *Patella* species. *J. Anim. Ecol.* 45 : 507–29.

———. 1984. Competition between marine organisms: Ecological and evolutionary implications. *Oceanography and Marine Biology Annual Review* 22 : 429–593.

Brusca, R. C. 1980. *Common intertidal invertebrates of the Gulf of California.* Tucson, Ariz.: University of Arizona Press.

Choat, J. H. 1977. The influence of sessile organisms on the population biology of three species of acmaeid limpet. *J. Exp. Mar. Biol. Ecol.* 26 : 1–26.

Connell, J. H. 1961. Effects of competition, predation by *Thais lapillus*, and other factors on natural populations of the barnacle *Balanus balanoides*. *Ecol. Monogr.* 31 : 61–104.

———. 1970. A predator-prey system in the marine intertidal region. I. *Balanus glandula* and several predatory species of *Thais*. *Ecol. Monogr.* 40 : 49–78.

———. 1972. Community interactions on marine rocky intertidal shores. *Annu. Rev. Ecol. Syst.* 3 : 169–92.

———. 1974. Ecology: Field experiments in marine ecology. In R. N. Mariscal (ed.), *Experimental marine biology*, pp. 21–54. New York: Academic Press.

Dawson, E. Y. 1966. *Marine algae in the vicinity of Puerto Peñasco, Sonora, Mexico.* Gulf of California Field Guide Series, No. 1. Tucson, Ariz.: University of Arizona.

Dayton, P. K. 1971. Competition, disturbance, and community organization: The provision and subsequent utilization of space in a rocky intertidal community. *Ecol. Monogr.* 41 : 351–89.

———. 1975. Experimental evaluation of ecological dominance in a rocky intertidal algal community. *Ecol. Monogr.* 45 : 137–59.

Denley, E. J., and A. J. Underwood. 1979. Experiments on factors influencing the settlement, survival, and growth of two species of barnacles in New South Wales. *J. Exp. Mar. Biol. Ecol.* 36 : 269–93.

Dethier, M. N., and D. O. Duggins. 1984. An "indirect commensalism" between marine herbivores and the importance of competitive hierarchies. *Amer. Naturalist* 124 : 205–19.

Dodson, S. I. 1970. Complementary feeding niches

sustained by size-selective predation. *Limnol. Oceanogr.* 15 : 131–47.

Dungan, M. L. 1984. Experimental analysis of processes underlying the structure of a rocky intertidal community in the northern Gulf of California. Ph.D. thesis, University of Arizona, Tucson.

———. 1985. Competition and the morphology, ecology, and evolution of acorn barnacles: an experimental test. *Paleobiology* 11 : 165–73.

———. 1986. Three-way interactions: barnacles, limpets, and algae in a Sonoran desert rocky intertidal zone. *Amer. Naturalist* 127 : 292–316.

Dungan, M. L., T. E. Miller, and D. A. Thomson. 1982. Catastrophic decline of a top carnivore in the Gulf of California rocky intertidal zone. *Science* 216 : 989–91.

Fairweather, P. G., A. J. Underwood, and M. J. Moran. 1984. Preliminary investigations of predation by the whelk *Morula marginalba*. *Marine Ecology Progress Series* 17 : 143–56.

Fawcett, M. H. 1984. Local and latitudinal variation in predation on an herbivorous marine snail. *Ecology* 65 : 1214–30.

Federer, W. T. 1975. The misunderstood split plot. In R. P. Gupta (ed.), *Applied statistics*, pp. 9–39. New York: American Elsevier.

Gaines, S. D., and J. Lubchenco. 1982. A unified approach to marine plant-herbivore interactions. II. Biogeography. *Annu. Rev. Ecol. Syst.* 13 : 111–38.

Garrity, S. D., and S. C. Levings. 1981. A predator-prey interaction between two physically and biologically constrained tropical rocky shore gastropods. *Ecol. Monogr.* 51 : 267–86.

Hart, D. D. 1985. Grazing insects mediate algal interactions in a stream benthic community. *Oikos* 44 : 40–46.

Hawkins, S. J., and R. G. Hartnoll. 1983. Grazing of intertidal algae by marine invertebrates. *Oceanography and Marine Biology Annual Review* 21 : 195–282.

Hurlbert, S. H. 1984. Pseudoreplication and the design of ecological field experiments. *Ecol. Monogr.* 54 : 187–211.

Jara, H. F., and C. A. Moreno. 1984. Herbivory and structure in a midlittoral community: a case in southern Chile. *Ecology* 65 : 28–38.

Jones, N. S. 1948. Observations and experiments on the biology of *Patella vulgata* at Port St. Mary, Isle of Man. *Proceedings and Transactions of the Liverpool Biological Society* 56 : 60–77.

Levine, S. 1976. Competitive interactions in ecosystems. *Amer. Naturalist* 110 : 903–10.

Lewis, J. R., and R. S. Bowman. 1975. Local habitat-induced variations in the population dynamics of *Patella vulgata* L. *J. Exp. Mar. Biol. Ecol.* 17 : 165–203.

Littler, M. M., and D. S. Littler. 1984. Relationships between macroalgal functional form groups and substrata stability in a subtropical rocky-intertidal system. *J. Exp. Mar. Biol. Ecol.* 74 : 13–34.

Lively, C. M. 1984. Competition, predation, and the maintenance of dimorphism in an acorn barnacle *(Chthamalus anisopoma)* population. Ph.D. thesis, University of Arizona, Tucson.

Lubchenco, J. 1978. Plant species diversity in a marine rocky intertidal community: Importance of herbivore food preference and algal competitive abilities. *Amer. Naturalist* 112 : 23–39.

———. 1979. Consumer terms and concepts. *Amer. Naturalist* 113 : 315–17.

Lubchenco, J., and J. Cubit. 1980. Heteromorphic life histories of certain marine algae as adaptations to variations in herbivory. *Ecology* 61 : 676–87.

Lubchenco, J., and S. D. Gaines. 1981. A unified approach to marine plant-herbivore interactions. I. Populations and communities. *Annu. Rev. Ecol. Syst.* 12 : 405–37.

Lubchenco, J., and B. A. Menge. 1978. Community development and persistence in a low rocky intertidal zone. *Ecol. Monogr.* 48 : 67–94.

Lubchenco, J., B. A. Menge, S. D. Garrity, P. J. Lubchenco, L. R. Ashkenas, S. D. Gaines, R. Emlet, J. Lucas, and S. Strauss. 1984. Structure, persistence, and the role of consumers in a tropical rocky intertidal community (Taboguilla Island, Bay of Panama). *J. Exp. Mar. Biol. Ecol.* 77 : 23–73.

Luckens, P. A. 1974. Removal of intertidal algae by herbivores in experimental frames and on shores near Auckland. *N. Z. J. Mar. Freshwater Res.* 8 : 637–54.

———. 1975. Predation and intertidal zonation of barnacles at Leigh, New Zealand. *N. Z. J. Mar. Freshwater Res.* 9 : 355–78.

Maluf, L. Y. 1983. Physical oceanography of the Gulf of California. In T. J. Case and M. L. Cody (eds.), *Island biogeography in the Sea of Cortez*, pp. 26–45. Los Angeles: University of California Press.

Malusa, J. R. 1983. Comparative reproductive ecology of two species of rocky intertidal barnacle. M. S. thesis, San Diego State University, San Diego, Calif.

———. 1985. Attack mode in a predatory gastropod: Labial spine length and the method of prey capture in *Acanthina angelica* Oldroyd. *The Veliger* 28 : 1–5.

McAuliffe, J. R. 1984. Competition for space, disturbance, and the structure of a benthic stream community. *Ecology* 65 : 894–908.

McCourt, R. M. 1984. Seasonal patterns of abundance, distributions, and phenology in relation to growth strategies of three *Sargassum* species. *J. Exp. Mar. Biol. Ecol.* 74 : 141–56.

Menge, B. A. 1976. Organization of the New England rocky intertidal community: Role of predation, competition, and environmental heterogeneity. *Ecol. Monogr.* 46 : 355–93.

———. 1978. Predation intensity in a rocky intertidal community: Effect of an algal canopy, wave action, and desiccation on predator feeding rates. *Oecologia* 34 : 17–35.

Menge, B. A., and J. Lubchenco. 1981. Community organization in temperate and tropical rocky intertidal habitats: Prey refuges in relation to consumer pressure gradients. *Ecol. Monogr.* 51 : 429–50.

Menge, B. A., and J. P. Sutherland. 1976. Species diversity gradients: Synthesis of the roles of predation, competition, and temporal heterogeneity. *Amer. Naturalist* 110 : 351–69.

Mercurio, K. S., A. R. Palmer, and R. B. Lowell. 1985. Predator-mediated microhabitat partitioning in two species of visually cryptic intertidal limpets. *Ecology* 66 : 1417–25.

Paine, R. T. 1966a. Food web complexity and species diversity. *Amer. Naturalist* 100 : 65–75.

———. 1966b. Function of the labial spine, composition of diet, and size of certain marine gastropods. *The Veliger* 9 : 17–24.

———. 1969a. The *Pisaster-Tegula* interaction: Prey patches, predator food preference, and intertidal community structure. *Ecology* 50 : 950–61.

———. 1969b. A note on trophic complexity and community stability. *Amer. Naturalist* 103 : 91–93.

———. 1971. A short-term experimental investigation of resource partitioning in a New Zealand rocky intertidal habitat. *Ecology* 52 : 1096–1106.

———. 1974. Intertidal community structure: Experimental studies on the relationship between a dominant competitor and its principle predator. *Oecologia* 15 : 93–120.

———. 1977. Controlled manipulations in the marine intertidal zone and their contributions to ecological theory. Academy of Natural Sciences, Philadelphia, Special Publications 12 : 245–70.

———. 1980. Food webs: Linkage, interaction strength, and community infrastructure. *J. Anim. Ecol.* 49 : 667–85.

———. 1981. Barnacle ecology: Is competition important? The forgotten roles of predation and disturbance. *Paleobiology* 7 : 553–60.

———. 1984. Ecological determinism in the competition for space. *Ecology* 65 : 1339–48.

Paine, R. T., and S. A. Levin. 1981. Intertidal landscapes: Disturbance and the dynamics of pattern. *Ecol. Monogr.* 51 : 145–78.

Paine, R. T., and T. H. Suchanek. 1983. Convergence of ecological processes between independently evolved competitive dominants: A tunicate-mussel comparison. *Evolution* 37 : 821–31.

Petraitis, P. S. 1983. Grazing patterns of the periwinkle and their effect on sessile intertidal organisms. *Ecology* 64 : 522–33.

Sousa, W. P. 1979. Experimental investigations of disturbance and ecological succession in a rocky intertidal algal community. *Ecol. Monogr.* 49 : 227–54.

————. 1984. Intertidal mosaics: patch size, propagule availability, and spatially variable patterns of succession. *Ecology* 65 : 1918–35.

Southward, A. J., and E. C. Southward, 1978. Recolonization of rocky shores in Cornwall after use of toxic dispersants to clean up the *Torrey Canyon* spill. *J. Fish. Res. Board Can.* 35 : 682–706.

Stephenson, T. A., and A. Stephenson. 1972. *Life between tidemarks on rocky shores.* San Francisco: Freeman and Co.

Stimson, J. S. 1970. Territorial behavior in the owl limpet, *Lottia gigantea. Ecology* 51 : 113–18.

Thomson, D. A., L. T. Findley, and A. N. Kerstich. 1979. *Reef fishes of the Sea of Cortez.* New York: Wiley and Sons.

Thomson, D. A., and C. E. Lehner. 1976. Resilience of a rocky intertidal fish community in a physically unstable environment. *J. Exp. Mar. Biol. Ecol.* 22 : 1–29.

Turk, M. J. 1979. Intertidal migration and formation of breeding clusters of labial spine morphs of the thaid gastropod, *Acanthina angelica.* M.S. thesis, University of Arizona, Tucson.

Underwood, A. J. 1979. The ecology of intertidal gastropods. *Advances in Marine Biology* 16 : 111–210.

————. 1980. The effects of grazing by gastropods and physical factors on the upper limits of intertidal macroalgae. *Oecologia* 46 : 201–13.

————. 1985. Physical factors and biological interactions: the necessity and nature of ecological experiments. In P. G. Moore and R. Seed (eds.), *The ecology of rocky coasts: Essays presented to J. R. Lewis,* pp. 372–90. London: Hodder and Stoughton.

Underwood, A. J., and E. J. Denley. 1984. Paradigms, explanations, and generalizations in models for the structure of intertidal communities on rocky shores. In D. Strong, D. Simberloff, L. G. Abele, and A. B. Thistle (eds.), *Ecological communities: Conceptual issues and the evidence,* pp. 151–80. Princeton, N.J.: Princeton University Press.

Underwood, A. J., E. J. Denley, and M. J. Moran. 1983. Experimental analyses of the structure and dynamics of mid-shore rocky intertidal communities in New South Wales. *Oecologia* 56 : 202–219.

Vandermeer, J. H. 1980. Indirect mutualism: variations on a theme by Stephen Levine. *Amer. Naturalist* 116 : 441–48.

Yensen, N. P. 1973. Limpets of the Gulf of California (Patellidae, Acmaeidae). M.S. thesis, University of Arizona, Tucson.

————. 1979. The function of the labial spine and the effect of prey size on switching polymorphs of *Acanthina angelica* (Gastropoda: Thaididae). Ph.D. thesis, University of Arizona, Tucson.

IV. Behavioral and Morphological Responses

14. Predators and Prey Lifestyles: An Evolutionary and Ecological Overview

Andrew Sih

Most studies of the effects of predators on prey emphasize lethal effects, i.e., direct predation. However, predators can also have far-reaching indirect influences on prey through their effects on prey life-styles; that is, on prey morphology, physiology, chemistry, life history, or behavior. This chapter provides an overview of (1) major categories of antipredator traits; (2) the evolution of such traits; and (3) the ecological effects that these traits have on interactions between prey, predators, competitors, and resources.

Prey can reduce predation pressure either by avoiding encounters with predators or by escaping after an encounter. Within these two main categories I distinguish several major subcategories (e.g., avoidance by hiding in refuges, avoidance by using ephemeral habitats) and suggest ways in which these subcategories affect overall prey life-style (e.g., avoidance by hiding often results in a "slow life-style," whereas the use of ephemeral habitats requires a "fast life-style"). The life-style evolved by a given species can be explained in terms of both cost/benefit analyses and in terms of constraints on selection.

Within any general life-style, prey often show short-term responses to predators, e.g., alterations in their habitat use, activity schedules, or movement patterns. Here, I list several predictions on optimal antipredator responses and review some studies that address these predictions. Many studies support the most obvious prediction, that greater predation risk should result in greater prey response. The other optimality-based predictions have rarely been addressed.

Notice, however, that although some prey show flexible responses to predators, other prey appear to show relatively fixed avoidance of predators; that is, they stay in refuges as a way of life despite fluctuations in local predation pressure. Here, I summarize a cost/benefit model that suggests that fixed predator avoidance might be due to the often prohibitive cost of gathering information about predation pressure. In essence, for prey that have a low chance of surviving an encounter with a predator, the likely cost of gathering information about predation risk is death. Without this information, adaptive shifts are not possible.

Finally, I explore some general ways in which antipredator defenses might affect species interactions. Studies of interactions among prey usually focus on competition for limited resources. Competition is often studied by examining niches. Here I discuss the consequences of the fact that prey niches (habitat use, diet, activity times) are often heavily affected by antipredator needs. Notably, predators can potentially increase competition among prey by forcing prey to share common refuges. Interactions among predators are also often couched in terms of competition for shared resources. My emphasis here is on mutual effects on prey availability not through exploitation but through effects on prey avoidance behaviors. Finally, I discuss behavioral "leapfrog" effects among three trophic levels. For example, carnivores can profoundly affect plants by influencing herbivore behavior.

Predation is generally thought to be one of the major forces influencing population dynamics (e.g., Murdoch and Oaten 1975; Hassell 1978; Taylor 1984), and community structure (e.g., Paine 1966; Connell 1975; Zaret 1980; Sih et al. 1985). The usual focus is on the direct lethal effects of predators. By killing prey, predators can control prey populations, drive some prey types extinct, and alter the relative and absolute abundances and species diversity of prey. Predators also, however, have potentially important indirect effects on prey through their shaping of prey life-styles; that is, through the effects that predators have on the morphology, physiology, chemistry, life history, and behavior of prey. In this and several subsequent chapters the focus will be on the evolution and ecological importance of predator effects on prey life-styles.

The goals of this chapter are to provide an overview on (1) major categories of antipredator traits; (2) the evolution of these traits; and (3) the ecological effects that these traits have on interactions between prey and predators, competitors, and resources.

PREDATORS AND PREY LIFE-STYLES

Prey show a dazzling range of antipredator traits (see reviews by Edmunds 1974, Morse 1980, Janzen 1981, Jeffries and Lawton 1984). In this section, I attempt to organize this diverse array of traits into several main categories and to discuss ways in which antipredator attributes might constrain the overall life-styles shown by prey.

Antipredator defenses can be divided into two main categories based on whether they act before or after a predator–prey encounter takes place. An encounter occurs when predators detect and recognize prey. Avoidance of predators occurs before encounters and tends to decrease the prey's encounter rate with predators. In contrast, escape occurs after prey encounter predators; escape acts to decrease the chance that prey will be attacked, captured, or consumed by predators. Some major categories of antipredator defense are listed in table 14.1 and discussed below.

Escape after Encountering Predators

Most studies of antipredator defenses examine traits that enhance escape success, i.e., traits that increase the probability that prey survive despite encounters with predators. Obvious examples of such defenses include: active flight and morphological or chemical defenses that make prey difficult to handle. The defenses themselves have been admirably reviewed (see Edmunds 1974 for examples of any of the escape behaviors discussed below). My focus here is on tradeoffs associated with these antipredator defenses. As the reader will see, al-though many potential tradeoffs can be identified, relatively few have been studied in any detail.

If prey are detected and recognized, their next line of defense is to reduce the probability that they will be attacked. Predators often avoid attacking prey that are clearly difficult to capture, handle, or digest. For example, many animals and perhaps most plants possess chemical defenses along with visual or chemical "warning" cues that signal their unpalatability (e.g., Edmunds 1974; Rosenthal and Janzen 1979; Denno and McClure 1983; this volume: Havel; Scrimshaw and Kerfoot). Although it is often assumed that these chemical defenses are costly (e.g., Feeny 1976; Rhoades 1979), in general, neither the costs of evolving a particular biochemical pathway nor the immediate costs of producing particular compounds have been quantified (e.g., Fox 1981; Futuyma 1983).

If prey are attacked, they next use defenses that reduce the probability of successful capture. Many prey prevent capture by attempting rapid escape or flight. Flight success is enhanced by staying near refuges, by using deimatic displays, and by using unpredictable (protean) escape patterns (see Edmunds 1974). Being built for speed can have costs in terms of reduced feeding or movement efficiency. A simple tradeoff, for example, exists between a full belly and fast movement (e.g., DeBenedictis et al. 1978). In a similar vein, in ungulates, a light body that is built for speed is incompatible with the extensive digestive systems required to process bulky plant foods efficiently (Jarman 1974). In addition, physiological or morphological traits that favor bursts of speed may be incompatible with traits required for efficient sustained movement (e.g., Gittelman 1974, Taigen et al. 1982, Giller 1982).

The other main defense used by prey to prevent capture is active defense, i.e., fighting. This recourse is, of course, only available to prey with appropriate weapons, e.g., claws, horns. Weapons are often also used by prey either to capture their own prey or for intraspecific combat (Vermeij 1982). The optimal shape and size of these weapons might differ for these different needs. It should be interesting to attempt to quantify the degree to which the evolution of weapons in response to one need reduces (or increases) efficiency relative to other needs.

Even after prey are "captured," they can prevent successful handling by various means.

Table 14.1. *Prey attributes that decrease predation risk and possible constraints associated with these traits*

Stage	Prey attribute	Possible constraint
ESCAPE AFTER ENCOUNTER		
Reduce probability of attack	Unpalatability	Cost of producing chemical
Reduce capture success	High speed of movement aided by deimatic and protean displays	Conflict between capacity for rapid and efficient movement
	Weapons for active defense	Conflict between optimal morphology for fighting vs. feeding
	Living in groups	Increased competition, intragroup aggression
Reduce handling success	Unwieldy size or shape	Conflict with optimal size or shape for feeding
AVOID ENCOUNTERS		
Reduce proximity to predators	Reduce activity, restrict activity in space or time	Slow life-style
	Use stressful habitats	Slow life-style
	Use ephemeral habitats	Fast life-style
Reduce probability of detection and recognition	Crypsis and reduced activity	Slow life-style

Prey have often evolved spines or armor (or other similar morphological traits) that tend to make them more difficult to handle. These defenses can be costly to produce, maintain and carry around so that prey possessing them often have lower feeding (photosynthetic), growth, or reproductive rates than similar prey without these defenses (e.g., Rhoades 1979; Dodson 1984; Havel, this volume). Prey also can reduce handling success by having a size or shape that is difficult to handle (see Zaret 1980; Jeffries and Lawton 1984; O'Brien, Power, Stemberger and Gilbert, this volume; for reviews). Body size and shape, however, also affect an organism's efficiency relative to many other needs (e.g., Hall et al. 1976, Peters 1983). Further analyses of optimal size or shape accounting for these conflicting needs should yield important insights.

Many of the antipredator mechanisms listed above are more effective when prey live in single-species or multispecies groups (see Bertram 1978; Pulliam and Caraco 1984; Folt, this volume, for reviews). Larger groups can detect predators earlier (e.g., Kenward 1978; Treherne and Foster 1980), confuse predators (e.g., Neill and Cullen 1974, Milinski 1979), and deter predators through group defense (Edmunds 1974). Bertram (1978) suggested that perhaps the strongest antipredator benefit of living in groups comes through dilution; that is, that if predators kill a limited number of prey, then even without the other advantages listed above, a prey individual can reduce the probability that it will be killed by living in a large group. Saturation of predators by synchronous emergence or reproduction (see Janzen 1981) is an extreme example of the use of the dilution effect.

Unlike most of the other defenses discussed above, the tradeoffs relevant to group living have been examined in some detail. Living near one's competitors presumably should increase competition and thus could result in reduced feeding efficiency; however, prey also might actually feed more efficiently in groups because groups might be more effective at finding resource patches, groups might forage more systematically, and individuals in groups might be able to devote more of their time to foraging (see the above reviews for details).

To summarize this section: behavioral ecologists have long known about a range of prey attributes that increase the prey's chances of surviving encounters with predators. Less attention has been devoted to quantifying and understanding the implications of the costs of these antipredator defenses; that is, the ways in which these defenses constrain the prey's overall life-style.

Avoiding Encounters

Prey can avoid encountering predators in two ways: they can either stay away from predators (avoid entering the predator's sensory field) or they can come close to predators but avoid detection or recognition as prey. Prey commonly use crypsis to avoid detection or recognition. To reduce proximity to predators, prey often either reduce their activity levels or limit activity to places (or times) where (or when) predators are inactive.

Crypsis (including transparency and mimicry of immobile objects) has been well studied (see Cott 1940; Edmunds 1974). It is usually only effective if prey are camouflaged and move relatively little. Crypsis thus is often associated with ambush foraging by prey, inactivity in the day, and restricted habitat use (to areas with the correct background). All of these behaviors tend to reduce encounter rates between prey and their own resources; that is, cryptic prey cannot take full advantage of the available food in the environment. Crypticity might thus be associated with relatively low metabolic, feeding, growth, developmental, and/or reproductive rates. This general life-style of low activity and low energetic rates I refer to as a slow life-style.

Noncryptic prey can persist locally with predators by limiting their activity to relatively predator-free situations. Reduced activity per se is a typical antipredator response that presumably reduces the frequency of contact with predators and the likelihood of detection and recognition as prey (e.g., Stein and Magnuson 1976, Sih 1979, 1981, 1982, Peckarsky 1980, Wright and O'Brien 1984, Dill and Fraser 1984; O'Brien, Power, Stemberger and Gilbert, this volume). Further, prey often restrict their activity to microhabitats where predators are less effective, i.e., use refuges (e.g., Charnov et al. 1976, Stein and Magnuson 1976, Wilson et al. 1978, Sih 1979, 1980, 1981, 1982, 1984, Peckarsky 1980, Cerri and Fraser 1983, Werner et al. 1983, Kotler 1984, Cooper 1984; Mittlebach and Chesson, Power, this volume); and restrict activity to times of the day (e.g., Gentry 1974, Wilson and Clark 1977, Polis 1980; and see Stein 1979 for a review) or seasons (e.g., Hairston; Stemberger and Gilbert, this volume) when predators are relatively inactive. These mechanisms for "coexisting in hiding"

are potentially costly to prey in terms of lost opportunities to feed or mate. Habitat use is a key niche component; and activity levels and movement patterns are key elements of an organism's foraging mode. The constraint imposed on these parameters by antipredator needs has been shown to reduce prey feeding or growth rates (e.g., Stein and Magnuson 1976, Murdoch and Sih 1978, Sih 1980, 1981, 1982, Werner et al. 1983, Dill and Fraser 1984; Lampert, Mittelbach and Chesson, Power, this volume). In essence, even when resource availability is high, an organism might not be able to take advantage of these resources because of the conflicting need to avoid predators. As a result, prey that coexist by hiding often appear to have evolved the slow life-style characterized by low energetic rates and the use of costly antipredator defenses. A diverse array of taxa, including many burrowing molluscs and annelids, leaf-litter arthropods, spiders, odonate larvae, antlions, stream salamander larvae, and late successional plants appear to fit this slow life-style.

Alternatively, prey can avoid predators by using habitats that predators are rarely found in either because the habitats are too ephemeral ("fugitive" prey) or too stressful ("stress-tolerant" prey) for predators.

Fugitive prey persist in habitats that are temporarily free from predators. These temporary refuges can either be habitats that are suitable for predators but that predators have simply not recolonized following a disturbance, or they can be habitats that are disturbed so frequently that predators cannot persist. Examples of fugitive prey include aquatic amphibian and insect larvae in temporary pools; terrestrial insect larvae in rotting fruits, carrion, or dung; and "weedy" early successional plants either on land or underwater.

Because of the temporary nature of the refuge, prey must both recolonize rapidly (through either dispersal or exit from diapause) and feed, grow, develop, and, if appropriate, reproduce rapidly. Rapid energetic rates are particularly important for prey that must attain a particular life stage before being able to escape from the habitat via either dispersal or diapause. In the above examples, amphibians and insects generally must reach the adult stage to disperse, and annual plants must produce seeds that then disperse or undergo dormancy. Because predators are rare or absent, prey need not devote time and energy to anti-predator responses that can be costly in terms of lost feeding opportunities. Prey are thus free to take advantage of available food supplies to enhance their chance of achieving the required high energetic rates. In sum, the fugitive strategy is associated with a "fast life-style" characterized by rapid energetic rates and weak anti-predator responses. This basic scheme has been noticed in several groups including amphibian larvae (e.g., Wilbur 1980, 1984, Woodward 1983), aquatic insects (e.g., Johnson and Crowley 1980, Pierce et al. 1985), zooplankton (e.g., O'Brien 1979) and both terrestrial (e.g., Feeny 1976, Rhoades 1979) and marine plants (e.g., Lubchenco and Gaines 1981; Hay 1984).

Stress-tolerant prey live in physically severe habitats that select for a very slow life-style in response to stress per se. These prey typically have low metabolic, foraging, growth, developmental, and reproductive rates, and late maturity (Grime 1977; Greenslade 1983). Physical stress often results in low predation pressure because predators are either rare or very inefficient (e.g., Connell 1975; Menge 1978). Prey then put little investment in antipredator defense. However, if predation pressure is not strongly reduced, then the low growth rates associated with stress select for high investment in antipredator defense (Coley et al. 1985). In plants, a negative association exists between inherent growth rates and amounts of chemical defense (Coley et al. 1985).

Many of the traits associated with the fugitive, coexist-in-hiding, and stress-tolerant strategies outlined above are similar to those of the oft-discussed r, K, and stress-tolerant strategies (e.g., Pianka 1970, Grime 1977, Greenslade 1983, Horn and Rubenstein 1984). The major difference is that while r and K traits are supposedly produced primarily in response to different competitive regimes, the same traits are here interpreted in terms of antipredator pressures. The r and K theory has fallen into disfavor because of the large number of exceptions. Many of the exceptions quite possibly can be explained in terms of antipredator needs (e.g., Wilbur et al. 1974). However, I do not mean to suggest that prey life-styles are exclusively molded by antipredator needs. The fugitive or stress-tolerant strategies often decrease encounter rates with predators; however, they also often decrease encounter rates

with competitors and other enemies. Generalities based on only one of many selective pressures are likely to yield incomplete explanations. My main point here is that the spectrum of prey life-styles often attributed to competitive pressures also can be partially, if not primarily, explained by antipredator needs.

But what evidence do we have that the use of ephemeral and stressful habitats are antipredator defenses? The answer depends on our definition of antipredator defense. Some might say that any trait that confers the benefit of reduced predation risk should be considered an antipredator defense. Gould and Vrba (1982) call these antipredator aptations (see Vermeij 1982, 1985; Sih 1985). By this definition, fugitive and stress-tolerant strategies are clearly antipredator defenses. Others might require that the trait evolved primarily for antipredator purposes; i.e., an antipredator adaptation (cf. Gould and Vrba 1982). Not only is this a much more restrictive definition, but it is much more difficult to gather evidence for. Evolutionary biologists will be most interested in identifying antipredator adaptations, whereas ecologists will be primarily interested in aptations with only a secondary interest in evolutionary pathways. As an evolutionary ecologist I take the tack of discussing aptations (e.g., fugitive and stress-tolerant strategies) because they are of ecological interest, while advising the reader to keep in mind that the evidence is unclear concerning whether they are adaptations or not.

To summarize this section: Prey can reduce predation pressure by avoiding encounters with predators. Four main strategies are identified. The fugitive strategy is associated with a fast life-style; crypsis and coexistence by hiding often require a slow life-style; the stress-tolerant strategy requires a very slow life-style.

EVOLUTION OF PREY LIFE-STYLES

What determines the particular life-style evolved by a given species? Two main classes of explanations exist: those based on natural selection versus those that emphasize the constraints on selection (e.g., Maynard Smith 1978; Gould and Lewontin 1979). I suggest that these mechanisms play a complementary role in shaping prey life-styles relative to predators. Using the adaptive landscape analogy

(Wright 1931), different prey life-styles can be represented as multiple adaptive peaks. The question then is, which peak will a species scale? Natural selection tends to push a species up the most accessible peak. If several peaks are similar in accessibility (i.e., if genetic, developmental, or phylogenetic constraints are not important barriers), then the costs and benefits of various strategies should determine the life-style followed.

Evaluating the costs and benefits associated with alternative life-styles is a complex multi-level problem. For example, suppose we wish to evaluate the relative benefit of evolving armor versus hiding as antipredator defenses. Measuring contemporary ecological costs and benefits (e.g., decreased mortality, decreased feeding rate) is not a trivial task; however, this difficulty pales beside that of evaluating possible genetic and developmental tradeoffs. An evolutionary analysis should also examine the cost of evolving armor in terms of all other required associated alterations. Further steps in analyzing the relative evolutionary costs and benefits of evolving new forms will require a deeper understanding of the links between genetics, developmental biology, and evolutionary ecology (e.g., Alberch 1982; Maynard Smith et al. 1985). Achieving this synthesis is surely a major challenge for modern evolutionary biology.

Environmental variability might play an important role in explaining why a given species relies more on behavioral versus morphological defenses against predators. Cost–benefit analyses usually focus on the mean fitness associated with a strategy. However, variance in fitness also can affect the evolution of traits (e.g., Real 1980; Caraco 1980). In general, theory says that variance in fitness decreases expected fitness. Inflexible traits, by definition, are incapable of responding to short-term alterations in selective pressures. In particular, inflexible traits might yield large benefits in reduced mortality when predators are present but also large costs in reduced feeding or mating when predators are absent. The result should be high variance in fitness as the predation regime varies. This suggests that (if all else is equal) fixed morphological traits should be favored in a stable environment, whereas flexible behavioral responses to predators should be favored in variable conditions. Whether this logic yields important insights into nature remains to be seen.

The alternative to cost–benefit explanations emphasizes the role of constraints on natural selection. Ancestry might often play a crucial role in determining current traits and life-styles. Returning to the adaptive landscape analogy, if one peak is far more accessible than the others, then a species will most likely scale that peak. In this scenario, a species' initial state (i.e., its ancestral condition) determines the path of natural selection, with both ancestry and selection having important influences on a species' current life-style.

A plausible example of this role of ancestral inertia was suggested by Kerfoot (1982) for explaining the evolution of antipredator adaptations in freshwater arthropods. Kerfoot (1982) noted that most species rely on either a tough body and chemical defenses or some combination of hiding, transparency, small size, and evasive maneuvers. The evolution of distastefulness via individual selection requires that prey individuals survive encounters with predators (see Vermeij 1982; Sih 1985). Species with ancestors that had hard or spongy bodies (mites, beetles, bugs) that could withstand being mouthed by fish evolved morphochemical defenses. In contrast, taxa lacking this "preadaptation" (mayflies, odonates, zooplankton) now rely on the latter set of defenses. In terms of adaptive landscapes, soft-bodied animals had too deep a valley to cross (the evolution of a hard or spongy body) to reach the morphochemical defense peak. More attempts at using a blend of ancestral constraints and natural selection to interpret differences in life-styles should prove rewarding.

SHORT-TERM ANTIPREDATOR RESPONSES

Within a given life-style, prey often show considerable flexibility in their antipredator traits. Individual prey often show behavioral flexibility, but other types of traits are also inducible (see Havel; Stemberger and Gilbert, this volume). Here I attempt to provide an evolutionary framework for understanding shifts in antipredator behavior depending on parameters such as predation risk, hunger level, mating opportunities, and patterns in current versus future reproductive success. I begin by summarizing some intuitively reasonable optimality-based predictions. I then review the existing data on these predictions and suggest some gaps requiring future study.

Optimality analyses have been criticized for requiring that organisms somehow be capable of gathering and processing the diverse information necessary to calculate the optimal behavior (e.g., Oaten 1977; Green 1980). Although it might be asking too much for prey to fit the quantitative predictions of optimality theory (but see Belovsky in press), prey do appear to generally fit the qualitative predictions of the theory (e.g., Pyke 1984; papers in Krebs and Davies 1984). In any case, the exercise of listing some predictions on optimal responses to predators along with a review of relevant empirical data should prove illuminating.

The variable of concern here is the degree of exposure to predators. Greater exposure has a cost in increased predation and usually a benefit in increased time devoted to other activities such as feeding or mating. Using terminology from life history theory, fitness is maximized by maximizing total expected reproductive success, which is the sum of current and future reproductive success; or in mathematical terminology, the optimal time spent exposed to predators (T^*), maximizes $b + pV$, where b = current reproductive success; p = survivorship to future reproduction, and V = future reproductive success (Schaffer 1974; Taylor et al. 1974).

Elsewhere (Sih, in preparation), I analyze a more detailed version of the above model. Here, I simply list some predictions:

1. An increase in potential predation risk decreases the optimal time spent exposed to predators, T^*.
1a. The decrease in T^* should result in decreased prey feeding, growth, or mating rates.
2. Increased mating opportunities should increase time spent exposed to predators.
3. Increased reproductive value of a current group of offspring should increase time spent protecting these offspring against predators.
4. Increased prey hunger level should increase time spent exposed to predators.
5. An increase in dV/dT (the marginal increase in future reproductive success with

additional exposure to predators) should increase time spent exposed to predators.
6. Increased future reproductive success per se should decrease the optimal degree of exposure to predators.

For predictions 2 through 6 corollary predictions are that when T^* increases, predator-caused mortality should increase, and vice versa.

The relative amount of time spent exposed to predators can be evaluated in several ways. For prey that rely heavily on avoidance of encounters with predators (e.g., by avoidance in time or space, or by reduced movement), the degree of exposure to predators is the proportion of time spent outside of refuges or the proportion of time spent moving. For prey that rely more on escape following an encounter, flight distance (the distance between predator and prey at which prey initiate an escape response) or intensity of flight can be used as an inverse measure of exposure, i.e., greater response corresponds to reduced exposure to predators.

What follows is a review of some studies that address these predictions. Unfortunately, only the more obvious prediction on the relationship between risk and exposure to predators has been examined in detail. The more subtle predictions on how other needs that conflict with antipredator needs might affect antipredator responses has received relatively little attention; and virtually no studies have examined how these conflicting needs affect predation rates.

Prediction 1: Greater Risk, Reduced Exposure

ONE PREY, VARYING PREDATION REGIMES. This prediction can be tested by comparing the responses of a given prey type to predation regimes that vary in risk. Numerous observations show that prey typically can evaluate predation risk at a gross level, e.g., they can tell predators from similar nonpredators (see Edmunds 1974 for a review), and they can tell if predators are actively foraging or not (e.g., Taylor and Chen 1969; Dayton et al. 1977; Shalter 1978). More interestingly, some prey appear to be capable of evaluating the relative danger associated with various types of predators (e.g., Walther 1969; Russel 1972; Hennessy and Owings 1978).

The above examples compare the responses of a given prey individual to different predators. Prey also have been shown to vary their responses to a given predator depending on factors that mediate risk (e.g., depending on the distance between the predator and the prey, or between the prey and a refuge). In general, prey show a more intense escape response when predators are closer or refuges are farther away. At the extreme, when predators are very close, so that escape appears impossible, prey feign death or fight. Elements of this spectrum of responses have been noted for moose (Mech 1970), gazelle (Schaller 1972), sparrows (Pulliam and Mills 1977), ducks (Sargeant and Eberhardt 1975), termites (Wilson and Clark 1977), backswimmers (Sih, unpublished data), aphids (Russel 1972), mayfly nymphs (Peckarsky 1980), and various marine invertebrates (see Edmunds 1974 for a review).

Table 14.2. *Examples of where more susceptible prey show stronger antipredator responses*

Prey	Predator	Relationship	Reference
Marine snails	Fish	Thin-shelled snails use shallow areas with fewer predators	Kitching and Lockwood 1975
Marine snails	Fish	Smaller snails avoid submergence or are less active when submerged	Bertness et al. 1981
Sea urchins	Seastars	Smaller prey respond more strongly to smell of predator	Duggins 1981
Seastars	Seastars	Smaller prey escape more actively	Van Veldhuizen and Oakes 1981
Freshwater zooplankton	Fish	Degree of vertical migration related to predation risk	Zaret and Suffern 1976
Marine zooplankton	Fish	More visible prey feed only at night	Hobson and Chess 1976
Crayfish	Fish	Smaller prey use areas with greater spatial heterogeneity	Stein and Magnuson 1976; Stein 1977
Grasshoppers	Birds	Smaller prey escape more actively	Schultz 1981
Lepidopteran larvae	Birds	More palatable prey are less active in day and stay on undersides of leaves	Heinrich 1979a
Juvenile backswimmers	Adult backswimmers	Smaller prey spend more time in refuge and are less active	Sih 1980, 1982
Aphids	Beetles	Smaller prey escape more actively	Russell 1972
Sticklebacks	Fish	Smaller prey with smaller, lighter spines spend more time in refuge	Wilz 1971
Mosquito fish	Fish	Smaller prey orient more strongly to shallow areas	Goodyear 1973
Bluegills	Bass	Smaller prey spend more time in vegetation	Werner et al. 1983
Armored catfish	Wading birds Fish	Larger prey use deeper waters Smaller prey use shallower waters	Power 1984
Fish	Fish	More conspicuous, less-armored prey show more elusive escape responses	Neill 1970
Frog larvae	Fish	Metamorphosing prey spend more time in refuge	Wassersug and Sperry 1977
Lizard	Various vertebrates	Gravid females are less active	Bauwens and Thoen 1981
Small rodents	Various vertebrates	Slower prey spend more time in vegetation	Kotler 1984
Marmots	Various vertebrates	Younger prey are more vigilant, stay closer to refuge	Holmes 1984
Moose	Wolf	Only cows with calves primarily use wolf-free habitats	Edwards 1983

Most of the studies cited above look at prey escape responses following an encounter with an individual predator (or hunting group). Few studies have examined adaptive shifts in avoidance behavior (refuge use). Also, perhaps because of the concentration on escape responses that tend to involve only one predator and one prey at a time, few studies have looked at prey responses to total predation risk (but see Dill and Fraser 1984; Sih 1984, 1986a). Total predation risk is affected by both the risk associated with a given predator individual (which is affected by species, size, hunger, etc.) and the density or encounter rate with different types of predators. It is total risk that should be most important to prey.

ONE PREDATOR, VARYING PREY SUSCEPTIBILITY. An alternative way of examining the relationship between risk and exposure is to compare the responses of different prey to a given predator. Prey that are more susceptible should show stronger responses and therefore reduced exposure to predators. This trend has been observed repeatedly (see table 14.2). Many of these examples show that, within a species, the prey size class that is preferred by predators also spends the least amount of time exposed to predators. Other examples follow the same basic trend but use interspecific comparisons.

Prediction 1a: Increased Risk, Decreased Exposure, Decreased Feeding, Growth or Mating Success

Prey often avoid predators by altering their habitat use, time of activity, or movement patterns (see earlier summary). If we assume that in the absence of predators prey behave in ways that result in high feeding efficiency, then it should often be the case that alterations in these behaviors due to the need to avoid predators will decrease feeding efficiency. Surprisingly few studies have documented the effects of predators on prey feeding efficiency. However, the existing studies confirm that antipredator responses do significantly decrease prey feeding or growth rates (e.g., Stein and Magnuson 1976; Murdoch and Sih 1978; Sih 1980, 1981, 1982; Grubb and Greenwald 1982; Werner et al. 1983; Dill and Fraser 1984; Holmes 1984). As might be expected, alterations in habitat use, time of activity and movement patterns due to predator presence also appear to alter

prey diets (e.g., Vance and Schmitt 1979; Krebs 1980; Dill 1983; Edwards 1983; Werner et al. 1983; Lima et al. 1985). The study of the effects of antipredator needs on both prey feeding rates and diets is a relatively untouched field in which further attention should prove rewarding.

Similarly, although it has been suggested that the need to avoid predators affects mating strategies (e.g., Strong 1973; Farr 1975; Endler 1980; Tuttle and Ryan 1981) and mating systems—e.g., antipredator needs might affect female group size and therefore the ability of males to monopolize females (Emlen and Oring 1977; Wittenberger 1980.)—few studies have experimentally quantified such effects (but see Endler 1980).

Prediction 2: More Mating Opportunities, Increased Exposure to Predators

Although numerous anecdotes exist on the disregard that mating animals have for danger, few quantitative studies exist on this point. Perhaps this relationship is so obvious that it need not be quantified; however, I suspect that more detailed studies would yield interesting insights.

Prediction 3: Increased Reproductive Value of Current Offspring, Increased Parental Defense of Young

The total reproductive value of a group of offspring is a function of both the number of offspring and the value of the average one. Up to the age of first reproduction, an individual offspring's reproductive value increases as it gets older. Several studies have shown that parents defend their young more vigorously either if they have more of them (e.g., Pressley 1981; Carlisle 1984) or if the offspring are older (e.g., Curio 1975; Patterson et al. 1981; Pressley 1981; Carlisle 1984).

Prediction 4: Greater Hunger, Greater Degree of Exposure to Predators

This pattern has often been observed (e.g., Griscom 1941; Morse 1970; Batten 1971; Zumeta and Holmes 1978; Milinski 1984; Metcalfe and Furness 1984), and has occasionally been corroborated experimentally (e.g., Milinski and Heller 1978; Sih 1981; Dill and Fraser 1984).

Prediction 5: Greater Marginal Increase in Future Reproductive Success with Exposure to Predators, Increased Exposure to Predators

This prediction can be addressed by comparing antipredator behaviors in conditions where increased feeding rate results in large versus small increases in future reproductive success. For example (suggested by R. Warner), male reproductive success is often highly size-dependent; the largest males enjoy disproportionately high breeding success. The relationship between size and success is thus often sigmoid. Males on the ascending portion of the size–success curve should have the largest value of dV/dT and might be most willing to take risks to gain extra food to reach more rapidly the large size required for high mating success. To my knowledge, this prediction has not been tested.

Prediction 6: Greater Future Reproductive Success, Decreased Exposure to Predators

Predictions 5 and 6 are not contradictory. Prediction 6 suggests that if you already have high expectations for future reproductive success, then you should avoid exposure to predators. In contrast, prediction 5 says that, independent of your current expectations for the future, if you can dramatically increase your future success by taking risks now, then you should do so. Again, few studies include enough information on both future reproductive success and antipredator behaviors to test prediction 6; however, some trends appear to fit. For example, older individuals often have lower future reproductive success, and older individuals sometimes expose themselves to greater predation risks (e.g., Sherman 1977). Of course, there may be confounding factors involved; for example, older individuals also might be larger and less susceptible to predation.

The corollary predictions for predictions 2 through 6 are that changes in time spent exposed to predators should produce corresponding changes in predation rates. Hungry prey, mating prey, prey protecting their young, and so on should have reduced survivorship. Although most behavioral ecologists no doubt trust that this is usually true, to my knowledge, few studies have quantified these effects.

To summarize this section: A cost–benefit approach was used to generate qualitative predictions on flexible prey responses to alterations in predation risk, prey hunger level, and various components of prey current and future reproductive success. Although, by and large, these predictions appear to be upheld, only the prediction on risk versus exposure has been subjected to numerous tests. The ecological consequences of these shifts in antipredator behavior also have been commented on anecdotally but have rarely been quantified or experimentally addressed.

FIXED BEHAVIORS

Behavioral ecologists tend to emphasize adaptive flexible behaviors that organisms show in response to alterations in their environment (e.g., Morse 1980; Krebs and Davies 1984; see also the previous section of this chapter). Many behaviors, however, are probably relatively fixed. Possible examples of fixed behaviors that aid in predator avoidance include diel vertical migration (e.g., Lampert, this volume), nocturnal foraging by prey, ambush foraging, and the permanent use of refuges. Although in most cases experimental verification has not been attempted (but see Allan 1984), most ecologists would probably guess that individual prey would not rapidly alter these behaviors in response to a reduction in predation pressure. As mentioned earlier, inflexibility can be inherently costly because it should increase the variance in fitness (for a given level of environmental variation) and should thereby often decrease expected mean fitness. Why then do prey show fixed antipredator behaviors?

A rather trivial explanation is that in some cases prey do not respond behaviorally because their behaviors are tied to morphologies that are fixed, i.e., prey do not respond because they cannot respond. For example, some prey have evolved an extreme ambush life-style that includes a drastic reduction in their ability to move around (e.g., many filter feeders). Even if the evolution of this life-style was originally partially in response to antipredator needs, a reduction in predation pressure is not likely to result in a flexible shift to an active foraging mode.

More interesting are cases where there are no obvious barriers to a behavioral shift (such as when closely related species behave in other ways, e.g., congeneric zooplankton that differ in their tendency to migrate vertically). Previous explanations of fixed behaviors either have emphasized the relatively low need for flexibility in constant environments (e.g., Levins 1968; Glasser 1979) or have suggested that fixed avoidance of predators is to be expected if the cost of fixed behavior is low (Stein 1979). These factors should certainly contribute to an overall cost–benefit analysis of fixed predator avoidance. Here I discuss an unexplored but potentially important consideration: the often prohibitive cost of gathering information about predation pressure.

Simple optimality theories implicitly assume that organisms have complete knowledge about their environments. The fact that real foragers must sample to gather information about various food types and locations has been recognized and studied (e.g., Krebs et al. 1978; Heinrich 1979b; Werner et al. 1981; Lima 1984). Sampling yields benefits in increased information but also costs in reduced efficiency while sampling. Prey also must sample to gather information about predation regimes if they are to show adaptive flexible antipredator responses. However, in contrast to the relatively low costs of sampling for food, sampling for predators can be a very dangerous business. For prey that have a low probability of surviving an encounter with a predator, the likely cost of gathering information about predation risk is death. Without this information, adaptive shifts in response are not possible. Fixed avoidance of predators might then be the best strategy even if the predation regime is highly variable (benefit of flexibility is high), and life in refuges is stressful (cost of avoidance is high). Incidentally, a conceptually similar idea was suggested by Levins and MacArthur (1969) for explaining feeding specialization in herbivorous insects. These insects might show fixed preferences because the cost of sampling toxic plants is too great.

Elsewhere, I develop some theory and discuss predictions on the relationship between prey uncertainty and fixed antipredator traits. Here I simply list a few intuitively reasonable predictions. Prey should sample more and po-

tentially show more flexible behavior if (1) predation risk is lower, (2) the benefit of exposure is greater, (3) the cost of refuge use is greater, and (4) the time required to gather information is lower. These predictions can be tested using animals that are not totally fixed but that respond relatively slowly to changes in the predation regime.

Finally, notice that although uncertainty might often cause prey to show fixed avoidance of predators, no analogous logic predicts that prey should show an inflexible lack of avoidance. In other words, an inherent asymmetry exists. An increase in predation pressure should produce rapid antipredator responses, but a decrease in predation pressure might have little immediate effect. This asymmetry in response has been noted in both this and other contexts (e.g., Palumbi 1984; Aronson 1985).

THREE PERSONAL OBSERVATIONS
Backswimmer Responses to Predatory Fish and Cannibalism

Backswimmers (*Notonecta*) are aquatic bugs that are commonly found in relatively still waters (lakes, ponds, quiet stream pools) worldwide. Some notonectid species appear to live the fugitive life-style in fishless temporary ponds and stream pools. Other species of *Notonecta* coexist with fish primarily by hiding in littoral vegetation. Cook and Streams (1984) showed that relative to notonectids that use ephemeral habitats, the species that coexist with fish are less valuable prey items, they are more difficult to detect, and perhaps most important, they have a greater tendency to remain motionless (i.e., live a slower life-style).

My studies have concentrated on a species that rarely lives with fish, *Notonecta hoffmanni*. In this species, the main source of predation on juveniles is cannibalism by older individuals (Fox 1975; Sih, unpublished data). Juvenile *N. hoffmanni* respond to cannibalistic adults by segregating from them in space, by reducing movement, and if attacked, by protean escape (Murdoch and Sih 1978; Sih 1980, 1981, 1982). As predicted, greater risk results in reduced exposure to predators (or alternatively, increased response to predators). For a given prey type, stronger responses accompany riskier situations. For escape responses, juveniles show stronger responses to adults than to less dangerous juveniles. I instar notonectids respond to an attack from another I instar by swimming away in a straight path, whereas they respond to an attacking adult with convoluted protean movement (Sih, unpublished data). For avoidance behaviors, I and II instars reduce their movement and use of open areas in direct proportion to their encounter rates with adults (Sih 1981). Relative to a given predator (adult notonectids), more susceptible prey (smaller juveniles) show stronger responses (Sih 1980, 1982). A detailed analysis suggested that juvenile notonectids balance the conflicting needs of feeding efficiently and avoiding predators in an adaptive way (Sih 1980).

Juvenile notonectids show asymmetrical responses to alterations in the predation regime (Sih, unpublished data). When predatory adults are added to a system, juvenile N. hoffmanni shift their habitat use and movement patterns within 30 minutes. In contrast, when adults are removed, juveniles remain in refuge for up to 12 hours. As theory would predict, the amount of time that juveniles remain in refuge after predator removal is related to predation risk. I instar juveniles are more susceptible to cannibalism than are III instars. Accordingly, relative to III instars, I instars stay in refuge after predator removal for a significantly greater amount of time. Most interestingly, I instars stay in refuge longer after predator removal if predator density had been higher. An alternative explanation for the difference between instars is that, relative to I instars, III instars simply realize more quickly that predators are gone; e.g., III instars might have superior sensory abilities. However, this explanation cannot account for the effect of predator density on a given instar. The observed patterns are thus consistent with the prediction that higher predation risk increases the cost of sampling the environment, so that prey must stay in refuge for a greater amount of time after predator removal.

Mosquito Responses to Notonectids

Most mosquito larvae appear to use the fugitive strategy. In these species, adults lay eggs and larvae develop in very temporary bodies of water, e.g., puddles, ditches, treeholes (Carpenter and LeCasse 1955). Some species, however, use somewhat more permanent habitats and commonly coexist with predators. The available evidence suggests that mosquitoes that co-occur with predators live a slower life-style (are less active, spend more time avoiding predators) than do fugitive mosquitoes. Relative to Aedes aegypti, a fugitive (treehole) mosquito, Culex pipiens that co-occur with predatory notonectids show much stronger reductions in movement and increases in refuge use in response to the presence of these predators [Sih, 1986a; Husbands (1978) observed a similar trend with congeners of the above species]. In essence, whereas A. aegypti show weak responses to notonectids, C. pipiens almost totally avoid open areas and completely stop moving when backswimmers are present.

As mentioned earlier, many studies look at prey behaviors as a function of the presence or absence of predators but do not examine more subtle responses to variations in predator density. Optimal antipredator responses should be proportional to predation risk. Predation risk (mortality rate caused by predators) is often correlated with predator density or encounter rate with predators. Thus, in general, prey responses should be proportional to predator density. In some situations, however, the above variables are only loosely correlated. For example, a high density of mobile predators (resulting in high encounter rate with predators) might not cause much mortality if the predators are satiated.

Both A. aegypti and C. pipiens respond to variations in predator density. Partial correlation analysis showed that the magnitude of antipredator response by C. pipiens is best explained by actual predation risk, whereas A. aegypti responses are best explained by the frequency of encounter with notonectids (Sih 1986a). These differences can be explained by the types of cues used by these species to gauge predation risk. A. aegypti show similar avoidance responses to crude disturbances and notonectids. Because the frequency of disturbance is only loosely correlated to actual predation risk, A. aegypti often show inappropriate antipredator responses. In contrast, C. pipiens apparently use chemical cues associated with the actual predation act to gauge predation risk. Higher predation rates result in a stronger smell of death and a stronger antipredator response. In sum: (1) the mosquito species that co-occurs with notonectids shows stronger and more appropriate anti-notonectid responses; (2) the differences between prey species can be explained by a fundamental difference in their means of gauging predation risk.

Salamander Larval Responses to Fish

Aquatic salamander larvae fall into two main categories. Those found in permanent stream pools coexist with predatory fish and show a slow life-style, whereas those found in ephemeral ponds are fugitives from predators and live a fast life-style. Stream salamander larvae appear to coexist with visual fish predators by hiding during the day, i.e., they forage actively only at night (e.g., Petranka 1984a). Even when "active," they move very little. Their generally low activity levels are associated with relatively slow growth and development. However, because their habitat is permanent, the cost of slow development is probably relatively small. In contrast, most pond salamander larvae live in ephemeral pools that very rarely contain fish. These larvae are fugitives from predatory fish and show some elements of the fast life-style. Most notably, they develop much more rapidly than do stream salamander larvae (e.g., Salthe and Mecham 1974; Petranka 1984b; Petranka and Sih, unpublished data). Unfortunately, few studies have examined the antipredator behavior of these larvae.

James Petranka and I have been studying larvae of the small-mouthed salamander, *Ambystoma texanum*, that live in ephemeral stream pools without fish (i.e., they are fugitives from predatory fish). Unlike most other stream salamanders these larvae live the fast life-style. They forage actively both day and night (Petranka and Sih, unpublished data) and develop very rapidly, taking only 6 to 8 weeks to go from hatching to metamorphosis (Petranka 1984b; Petranka and Sih, unpublished data). Interestingly, contrary to the generality that fugitive prey should have weak antipredator responses, the initial responses of *A. texanum* to an experimental addition of fish were anything but weak (fig. 14.1). Within minutes after fish addition, virtually all of the larvae were in hiding. Furthermore, their refuge use was at least partially mediated by chemical cues emitted by fish, i.e., *A. texanum* larvae were alarmed by the smell of predatory fish (Sih and Petranka, unpublished data). Despite this strong initial response, *A. texanum* did not persist with fish predators (fig. 14.2). Detailed observations showed that after a few hours, these salamander larvae apparently got hungry and emerged to forage. Fish were quick to attack.

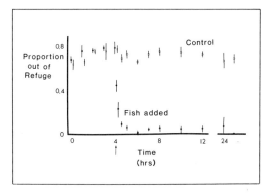

Figure 14.1. *The effects of adding fish on the proportion of time that larval Ambystoma texanum spent outside of refuges. Shown are means and standard errors. The time at which fish were added is shown by the arrow.*

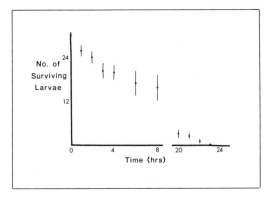

Figure 14.2. *The number of surviving Ambystoma texanum larvae as a function of time after fish were added. Shown are means and standard errors.*

The proportion of larvae out in the open (fig. 14.1) stayed low only because the larvae were either immediately killed or chased back into refuge. In sum, if fish are present in ephemeral pools, then salamander larvae are stuck between a rock and a hard place. If they stay in refuge, they die when the pool dries up; whereas if they emerge to feed, they fall victim to fish predation.

Why do the larvae of *A. texanum* live the fast life-style in ephemeral stream pools while most other stream salamanders show the slow life-style in permanent pools? We suggest that the answer lies in the ancestral inertia discussed earlier. *A. texanum* now breed in streams but they are almost certainly derived from pond ancestry (almost all other *Ambystoma* species breed exclusively in ponds). The most direct ancestor of stream breeding *A. texanum* was a still extant pond-breeding form of the same species. Like most *Ambystoma*, the pond form of *A. texanum* lives a fast life-style in ephemeral pools without fish. The basic life-style now shown by the stream form of *A. texanum* is thus most likely explained by the "pre-adaptations" of its ancestor at the time of colonization into streams. We are currently doing field experiments to test this hypothesis.

ECOLOGICAL IMPLICATIONS

In this final section, I explore some general ways in which antipredator defenses might affect interactions among species and thus the distribution, abundance, and diversity of prey.

Antipredator defenses, particularly those that decrease encounter rates with predators, can have profound effects on the prey's general life-style (e.g., fast versus slow life-styles). Prey life-style can strongly constrain the local distribution of prey. For example, prey that coexist with predators by hiding often have slow energetic rates that can make it impossible for them to complete development in ephemeral habitats or can simply make them inferior competitors in habitats with little predation pressure. Conversely, fugitive or stress-tolerant prey that rarely see predators typically have ineffective antipredator defenses and high activity levels that make them extremely conspicuous and vulnerable to predators. In essence, the life-styles favored in the various habitats are different enough that few organisms can switch between them. This basic scheme has been noted for a variety of animals (e.g., Woodward 1983; Pierce et al. 1985; Sih 1986a; Sih and Petranka, unpublished data) and plants (e.g., Lubchenco and Gaines 1981; Hay 1984).

Antipredator responses also can have more subtle effects on prey dynamics. As discussed above, prey often attempt to avoid encountering predators by altering their own habitat use, activity schedule, and foraging mode. These are key elements of the prey's niche that affect encounter rates, not only between prey and a given predator but also between prey and other predators, other prey, and the resources that the prey require. The interactions affected can thus be either within one trophic level or can involve two or three trophic levels. Within a

trophic level, the interactions affected can be among prey or among predators.

Interactions among Prey

Most studies that examine interactions within one trophic level focus on competition for limited resources (for reviews see Connell 1983; Schoener 1983). The importance of competition can depend on the intensity of predation. Although many studies have looked at the effects of predators on competitors, the emphasis generally has been on lethal effects that decrease the intensity of competition and thus enhance the likelihood of coexistence among prey (e.g., Paine 1966; Connell 1975; Lubchenco 1978; Caswell 1978; Sih et al. 1985). Here I emphasize that predators can have profound effects on interactions among prey via nonlethal effects of predators on prey niches.

Interspecific competitive interactions have often been interpreted in terms of resource niches (for reviews see Pianka 1981; Abrams 1984). Niche theory typically assumes that competition determines the position and breadth of niches and that niche overlaps measure the intensity of competition. The simple fact that niches are influenced not only by competitive but also by antipredator needs disrupts the logic of competition-based niche theory.

Imagine a system where competition exists and niche overlap accurately estimates the degree of competition, but where niches are determined by both competitive and antipredator needs. In such a system, predators can have profound effects on prey competition independent of any actual predation. The nonlethal effect of predators on prey niches can either decrease or increase the intensity of competition, depending on whether prey use the same or different refuges. Competition is decreased if prey that have similar niches in the absence of predators are forced into different niches when predators are present; whereas competition can be intensified if prey that use different niches in the absence of predators are forced into a common refuge. I suspect that the latter scenario is very common; however, to date it has rarely been studied (but see Mittelbach 1984; Mittelbach and Chesson, this volume).

When might we expect prey to use different versus similar refuges? Or more generally, when should prey use different versus similar defense mechanisms? The answer is trivial if one type of refuge or defense is far superior to all others. Then, despite overcrowding within that refuge all prey are forced to share. A more interesting case arises when alternative defenses are potentially roughly equal in effectiveness. Then, whether prey should diverge or converge in their antipredator traits might depend on whether prey show "apparent competition" or "apparent mutualism" (cf. Holt 1977).

Apparent competition occurs when prey that do not compete for resources nonetheless have negative effects on one another through their effects on a common predator. For example, the presence of prey A might result in an increase in the density of a predator that then reduces the abundance of prey B, and vice versa. More generally, total predation rate is determined by the product of per-predator predation rates and the number of predators. Per-predator predation rates depend on encounter rates and the conditional probabilities of detecting, recognizing, attacking, and killing prey. Apparent competition exists if the presence of prey A increases these factors for prey B (and vice versa). However, it is also conceivable that the presence of prey A decreases predation on prey B. For example, time spent handling A should decrease time spent searching for B and thus could decrease predation rates on B and vice versa. See Abrams (this volume) for a detailed discussion of the rich range of possible indirect effects between prey that share a predator.

If apparent competition occurs, then prey should diverge in their defense mechanisms to minimize negative effects. These niche shifts are analogous to those predicted by competition theory; however, they are motivated by the need to avoid not competitors but predators, and they should result in the partitioning not of resource space but "enemy free space." For example, it has been suggested that if predators form search images, then prey can minimize predation risk by being different from other prey. The difference can be morphological (e.g., Blest 1963; Ricklefs and O'Rourke 1975) or behavioral (e.g., Schall and Pianka 1980). The result is prey aspect diversity. Alternatively, if prey show "apparent mutualism," they should converge in their antipredator de-

fense mechanisms. This basic set of ideas has been explored in the plant–herbivore literature (e.g., Feeny 1976; Atsatt and O'Dowd 1976; Rhoades 1979; Price et al. 1980) but has not received much attention from animal predator–prey ecologists.

The analogy between niche theories based on resource competition versus apparent competition can be taken further. Jeffries and Lawton (1984) suggested that prey might show resource partitioning and limiting similarity along an "enemy free space" axis. A full range of community parameters (e.g., species diversity and patterns of prey abundance as well as predator/prey abundance ratios) might be explainable in terms of these indirect antipredator effects. Jeffries and Lawton (1984) caution, however, that just as niche theory should not be based entirely on competitive needs, it also should not be based entirely on antipredator needs. The suggestion is that further theory and empirical work examining niches and coexistence should consider both competitive and antipredator needs.

Interactions among Predators

Prey responses to predators also can affect interactions among predators by affecting prey availability. Prey availability is a function of both prey density and the susceptibility of the average prey individual. It is usually assumed that predators reduce one another's prey availability by reducing prey density, i.e., by consuming prey. However, predators also can either depress or enhance each other's resource availabilities by affecting prey behavior and thus prey susceptibility to other predators.

If a prey response to one predator also reduces that prey's susceptibility to other predators, then predators have reciprocal negative effects on prey availability. This probably occurs often when two predators (of the same or different species) have similar foraging adaptations, so that a prey defense against one is also effective against the other. This phenomenon has been called resource depression (Charnov et al. 1976) and mutual interference through prey responses to predators (Sih 1979).

Prey responses to one predator also can make prey more susceptible to other predators. For example, the act of escaping from one predator can make prey more conspicuous and more likely to be attacked by other predators (e.g., Rand 1954; Moynihan 1962; Sih 1979). Similarly, a shift in habitat use or activity time to avoid one predator often might increase the likelihood of attack from another type of predator. Charnov et al. (1976) termed this phenomenon resource enhancement.

Perhaps most interestingly, whether predators show enhancement or depression of each other's resources can depend on predator density. Sih (1979) showed with predatory backswimmers (*Notonecta hoffmanni*) and mosquito larvae prey (*Culex quinquefasciatus*) that at low predator densities enhancement occurs, whereas at high predator densities interference occurs. Overall, although both effects have often been suggested, they have rarely been studied in detail (see Charnov et al. 1976; Sih 1979).

Interactions between Two Trophic Levels

The vast majority of the literature on predator–prey coexistence deals with predator characteristics, e.g., functional, developmental, numerical and aggregative responses, switching, and mutual interference among predators (for reviews see Murdoch and Oaten 1975; Hassell 1978; Taylor 1984). Curiously, little attention has been paid to the importance of prey responses to predators. Here I have discussed some adaptive trends in antipredator responses. Most notably, prey refuge use is usually positively related to predation risk. An increase in refuge use, by definition, reduces prey availability to predators. The usual trend is thus that an increase in predator density or predation risk results in a decrease in prey availability and thus a decrease in per-predator predation rates. The result is mutual interference. Numerous papers have shown that mutual interference has a stabilizing effect on the predator prey interaction (e.g., Hassell 1978; Sih 1979, 1986b).

The implications of the other predicted trends in refuge use (e.g., decreased refuge use when prey are hungrier) for the likelihood of predator–prey coexistence have not been explored.

Leapfrog Effects among Three Trophic Levels

Most communities include at least three trophic levels (Pimm 1982). I will refer to the three levels as predators, foragers, and prey; predators eat foragers and foragers eat prey. In some cases, predators also can eat prey. Behavioral "leapfrog" effects occur when two nonadjacent trophic levels affect each other through their effects on the behavior of a middle trophic level. Predators clearly can affect forager behavior, feeding rates, and diet choice. Thus, predators can have important effects on the total abundance, relative abundance, and species diversity of prey that they do not consume. In general, predators reduce forager feeding rates so that predators have a positive effect on prey. Indeed, it has been suggested that plant distribution and abundance might be heavily influenced by the restriction that predators place on the habitat use of herbivores (e.g., Holmes 1984; Power, this volume). Prey might, in turn, have a positive effect on predators. Greater prey abundance outside of refuges can induce foragers to spend more time outside of refuges, with the result that predation rates increase. In this scenario predators and prey are indirect mutualists. In fact, Abrams (1984) showed that, in theory, these indirect mutualistic effects should often be of equal or greater magnitude than the direct effects of adjacent trophic levels. To date, only a few theoretical (Abrams 1984) and empirical studies (e.g., Price et al. 1980; Lampert, Power, this volume) have begun to explore the potentially fruitful field of indirect three-trophic-level behavioral interactions.

A special case of a behaviorally mediated three-trophic-level effect occurs in age-structured populations of predators. For example, adult predatory backswimmers (*Notonecta*) and adult water striders (*Gerris*) are both cannibalistic and capable of mutual predation on juveniles of either species. Juveniles of both species avoid adults of both species (Murdoch and Sih 1978; Sih 1980, 1981, 1982, unpublished data). This avoidance can reduce juvenile feeding rates by up to 90%. The reduction in juvenile feeding rates is greater when alternative prey are less abundant (Sih 1981). The result is a strong stabilizing effect of predators on the forager–prey interaction. When prey are scarce, predators consume foragers and reduce forager feeding rates. The latter decreases forager developmental rates, which further decreases future forager feeding rates. In contrast, when prey get more abundant, a storehouse of potential feeding is released. Because alternative prey are available, predators neither consume nor interfere with foragers. Forager feeding rate, developmental rate, and survival increase. All of this results in a disproportionate increase in feeding rate on prey, i.e., in stabilizing density-dependent predation.

Ecology of Fixed Behaviors

Many prey might show relatively fixed avoidance of predators. As discussed earlier, fixed behaviors are expected in many situations where the cost of flexibility might be greater than the benefit. Even where fixed behaviors are optimal (in a constrained sense), they are costly to prey simply because they limit the prey's ability to take advantage of short-term patchiness (in space and time) in predation pressure. The larger ecological consequence of this is that the effects of short-term antipredator responses on population interactions discussed above no longer apply. If prey do not show rapid behavioral responses to predators, then, for example, interactions among predators will not be affected by behavioral resource depression (see Sih 1979), and the stabilizing effect of prey behaviors on the coexistence of predators and prey is negated. In contrast, the idea that prey community structure is best understood in terms of niche partitioning along an "enemy free space" axis should be just as important and, in fact, easier to analyze if prey use of enemy free space is fixed. Finally, note that fixed versus flexible behaviors are only endpoints on a spectrum. Behaviors are probably usually flexible, but they vary in their lag time between the environmental change and the trait change. Analyses of how varying time lags influence the effect of antipredator defenses on community dynamics should be interesting.

CONCLUSIONS, SUGGESTIONS FOR FUTURE STUDY

1. Many studies show that the magnitude of prey response to predators is proportional to predation risk. Relatively few studies quantitatively examine the effects of other needs, such as feeding and mating needs, on antipredator behaviors. Ideally, we should look at the interacting effects of these other needs and predation risk in determining antipredator traits.

2. Most studies of antipredator behaviors emphasize adaptive shifts in behavior. Many behaviors, however, may be fixed or at least relatively slow to respond to alterations in the environment. More theoretical and empirical attention should be paid to the evolution of fixed behaviors.

3. In general, attempts to understand antipredator traits emphasize apparent adaptation to current environmental conditions. The assumption is thus that recent natural selection plays the dominant role in determining observed traits. The influence of ancestral inertia, however, should be given greater consideration.

4. Effective antipredator defense can place constraints on a broad range of traits that are not usually thought of as antipredator traits. The result can be suites of associated traits (life-styles) shaped partially or primarily by antipredator needs. Generalities about these life-styles and the role that predators have in molding them need to be identified and tested.

5. Theoretical and empirical studies should be done to elucidate the nonlethal effects that predators have on population interactions among prey, among predators, between predators and prey, and among three trophic levels. The community effects of time lags in antipredator response also should be investigated.

ACKNOWLEDGMENTS

The ideas in this paper gradually evolved over many years, aided by discussions with countless colleagues. In particular, recent discussions with Phil Crowley, Marie-Sylvie Baltus, and Loric Sih helped to clarify previous misconceptions. Financial aid was provided by a grant from the National Science Foundation.

REFERENCES

Abrams, P. A. 1984. Foraging time optimization and interactions in food webs. *Amer. Naturalist* 124 : 80–96.
Alberch, P. 1982. Developmental constraints in evolutionary processes. In J. T. Bonner (ed.), *Evolution and development*, pp. 313–32. New York: Springer-Verlag.
Allan, J. D. 1984. The size composition of invertebrate drift in a Rocky Mountain stream. *Oikos* 43 : 68–76.
Aronson, R. 1985. Ecological release in a Bahamian salt water lake: *Octopus briareus* (Cephalopoda)

and *Ophiothrix oerstedii* (Ophiuroidea). Ph.D. thesis, Harvard University.

Atsatt, P. R., and D. J. O'Dowd. 1976. Plant defense guilds. *Science* 193 : 24–29.

Batten, L. A. 1971. Bird population changes on farmland and in woodland for the years 1968–69. *Bird Study* 18 : 1–8.

Bauwens, D., and C. Thoen. 1981. Escape tactics and vulnerability to predation associated with reproduction in the lizard *Lacerta vivpara. J. Anim. Ecol.* 50 : 733–43.

Belovsky, G. E. 1986. Optimal foraging and community structure: Implications for a guild of generalist grassland herbivores. *Oecologia* (in press).

Bertness, M. D., S. D. Garrity, and S. C. Levings. 1981. Predation pressure and gastropod foraging: A tropical-temperate comparison. *Evolution* 35 : 995–1007.

Bertram, B. C. 1978. Living in groups: Predators and prey. In J. R. Krebs and N. B. Davies (eds.), *Behavioural ecology: An evolutionary approach*, pp. 64–96. Oxford: Blackwell.

Blest, A. D. 1963. Longevity, palatability and natural selection in five species of New World saturniid moth. *Nature* 197 : 1183–86.

Caraco, T. 1980. On foraging time allocation in a stochastic environment. *Ecology* 61 : 119–28.

Carlisle, T. R. 1984. Parental response to brood size in a cichlid fish. *Anim. Behav.* 33 : 234–38.

Carpenter, S. J., and W. J. Lacasse. 1955. *Mosquitoes of North America*. Berkeley, Calif.: University of California Press.

Caswell, H. 1978. Predator-mediated coexistence: A nonequilibrium model. *Amer. Naturalist* 112 : 127–54.

Cerri, R. D., and D. F. Fraser. 1983. Predation and risk in foraging minnows: balancing conflicting demands. *Amer. Naturalist* 121 : 552–61.

Charnov, E. L., G. H. Orians, and K. Hyatt. 1976. Ecological implications of resource depression. *Amer. Naturalist* 110 : 247–59.

Coley, P. D., J. P. Bryant, and F. S. Chapin III. 1985. Resource availability and plant antiherbivore defense. *Science* 230 : 895–99.

Connell, J. H. 1975. Some mechanisms producing structure in natural communities. In M. L. Cody and J. M. Diamond (eds.), *Ecology and evolution of communities*, pp. 460–90. Cambridge, Mass.: Belknap Press.

———. 1983. On the prevalence and relative importance of interspecific competition: evidence from field experiments. *Amer. Naturalist* 122 : 661–96.

Cook, W. L., and F. A. Streams. 1984. Fish predation on *Notonecta* (Hemiptera): Relationship between prey risk and habitat utilization. *Oecologia* 64 : 177–83.

Cooper, S. D. 1984. The effects of trout on water striders in stream pools. *Oecologia* 63 : 376–79.

Cott, H. B. 1940. *Adaptive coloration in animals*. London: Methuen.

Curio, E. 1975. The functional organization of antipredator behaviour in the pied flycatcher: A study of avian visual perception. *Anim. Behav.* 23 : 1–115.

Dayton, P. K., R. J. Rosenthal, L. C. Mahen, and T. Antezana. 1977. Population structure and foraging biology of the predaceous Chilean asteroid *Meyenaster gelatinosus* and the escape biology of its prey. *Mar. Biol.* 39 : 361–70.

DeBenedictis, P. A., F. B. Gill, F. R. Hainsworth, G. H. Pyke, and L. L. Wolf. 1978. Optimal meal size in hummingbirds. *Amer. Naturalist* 112 : 301–16.

Denno, R. F., and M. S. McClure. 1983. Variable plants and herbivores in natural and managed systems. New York: Academic Press.

Dill, L. M. 1983. Adaptive flexibility in the foraging behavior of fishes. *Can. J. Fisheries Aquat. Sci.* 40 : 398–408.

Dill, L. M., and A. H. G. Fraser. 1984. Risk of predation and the feeding behavior of juvenile coho salmon (*Oncorhynchus kisutch*). *Behav. Ecol. Sociobiol.* 16 : 65–71.

Dodson, S. I. 1984. Predation of *Heterocope septentrionalis* on two species of *Daphnia*: Morphological defenses and their cost. *Ecology* 65 : 1249–57.

Duggins, D. O. 1981. Interspecific facilitation in a guild of benthic marine herbivores. *Oecologia* 48 : 157–63.

Edmunds, M. 1974. *Defence in animals*. New York: Longman.

Edwards, J. 1983. Diet shifts in moose due to predator avoidance. *Oecologia* 60 : 185–89.

Emlen, S. T., and L. W. Oring. 1977. Ecology, sexual selection, and the evolution of mating systems. *Science* 197 : 215–23.

Endler, J. A. 1980. Natural selection on color patterns in *Poecilia reticulata. Evolution* 34 : 76–91.

Farr, J. A. 1975. The role of predation in the evolution of social behavior of natural populations of the guppy, *Poecilia reticulata* (Pisces: Poeciliidae). *Evolution* 29 : 151–58.

Feeny, P. O. 1976. Plant apparency and chemical defense. *Recent Advances in Phytochemistry* 10 : 1–40.

Fox, L. R. 1975. Some demographic consequences of food shortage for the predator, *Notonecta hoffmanni. Ecology* 56 : 868–80.

———. 1981. Defense and dynamics in plant-herbivore systems. *Amer. Zool.* 21 : 853–64.

Futuyma, D. J. 1983. Evolutionary interactions among herbivorous insects and plants. In D. J. Futuyma and M. Slatkin (eds.), *Coevolution*, pp. 207–31. Sunderland, Mass.: Sinauer Assoc.

Gentry, J. B. 1974. Response to predation by colonies of the Florida harvester ant, *Pogonomyrmex badius. Ecology* 55 : 1328–38.

Giller, P. S. 1982. Locomotor efficiency in the predation strategies of the British *Notonecta* (Hemiptera: Heteroptera). *Oecologia* 52 : 273–77.

Gittelman, S. H. 1974. Locomotion and predatory strategies in backswimmers (Hemiptera: Notonectidae). *Amer. Midland Naturalist* 92 : 496–500.

Glasser, J. W. 1979. The role of predation in shaping and maintaining the structure of communities. *Amer. Naturalist* 113 : 631–41.

Goodyear, C. P. 1973. Learned orientation in the predator avoidance behavior of mosquito fish, *Gambusia affinis*. *Behaviour* 45 : 191–220.

Gould, S. J., and R. C. Lewontin. 1979. The spandrels of San Marco and the Panglossian paradigm: a critique of the adaptationist programme. *Proc. Roy. Soc. London [Biol.]* 205 : 581–98.

Gould, S. J., and S. Vrba. 1982. Exaptation—a missing term in the science of form. *Paleobiology* 8 : 4–15.

Green, R. F. 1980. Bayesian birds: A simple example of Oaten's stochastic model of optimal foraging. *Theor. Pop. Biol.* 18 : 244–56.

Greenslade, P. J. M. 1983. Adversity selection and the habitat templet. *Amer. Naturalist* 122 : 352–65.

Grime, J. P. 1977. Evidence for the existence of three primary strategies in plants and its relevance to ecological and evolutionary theory. *Amer. Naturalist* 111 : 1169–94.

Griscom, L. 1941. The recovery of birds from disaster. *Audubon Magazine* 43 : 191–96.

Grubb, T. C., and L. Greenwald. 1982. Sparrows and a brushpile: Foraging responses to different combinations of predation risk and energy cost. *Anim. Behav.* 30 : 637–40.

Hall, D. J., S. T. Threlkeld, C. W. Burns, and P. H. Crowley. 1976. The size-efficiency hypothesis and the size-structure of freshwater zooplankton communities. *Annu. Rev. Ecol. Syst.* 7 : 177–208.

Hassell, M. P. 1978. The dynamics of arthropod predator-prey systems. Princeton, N.J.: Princeton University Press.

Hay, M. E. 1984. Predictable spatial escapes from herbivory: How do these affect the evolution of herbivore resistance in tropical marine communities? *Oecologia* 64 : 396–407.

Heinrich, B. 1979a. Foraging strategies of caterpillars. Leaf damage and possible predator avoidance strategies. *Oecologia* 42 : 325–37.

———. 1979b. "Majoring" and "minoring" by foraging bumblebees, *Bombus vagrans*: An experimental analysis. *Ecology* 60 : 245–55.

Hennessy, D. F., and D. H. Owings. 1978. Snake species discimination and the role of olfactory cues in the snake-directed behavior of the California ground squirrel. *Behaviour* 65 : 115–24.

Hobson, E. S., and J. R. Chess. 1976. Trophic interactions among fishes and zooplankters near shore at Santa Catalina Island, California. *Fish. Bull.* 74 : 567–98.

Holmes, W. G. 1984. Predation risk and foraging behavior of the hoary marmot in Alaska. *Behav. Ecol. Sociobiol.* 15 : 293–301.

Holt, R. D. 1977. Predation, apparent competition and the structure of prey communities. *Theor. Pop. Biol.* 12 : 197–229.

Horn, H. S., and D. I. Rubenstein. 1984. Behavioural adaptations and life history. In J. R. Krebs and N. B. Davies (eds.), *Behavioural ecology: An evolutionary approach*, pp. 279–98. 2nd ed. Sunderland, Mass.: Sinauer.

Husbands, R. C. 1978. The influence of mosquito larvae behavior on predator efficiency in a natural habitat. *California Mosquito Vector Control Association Biological Briefs* 4 : 2.

Janzen, D. H. 1981. Evolutionary physiology of personal defense. In C. R. Townsend and P. Calow (eds.), *Physiological ecology: An evolutionary approach to resource use*, pp. 145–64. Oxford: Blackwell.

Jarman, P. J. 1974. The social organization of antelope in relation to their ecology. *Behaviour* 59 : 215–67.

Jeffries, M. J., and J. H. Lawton. 1984. Enemy free space and the structure of ecological communities. *Biol. J. Linn. Soc.* 23 : 269–86.

Johnson, D. M., and P. H. Crowley. 1980. Odonate "hide-and-seek": habitat-specific rules? In W. C. Kerfoot (ed.), *Evolution and ecology of zooplankton communities*, pp. 569–79. Hanover, N.H.: University Press of New England.

Kenward, R. E. 1978. Hawks and doves: Factors affecting success and selection in goshawk attacks on wood pigeons. *J. Anim. Ecol.* 47 : 449–60.

Kerfoot, W. C. 1982. A question of taste: Crypsis and warning coloration in freshwater zooplankton communities. *Ecology* 63 : 538–54.

Kitching, J. A., and J. Lockwood. 1975. Observations on shell form and its ecological significance in thaisid gastropods of the genus *Lepsiella* in New Zealand. *Mar. Biol.* 28 : 131–44.

Kotler, B. P. 1984. Risk of predation and the structure of desert rodent communities. *Ecology* 65 : 689–701.

Krebs, J. R. 1980. Optimal foraging, predation risk and territorial defence. *Ardea* 68 : 83–90.

Krebs, J. R., and N. B. Davies. 1984. *Behavioural ecology: An evolutionary approach*. 2nd ed. Sunderland, Mass. Sinauer.

Krebs, J. R., A. Kacelnik, and P. J. Taylor. 1978. Tests of some optimal sampling by foraging great tits. *Nature* 275 : 27–31.

Levins, R. 1968. Evolution in changing environments. Princeton, N.J.: Princeton University Press.

Levins, R., and R. MacArthur. 1969. An hypothesis to explain the incidence of monophagy. *Ecology* 50 : 910–11.

Lima, S. L. 1984. Downy woodpecker foraging behavior: Efficient sampling in simple stochastic environments. *Ecology* 65 : 166–74.

Lima, S. L., T. J. Valone, and T. Caraco. 1985. Foraging efficiency-predation risk trade-off in the grey squirrel. *Anim. Behav.* 33 : 155–65.

Lubchenco, J. 1978. Plant species diversity in a marine intertidal community: Importance of herbivore food preferences and algal competitive abilities. *Amer. Naturalist* 112 : 23–39.

Lubchenco, J., and S. D. Gaines. 1981. A unified approach to marine plant-herbivore interactions. I. Populations and communities. *Annu. Rev. Ecol. Syst.* 12 : 405–37.

Maynard Smith, J. 1978. Optimality theory in evolution. *Annu. Rev. Ecol. Syst.* 9 : 31–56.

Maynard Smith, J., R. Burian, S. Kauffman, P. Alberch, J. Campbell, B. Goodwin, R. Lande, D. Raup, and L. Wolpert. 1985. Developmental constraints and evolution. *Q. Rev. Biol.* 60 : 265–87.

Mech, L. D. 1970. The wolf: The ecology and behaviour of an endangered species. New York: Natural History Press.

Menge, B. A. 1978. Predation intensity in a rocky intertidal community. Relationship between predator foraging activity and environmental harshness. *Oecologia* 34 : 1–16.

Metcalfe, N. B., and R. W. Furness. 1984. Changing priorities: The effect of pre-migratory fattening on the trade-off between foraging and vigilance. *Behav. Ecol. Sociobiol.* 15 : 203–206.

Milinski, M. 1979. Can an experienced predator overcome the confusion of swarming prey more easily? *Anim. Behav.* 27 : 1122–26.

———. 1984. A predator's costs of overcoming the confusion effect of swarming prey. *Anim. Behav.* 32 : 1157–62.

Milinski, M., and R. Heller. 1978. Influence of a predator on the optimal foraging behaviour of sticklebacks (*Gasterosteus aculeatus* L.). *Nature* 275 : 642–44.

Mittelbach, G. G. 1984. Predation and resource partitioning in two sunfishes (Centrarchidae). *Ecology* 65 : 499–513.

Morse, D. H. 1970. Ecological aspects of some mixed-species foraging flocks of birds. *Ecol. Monogr.* 40 : 119–68.

———. 1980. Behavioral mechanisms in ecology. Cambridge, MA: Harvard University Press.

Moynihan, M. 1962. The organization and possible evolution of some mixed species flocks of neotropical birds. *Smith. Misc. Coll.* 143 : 1–140.

Murdoch, W. W., and A. Oaten. 1975. Predation and population stability. *Adv. Ecol. Res.* 9 : 1–131.

Murdoch, W. W., and A. Sih. 1978. Age-dependent interference in a predatory insect. *J. Anim. Ecol.* 47 : 581–92.

Neill, S. R. 1970. A study of antipredator adaptations in fish with special reference to silvery camouflage and shoaling. Ph.D. thesis, Oxford University.

Neill, S. R., and J. M. Cullen. 1974. Experiments on whether schooling by their prey affects the hunting behaviour of cephalopods and fish predators. *J. Zool.* 172 : 549–569.

Oaten, A. 1977. Optimal foraging in patches: A case for stochasticity. *Theor. Pop. Biol.* 12 : 263–85.

O'Brien, W. J. 1979. The predator-prey interaction of planktivorous fish and zooplankton. *American Scientist* 67 : 572–81.

Paine, R. T. 1966. Food web complexity and species diversity. *Amer. Naturalist* 100 : 65–75.

Palumbi, S. R. 1984. Tactics of acclimation: Morphological changes of sponges in an unpredictable environment. *Science* 225 : 1478–80.

Patterson, T. L., L. Petrinovitch, and D. K. James. 1980. Reproductive value and appropriateness of response to predators by white-crowned sparrows. *Behav. Ecol. Sociobiol.* 7 : 227–31.

Peckarksy, B. L. 1980. Predator-prey interactions between stoneflies and mayflies: Behavioral observations. *Ecology* 61 : 932–43.

Peters, R. H. 1983. Ecological implications of body size. New York: Cambridge University Press.

Petranka, J. W. 1984a. Ontogeny of the diet and feeding behavior of *Eurycea bislineata* larvae. *J. Herpetol.* 18 : 48–55.

———. 1984b. Sources of intrapopulational variation in growth responses of larval salamanders. *Ecology* 65 : 1857–65.

Pianka, E. R. 1970. On r- and K-selection. *Amer. Naturalist* 104 : 592–97.

———. 1981. Competition and niche theory. In R. M. May (ed.), *Theoretical ecology*, pp. 167–96. 2nd ed. Oxford: Blackwell.

Pierce, C. L., P. H. Crowley, and D. M. Johnson. 1985. Behavior and ecological interactions of larval Odonata. *Ecology* 66 : 1504–12.

Pimm, S. L. 1982. *Food webs.* London: Chapman and Hall.

Polis, G. A. 1980. The effect of cannibalism on the demography and activity of a natural population of desert scorpions. *Behav. Ecol. Sociobiol.* 7 : 25–35.

Power, M. E. 1984. Depth distributions of armored catfish: predator-induced resource avoidance? *Ecology* 65 : 523–28.

Pressley, P. H. 1981. Parental effort and the evolution of nest-guarding tactics in the threespine stickleback, *Gasterosteus aculeatus* L. *Evolution* 35 : 282–95.

Price, P. W., C. E. Bouton, P. Gross, B. A. McPherson, J. N. Thompson, and A. E. Weis. 1980. Interactions among three trophic levels: Influence of plants on interactions among insect herbivores and natural enemies. *Annu. Rev. Ecol. Syst.* 11 : 41–65.

Pulliam, H. R., and T. Caraco. 1984. Living in groups: Is there an optimal group size? In J. R.

Krebs and N. B. Davies (eds.), *Behavioural ecology: An evolutionary approach*, pp. 122–47. 2nd ed. Sunderland, Mass.: Sinauer.

Pulliam, H. R., and G. S. Mills. 1977. The use of space by wintering sparrows. *Ecology* 58 : 1393–99.

Pyke, G. H. 1984. Optimal foraging theory: A critical review. *Annu. Rev. Ecol. Syst.* 15 : 523–75.

Rand, A. L. 1954. Social feeding behavior of birds. *Fieldiana: Zoology* 36 : 1–71.

Real, L. 1980. Fitness, uncertainty and the role of diversification in evolution and behavior. *Amer. Naturalist* 115 : 623–38.

Rhoades, D. F. 1979. Evolution of plant chemical defense against herbivores. In G. A. Rosenthal and D. H. Janzen (eds.), *Herbivores: Their interactions with secondary metabolites*, pp. 3–54. New York: Academic Press.

Ricklefs, R. E. and K. O'Rourke. 1975. Aspect diversity in moths: A temperate-tropical comparison. *Evolution* 29 : 313–24.

Rosenthal, G. A. and D. H. Janzen. 1979. Herbivores: Their interactions with secondary metabolites. New York: Academic Press.

Russell, R. J. 1972. Defensive responses of the aphid *Drepanosiphum platanoides* in encounters with the bug *Anthocoris nemorum*. *Oikos* 23 : 264–67.

Salthe, S. N., and J. S. Mecham. 1974. In B. Lofts (ed.), *Physiology of the Amphibia*, vol. 2, pp. 309–521. New York: Academic Press.

Sargeant, A. B., and L. E. Eberhardt. 1975. Death feigning by ducks in response to predation by red foxes (*Vulpes fulva*). *Amer. Midland. Naturalist* 94 : 108–19.

Schaffer, W. M. 1974. Optimal reproductive effort in fluctuating environments. *Amer. Naturalist* 108 : 783–90.

Schall, J. J., and E. R. Pianka. 1980. Evolution of escape behavior diversity. *Amer. Naturalist* 115 : 551–66.

Schaller, G. B. 1972. *The Serengeti lion.* Chicago: University of Chicago Press.

Schoener, T. W. 1983. Field experiments on interspecific competition. *Amer. Naturalist* 122 : 240–85.

Schultz, J. C. 1981. Adaptive changes in antipredator behavior of a grasshopper during development. *Evolution* 35 : 175–79.

Shalter, M. D. 1978. Effect of spatial context on the mobbing reaction of pied flycatchers to a predator model. *Anim. Behav.* 26 : 1219–21.

Sherman, P. W. 1977. Nepotism and the evolution of alarm calls. *Science* 197 : 1246–53.

Sih, A. 1979. Stability and prey behavioural responses to predator density. *J. Anim. Ecol.* 48 : 79–89.

———. 1980. Optimal behavior: Can foragers balance two conflicting demands? *Science* 210 : 1041–43.

———. 1981. Stability, prey density and age-dependent interference in an aquatic insect predator, *Notonecta hoffmanni*. *J. Anim. Ecol.* 50 : 625–36.

———. 1982. Foraging strategies and the avoidance of predation by an aquatic insect, *Notonecta hoffmanni*. *Ecology* 63 : 786–96.

———. 1984. The behavioral response race between predator and prey. *Amer. Naturalist* 123 : 143–50.

———. 1985. Evolution, predator avoidance, and unsuccessful predation. *Amer. Naturalist* 125 : 153–57.

———. 1986a. Antipredator responses and the perception of danger by mosquito larvae. *Ecology* 67 : 434–41.

———. 1986b. Prey refuges and predator-prey stability. *Theor. Pop. Biol.* (in press).

Sih, A., P. H. Crowley, M. A. McPeek, J. W. Petranka, and K. Strohmeier. 1985. Predation, competition and prey communities: a review of field experiments. *Annu. Rev. Ecol. Syst.* 16 : 269–311.

Stein, R. A. 1977. Selective predation, optimal foraging and the predator-prey interaction between fish and crayfish. *Ecology* 58 : 1237–53.

———. 1979. Behavioral response of prey to fish predators. In R. H. Stroud and H. Clepper (eds.), *Predator-prey systems in fisheries management*, pp. 343–53. Washington, D.C.: Sport Fishing Institute.

Stein, R. A., and J. J. Magnuson. 1976. Behavioral response of crayfish to a fish predator. *Ecology* 57 : 751–61.

Strong, D. R. 1973. Amphipod amplexus, the significance of ecotypic variation. *Ecology* 54 : 1383–88.

Taigen, T. L., S. B. Emerson, and F. H. Pough. 1982. Ecological correlates of anuran exercise physiology. *Oecologia* 52 : 49–56.

Taylor, H. M., R. S. Gourley, and C. E. Lawrence. 1974. Natural selection of life history attributes. *Theor. Pop. Biol.* 5 : 104–22.

Taylor, P. B., and L. C. Chen. 1969. The predator-prey relationship between the octopus (*Octopus bimaculatus*) and the California scorpionfish (*Scorpaena guttata*). *Pac. Sci.* 23 : 311–16.

Taylor, R. J. 1984. *Predation.* New York: Chapman and Hall.

Treherne, J. E., and W. A. Foster. 1980. The effects of group size on predator avoidance in a marine insect. *Anim. Behav.* 28 : 1119–22.

Tuttle, M. D., and M. J. Ryan. 1981. Bat predation and the evolution of frog vocalizations in the neotropics. *Science* 214 : 677–78.

Vance, R. R., and R. J. Schmitt. 1979. The effect of the predator-avoidance behavior of the sea urchin, *Centrostephanus coronatus*, on the breadth of its diet. *Oecologia* 44 : 21–25.

Van Veldhuizen, H. D., and V. J. Oakes. 1981. Behavioral response of seven species of asteroids to the asteroid predator, *Solaster dawsoni*. *Oecologia*

48 : 214–20.

Vermeij, G. J. 1982. Unsuccessful predation and evolution. *Amer. Naturalist* 120 : 701–20.

————. 1985. Aptations, effects, and fortuitous survival: Comment on a paper by A. Sih. *Amer. Naturalist* 125 : 470–72.

Walther, F. R. 1969. Flight behaviour and avoidance of predators in Thomson's gazelle (*Gazella thomsoni* Guenter 1884). *Behaviour* 34 : 184–221.

Wassersug, R. J., and D. G. Sperry. 1977. The relationship of locomotion to differential predation on *Pseudacris triseriata* (Anura: Hylidae). *Ecology* 58 : 830–39.

Werner, E. E., J. F. Gilliam, D. J. Hall, and G. G. Mittelbach. 1983. An experimental test of the effects of predation risk on habitat use. *Ecology* 64 : 1540–48.

Werner, E. E., G. G. Mittelbach, and D. J. Hall. 1981. The role of foraging profitability and experience in habitat use by the bluegill sunfish. *Ecology* 62 : 116–25.

Wilbur, H. M. 1980. Complex life cycles. *Annu. Rev. Ecol. Syst.* 11 : 67–93.

————. 1984. Complex life cycles and community organization in amphibians. In P. W. Price, C. N. Slobodchikoff, and W. S. Gaud (eds.), *A new ecology: Novel approaches to interactive systems*, pp. 195–226. New York: John Wiley & Sons.

Wilbur, H. M., D. W. Tinkle, and J. P. Collins. 1974. Environmental uncertainty, trophic level and resource availability in life history evolution. *Amer. Naturalist* 108 : 805–17.

Wilson, D. S., and A. B. Clark. 1977. Above ground predator defence in the harvester termite, *Hodotermes mossambicus* (Hagen). *J. Entomol. Soc. S. Afr.* 40 : 271–82.

Wilson, D. S., M. Leighton, and D. R. Leighton. 1978. Interference competition in a tropical ripple bug (Hemiptera: Veliidae). *Biotropica* 10 : 302–306.

Wilz, K. J. 1971. Comparative aspects of courtship behavior in the ten-spined stickleback, *Pygosteus pungitius* (L.). *Z. Tierphysiol.* 34 : 360–70.

Wittenberger, J. F. 1980. Group size and polygamy in social mammals. *Amer. Naturalist* 115 : 197–222.

Woodward, B. D. 1983. Predator-prey interactions and breeding-pond use of temporary-pond species in a desert anuran community. *Ecology* 64 : 1549–55.

Wright, D. I., and W. J. O'Brien. 1984. The development and field test of a tactical model of the planktivorous feeding of white crappie (*Pomoxis annularis*). *Ecol. Monogr.* 54 : 65–98.

Wright, S. W. 1931. Evolution in Mendelian populations. *Genetics* 16 : 97–159.

Zaret, T. M. 1980. Predation and freshwater communities. New Haven, Conn.: Yale University Press.

Zaret, T. M., and J. S. Suffern. 1976. Vertical migration in zooplankton as a predator avoidance mechanism. *Limnol. Oceanogr.* 21 : 804–13.

Zumeta, D. C., and R. T. Holmes. 1978. Habitat shift and roadside mortality of scarlet tanagers during a cold wet New England spring. *Wilson Bulletin* 90 : 575–86.

A. Evolutionary Responses
   Mediated by Chemicals

15. Defenses of Planktonic Rotifers Against Predators

Richard S. Stemberger

John J. Gilbert

Planktonic rotifers form an important constituent in the diets of many aquatic predators, such as other rotifers (*Asplanchna*), insect larvae (*Chaoborus*), cyclopoid and calanoid copepods (*Diacyclops, Epischura*), malacostracans (*Mysis*), and particulate-feeding and filter-feeding fish. Despite this, they often are important members of freshwater, zooplankton communities.

Rotifer defenses fall into four general functional categories: (1) morphology or behavior that reduces the probability of capture or ingestion—spines, stiffened lorica, mucus sheaths, large body size, coloniality, body turgidity, and rapid escape responses; (2) morphology or behavior that reduces detection of the prey by the predator—small body size, deadman response, and slow, smooth swimming motion minimizing water disturbance; and less direct modes of defense, which include (3) seasonal or spatial distributions that minimize contact with major predators; (4) life-history traits, such as high intrinsic growth rates, which permit some species to coexist with their predators. The majority of these defenses appear to be effective against small, invertebrate predators that use mechanoreceptors or contact chemoreceptors for prey detection. Consequently, the influence of selective predation on rotifer community structure should be most pronounced in systems where these small predators predominate.

Limnoplanktonic rotifers (200 or more species generally 60–200 μm in length) have many different types of morphological, behavioral, and life-history features that can protect them from a wide variety of predators—fish, dipteran larvae (*Chaoborus*), mysids, copepods, and predatory rotifers (*Asplanchna*). Such features must partially explain the now well-recognized importance of rotifers as primary consumers (Pace and Orcutt 1981; Bogdan and Gilbert 1982 and references therein) and agents of nutrient cycling (Markarewicz and Likens 1979) in many freshwater zooplankton communities.

Planktonic rotifers clearly evolved some defensive adaptations in direct response to predation pressure. Such adaptations include active escape responses in *Polyarthra, Hexarthra*, and *Filinia*, and predator-induced spination and body growth in some species of *Brachionus* and *Keratella*. On the other hand, many features providing some defense against predators may have evolved for entirely different reasons. For example, mucus sheaths in *Conochiloides* and *Ascomorpha* and body spines in nonpolymorphic species may have evolved to increase

buoyancy, and therefore to decrease respiration costs (Stemberger and Gilbert 1985). Similarly, smallness in body size, which may reduce conspicuousness to both vertebrate and invertebrate predators, may have evolved to permit reproduction at low food levels. In planktonic rotifers, body size is directly related to the concentration of food required for the intrinsic rate of natural increase (r_m) to equal zero, the threshold food concentration (Stemberger and Gilbert 1985).

Upper size limits in planktonic rotifers also may be set by the inefficiency of ciliary locomotion (Sleigh and Blake 1977; Epp and Lewis 1984) and by the number of cilia that can be located on the coronal margin. Body mass increases as a cubic function of body length, but the number of cilia on the coronal margin increases as a linear function of body length. Therefore, at some body size there will be insufficient numbers of cilia for the rotifer to swim effectively. The largest free-swimming rotifers, campanulate morphs of polymorphic *Asplanchna* species have body lengths of ˜1600 μm (˜7 μg dry mass) and have the slowest rela-

227

tive swimming speeds (0.2 body lengths/s) reported for rotifers (Epp and Lewis 1984; Gilbert and Stemberger 1985a; Stemberger and Gilbert 1984a).These giant morphs have a greatly enlarged corona, which may have evolved to increase the number of coronal cilia as well as encounter rates with prey (Gilbert and Stemberger 1985b).

Because the small size of most planktonic rotifers reduces the impact of predation by visually feeding fish predators, we suggest that defensive structures and swimming behavior are primarily effective against small invertebrate predators. The growing number of rotifers that show developmental polymorphisms controlled by *Asplanchna* and copepods supports this contention (Gilbert 1980a; Stemberger and Gilbert 1984b; Gilbert and Stemberger 1985c).

The purpose of this chapter is to review and discuss the many and diverse features of planktonic rotifers that may limit mortality from vertebrate, and especially invertebrate, predators. The extent to which predation may have provided the selective pressure for the evolution of many of these features is uncertain. However, these features are undoubtedly important in permitting rotifers to co-occur with a diversity of predators.

We first discuss the basic feeding mechanisms of the major predators on planktonic rotifers. We then consider the different features of planktonic rotifers that may protect them against these predators. We explain the mechanistic bases for the defenses, suggest possible costs for these defenses, and offer some alternative explanations for the evolution of these defenses. Our thoughts on these matters are summarized in table 15.1.

PREDATORS AND THEIR FEEDING MECHANISMS

Fish

Zooplanktivorous fish have two basic modes of feeding. Visually feeding species capture prey using suction by rapidly expanding their buccal cavities. Filter-feeding species, which do not use vision, actively pump and strain water across their sievelike gill rakers or swim with open mouths forcing water through them. The size of the particles retained on the gill rakers may not be a simple function of interraker distances (Wright et al. 1983). A fish's mode of

feeding may change with its developmental stage or body size, with prey size and density, or in response to light conditions (O'Brien 1979; Janssen 1980; Drenner et al. 1982). Some filter-feeders, like the gizzard shad, are highly efficient at capturing small rotifers and even large phytoplankters (Drenner et al., 1982, 1984). The impact of selective predation by such species may favor survivorship of large, strong swimming zooplankton species that can avoid capture (Drenner et al., 1982). In contrast, visual-feeders tend to select larger-bodied crustaceans and planktonic insect larvae (*Chaoborus*). This latter type of predation favors communities dominated by rotifers and small crustaceans (Hrbacek 1962; Brooks and Dodson 1965; Lynch and Shapiro 1981). However, early larvae of many species of fish select small zooplankton, like rotifers (Siefert 1972; Guma'a 1978; Crecco and Blake 1983). The impact of fish larvae on rotifer community structure has not been adequately assessed.

Cladocerans

Cladocerans, such as *Daphnia galeata mendotae* and *D. rosea*, can injure and kill small rotifers like *Keratella cochlearis* during normal filter-feeding for algae. We have called this interaction interference competition (Gilbert and Stemberger 1985a). The mortality rate on small *Keratella cochlearis* by interference competition is a function of *Daphnia* body size (Burns and Gilbert 1986) and may represent an important mechanism controlling some rotifer populations in nature.

Copepods

Cyclopoid and calanoid copepods are abundant and widely distributed crustaceans in the plankton of freshwater lakes. Many species are herbivorous in naupliar and early copepodite stages but are carnivorous or omnivorous as late-stage copepodites and adults (Jamieson 1980; Wyngaard and Chinnappa 1982). They possess mechanoreceptors that detect vibrations and water displacements created by swimming prey (Strickler and Bal 1973). Calanoid copepods also are known to possess contact chemoreceptors enabling them to discriminate among different algal prey types (Friedman 1980; Poulet and Marsot 1980). Copepods display varied and complex swimming behavior associated with feeding, such as looping, which probably increases encounters with the prey (Fryer 1957; Williamson 1981).

Raptorially feeding cyclopoid copepods capture prey by clasping them with paired, well-developed, spined maxillae. Other mouth parts manipulate, shred, and push the prey into the mouth (Fryer 1957). Calanoid copepods can feed raptorially or switch to a filter-feeding mode (Poulet and Marsot 1980). Particle capture in marine herbivorous species is a very complex and relatively unknown process (see review in Koehl 1984). However, the paired second maxillae, which vary greatly in structure among different species, are used to capture and manipulate prey. In *Epischura* they form a coarsely spined, cagelike structure that is used for grasping large prey and may completely envelop small prey like rotifers.

Mysidacea

Mysids are opportunistic omnivores capable of both filter- and raptorial-feeding modes (Grossnickle 1982). They are important grazers and predators of the plankton of fresh- and brackish-water systems. Although mysids select cladocerans and copepods, rotifers have been reported in mysid guts—particularly *Kellicottia longispina*, *Keratella*, and *Notholca* (Lasenby and Langford 1973; Threlkeld et al., 1980; Siegfried and Kopache 1980; Murtaugh 1981). Soft-bodied rotifers probably are easily consumed by mysids but would be much more difficult to identify in mysid guts because only their mouth parts, or portions thereof, would likely remain intact. In deep, stratified lakes rotifers often form dense bands (>1,000 individuals per liter) in the vicinity of the metalimnion and may be extensively preyed on in this region by diurnally migrating mysids (Stemberger 1974; Bowers and Grossnickle 1978). The relationship of mysid predation to mortality of rotifers in nature needs to be more thoroughly investigated.

Insects

The four larval instars of the dipteran *Chaoborus* are the most important insect predators in the plankton of freshwater lakes. The early instars often feed extensively on rotifers (Lewis 1977; Saunders 1980; Chimney et al. 1981; Neill 1985; M. M. Rodenhouse, unpublished data) and may depend on them for survival to the later instars, which prefer larger, crustacean prey (Neill and Peacock 1980; Neill 1985). They have compound eyes but use mechanoreception as the primary means of locating prey (Giguere and Dill 1979). They also possess contact chemoreceptors on the labrum (Pastorok 1980), which may allow them to discriminate prey on the basis of taste. *Chaoborus* larvae can regulate their buoyancy and therefore their position in the water column. They remain motionless and wait for prey to swim within the radius of detection, which may extend to about one-third of their body length for large, active prey. In an attack, the hinged prehensile antennae fling out and sweep the prey into a traplike structure created by the mandibular fan where it is held. The prey is swallowed and enters a muscular pharynx where it is crushed and partially digested. The liquid portion passes into the gut, and the undigested remains are egested out the mouth.

Rotifers

A wide variety of predatory rotifers occur in the plankton. However, only *Asplanchna*, and to a lesser extent *Ploesoma*, are known to be important predators on smaller rotifers. In *Asplanchna* the prey must contact coronal receptors near the mouth to elicit a feeding response (Gilbert 1980a). If stimulated, the muscularized pharynx rapidly expands, creating a suction that draws the prey into the expanded pharyngeal cavity in about 50 msec; the paired, pincer-like trophi then manipulate and push the prey into the stomach (Gilbert and Stemberger 1985c).

The feeding habits of *Ploesoma* are poorly known. *Ploesoma truncatum* is an effective predator on *Polyarthra vulgaris* but may be primarily herbivorous (Stemberger et al. 1979). The large *P. hudsoni* may be the most predaceous species of the genus; a variety of rotifer trophi have been identified from stomach contents of field collected specimens, including those of *Keratella*, *Synchaeta*, and *Polyarthra* (Guiset 1977; Stemberger, unpublished data). *Ploesoma* have specialized virgate mouth parts that probably grasp and tear the integument of their prey; then rapid expansions of the muscularized pharynx probably suck out the soft, internal tissues.

Table 15.1. *Summary of major types of morphological, behavioral, physiological, and reproductive characteristics of rotifers which provide protection against predators*

Means of defense	Primary predators	Point of predation cycle interruption	Potential benefits	Potential disadvantages	Examples
MORPHOLOGICAL FEATURES					
Spines and lorica	Copepods, Asplanchna	Capture, ingestion	Reduced sinking rates	High threshold food requirements	Keratella Kellicottia Notholca Brachionus Trichocerca
Mucus	Copepods, Asplanchna	Attack, capture	Reduced sinking rates; lower threshold food requirements; reduced swimming costs	Reduced encounter rates with algae	Ascomorpha Collotheca Conochiloides
Large body size	Copepods, Asplanchna Chaoborus Mysis Fish larvae	Capture, manipulation	High r_{max}	Fish predation; high threshold food requirements; increased detection; high sinking rates	Asplanchna Synchaeta grandis
Small body size	Copepods; Asplanchna Mysis	Reduced detection	Low threshold food requirements	Decreased r_{max}; interference competition with cladocerans	Keratella Synchaeta Anuraeopsis
Coloniality	Copepods; Asplanchna Chaoborus Fish larvae	Capture; manipulation	Reduced swimming costs	Fish predation; increased sinking rates	Conochilus
Transparency	Fish larvae	Detection	?	Damage by short wave radiation	Asplanchna
ESCAPE RESPONSES					
Active escape	Copepods, Asplanchna Chaoborus	Encounter; pursuit; capture	?	Increased threshold food requirements	Polyarthra Hexarthra Filinia
Passive escape (deadman)	Copepods, Asplanchna	Detection	No swimming costs; protects corona	Cessation of feeding	Keratella Synchaeta Asplanchna
Body turgor	Copepods, Asplanchna	Detection, capture, manipulation	Low energetic cost?; protects corona when body contracted	Sink to poor food environment	Synchaeta Asplanchna

Table 15.1 (*continued*)

Means of defense	Primary predators	Point of predation cycle interruption	Potential benefits	Potential disadvantages	Examples
REPRODUCTIVE FEATURES					
High r_{max}	Copepods; Asplanchna	High birth rate to offset death rate	Opportunistic species	Vulnerable to most predators	Synchaeta Brachionus
Protection of eggs	Copepods; Asplanchna Cladocera Chaoborus		—		
Eggs attached or free within mucus sheath		—	Increased body size	Hydrodynamic drag	Keratella Ascomorpha Collotheca Conochiloides Brachionus
Eggs carried but easily detached		—	—	Sink to bottom; egg ingestion by cladocerans	Poluarthra
Eggs free-floating		—	Reduced visibility of adults; no hydrodynamic drag	Sink to bottom; egg ingestion by cladocerans copepods and Asplanchna	Synchaeta
Live-bearing		—	Newborn escape predation by small copepods and cannalbalism in Asplanchna	Higher maternal costs	Asplanchna
Protection by egg mass	Asplanchna	Capture	?	Increased hydrodynamic drag, sinking rates and visibility	Brachionus some Keratella
DISTRIBUTION IN SPACE AND TIME					
Limited seasonal distribution	Copepods; Asplanchna; fish	Encounter	Present during periods of resource abundance	Absent during periods of resource abundance	Synchaeta
Behavioral avoidance	Copepods mysids cladocerans	Encounter	?	?	Probably many genera
Physiological tolerance to low dissolved oxygen	Copepods; Asplanchna; fish	Encounter	Reduced metabolic costs at lower temperatures	Cost of special pigments	Keratella Kellicottia Filinia Brachionus

ROTIFER DEFENSES
Morphological Features

SPINES AND LORICA. Spines are the most common form of morphological structure protecting rotifers against predators. Spines, which are evaginations of the lorica (body wall, or integument, containing various amounts of cuticle) are most highly developed in the Brachionidae, the family containing some of the most important genera of eulimnetic rotifers—*Keratella, Kellicottia, Notholca,* and *Brachionus.* The effectiveness of spines against a potential predator depends on the length of the spines as well as on the stiffness and thickness of the skeletal material in the body wall.

Generally, spines interfere with a predator's ability to capture or manipulate prey for ingestion. For example, the single posterior spine and, to a lesser extent, the six anterior spines of the small-bodied *Keratella cochlearis* f. *typica* lodge in the soft tissue of the pharynx of *Asplanchna*, rendering manipulation by the trophi ineffective (Stemberger and Gilbert 1984a). This results in a high rate of rejection. For example, *K. cochlearis* possessing a posterior spine are four times more likely to be rejected after capture by *Asplanchna girodi* (Stemberger and Gilbert 1984b). The pharyngeal cavity rapidly expands and contracts, and the inhaled and exhaled water eventually dislodges the rotifer from the pharynx and expels it through the mouth. Most of these rejected prey swim away unharmed (Stemberger and Gilbert 1984b). On the other hand, the linear dimensions of long-spined morphs of the much larger *Keratella slacki* exceed the width of the mouth opening of *Asplanchna*, usually preventing their capture by this predator. If they enter the mouth in a lengthwise position, only a portion of the body fits into the expanded pharyngeal cavity, the spines stick in the pharynx, and the rotifer is rejected as previously described (Gilbert and Stemberger 1984).

Brachionus calyciflorus has a weakly developed lorica, but forms with long posterolateral spines still are well defended against *Asplanchna*. On contacting *Asplanchna*, *Brachionus* retracts its corona; this behavior increases body turgor and causes the paired posterior spines to spread apart at their basal articulation points. These movable spines increase the effective size of the rotifer; they also lodge in the pharynx of *Asplanchna*, greatly reducing the probability of capture by *Asplanchna* (Gilbert 1967, 1980b).

Although *Asplanchna* predation may control rotifer populations in some lakes (Magnien 1983; Hofmann 1983), copepods are probably the most important invertebrate predators affecting rotifer population dynamics and community structure (Stemberger and Evans, 1984). Spination can be an effective deterrent against copepods if the lorica is stiffened. Because *Brachionus calyciflorus* has a thin lorica, it is easy prey for many predaceous copepods, such as *Mesocyclops edax*, even if it possesses long posterolateral spines (Gilbert 1980b; Williamson and Gilbert 1980). In contrast, similarly sized and even smaller-bodied rotifer species with well-developed loricae, like *Keratella crassa* and *K. cochlearis*, are much better protected against copepod predators than are *B. calyciflorus* and other species with soft bodies like *Synchaeta* (Stemberger 1985). The protection provided by a rigid lorica probably reduces damage inflicted by the stoutly spined feeding appendages. The loricae of *Keratella cochlearis*, which have been eaten by *Diacyclops thomasi*, often have puncture wounds, bite-sized pieces removed, or large portions torn off (Stemberger 1985).

Thus, the features that protect rotifers from copepods are somewhat more complex than those that protect them from *Asplanchna*. Greater spination, together with a relatively large body size are effective deterrents against *Asplanchna*. However, a rigid lorica must be added to these features for them to be effective against copepods.

For rotifers with rigid loricae, strength of spines gives considerable additional protection against copepod predation. For example, *Keratella cochlearis* f. *tecta*, a form without a posterior spine, is four times more susceptible to predation by the small-bodied *Tropocyclops prasinus* (500 μm body length) than *K. cochlearis* f. *typica*, a form of identical body size but possessing a single posterior spine (Stemberger and Gilbert 1984b).

Any protection conferred by a stiff lorica and spines may be reduced or lost if predator body size is relatively great in comparison to prey body size. For example, large *Mesocyclops edax* (1500-μm body length) selects equally for spined and unspined morphs of *K. cochlearis* (Stemberger and Gilbert 1984b) and is an effective predator on the larger, well-defended *K. cochlearis [crassa]* (Gilbert and Williamson 1978). However, *Diacyclops thomasi* (1,100-μm body size) preys on *K. cochlearis* about three

Table 15.2. *Known cases of predator-induced polymorphisms in planktonic rotifers*

Polymorphic rotifer	Inducing predator(s)	Reference(s)
Brachionus calyciflorus	*Asplanchna*	Beauchamp, 1952; Gilbert 1966, 1967, 1980b; Halbach, 1970
B. bidentata	*Asplanchna*	Pourriot, 1974
B. urceolaris sericus	*Asplanchna*	Pourriot, 1964
Filinia mystacina	*Asplanchna*	Pourriot, 1964
Keratella cochlearis	*Asplanchna* *Mesocyclops* *Tropocyclops*	Stemberger and Gilbert, 1984b
K. slacki	*Asplanchna*	Gilbert and Stemberger, 1984
K. testudo	*Asplanchna*	Stemberger and Gilbert, in press
	Tropocyclops	Ibid.
	Epischura	Ibid.
	Daphnia	Ibid.

times more effectively than *K. crassa* (Stemberger 1985).

The effect of the relative size of predator and prey suggests that the earlier and smaller predaceous instars of large copepods like *M. edax* may be more selective on small-spined and unspined *K. cochlearis* than are the adults. This prediction has not been tested but has important ecological consequences for predation-mediated changes in the species structure of rotifer communities.

Some rotifer species exhibit predator-controlled polymorphisms in which soluble factors secreted by both copepods (*Mesocyclops*, *Tropocyclops*, and *Epischura*) and *Asplanchna* induce a lengthening and/or *de novo* development of spines and also, in some cases, an increase in body size (table 15.2 and references therein; fig. 15.1). The adaptive significance of spined, predator-induced phenotypes is clear: to reduce mortality from the predators. These polymorphisms undoubtedly evolved as defense mechanisms against the predators that control them. However, the reason for the prevalence of the less spinous and sometimes smaller-bodied basic phenotypes in the absence of these predators is not clear. Presumably, the basic phenotypes are more fit than the exuberant ones in the absence of predators because of either greater longevities or reproductive potentials. Unless there is some cost involved with increased spination and body size, it is uncertain why natural selection would have led to the evolution of these polymorphisms. In light of the relationship between food concentration, population growth rates, and roti-

fer body size, basic phenotypes should have lower-threshold food requirements than exuberant ones (Stemberger and Gilbert 1985). This prediction has not yet been experimentally verified and is critical to assessing the cost of polymorphic responses in rotifers.

MUCUS SHEATHS. Many species of planktonic rotifers secrete a layer of mucus that partially or completely surrounds the body of the rotifer. Such sheaths commonly occur in soft-bodied planktonic species that are slow swimmers and are without behavioral or physical means of defense. Mucus sheaths vary greatly in function and extent to which they cover the body of the rotifer. Some rotifers, like *Ascomorpha ecaudis*, produce a spherical mucus coat that extends out to a diameter that may exceed 1 mm. Algae contacting the sticky surface of the coat are trapped and fed on by the rotifer within (Stemberger 1985). In *Conochilus* and *Conochiloides* the mucus surrounds only the lower portion of the body and foot. When disturbed, these rotifers contract their bodies within the sheath, thus protecting their delicate coronae. Experimental laboratory studies indicate that mucus sheaths in species like *Ascomorpha ecaudis* and *Conochilus* are very effective structures against predation by predatory copepods (Stemberger 1985; Williamson 1983). The mucus adheres to the feeding appendages of the copepod, inhibiting it from making further attacks on the rotifer (Stemberger 1985). The mucus sheaths of *Conochilus*, *Conochiloides*, and *Ascomorpha* also are very effective against predation by *Asplanchna*. *Asplanchna*

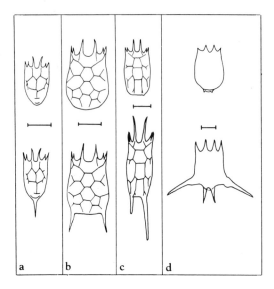

Figure 15.1. *Basic (top row) and exuberant (bottom row) phenotypes of four polymorphic rotifers:* **(a)** Keratella cochlearis, **(b)** K. testudo, **(c)** K. slacki, *and* **(d)** Brachionus calyciflorus. *Scalar = 50 μm.*

BODY SIZE AND COLONIALITY. Body size relative to the predator is an important variable affecting a species' vulnerability to predation. This has already been alluded to above in the section on spines and lorica. A rotifer that is large relative to its predator may be easily detected but difficult to capture. For example, *Diacyclops thomasi* cannot capture adult *Asplanchna priodonta* but easily captures and ingests the newborn of the species (Stemberger 1985). Likewise, adult females of *Mesocyclops edax* eat the small males of *Asplanchna girodi* much more readily than the females of this rotifer (Williamson 1981a). Similarly, there is a direct relationship between the size of *Asplanchna* and the size of the rotifer prey this predator can eat (Gilbert 1980a; Gilbert and Stemberger 1985c). For this reason, large *Asplanchna* are better able to capture long-spined *Brachionus calyciflorus* of a given size than are smaller ones (Gilbert 1967). On the other hand, small body size may reduce a prey's probability of detection because of weak disturbances it produces while swimming.

Coloniality increases the effective size of a species relative to the predator and therefore should protect individuals in the colony from many kinds of small invertebrate predators. Coloniality in *Conochilus unicornis*, for example, protects individuals from predation by copepods and *Asplanchna*, as discussed above in the section on mucus. However, large and colonial species may be more susceptible to visually and filter-feeding fish. Most large-bodied planktonic rotifers, such as *Asplanchna*, and colonial forms, though, possess clear, translucent bodies that may reduce their visibility to visually feeding fish. The most prevalent colonial species, *C. unicornis* and *C. hippocreppis*, are soft-bodied and therefore should be vulnerable to invertebrate predators only when they occur in solitary form. Indirect evidence for the effectiveness of coloniality as a predator defense is the fact that populations of *Conochilus* colonies often persist in nature in the presence of strong predators like *Epischura* and *Diacyclops* (Stemberger and Evans 1984; Stemberger, personal observation).

girodi cannot capture individuals of *Conochilus unicornis* or *Conochiloides dossuarius* when they are surrounded by mucus but readily eats them when the mucus is digested away by a proteolytic enzyme (Gilbert 1980c). Similarly, *Asplanchna girodi* eats *Ascomorpha ecaudis* only when the latter are not surrounded by their mucus sheath (Gilbert, unpublished data).

Mucus sheaths also may reduce a species' specific gravity and therefore confer a substantial energy savings for maintaining position in the water column. This function is supported by the occurrence of mucus in sit-and-wait predators like *Collotheca mutabilis* and *C. pelagica*. These species are weak swimmers and spend much of their time in a quiescent state. The saclike mucus sheath surrounds the body and has a single anterior opening from which it protrudes most of its body when feeding or swimming. However, if disturbed, the rotifer retracts its body into the mucus sheath (Stemberger, personal observation). Species with well-developed mucus sheaths lay their eggs within the mucus cavity, thus protecting them from predators such as copepods, *Asplanchna*, and early instars of *Chaoborus* (M. Moore Rodenhouse 1986). Finally, the very low swimming activity of these species likely reduces their detection by predators that use mechanoreception to locate prey.

Escape Responses

ACTIVE ESCAPE. Three different genera (*Polyarthra, Filinia, Hexarthra*) have evolved highly specialized morphological and behavioral adaptations to escape from predators. In each genus, physical contact with a predator initiates escape responses that propel the body away from the predator. These responses are mediated by muscle-controlled, tubular, or bladelike extensions of the body wall. These jumping or skipping responses also may occur spontaneously, without contact with a predator or another object, apparently especially in *Hexarthra* (Hudson 1871). The ecological significance of such spontaneous responses is unknown.

All species of *Polyarthra* possess four triplets of anterolaterally inserted, paddle-like apendages. In *P. vulgaris*, up to three cycles of asynchronous, up-and-down movements of these 12 paddles cause the body to tumble an average distance of 1.95 mm (15 body lengths) at an average velocity of 35.7 mm/sec (270 body lengths/sec) (Gilbert 1985). This response may be initiated as soon as 7 msec after contact with a predator (Gilbert 1985). Direct observations and experiments have shown that such escape responses are effective in limiting the capture of various *Polyarthra* species by *Asplanchna* (Gilbert and Williamson 1978; Gilbert 1980a), *Diacyclops* (Stemberger 1985) and early instar larvae of *Chaoborus* (M. Moore Rodenhouse 1986).

Very little is known about the mechanics and effectiveness of the escape responses of *Filinia* and *Hexarthra*. *Filinia* species possess three or four long, thin appendages, two or three of which are movable. *Hexarthra* has six large, conical appendages that have setous distal ends and through which muscles extend. Direct observations have shown that the escape response of *Hexarthra mira* is very effective in preventing capture by *Asplanchna* (Gilbert, unpublished data), and a study in a Javanese sewage pond showed that the frequency of *Hexarthra* species in the gut of *Asplanchna brightwelli* was less than that in the environment (Green and Lan 1974). Interactions between *Filinia* and potential predators have not been directly observed; however, one study suggests that both *Filinia opoliensis* and *Hexarthra intermedia* may avoid predation by *Chaoborus* larvae (Lewis 1977).

PASSIVE ESCAPE. Many slow-swimming rotifer species display a dead-man sinking response when contacted by a predator or large cladoceran or when disturbed during routine pipetting in the laboratory. During this response the corona is retracted into the body, and the rotifer sinks passively. Passive sinking reduces swimming disturbances and hence may reduce the ability of the predator to relocate the prey.

When some soft-bodied forms, such as *Synchaeta* and *Asplanchna*, are contacted by predators, they retract their corona, making their bodies swollen and turgid. This turgor makes them difficult to grasp and hence capture by some copepods (Williamson 1983; Stemberger 1985). Interestingly, the soft-bodied *Polyarthra*, which has an active escape response, does not retract its corona after being contacted by a predator (Stemberger, personal observation).

Reproductive Features

Defensive structures and maximal reproductive rates may be inversely correlated. For example, *Synchaeta pectinata* and *Brachionus calyciflorus* have very high maximal rates of population growth (r_m = 0.8/day at 20°C) (Stemberger and Gilbert 1985) but have thin loricae and are very vulnerable to predaceous copepods (Gilbert and Williamson 1978; Williamson 1983; Stemberger 1985). The lack of a stiffened lorica may permit high reproductive rates relative to defended species of similar body mass. High intrinsic rates of population growth also may offset high mortality rates from predation thus permitting coexistence of the prey. In contrast, *Keratella crassa* has a lower maximal rate of population growth than predicted on the basis of body size (Stemberger and Gilbert 1985) but has a very thick lorica that protects it against copepod predation (Stemberger 1985) (see section on spines and lorica above).

There is a great deal of variation in the way that rotifers oviposit, and the oviposition characteristics of a species may affect the vulnerability of both the eggs and the parent to predators (Gilbert 1983). Vulnerable species like *Synchaeta* often release their eggs directly into the water rather than carry them attached to their bodies. Such behavior may reduce the probability of the eggs being eaten with the parental female. At low water temperatures and when predators are absent from the plankton, some populations of species, like *Synchaeta lackowitziana*, may carry their eggs attached by a mucus thread (Stemberger, personal observation). In contrast, species of rotifers that are well defended against copepods, like *Keratella coch-*

learis, *K. crassa*, and *Kellicottia*, carry their eggs firmly attached to the lorica. The weakly loricate species *B. calyciflorus* often carries more than one egg. Such attached eggs increase the effective size of the rotifer and significantly reduce predation of the parental female by *Asplanchna* (Gilbert 1980a). *Polyarthra* carry eggs that are easily detached, and hence probably spared, when the mother is attacked or captured by a copepod (Stemberger, personal observation). However, detached eggs of *Brachionus* and *Kellicottia* have been reported in the crops of early instars of field-collected *Chaoborus* (Saunders 1980; M. Moore Rodenhouse, 1986). Presumably, these eggs were detached from the mother's body in unsuccessful capture attempts. Species with a mucus sheath lay their eggs directly into the free, protected space of the sheath. *Ascomorpha* may temporarily carry its eggs but later sheds them into the mucus cavity. In *Conochilus*, eggs are attached to the mother's foot, and the larvae migrate and attach near the distal part of the foot; there they grow to adult size, being protected by the colony as well as contributing to the colony diameter.

Eggs oviposited free in the environment could sink to deeper waters that are low in dissolved oxygen or have low concentrations of food; such sedimentation might reduce the survivorship of newborns. The colder temperatures of deeper waters also may increase egg development times. Additionally, freely suspended eggs could be susceptible to *Asplanchna* predation and to ingestion or damage by filter-feeding cladocerans (Gilbert and Stemberger 1985a). Thus, it may be advantageous for some rotifers to carry their eggs. Possible disadvantages of this habit may be increased swimming costs to the parent due to viscous drag, increased sinking rates of the parent, or increased egg mortality.

Live-bearing rotifers may increase survival of the young by giving birth to relatively large-size, and hence well-protected, individuals. The newborn of the predaceous, soft-bodied, ovoviviparous rotifers *Asplanchna* and *Asplanchnopus* are more than half of the adult body length. For example, newborn of *Asplanchna priodonta* (400 μm body length) are still vulnerable to predation by copepods but within 48 hours they attain adult size ($^\sim$600 μm) and are much less vulnerable to such predation (Stemberger 1985; Stemberger et al., 1979).

Escape in Time or Space

The pattern of seasonal succession of planktonic rotifers is often very pronounced and predictable (Carlin 1943; Stemberger 1974; Nauwerck 1978; Stemberger and Evans 1984). Resource quantity and quality as well as selective predation influence the succession. Species that coexist with their predators must be defended against them or possess some means of avoiding them in space or time. Undefended species of soft-bodied *Synchaeta* colonize the plankton early in the spring before cohorts of late-stage, predaceous *Diacyclops* move into the plankton (Stemberger and Evans 1984). *Synchaeta* populations rapidly decline with the appearance of these later instars, suggesting that *Synchaeta* rely on escape in time as a means to establish short-lived but often dense populations. *Synchaeta's* ability to develop populations before predation becomes intense probably is due to its ability to reproduce under conditions unsuitable for the reproduction of its predators and to its very high reproductive potential (Stemberger and Gilbert 1985) (see section on reproductive features above).

Some field studies suggest that rotifers may escape predation in space by occupying strata in the column where predator densities are low or absent. For example, rotifers may show well-defined, small-scale movements of less than 1 m, over a time period of several hours, which are negatively and significantly correlated with the diurnal migratory movements of some cladocerans and predaceous copepods (Dumont 1972; Fairchild et al., 1977). Such small-scale, short-lived, vertically oriented patches may be generated by tactic or kinetic responses of the rotifers to migrating, densely stratified crustacean populations. For example, behavioral avoidance, such as passive sinking, could cause the reverse vertical migrations occasionally reported in natural populations of rotifers (Dumont 1972; George and Fernando 1970).

A variety of rotifers tolerate, and even thrive in, water with very low dissolved-oxygen concentrations (Ruttner-Kolisko 1975, 1980; Miracle and Vicente 1983). The ability to occupy oxygen-depleted strata may provide a spatial refuge from most invertebrate predators and filter-feeding fish. Such physiological adaptations also could have evolved to take advantage of dense food resources often associated with oxygen-depleted zones (Miracle and Vicente 1983).

CONCLUSIONS

Rotifers have a variety of morphological, behavioral, physiological, and reproductive features that may protect them against certain predators. Some of these features seem to have evolved especially for defense; others may have evolved to maximize ingestion and or minimize respiration. Possibly some features represent adaptations to both predators and optimum energetics. The specific mechanism by which a species is protected against a particular predator is generally poorly understood, even in some cases where predator–prey interactions have been studied. Interactions of rotifers with calanoid copepods, cladocerans, mysids, and *Chaoborus* especially need to be further studied. More detailed studies of the mechanisms of predator defense in rotifers should provide a fruitful area of future research.

ACKNOWLEDGMENTS

We thank M. Moore Rodenhouse for comments that improved the manuscript. This work was supported by NSF grant BSR-8500561.

REFERENCES

Beauchamp, P. de. 1952. Un facteur de la variabilite chez les rotiferes du genre *Brachionus*. *C.R. Acad. Sci., Paris* 234 : 573–75.

Bogdan, K. G., and J. J. Gilbert. 1982. Seasonal patterns of feeding by natural populations of *Keratella, Polyarthra* and *Bosmina*: Clearance rates, selectivities, and contributions to community grazing. *Limnol. Oceanogr.* 27 : 918–34.

Bowers, J. A., and N. E. Grossnickle. 1978. The herbivorous habits of *Mysis relicta* in Lake Michigan. *Limnol. Oceanogr.* 23 : 767–76.

Brooks, J. L., and S. I. Dodson. 1965. Predation, body size, and composition of the plankton. *Science* 150 : 28–35.

Burns, C. W., and J. J. Gilbert. 1986. Effects of daphnid size and density on interference between *Daphnia* and *Keratella cochlearis*. *Limnol. Oceanogr.* 31 : 848–58.

Carlin, B. 1943. Die Plankton rotatorien des Motalastrom Medd. Lunds Univ. Limn. Inst. 5 : 1–255.

Chimney, M. J., R. W. Winner, and S. K. Seilkop. 1981. Prey utilization by *Chaoborus puntipennis* Say in a small, eutrophic reservoir. *Hydrobiology* 85 : 193–99.

Crecco, V. A., and M. M. Blake. 1983. Feeding ecology of coexisting larvae of American Shad and Blueback Herring in the Connecticut River. *Trans. Amer. Fish. Soc.* 112 : 498–507.

Drenner, R. W., F. de Noyelles, and D. Kettle. 1982. Selective impact of filter-feeding gizzard shad on zooplankton community structure. *Limnol. Oceanogr.* 27 : 965–68.

Drenner, R. W., J. R. Mummert, F. de Noyelles, Jr., and D. Kettle. 1984. Selective particle selection by filter-feeding fish and its impact on phytoplankton community structure. *Limnol. Oceanogr.* 29 : 941–48.

Dumont, H. J. 1972. A competition-based approach to the reverse vertical migration in zooplankton and its implications, chiefly based on a study of the interactions of the rotifer *Asplanchna priodonta* (Gosse) with several Crustacea Entomostraca. *Int. Rev. ges. Hydrobiol.* 57 : 1–38.

Epp, R. W., and W. M. Lewis, Jr. 1984. Cost and speed of locomotion for rotifers. *Oecologia* 61 : 289–92.

Fairchild, G. W., R. S. Stemberger, L. C. Escamp, and H. A. Debaugh. 1977. Environmental variables affecting small scale distribution of five rotifer species in Lancaster Lake, Michigan. *Int. Rev. ges. Hydrobiol.* 62 : 511–21.

Friedman, M. M. 1980. Comparative morphology and functional significance of copepod receptors and oral structures. In W. C. Kerfoot (ed.), *Evolution and ecology of zooplankton communities*, pp. 185–97. Hanover, N.H.: University Press of New England.

Fryer, G. 1957. The feeding mechanism of some freshwater cyclopoid copepods. *Proc. Zool. Soc. London.* 129 : 1–25.

George, M. G., and C. H. Fernando. 1970. Diurnal migration in three species of rotifers in Sunfish Lake, Ontario. *Limnol. Oceanogr.* 15 : 218–33.

Giguere, L. A., and L. M. Dill. 1979. The predatory response of *Chaoborus* larvae to acoustic stimuli, and the acoustic characteristics of their prey. *Z. Tierpsychol.* 50 : 113–23.

Gilbert, J. J. 1966. Rotifer ecology and embryological induction. *Science* 151 : 1234–37.

———. 1967. *Asplanchna* and postero-lateral spine production in *Brachionus calyciflorus*. *Arch. Hydrobiol.* 64 : 1–62.

———. 1980a. Feeding in the rotifer *Asplanchna*: Behavior, cannibalism, selectivity, prey defenses, and impact on rotifer communities. In W. C. Kerfoot (ed.), *Evolution and ecology of zooplankton communities*, pp. 509–17. Hanover, N.H.: University Press of New England.

———. 1980b. Further observations on developmental polymorphism and its evolution in the rotifer *Brachionus calyciflorus*. *Freshwater Biol.* 10 : 281–94.

———. 1980c. Observations on the susceptibility of

some protists and rotifers to predation by *Asplanchna girodi. Hydrobiology* 73 : 87–91.

———. 1983. Rotifera. Oogenesis, Oviposition, and Oosorption. In K. G. and R. G. Adiyodi (eds.), *Reproductive biology of invertebrates*, vol. 1, pp. 181–209. New York: John Wiley and Sons.

———. 1985. Escape response of the rotifer *Polyarthra*: a high-speed cinematographic analysis. *Oecologia* 66 : 322–31.

Gilbert, J. J., and R. S. Stemberger. 1984. *Asplanchna*-induced polymorphism in the rotifer *Keratella slacki. Limnol. Oceanogr.* 29 : 1309–16.

———. 1985a. Control of *Keratella* populations by interference competition by *Daphnia. Limnol. Oceanogr.* 30 : 180–88.

———. 1985b. The costs and benefits of gigantism in polymorphic species of the rotifer *Asplanchna. Arch. Hydrobiol. Beih.* 21 : 185–92.

———. 1985c. Prey capture in the rotifer *Asplanchna girodi. Verh. Int. Ver. Limnol.* 22 : 2997–3000.

Gilbert, J. J., and C. E. Williamson. 1978. Predator-prey behavior and its effect on rotifer survival in associations of *Mesocyclops edax, Asplanchna girodi, Polyarthra vulgaris*, and *Keratella cochlearis. Oecologia* 26 : 13–22.

Green, J., and O. B. Lan. 1974. *Asplanchna* and the spines of *Brachionus calyciflorus* in two Javanese sewage ponds. *Freshwater Biol.* 4 : 223–26.

Grossnickle, N. E. 1982. Feeding habits of *Mysis relicta*—an overview. *Hydrobiology* 93 : 101–107.

Guiset, A. 1977. Stomach contents in *Asplanchna* and *Ploesoma. Arch. Hydrobiol. Beih.* 8 : 126–29.

Guma'a, S. A. 1978. The food and feeding habits of young perch, *Perca fluviatilis* in Windermere. *Freshwater Biol.* 8 : 177–87.

Halbach, U. 1970. Die Ursachen der Temporal variation von *Brachionus Pallas* (Rotatoria). *Oecologia* 4 : 62–318.

Hofmann, W. 1983. Interactions between *Asplanchna* and *Keratella cochlearis* in the PluBsee. *Hydrobiology* 104 : 363–65.

Hrbáček, J. 1962. Species composition and the amount of zooplankton in relation to the fish stocks. *Rozpr. Česk. Akad. Věd.* 72 : 1–116.

Hudson, C. T. 1871. On a new rotifer. *Mon. Microscopical J.* 6 : 121–24.

Jamieson, C. D. 1980. The predatory feeding of copepodite stages III to adult *Mesocyclops leuckarti* (Claus). In W. C. Kerfoot (ed.), *Evolution and ecology of zooplankton communities*, pp. 518–37. Hanover, N.H.: University Press of New England.

Jannsen, J. 1980. Alewives (*Alosa pseudoharengus*) and ciscoes (*Coregonus artedii*) as selective and non-selective planktivores. In W. C. Kerfoot (ed.), *Evolution and ecology of zooplankton communities*, pp. 580–86. Hanover, N.H.: University Press of New England.

Koehl, M. A. R. 1984. Mechanisms of particle capture by copepods at low Reynolds numbers: pos-

sible modes of selective feeding. In D. G. Myers and J. R. Strickler (eds.), *Trophic interactions within aquatic ecosystems*, pp. 135–66. AAAS Selected Symposium 85. Boulder, Colo.: Westview Press.

Lasenby, D. C., and R. R. Langford. 1973. Feeding and assimilation of *Mysis relicta. Limnol. Oceanogr.* 18 : 280–85.

Lewis, W. M., Jr. 1977. Feeding selectivity of a tropical *Chaoborus* population. *Freshwater Biol.* 7 : 311–25.

Lynch, M., and J. Shapiro. 1981. Predation, enrichment, and phytoplankton community structure. *Limnol. Oceanogr.* 26 : 86–102.

Magnien, R. E. 1983. Analysis of zooplankton dynamics: Case studies of temporal succession controlled by the predator, *Asplanchna*, and of diel cycles in *Keratella*. Ph.D. diss., Dartmouth College. Hanover, N.H.

Markarewicz, J. C., and G. E. Likens. 1979. Structure and function of the zooplankton community of Mirror Lake, New Hampshire. *Ecol. Monogr.* 49 : 109–27.

Miracle, M. R., and E. Vicente. 1983. Vertical distribution and rotifer concentrations in the chemocline of meromictic lakes. *Hydrobiology* 104 : 259–67.

Moore Rodenhouse, M. 1986. Age-specific predation risk of *Chaoborus puctipennis*. Ph.D. diss., Dartmouth College. Hanover, N.H.

Murtaugh, P. A. 1981. Selective predation by *Neomysis mercedis* in Lake Washington. *Limnol. Oceanogr.* 26 : 445–53.

Nauwerck, A. 1978. Notes on the planktonic rotifers of Lake Ontario. *Arch. Hydrobiol.* 84 : 269–301.

Neill, W. E. 1985. The effects of herbivore competition upon the dynamics of *Chaoborus* predation. *Arch. Ergebn. Limnol. Hydrobiol. Beih.* 21 : 483–91.

Neill, W. E., and A. Peacock. 1980. Breaking the bottleneck: interactions of invertebrate predators and nutrients in oligotrophic lakes. In W. C. Kerfoot (ed.), *Evolution and ecology of zooplankton communities*, pp. 715–24. Hanover, N.H.: University Press of New England.

O'Brien, W. J. 1979. The predator-prey interaction of planktivorous fish and zooplankton. *American Scientist* 67 : 572–81.

Pace, M. L., and J. D. Orcutt, Jr. 1981. The relative importance of protozoans, rotifers, and crustaceans in a freshwater zooplankton community. *Limnol. Oceanogr.* 26 : 822–30.

Pastorok, R. A. 1980. Selection of prey by *Chaoborus* larvae: A review and new evidence for behavioral flexibility. In W. C. Kerfoot (ed.), *Evolution and ecology of zooplankton communities*, pp. 538–54. Hanover, N.H.: University Press of New England.

Poulet, S. A., and P. Marsot. 1980. In W. C. Kerfoot (ed.), *Evolution and ecology of zooplankton communities*, pp. 198–218. Hanover, N.H.: University Press of New England.

Pourriot, R. 1964. Etude experimentale de variations morphologiques chez certaines especes de rotiferes. *Bull. Soc. Zool. Fr.* 89 : 555–61.

———. 1974. Relations predateur-proie chez les rotiferes: influence du predateur *(Asplanchna brightwelli)* sur la morphologie de la proie *(Brachionus bidentata)*. *Ann. Hydrobiol.* 5 : 43–55.

Ruttner-Kolisko, A. 1975. The vertical distribution of plankton rotifers in a small alpine lake with a sharp oxygen depletion (Lunzer Obersee). *Verh. Internat. Verein. Limnol.* 19 : 1286–94.

———. 1980. The abundance and distribution of *Filinia terminalis* in various types of lakes as related to temperature, oxygen, and food. *Hydrobiology* 73 : 169–75.

Saunders, J. F. 1980. The role of predation as a mechanism controlling planktonic herbivore populations in Lake Valencia, Venezuela. Ph.D. thesis, University of Colorado, Boulder, Colo.

Siefert, R. E. 1972. First food of larval yellow perch, white sucker, bluegill, emerald shiner and rainbow smelt. *Trans. Amer. Fish. Soc.* 101 : 219–25.

Siegfried, C. A., and M. E. Kopache. 1980. Feeding of *Neomysis mercedis* (Holmes). *Biol. Bull.* 159 : 193–205.

Sleigh, M. A., and J. R. Blake. 1977. Methods of ciliary propulsion and their size limitations. In J. R. Pedly (ed.), *Scale effects in animal locomotion*, pp. 243–56. New York: Academic Press.

Stemberger, R. S. 1974. Temporal and spatial distributions of planktonic rotifers in Milwaukee Harbor and adjacent Lake Michigan. *Proceedings of the 17th Conference on Great Lakes Research, International Association Great Lakes Research*, pp. 120–34.

———. 1985. Prey selection by the copepod *Diacyclops thomasi*. *Oecologia* 65 : 492–97.

Stemberger, R. S., and M. S. Evans. 1984. Rotifer seasonal succession and copepod predation in Lake Michigan. *J. Great Lakes Res.* 10 : 417–28.

Stemberger, R. S., D. R. Fuller, and A. M. Beeton. 1979. *The role of predaceous rotifers in Great Lakes plankton dynamics*. A final report to the National Oceanic and Atmospherics Administration, Great Lakes Research Division of the Great Lakes and Marine Waters Center, University of Michigan, Ann Arbor.

Stemberger, R. S., and J. J. Gilbert. 1984a. Body size, ration level, and population growth in *Asplanchna*. *Oecologia* 64 : 355–59.

———. 1984b. Spine development in the rotifer *Keratella cochlearis*: Induction by cyclopoid copepods and *Asplanchna*. *Freshwater Biol.* 14 : 639–47.

———. 1985. Body size, food concentration, and population growth in planktonic rotifers. *Ecology* 66 : 1151–59.

———. In press. Multiple species induction of morphological defenses in the rotifer *Keratella testudo*. *Ecology* 68.

Strickler, J. J., and A. K. Bal. 1973. Setae of the first antennae of the copepod *Cyclops scutifer* (Sars): Their structure and importance. *Proc. Nat. Acad. Sci. (U.S.A.)* 70 : 2656–59.

Threlkeld, S. T., J. T. Rybock, M. D. Morgan, C. L. Folt, and C. R. Goldman. 1980. The effects of an introduced invertebrate predator and food resource variation on zooplankton dynamics in an ultraoligotrophic lake. In W. C. Kerfoot (ed.), *Evolution and ecology of zooplankton communities*, pp. 555–68. Hanover, N.H.: University Press of New England.

———. 1981b. Foraging behavior of a freshwater copepod: Frequency changes in looping behavior at high and low prey densities. *Oecologia* 50 : 332–36.

Williamson, C. E. 1981. The feeding ecology of the freshwater cyclopoid copeped *Mesocyclops edax*. Ph.D. thesis, Dartmouth College, Hanover, N.H.

———. 1983. Invertebrate predation on planktonic rotifers. *Hydrobiology* 104 : 385–96.

Williamson, C. E., and J. J. Gilbert. 1980. Variation among zooplankton predators: the potential of *Asplanchna*, *Mesocyclops* and *Cyclops* to attack, capture and eat various rotifer prey. In W. C. Kerfoot (ed.), *Evolution and ecology of zooplankton communities*, pp. 509–17. Hanover, N.H.: University Press of New England.

Wright, D. J., W. J. O'Brien, and C. Lueke. 1983. A new estimate of zooplankton retention by gill rakers and its ecological significance. *Trans. Amer. Fish. Soc.* 112 : 638–46.

Wyngaard, G. A., and C. C. Chinnapa. 1982. General biology and cytology of cyclopoids. In *Developmental biology of freshwater invertebrates*, pp. 485–533. New York: Alan R. Liss.

16. Chemical Defenses of Freshwater Organisms: Beetles and Bugs

Steve Scrimshaw

W. Charles Kerfoot

Numerous freshwater organisms seem protected against predators by distasteful or toxic substances. One of the most intensively studied groups, the aquatic insects, includes excellent examples in the orders Coleoptera (beetles) and Hemiptera (bugs). Here we review the kinds of substances carried by species, the morphology of the glands that produce and deliver the compounds, and the relative effectiveness of these defenses against dominant aquatic predators.

We argue that chemical defenses are especially common in active, large-bodied insects that are exposed to risk from fish predation. Identified compounds range from vertebrate hormones to acids or active substances (e.g., hydrogen peroxide) that may cause local cell damage. Delivery of compounds varies from coating of external surfaces to active expulsion of glandular contents. Both the intended targets (fish) and the modifying influences of the medium (water) offer some interesting twists to conventional predator–prey interactions.

Contrasting with the relatively small size of most pelagic organisms, the plants and animals of the littoral zone more closely resemble their terrestrial counterparts in both size and taxonomic composition. Large plants are common, as are a wide variety of large-bodied and conspicuous insects. Cover has improved, yet as in terrestrial fields or woodlands, these insects find themselves at risk from a variety of vertebrate and invertebrate predators because of their size and potential food value. Along the shoreline, predators can be similar to terrestrial taxa, yet in the submerged portion of lakes and ponds, the predominant predators are an assortment of visually feeding fish. Prey species do show some behavioral or morphological adjustments that lower risk to these predators. Some of the more commonly observed adaptations include (1) crypsis or protective concealment, (2) nocturnal habits, (3) swift evasive movements or extreme agility, and (4) spines, claws, or thick chitinous coats. For instance, caddisfly larvae construct durable, protective houses from background materials, blending with the colors and texture of the substrate. Larval dragonflies or damselflies fall into two groups: exposed species that are protectively colored and concealed species that bury themselves in loose sediment (Benke and Benke 1975; Johnson and Crowley 1980; Morin 1984). Finally, whirligig beetles and water striders can sense the slightest agitation of the surface film, reacting with immediate darts or jumps. Yet are these attributes the only defenses of aquatic insects to patrolling or ambush predators?

We assert that many freshwater organisms rely on chemical defenses against potential consumers, in particular fishes. This defense is used either as a primary deterrent (water mites) or, most often, in conjunction with an assortment of behavioral traits or morphological adaptations (beetles, bugs). However, published information on chemical defenses is so scattered throughout the literature and unavailable to the casual reader that it gives a misleading impression, i.e., that chemical defenses are rare and unusual in aquatic organisms.

To our knowledge, there is no comprehensive review of the evidence for chemical defenses and little discussion of its significance. Moreover, there is considerable dispute over the broad function of the chemical substances produced by several taxa. We find that the specific actions of many compounds are poorly related to field conditions and that judgments of effects are often based on subjective criteria. This is unfortunate for two reasons: (1) contrasts between poorly adapted and well-adapted aquatic species offer interesting comparisons, and (2) both the medium and the kinds of predators differ for insects found on emergent, neustonic, or fully submerged substrates. In particular, the physical and chemical properties of water significantly influence the options for delivery and dispersion of toxic or foul-tasting compounds. Moreover, fish may differ from typical terrestrial predators in physiology and prey-handling characteristics.

In this chapter we attempt a broad review of the published literature on two groups (beetles, bugs), where evidence is the most detailed and convincing. Here chemical compounds and glandular organs have been characterized extensively and their effects on various predators determined in specific bioassays. Despite our caution with certain cases, we feel that the current state of knowledge justifies a strong conclusion: the use of chemical defenses against fish is not peculiar to a few species but is widespread among the larger, active, and/or more exposed organisms of the littoral environment.

In the following discussion, we emphasize three types of evidence: (1) the nature of glands and delivery devices, (2) the isolation and identification of compounds produced by those glands, and (3) laboratory and field observations that suggest the action of the secreted compounds. At last count, approximately 80 chemical compounds had been identified from 68 species in 11 families of aquatic arthropods (for recent reviews, see Blum 1981; Bettini 1978). Of these, 56 compounds have been isolated from 70 different species of Coleoptera, primarily within the Dytiscidae and Gyrinidae. Thus, previous work has centered on relatively few taxa. Our own surveys have uncovered a few new families in which palatability tests indicate noxious substances. Other cases include instances where gland systems have been described but where no attempt has been made to identify or evaluate the effect of secretions.

BEETLES (COLEOPTERA): EVIDENCE FOR CHEMICAL DEFENSES

In north temperate ponds and lakes, approximately five families of beetles are both com-

mon and specialized enough to be considered a typical part of the littoral fauna. These families include predaceous diving beetles (Dytiscidae), whirlgig beetles (Gyrinidae), crawling water beetles (Haliplidae), elmids (Elmidae), and water scavenger beetles (Hydrophilidae) (fig. 16.1). The first three families fall within the suborder Adephaga; the latter two are in the Polyphaga. Dytiscids and gyrinids are especially good divers and swimmers, whereas elmids usually crawl about on the bottom substrate.

Predaceous Diving Beetles (Dytiscidae)

Within the aquatic Coleoptera, the family Dytiscidae has received by far the most attention, with respect not only to the isolation and identification of chemicals but also to the effects of substances on suspected targets (Weatherston and Percy 1978a). Roughly 41 compounds have been identified from 58 species in four subfamilies. Heightened interest in dytiscids comes from (1) evidence that species synthesize "vertebrate" hormones as a defense against fishes, and (2) the opportunity for comparisons, dytiscids being closely related to the large terrestrial family Carabidae (both fall within the same superfamily, i.e., Caraboidea). The superfamily Caraboidea includes the terrestrial Carabidae and several aquatic families: Dytiscidae, Gyrinidae, Haliplidae, Noteridae, Hygrobiidae, and Amphizoidea. The dytiscids, however, are the most specialized aquatic branch.

The 41 compounds come from two major gland systems that open externally: the prothoracic and pygidial glands (fig. 16.2). All dytiscids possess prothoracic glands. These glands are absent from all other Caraboidea except the Hygrobiidae. In the Dytiscidae, the thoracic glands open close to the anterolateral angle of the prothorax, whereas in the Hygrobiidae they open near the posterolateral angle of the prothorax. Additionally, in the dytiscids there are no intrinsic muscles covering the central reservoir, but hygrobids possess such layers. These differences prompted Forsyth (1970) to suggest that prothoracic glands evolved independently in the two families, despite an occurrence that superficially argues for close affinity.

Prothoracic glands have been described in detail for several species of dytiscids (Casper 1913; Blunck 1917; Korschfelt 1924; Forsyth

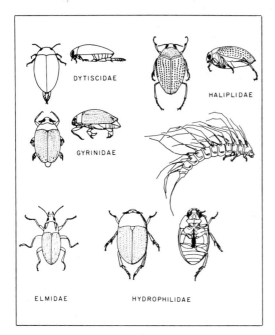

Figure 16.1. *Examples of beetles (Coleoptera) that are rejected by fish. (Redrawn from Doyen and Ulrich 1978).*

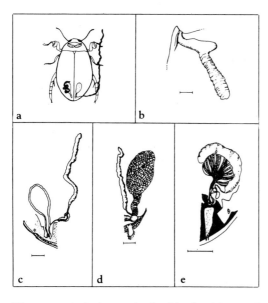

Figure 16.2. *Defensive glands of beetles:* **(a)** *general location of prothoracic and pygidial glands;* **(b)** *right prothoracic gland of* Hyphydrus *(Dytiscidae); pygidial glands of* **(c)** Deronectes *(Dytiscidae),* **(d)** Gyrinus *(Gyrinidae), and* **(e)** Haliplus *(Haliplidae).* (b–e *redrawn from Forsyth 1968).*

1968). In most instances, a simple, elongate reservoir passes laterally along the body midline, then veers posteriorly for about an equal distance. Secretory cells form a lobe draped around the reservoir, a more or less continuous sheet of tissue about one cell deep. Secretory cell ducts open individually into the chitinous collecting reservoir through circular sieve plates distributed haphazardly over the basal half of the vessel (Forsyth 1968). There are no intrinsic muscles surrounding the reservoir. Instead, the secretion is forced out partly by turgor pressure and partly by the indirect action of the nearby tergosternal muscles. A single muscle regulates discharge by controlling the opening valve of the reservoir (fig. 16.2).

The pygidial glands are paired structures located in the posterior part of the abdomen, behind the hindgut and above the reproductive organs (fig. 16.2). Each gland connects to a reservoir that opens externally behind the eighth abdominal tergite (Forsyth 1968). Most families in the Caraboidea possess pygidial bladders. Based on comparative morphology, all of the glands are fairly similar, leading Forsyth (1968) to suspect that pygidial defense glands evolved only once in the Caraboidea.

In all dytiscids, the pygidial glands consist of a long, stringy secretory lobe, a collecting canal, a round reservoir, an efferent duct, and an opening valve. In some groups (e.g., the subfamily Hydroporinae), there may be an accessory gland. The secretory cells of the lobe are arranged into an elongate, cylindrical tube, often many times the length of the reservoir. Their secretions pass into a central canal, which discharges into the efferent duct of the reservoir. The walls of the collecting reservoir are thin cuticle, surrounded by muscle one to four layers deep. When the reservoir valve opens, direct muscular contraction expels the clear or yellowish contents of the pygidial bladders. The secretions ooze out as a viscous fluid, for the dytiscids have lost the ability of the carabid beetles suddenly to discharge gland products. Instead, beetles use their rear legs as brushes to spread the pale or yellowish fluids over the rear of the body and wings.

SECRETIONS OF PROTHORACIC GLANDS. The dominant kinds of compounds produced by the prothoracic glands of dytiscids are steroids, although an alkaloid, a diterpene, and a nucleoprotein also have been isolated and identi-

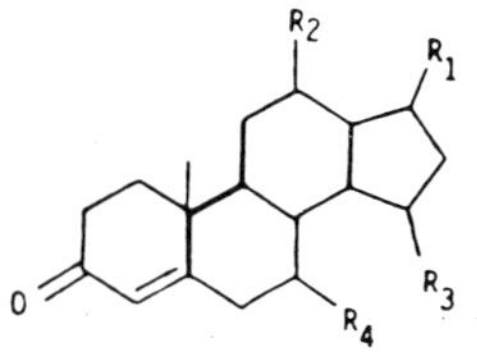

		R_1	R_2	R_3	R_4
1	4-pregnene-21-ol-3,20-dione	$-COCH_2OH$	H	H	H
2	4-pregnene-20α-ol-3-one	$-CHOHCH_3$	H	H	H
3	4-pregnene-20β-ol-3-one	$-CHOHCH_3$	H	H	H
4	4-pregnene-20β,21-diol-3-one	$-CHOHCH_2OH$	H	H	H
5	4-pregnene-15α,20β-diol-3-one	$-CHOHCH_3$	H	OH	H
6	4-pregnene-12β-ol-3,20-dione	$-COCH_2$	OH	H	H
7	4-pregnene-12β-ol-3,20-dione-pentetate	$-COCH_3$	$-OCOCH_2CHCHCH_3$	H	H
8	4-pregnene-15α-ol-3,20-dione-7α-isobutyrate	$-COCH_3$	H	OH	$-OCOC_3H_7$
9	4-pregnene-15α-ol-3,20-dione-7α-hydroxyisobutyrate	$-COCH_3$	H	OH	$-OCOC_3H_7O$

Figure 16.3. *Pregnene derivatives in the prothoracic glands of dytiscid water beetles.*

fied (figs. 16.3–16.5; tables 16.1, 16.2). Overall, 22 varieties of steroids have been isolated and characterized. Because both the glands and their secretions are unique in that certain compounds are identical with vertebrate steroids (Schildknecht et al. 1966; Schildknecht et al. 1967; Schildknecht and Hotz 1967; Schildknecht 1971), prothoracic gland secretions have received more attention than any other class of arthropod defensive substance. An additional contributing historical factor was the early medical use of fish as bioassay organisms for the anesthetic properties of steroids (Selye and Heard 1943). These experiments not only underscored the sensitivity of fish to steroids but also established a protocol for later physiological tests.

Most of the compounds characterized so far from prothoracic glands come from the two subfamilies Dytiscinae and Colymbetinae. These substances fall into four broad groupings: pregnene derivatives, pregnediene derivatives, other steroids, and nonsteroids (table 16.1). The more exotic compounds include cortexone, estrone, estradial, and testosterone (figs. 16.3–16.5). Steroids are especially concentrated in the Dytiscinae, whereas some of the Colymetinae contain substantial amounts of alkaloids. For example, the main component in the prothoracic glands of *Colybius fenestratus* is the alkaloid, 8-hydroxy-quinoline-2-carboxylate (fig. 16.5) (Schildknecht 1971).

SECRETIONS OF PYGIDIAL GLANDS. The most common compounds isolated from dytiscid pygidial glands are aromatic aldehydes, esters,

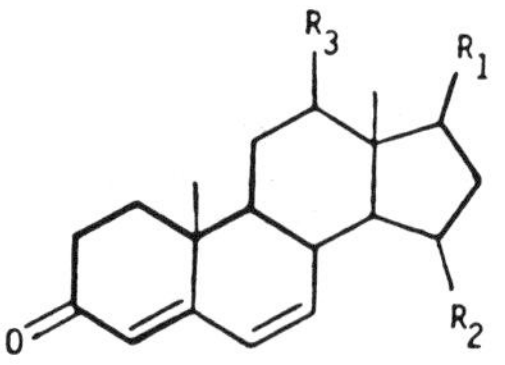

Figure 16.4. *Pregnediene derivatives found in the prothoracic glands of dytiscid water beetles.*

		R_1	R_2	R_3
10	4,6-pregnadiene-21-ol-3,20-dione	-COCH$_2$OH	H	H
11	4,6-pregnadiene-20α-ol-3-one	-CHOHCH$_3$	H	H
12	4,6-pregnadiene-3,20-dione	-COCH$_3$	H	H
13	21-hydroxy-4,6-pregnadiene-3,20-dione	-COCH$_2$OH	H	H
14	4,6-pregnadiene-12β-ol-3,20-dione	-COCH$_3$	H	OH
15	4,6-pregnadiene-12β,20α-diol-3-one	-CHOHCH$_3$	H	OH
16	4,6-pregnadiene-15α-ol-3,20-dione	-COCH$_3$	OH	H
17	4,6-pregnadiene-15α-ol-3,20-dione-isobutyrate	-COCH$_3$	-OCOCH(CH$_3$)$_2$	H
18	4,6-pregnadiene-15α-20β-diol-3-one-20-isobutyrate	-CHCH$_3$ OCOCH(CH$_3$)$_2$	OH	H

and acids (fig. 16.6, table 16.3). Sixteen compounds have been identified from 54 species in four subfamilies, with much of the work published recently (Dettner 1979; Newhardt and Mumma 1979a; Dettner and Schwinger 1980; Classen and Dettner 1983). These compounds differ from those characteristic of terrestrial carabids and show distinctions between dytiscid subfamilies. For example, most carabid beetles (e.g., subfamilies Carabinae, Harpalinae) possess pygidial secretions that contain either formic acid or a mixture of 95% methacrylic acid and 5% tiglic acid (Schildknecht 1970). In contrast, the dytiscid subfamilies Dytiscinae and Colymbetinae possess only one major carboxylic acid, i.e., benzoic acid. The other principal components of their pygidial paste are phenols. The most prominent phenolic compound is *p*-hydroxybenzaldehyde, usually accompanied by *p*-hydroxybenzoic acid methyl ester (Schildknecht 1970). The Colymbetinae, which are far more amphibious than the highly aquatic Dytiscidae, possess a quinone (hydroquinone) more reminiscent of terrestrial carabids. In contrast, the compounds found in the subfamily Hydroporinae are principally α-hydroxycarbocyclic acids and a lactide, e.g., forms of acetic or phenylpyruvic acids (fig. 16.6, table 16.3). Dettner and Schwinger (1980) felt that the large degree of non-overlap between compounds in the Hydroporinae and the former two dytiscid subfamilies was more a reflection of distant phylogeny than of immediate adaptive significance.

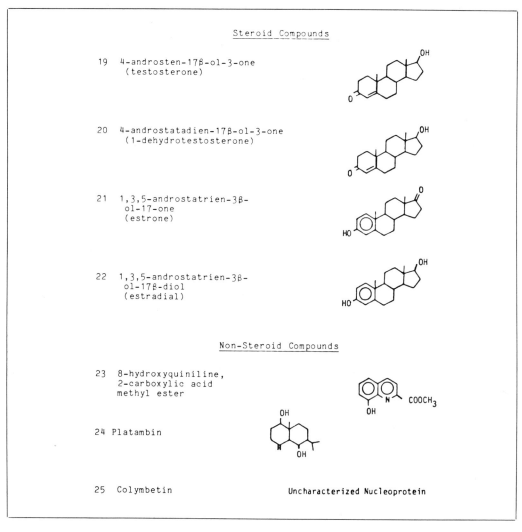

Figure 16.5. *Miscellaneous steroid and nonsteroid compounds found in the prothoracic glands of dytiscid water beetles. (Sources: 19–22, Schildknecht and Bir-ringer 1969; 23, 24, Schildknecht 1976; Schildknecht and Tacheci 1971.)*

PHYSIOLOGICAL EFFECTS OF GLAND SECRETIONS ON PREDATORS, PARASITES, AND PATHOGENS. The extreme distastefulness of dytiscid secretions to fish has been known for some time. In 1721, Frisch (Schildknecht et al. 1966; Schildknecht 1970) stated that these beetles produce a "foul drop that affords torment to a pike or other animal that swallows it." Subsequently, Blunck (1917) force-fed beetles and their prothoracic secretions to a variety of fish and anurans. In all cases, the fish responded with immediate "agitation" or at high doses lapsed into an "anesthetic" state. Anurans vomited the beetles and began mouth frothing and cleaning activity.

Our own experiments and those of Schildknecht (Schildknecht 1970, 1971; Schildknecht et al. 1966) confirm the repugnancy of dytiscids to fish. Capture releases contents of both the prothoracic and pygidial bladders. Most often this results in immediate rejection, followed by buccal flushing by the fish and an erratic and evasive escape reaction on the part of the prey. Thus, there is no doubt about the

Table 16.1. Compounds found in the prothoracic glands of dytiscid water beetles

	Pregnene derivatives									Pregnediene derivatives									Other steroids				Non-steroids		
	1	2	3	4	5	6	7	8	9	10	11	12	13	14	15	16	17	18	19	20	21	22	23	24	25
COLYMBETINAE																									
Agabus bipustulatus	+																								
A. sturmi					+			+	+							+	+	+							
A. seriatus	+		+																						
Colymbetes fuscus																									+
Illyabus fuliginosis																			+						
I. fenestratus			+	+	+														+	+	+	+	+		
Platambus maculatus	+				+															+	+	+		+	
DYTISCINAE																									
Acilus sulcatus	+	+								+	+	+	+												
A. semisulcatus	+	+								+	+	+													
A. mediatus	+																								
A. seriates	+																								
A. obtusus	+	+																							
Dytiscus marginalis	+	+									+														
Cybister confusus	+																								
C. lateralimarginalis	+					+	+			+	+	+	+	+	+										
C. tripunctatus			+							+	+	+		+											
C. limbatus	+	+				+				+	+			+											
Graphoderus cinereus	+						+																		
G. liberus	+																								

Table 16.2. *Pregnene and pregnadiene derivatives in the prothoracic glands of dytiscid water beetles*

	Compound	Beetle	μg/beetle	References
1	4-pregnene-21-ol-3,20-dione	*Acilius sulcatus*	19	Schildknecht (1971)
		A. semisulcatus	M	Miller and Mumma (1976a)
		Agabus seriatus	40	Miller and Mumma (1973, 1974)
		A. bipustulatus	M	Schildknecht and Hotz (1967, 1970)
		Cybister confusus	M	Chadha et al. (1970)
		C. lateralimarginalis	3	Schildknecht (1971)
		C. limbatus	133[a]	Simpahimalani et al. (1970)
		C. tripunctatus	143	Chadha et al. (1970)
		Dytiscus marginalis	400	Schildknecht (1971)
		Graphoderus liberus	20	Miller and Mumma (1973)
2	4-pregnene-20-ol-3-one	*Acilius sulcatus*	1	Schildknecht (1971)
		Cybister limbatus	8	Simpahimalani et al. (1970)
		Dytiscus marginalis	m	Schildknecht and Hotz (1967)
3	4-pregnene-20-ol-3-one	*Cybister tripunctatus*	100	Chadha et al. (1970)
		Ilyabus fenestrus	1	Schildknecht and Birringer (1969)
4	4-pregnene-20,21-diol-3-one	*Ilyabus fenestrus*	4	Schildknecht (1971)
5	4-pregnene-15,20-diol-3-one	*Ilyabus fenestrus*	7	Schildknecht (1971)
		Platambus maculatus	7	Schildknecht (1971)
6	4-pregnene-12-cl-3, 20-dione	*Cybister limbatus*	80	Chadha et al. (1970)
		C. lateralimarginalis	28	Schildknecht (1971)
7	4-pregnene-12-ol-3,20-dione-pentetate	*Cybister lateralimarginalis*	4	Schildknecht (1971)
8	4-pregnene-15-ol-3,20-dione-7-isobutyrate	*Agabus sturmi*	—	Schildknecht and Hotz (1971)
9	4-pregnene-15-ol-3,20-dione-7-hydroxyisobutyrate	*Agabus sturmi*	—	Schildknecht and Hotz (1971)

PREGNADIENE

	Compound	Beetle	μg/beetle	References
10	4,6-pregnadiene-21-ol-3,20-dione	*Acilus sulcatus*	56	Schildknecht (1970, 1971)
		Cybister limbatus	13	Sipahimalani et al. (1970)
		C. lateralimarginalis	90	Schildknecht (1971)
		C. tripunctatus	1	Schildknecht (1971)
11	4,6-pregnadiene-20-ol-3-one	*Acilus sulcatus*	7	Schildknecht (1971)
		Cybister limbatus	17	Sipahimalani et al. (1970)
		C. lateralimarginalis	140	Schildknecht (1971)
		Dytiscus marginalis	—	Schildknecht and Hotz (1967)
12	4,6-pregnadiene-3,20-dione	*Acilus sulcatus*	6	Schildknecht (1971)
		Cybister lateralimarginalis	—	Miller and Mumma (1976a,b)
13	21-hydroxy-4,6-pregnadiene-3,20-dione	*Acilus sulcatus*	—	Schildknecht (1976)
		Cybister lateralimarginalis	—	Schildknecht (1976)
14	4,6-pregnadiene-12-ol-3,20-dione	*Cybister limbatus*	37	Chadha et al. (1970)
		C. lateralmarginalis	6	Schildknecht (1970)
		C. tripunctatus	1050	Schildknecht (1971, 1976)
15	4,6-pregnadiene-12,20-diol-3-one	*Cybister lateralmarginalis*	36	Schildknecht (1971)
16	4,6-pregnadiene-15-ol-3,20-dione	*Agabus sturmi*	—	Schildknecht and Hotz (1971)

Table 16.2 (continued)

	Compound	Beetle	μg/beetle	References
17	4,6-pregnadiene-15-ol-3,20-dione-isobutyrate	*Agabus sturmi*	—	Schildknecht and Hotz (1971)
18	4,6-pregnadiene-15-20-diol-3-one-20-isobutyrate	*Agabus sturmi*	—	Schildknecht and Hotz 1971

M, major constituent; m, minor constituent. [a]Up to 1 μg per beetle (Schildknecht, 1971).

Table 16.3. *Pygidial gland constituents of dytiscid water beetles*

	Compound															
	1	2	3	4	5	6	7	8	9	10	11	12	13	14	15	16
DYTISCINAE																
Acilius mediatus			n	n	n											
A. semisulcatus			n	n	n											
A. sulcatus		+	+	+	+											
A. sylvanus			n	n	n											
C. lateralimarginalis			+	+	+	+	+									
C. tripunctatus			+	+	+											
Dytiscus lattissimus			+	+	+											
D. marginalis			+	+	+			+								s
Graphoderus cinereus	+		+	+	+	+										
G. liberus				m	+											
Hydaticus seminiger			+	+	+						+	+				
COLYMBETINAE																
Agabus affinus		+	+	+	+	+										
A. bipusulatus	c	c	c	c	c	c										
A. chalconotus		+	+	+	+	+										
A. congener		+	+	+												
A. didymus		+	+	+												
A. guttatus		+	+	+	+	+										
A. labiatus	+	+	+	+	+											
A. melanaris		+	+	+	+	+										
A. nebulosus		+	+	+	+											
A. paludosus	c	c	c	c	c	c										
A. seriatus			m													
A. solieri		+	+	+	+	+										
A. sturmi		+	+	+	+											
A. wasastjernae			+	+												
Copelatus ruficollis			s	s												
C. haemorrhoidalis		+		+	+								+			
Colymbetes fuscus	s	+	+	+	+	s										
Illyabus ater	+	+	+	+	+				+							
I. crassus		+	+	+	+	+										
I. fenestratus	+	+	+	+	+	+										
I. fuliginosis		+	+	+	+											
I. guttiger	+	+	+	+	+				+							
Platambus maculatus		+	+	+	s											
Rhantus exoletus	+	+	+	+	+											
R. pulverosus		+	+	+	+	+										

Table 16.3 (*continued*)

								Compound								
	1	2	3	4	5	6	7	8	9	10	11	12	13	14	15	16
HYDROPORINAE																
Graptopdytes pictus											+	+	+	+	+	
Guignotus pusillus											+	+	+			
Hydroporus angustatus											+	+	+		+	
H. dicretus											+	+			+	
H. dorsalis												+		+		
H. ferrugineus												+	+	+		
H. marginatus											+	+	+		+	
H. melanaris				+										+		
H. obscurus												+	+	+		
H. palustris				+							+	+		+		
H. planus												+	+	+		
H. tristus												+	+	+		
Hygrotus inaequalis											+	+	+	+	+	
Hyphydrus ovatus												+	+	+		
Potamonectes depressus												+	+	+		
Scarodytes halensis												+	+	+	+	
Stictotarsus												+	+	+	+	
duodecimpustulatus												+	+	+	+	
LACCOPHILANAE																
Laccophilus hylanus								+								

+, Dettner (1979); c, Classen and Dettner (1983); m, Miller and Mumma (1973); n, Newhardt and Mumma (1979a); s, Schildknecht (1970). References represent the most recent original source article reporting the presence of a compound in particular species.

"defensive" nature of the substances nor their effectiveness on fish. However, the presence of two such organs has raised questions about relative roles and targets.

The standard bioassays of Selye and Heard (1943) and follow-up assays by Miller and Mumma (1976a,b) have established the strong physiological impacts that prothoracic steroids have on fish. Miller and Mumma consider the dytiscid steroids among the more highly anesthetic steroids known on a per weight basis. Yet the quantity that certain beetles can deliver is also great. As pointed out by Schildknecht et al. (1966), a single *Dytiscus* can store up to 0.4 mg of cortexone, an amount equivalent to that isolated from more than 1,000 ox suprarenal glands. Because it is a mineralocortical hormone of vertebrates, exposure to such concentrations could have strong effects, aside from interfering with the sodium-potassium balance of exchange surfaces. On the other hand, because cortexone is a hormone foreign to the beetles, it has no corresponding

hormonal impact on the beetle. Known concentrations of certain steroids in the prothoracic glands of dytiscids are listed in table 16.2.

Steroids are known to have dramatic impacts on fish, but what are the primary physiological effects of contact with secretions or ingestion of organisms: local cell damage to exchange tissues (gills), general anesthesia, or simply a noxious taste? Although the ultimate results probably depend on exposure, Miller and Mumma (1976a) suggested that the prothoracic steroids primarily affect fish via membrane lysis, with toxic effects following if fish experience prolonged narcosis. To test the effects of concentration, fish were placed in water that contained either pure steroids or whole extracts from prothoracic glands. For example, the most commonly reported steroid in dytiscid prothoracic secretions is 4-pregnene-21-ol-3,20-dione (deoxycorticosterone, DOC). Miller and Mumma (1976a) found this compound to be one of the most potent on *Pimephales promales*,

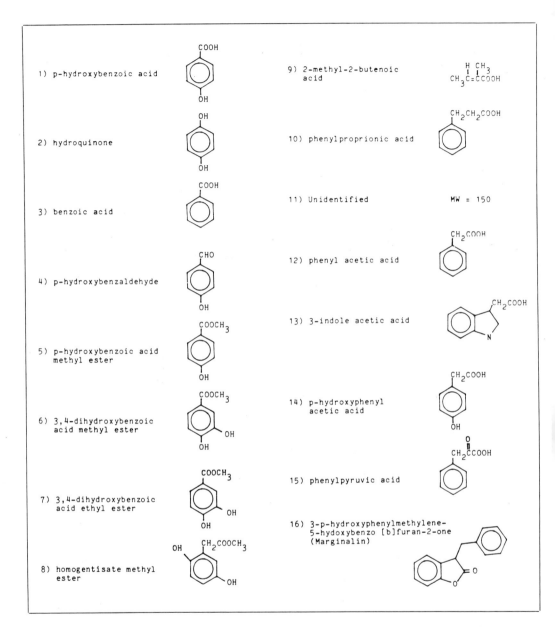

Figure 16.6. *Structures of compounds found in the pygidial gland secretions of dytiscid water beetles.*

with an LD_{50} of 300 nmoles/g fish, or approximately 100 μg/g fish. The minnow absorbed 80% of the steroids through the gills and 20% through the skin, although actual uptake rates varied with external concentration (Miller and Mumma 1976b). However, even if 100% of the prothoracic gland contents were absorbed, these tolerance levels make it very doubtful whether the ingestion of even a single large beetle would cause fish death. Rather, Miller

and Mumma (1976b) concluded that anesthetic properties, which are effective at concentrations an order of magnitude less, are probably far more important than toxicity per se.

The primary effect of exposure to steroids at low concentrations may be even more subtle. Hara (1967) and Oshima and Gorbman (1968) have shown that steroids affect the spontaneous and evoked neural discharges from the olfactory systems of fish. This finding led Clayton (1964) to speculate that steroids could act as repellents independent of their anesthetic properties. Miller and Mumma (1976b) also suggested that possibility.

Although only a fraction of the prothoracic gland's contents are discharged with each encounter, information is available on synthesis and renewal rates. As noted by Clayton (1964), Blum (1981), and Fescemeyer and Mumma (1983), all insects require an exogenous source of sterols for normal growth, development, and reproduction because they are unable to synthesize steroids from simple precursors. Both Schildknecht (1970) and Fescemeyer and Mumma (1983) found that radioactively labeled cholesterol was incorporated into the prothoracic steroids, suggesting that cholesterol was the natural precursor in aquatic environments. The latter two workers also found that *Agabus seriates* and *A. obtusus* were able to regenerate 80% of their prothoracic steroids by the end of 2 weeks.

The presence of alkaloids in the Colymbetinae may relate to their more amphibious habits. The main component, 8-hydroxyquinoline-2-carboxylate, is not toxic to fish or amphibians but apparently produces spasms in small mammals, especially mice (Schildknecht 1971).

In contrast to the acknowledged defensive nature of prothoracic secretions, early investigators ascribed an antimicrobial or hydrophobic action to pygidial secretions (Dierckx 1899; Maschiwitz 1967; Schildknecht 1970, 1971). This exclusive interpretation has been questioned by Eisner (1970), Hepburn et al. (1973), and Blum (1981). Eisner (1970) noted the unpalatability of adult beetles to fish and stated that the "pygidial glands . . , despite the unconvincing claim that they are antimicrobial, . . may also be protective against predators." Subsequent experiments have provided direct evidence for protection against certain vertebrate predators. For example, when two of the three most common components (*p*-hydroxybenzaldehyde, *p*-hydroxybenzoic acid methyl ester) were mixed in low concentration (6×10^{-6} moles) with palatable media, they rendered the food unpalatable to *Tilapia sparminni* (Hepburn et al. 1973). Likewise, 2-methyl-2-butenoic acid, found in the secretions of *Ilbius* (Colymbetinae) is a strong cell poison characteristic of carabid secretions (Dettner and Schwinger 1980). However, a number of the other secretions have unclear functions. Phenylacetic acid and 3-indoleacetic acid are phytohormones, whereas *p*-hydroxyphenylacetic acid is a main component in the venom of a spider. Phenylpropionic acid has both a strong odor and antibiotic properties (Dettner and Schwinger 1980). Like many other familiar terrestrial secretions (e.g., ant or termite), it is possible that pygidial products serve several functions: deterrence of large predators, antiseptic actions, and conditioning of tissues associated with deposition of eggs on plant stems.

Whirligig Beetles (Gyrinidae)

True to their common name, whirligig beetles are highly specialized swimmers, capable of executing rapid evasive movements to approaching predators. Hind legs can stroke 50 to 60 times per second at the water surface, propelling the beetles at speeds up to 1 m/s or throwing them into an erratic series of tight loops, zig-zags, or spirals. Eyes are positioned to receive images from either above or below the water surface, and echo location is used to locate solid objects (Tucker 1969). If danger threatens at the water surface, the beetles can also dive, although speeds here are only about one-tenth the surface values. As with the dytiscids and most other beetles, the body is covered by a hard exoskeleton (fig. 16.1).

In addition to these highly specialized neustonic adaptations, whirligig beetles also possess paired pygidial glands that secrete an odoriferous, volatile fluid (Forsyth 1968; Benfield 1972; Meinwald et al. 1973). The structure of the pygidial glands is similar to that in other families in the Caraboidea (fig. 16.2), except in the external opening. The opposing dorsal and ventral surfaces of the opening are extended; they touch at rest and return after displacement, suggesting additional use as a valve (Forsyth 1968). Benfield (1972) noted that the anatomy of the glands is such that visible amounts of secretion are noticeable only when beetles are squeezed, leading him to suspect that gyrinids conserve the emission of glandular products.

In contrast to the actions of dytiscid pygidial secretions, those of gyrinids are well studied. Gyrinid beetles produce four varieties of norsesquiterpene (fig. 16.7) that are strongly repellent to fish, birds, and small mammals. The effectiveness of compounds has been demonstrated in taste tests (Benfield 1972; Meinwald

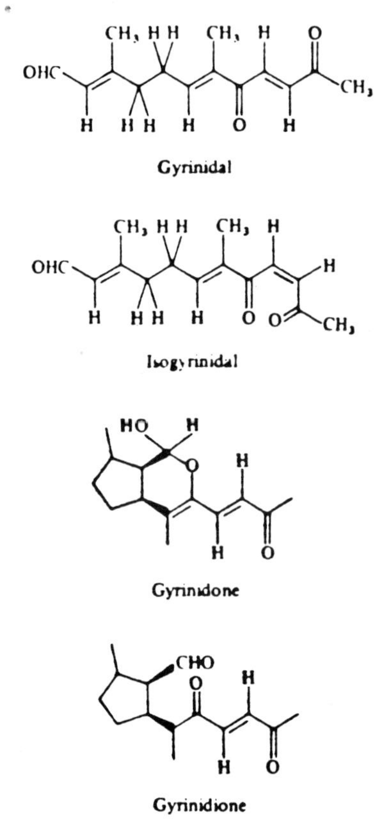

Figure 16.7. *Structures of four major compounds found in the pygidial glands of gyrinid water beetles (from Weatherston and Percy 1978).*

et al. 1973) and in bioassays of refined or synthesized components (Meinwald et al. 1973; Miller et al. 1975; Miller and Mumma 1976a). For example, Benfield (1972) assayed both whole beetles and the contents of pygidial glands on bluegill sunfish (*Lepomis macrochirus*), rainbow trout (*Salmo gairdneri*), and newts (*Notopthalmus viridescens*). Although there were occasional ingestions of whole beetles or of palatable items coated with the contents of pygidial glands, the overwhelming response was immediate rejection followed by buccal flushing. In all cases where a beetle was attacked and rejected, the individual survived the attack with no noticeable damage. The amounts of secretion necessary for rejection were small relative to the amounts carried in glands (table 16.4). Meinwald et al. (1972) de-

termined that, when applied topically to mealworms, as little as 0.3 to 0.5 μg of the norsesquiterpene gyrinidal was sufficient to cause rejection by largemouth bass (*Micropterus salmoides*).

The kinds and amounts of secreted substances vary somewhat among species of the two gyrinid genera *Gyrinus* and *Dinuetes* (fig. 16.7). The norsesquiterpene gyrinidal is the dominant component in *Gyrinus ventralis*, *G. frosti*, *Dinuetes hornii*, and *D. serralates* (Meinwald et al. 1972, 1973; Newhardt and Mumma 1979b). Both gyrinidal and gyrinidione are important in the pygidial secretions of *D. assimilus* and *D. nigrior*.

As in the case of dytiscid steroids, Miller and Mumma (1976a) were doubtful whether sufficiently high concentrations of compounds could be established to permit toxic gill uptake. As in the case of steroids, when the minnow *P. promales* was placed in water that contained norsesquiterpene or whole extracts from pygidial bladders, 80% of the chemicals were absorbed through the gills, whereas 20% passed through the skin. The lowest concentration that caused 100% mortality (LC_{100}) was approximately 3 μg/ml for gyrinidione, approximately 2 μg/ml for gyrinidal, 15 μg/ml for gyrinidone, and 90 μg/ml for isogyrinidal. In an independent series of trials, Miller and Mumma (1976b) determined weight-specific dosage levels necessary for fish kill. As before, fish were far more sensitive to gyrinidal (45 μg/g fish) than to gyrinidone (120 μg/g fish). Unlike the steroids of dytiscids, the anesthetic dosages were very close to lethal limits. As in the case of steroids, repellency seemed more a contact or taste phenomenon, with high sensitivity to compounds at relatively low concentrations.

Crawling Water Beetles (Haliplidae) and Other Minor Related Families (Hygrobidae, Noteridae)

Beetles in the families Haliplidae, Hygrobidae, and Noteridae also possess pygidial glands, from which some compounds have been isolated and identified (figs. 16.1, 16.2, table 16.5). Among the three families, the haliplids are of particular interest for three reasons: (1) they are among the most common aquatic coleopterans in the littoral zones of larger water bodies, often in regions of high fish abundance

Table 16.4. *Compounds isolated from gyrinid pygidial glands (in μg ± 1SD per beetle[-1])*

Beetle	Compound	μg/beetle ♂	♀	References
Dineutes assimilis	Gyrinidal	99.7 ± 29.2	136.5 ± 30.1	Miller et al. (1975)
	Isogyrinidal	14.3 ± 5.9	14.3 ± 5.9	
	Gyrinidione	80.9 ± 22.7	110.1 ± 24.9	
	Gyrinidone	17.0 ± 9.5	27.6 ± 9.5	
D. discolor	Gyrinidone	Major component		Wheeler et al. (1972)
	Octanal	Minor component		Wheeler in Blum (1981)[c]
D. horneii	Gyrinidal	Major component		Meinwald et al. (1972, 1973)
D. nigrior	Gyrinidal	71.3 ± 25.8	61.1 ± 24.5	Miller et al. (1975)
	Isogyrinidal	10.6 ± 5.7	6.3 ± 3.4	
	Gyrinidione	58.2 ± 24.6	48.9 ± 26.7	
	Gyrinidone	10.8 ± 6.1	13.5 ± 9.4	
D. serrulates	Gyrinidal	Major component		Meinwald et al. (1972, 1973)
Gyrinus frosti	Gyrinidal	23.4 ± 4.0 – 19.8 ± 4.6[a]		Newhardt and Mumma (1979b)
	Isogyrinidal	0.6 ± 0.08 – 6.9 ± 0.4[a]		
G. minutus	Gyrinidal	Major component		Schildknecht (1976); Blum (1981)
G. natator	Gyrinidal			Schildknecht (1976); Blum (1981)
	3-Methyl butanal	2.5		
	3-Methyl buta-1-ol	0.2		
G. substriatus	Gyrinidal			Schildknecht (1976); Blum (1981)
G. ventralis	Gyrinidal	[b]		Meinwald et al. (1972, 1973)

[a]Seasonal variation; June and August values are given and represent the extremes in variation.
[b]Major component in an 80 μg/beetle extract.
[c]Personal communication cited by Blum (1981).

Table 16.5. *Pygidial gland constituents of water beetles in the families Haliplidae, Noteridae, and Hygrobidae (from Dettner 1979)*

Beetle	Compound
HALIPLIDAE (CRAWLING WATER BEETLES)	
Haliplus lineaticollis	phenylacetic acid
	unidentified compound, MW 150
NOTERIDAE (BURROWING WATER BEETLES)	
Noterus clavicornis	benzoic acid
	phenylacetic acid
	phenylpyruvic acid
	p-hydroxyphenyl acetic acid
N. crassicornis	phenylacetic acid
	phenylpyruvic acid
HYGROBIIDAE	
Hygrobia herminii	2-hydroxyhexanoic acid
	2-hydroxy-4-thiomethyl butyric acid
	1-carboxypentyl-2-hydroxyhexanoate
	1-carboxy-4-thiobutyl; 2-hydroxyhexanoate
	1-carboxy-4-thiobutyl; 2-hydroxy-5-thiohexanoate
	3, 6-bis(3-thiobutyl)-1,4-dioxane-2, 5-dione
	3-butyl-6-(3-thiobutyl)-1,4-dioxane-2, 5-dione

(Wilson 1923; Hutchinson, personal communication; our own observations), (2) this abundance relates in part to their food habits, for unlike most aquatic coleopterans, both larval and adult stages are herbivorous and often strongly associated with particular plants (e.g., *Chara*); and (3) adults *and larvae* are almost always rejected by fishes.

The pygidial glands of crawling water beetles are situated near the anus, as in other Caraboidea (fig. 16.2). Dettner (1979) isolated two compounds from these glands, but was able positively to identify only one, phenylacetic acid (table 16.5). The other compound had a molecular weight of 150. The former compound had also been isolated from the pygidial secretions of the dytiscid subfamily Hydroporinae (table 16.3, fig. 16.6) and is known as a phytohormone.

In taste tests that we conducted, haliplids were almost always rejected by fish, rivaling dytiscids and gyrinids. Individual beetles usually survived encounters with little to no damage, whereas fish soon lost interest. It is thus interesting that in 1888 Forbes (as reported in Wilson 1923), found haliplid beetles in the guts of only 2 species of fish out of 36 species studied, despite a much greater incidence of other beetles in the stomach contents. Unfortunately, there were no objective measurements of prey densities, exposure values, or rejections.

Because most aquatic beetles are carnivorous and their larvae carnivorous or detritivorous, the food habits of haliplids stand in marked contrast to other beetles. Rejection of "bristly" *Peltodytes* larvae (fig. 16.1) by sunfish (*Lepomis*) in our taste tests suggests either the use of noxious compounds or mechanical irritation by the protuberances. Because larvae and adults are herbivorous, the exciting possibility arises that they could sequester secondary plant products, although we have no direct evidence for this intriguing option.

The noterids are closely related to the dytiscids (treated by some as a subfamily) but differ by the lack of prothoracic glands. These are relatively dull-colored, small beetles (1.2–5.5 mm length) occasionally found in the littoral zones of ponds and lakes (Merrit and Cummins 1978; Borror et al. 1976). Four compounds have been isolated from the pygidial glands of noterids, all of which also occur in dytiscids (table 16.5). One of these compounds,

benzoic acid, is known to be distasteful to the cichlid *Tilapia sparminni* (Hepburn et al. 1973). Unfortunately, benzoic acid titers have not been directly established for species in the Noteridae, nor have beetles been tested directly against fish or other predators.

The Hygrobidae are of interest because they are the only aquatic beetle family other than the Dytiscidae that possess prothoracic glands in addition to pygidial glands. This is a small family with only one genus and four species (one each in Europe and China, two in Australia). Seven compounds have been isolated from the pygidial glands of *Hygrobia herminii* (table 16.5), all hydroxycarboxylic acids and their lactides. No compounds have been described from the prothoracic glands, nor to our knowledge have any workers described palatability tests.

It is interesting to note that hygrobids possess a stridulatory apparatus. Balfour-Browne (1940) reported that individuals emitted loud "squeaks" when caught in nets or when handled, raising the possibility that sudden noise might be used to deter predators. We have discovered this phenomenon in another group, the hydrophilids.

Riffle (Elmidae) and Water Scavenger (Hydrophilidae) Beetles

Both the elmids and hydrophilids belong in the subclass Polyphaga. Riffle beetles are very poor swimmers, whereas some water scavenger beetles are highly streamlined and adapted for swimming (fig. 16.1). Neither family possesses prothoracic or pygidial glands, and yet both are strongly rejected by fish.

Riffle beetles are most characteristic of running waters, but a few species are common in ponds or swamps. As noted earlier, these beetles do not swim. Rather, they crawl about slowly using their tarsal claws to pull the body along the substrate (Pennak 1953). Our taste tests show that adult *Stenelmis crenata* are rejected by *Lepomis*. Independent taste tests by D. White (University of Michigan; unpublished data) confirm the general unpalatability of elmids to fish. Adult *Ancyronyx variegata*, *Dubiraphia vittata*, *Hexacylloepus ferrugineus*, *Macronychus glabratus*, *Microcylloepus pusillus*, *Optioservus trivittatus*, and *Stenelmis crenata* were taste-tested against 3 to 5 fish (e.g., *Lepomis*, *Micropterus*, *Ictalurus*, *Perca*). In 98% of the offerings (N = 105 presentations), beetles were rejected. Forty-five trials with adults of other stream beetles (families Dryopidae, Limnichidae, Eubriidae, and Psephenidae) all showed acceptance by fish. Unfortunately, there are no known descriptions of secretory glands, and thus the precise mechanism causing rejection is unclear.

The final family that we consider, the hydrophilids, originally presented an enigma. These beetles were rejected by fish, yet we could find no published evidence for glands or glandular secretions. Moreover, efforts by R.O. Mumma (personal communication) to elicit secretions among several species have been unsuccessful. The Hydrophilidae, however, are characterized by a peculiar ventral keel along the ventral midline, the intercoxal process (fig. 16.1). On handling, *Hydrophilus* emits a loud noise, a "squeak," which must startle unsuspecting predators. This defense, in conjunction with their very hard exoskeleton, may provide the answer for high rejection rates.

BUGS (HEMIPTERA): CONTRASTS, GLANDS, AND SECRETED SUBSTANCES

Given the volumes devoted to taxonomic or life history surveys, there is relatively little published evidence for chemical defenses within the aquatic Hemiptera. This lack of attention is unfortunate, for numerous species have glands that secrete a diversity of odoriferous and potentially protective compounds.

On a taxonomic level, the Hemiptera are divided into two suborders, the Heteroptera and the Homoptera. Although some species of homopterans might be considered marginally semiaquatic, the vast majority of semiaquatic or truly aquatic hemipterans are in the Heteroptera (Merrit and Cummins 1978). Despite the abundance of macrophytes in aquatic environments, it is peculiar that there is no strict analog of the aphid in submerged portions of the littoral region. Moreover, while most terrestrial heteropterans are phytophagous, using their sucking mouth parts to pierce and withdraw plant fluids or partially digested tissues, the vast majority of aquatic families are predatory, feeding on small insects, amphibians, or fish. These distinctions suggest a major decline in the role of plants in trophic dynamics.

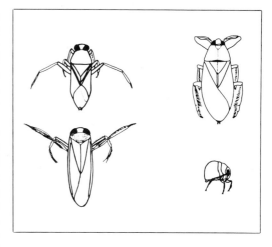

Figure 16.8. *Examples of bugs (Hemiptera) that secrete chemical compounds. From upper left clockwise: backswimmer (Notonecta), giant water bug (Lethocerus), pigmy backswimmer (Neoplea), and water boatman (corixid). (Redrawn from Pennak 1953).*

Yet some similarities do persist. Almost every terrestrial heteropteran species has specialized glands that emit pungent, protective secretions against potential predators. At least 8 of 10 aquatic families possess glands in the metasternal region of the prothorax (figs. 16.8, 16.9). These glands have been interpreted as defensive (Weatherston and Percy 1978b), although judgments are based on analogy of glands or chemicals with those of terrestrial species. The gland system of adults consists of paired glandular lobes that connect to a saclike reservoir through a collecting duct (fig. 16.9). Short efferent ducts lead from the reservoirs to lateral orifices. The secretory substances are not discharged forcibly, but ooze onto the cuticle surface, then along grooves or into a bristle field.

Heteropteran nymphs also possess glands, only these are located dorsally in the abdominal region. The so-called scent glands are found between segments 3/4, 4/5, and/or 5/6 (Weatherston and Percy 1978b). Each gland opens through two pores near the intersegmental membranes. As with adults, there are no muscles directly associated with the reservoir.

The glandless exceptions are especially instructive. These three groups, the two families Hydrometridae and Nepidae and one subfamily in the Belastomatidae (Belastomatinae), are

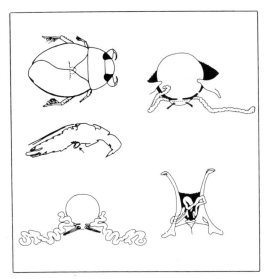

Figure 16.9. *Metathoracic glands of bugs: (a) location and structure of metathoracic gland in* Ilyocoris *(Naucoridae); (b) comparison of fully developed naucorid gland* (top) *with diminutive "scent" gland of female* Lethocerus *(Belostomatidae)* (bottom). *Parts a and b* (top) *from Staddon and Thorne 1973; b* (bottom) *from Pattenden and Staddon 1970.*

all relatively inactive and inconspicuous. The Hydrometridae (water measurers) are small (ca. 8 mm), very slender bugs that move slowly across emergent vegetation (Borror et al. 1976). Likewise, the Nepidae (water scorpions) are sluggish, sit-and-wait predators. Both groups are so sticklike and move so slowly that the fishes in our taste tests failed to recognize either species as edible items.

Of the eight families known to possess metasternal glands, the Gelatotrichidae are semi-aquatic, the Velidae and Gerridae are surface-dwelling skaters, and the Notonectidae, Corixidae, Pleidae, Belastomatidae, and Naucoridae are truly aquatic swimmers. Only the latter seven families are treated here. Chemical compounds have been isolated and identified from eight species in six families (table 16.6). The suspected functions of these compounds are multiple (e.g., as sexual attractants, alarm substances, or defensive compounds). In addition, species in the Belastomatidae, Naucoridae, and Notonectidae also can inflict painful oral stings to humans (Weatherston and Percy 1978b).

Water Striders and Skaters (Velidae and Gerridae)

As surface-dwelling skaters, velids and gerrids must be wary of attack from above or below the water surface. Both groups are extremely agile, sensitive to the slightest pressure waves, and able to hop or jump. They also school, glide across the surface film with only minimal disturbance, and have countershading to reduce shadows (Gerridae) or motionless responses (Veliidae [Brönmark et al. 1984]) to frustrate visual or acoustic search. In sum, both of these families are highly adapted to a neustonic existence. Despite this impressive arsenal of tactics, both possess chemical defenses, although protection appears less complete than in the case of beetles.

Although metasternal glands have been described for veliids, no chemicals have been isolated. Despite this, predator–prey experiments have disclosed distasteful compounds. In palatability tests, Brönmark et al. (1984) found that 90% of *Velia caprai* were rejected alive after capture by brown trout (*Salmo trutta*). Pellets made from *Velia* also were rejected, whereas control pellets made from the amphipod *Gammarus pulex* were all eaten.

Paired metathoracic glands are present in adult Gerridae. The reservoir discharges through a single opening located on the midline near the posterior margin of the metasternum (Brindley 1930). The oriface of the opening is surrounded by many setae, prompting Brinkhurst (1960) to term the region a "hair pile." Secretions saturate the region, then evaporate. In our tests, bluegills (*Lepomis macrochirus*) rejected some adult veliids and gerrids, yet rejection levels were often only slightly higher than expected for hard-coated objects.

Backswimmers and Water-Boatmen (Notonectidae and Corixidae)

Glands are well described for both notonectids and corixids. In the family Notonectidae, scent glands may be lacking in certain members of the subfamily Anisopinae (e.g., *Buenoa* and *Anisops*) but probably are found in most members of the subfamily Notonectinae (Staddon and Thorne 1974). For example, in *Notonecta* the paired tubular glands are very long, between 6 and 9 mm, thrown into loops held in

Table 16.6. *Compounds found in glands opening externally in aquatic hemipterans*

Taxon	Compound	Notes
NOTONECTIDAE		
Notonecta glauca	p-hydroxybenzaldehyde p-hydroxybenzoic acid methyl ester	A 7/3 solution found, study carried out on adult males and females. (Pattendon and Staddon, 1968)
CORIXIDAE		
Sigari falleni	trans-4-oxo-hex-2-enal	Composed 95% of material in gland. Total amount of secretion was approximately 30–50 μg/individual. (Pinder and Staddon (1965)
Corixa dentipes	trans-4-oxo-hex-2-enal	Secretion was indistinguisable from that of *Sigari falleni*. (Pinder and Staddon 1965)
BELASTOMATIDAE		
Lethocerus indicus	trans-hex-2-enyl acetate trans-hex-2-enyl butyrate	Nine compounds found in a secretion dominated by trans-hex-2-enyl acetate. Trans-hex-2-butyrate was the second most prevalent compound. (Devakul and Maarse 1964)
L. cordofanus	trans-hex-2-enyl acetate	Comprised 98% of secretion. Composition of constituents in males and females was approximately equal. Study performed on males, four other minor constituents. (Pinder and Staddon 1965)
NAUCORIDAE		
Ilycoris cimicoides	p-hydroxybenzaldehyde p-hydroxytenzoic acid methyl ester	These two compounds formed the major portion of the secretion. There were 4 other minor constituents. (Staddon and Weatherston 1967)
PLEIDAE		
Plea leachi	hydrogen peroxide	Found in a 10–15% solution. (Maschiwitz, 1971)
GELASTOTRICHADAE		
Gelastotricus oculatus	trans-4-oxo-hex-2-enal octenal[a] 4-oxo-octenal[a] hexal[a]	These compounds formed 63%, 25%, 9% and 3% respectively of the secretion. (Staddon 1973)

[a]Tentative identification.

position by branches from the metathoracic spiracle. These empty into a median reservoir, a membranous bilobed sac. Secretions are discharged through two laterally placed openings, accumulating on the external boundary between the mesothoracic epimeron and metathoracic episternum (Staddon and Thorne 1974). The secretion is an odorless, brownish fluid. Characterization by Pattenden and Staddon (1968) disclosed two major components: p-hydroxybenzaldehyde and methyl p-hydroxybenzoate (table 16.6). Recall that these compounds were also found in the pygidial glands of dysticid beetles, and that Hepburn et al.

(1973) found p-hydroxybenzaldehyde repellent to the cichlid *Tilapia* at a concentration of 6 × 10^{-7} moles. Pattenden and Staddon (1968) suggest defensive functions for both compounds.

Adult corixids also possess a pair of tubular glands that connect to a median saclike reservoir (Betten 1943). The external openings are on the metacoxa, and there is a closing mechanism. The pale yellow secretion appears forced out by the elasticity of the reservoir, as there are no intrinsic muscles (Pinder and Staddon 1965). This oillike liquid has a pleasant aldehyde odor. It has been characterized as *trans-4-*

oxo-hex-2-enal in the species *Corixa dentipes* and *Sigari falleni* (Pinder and Staddon 1965) (table 16.6).

The reports of protection against fish are ambiguous. Based on the results of examining 15,000 guts from 18 fish species, Frost and Macan (1948) concluded that the number of corixids eaten by fish was small relative to the available fauna. Noting the presence of metasternal glands, they suggested that chemicals secreted by these glands might be noxious to fish. However, in later papers Macan (1965, 1976, 1977) ascribed an important role to fish in influencing both the distribution and abundance of corixids. In particular, Macan (1965) suggested that corixid species may differ markedly in their susceptibility to fish predation. More recently, Henrikson and Oscarson (1978) have noted the importance of fish in influencing the distribution and abundance of *Glaenocorisa propinqua.*

Pigmy Backswimmers (Pleidae)

The metasternal glands of pleids are undescribed, but their reservoir contents open laterally through a pair of pores. The glands of these small backswimmers contain hydrogen peroxide. Maschiwitz (1971) discovered a 15 to 17% solution of hydrogen peroxide in the secretions of *Plea leachi.* Whereas Maschiwitz suggested an antimicrobial function, we have found the secretions very effective against small fish (*Gambusia, Neoheterandria*) (Kerfoot 1982) and moderately effective against small *Lepomis.* When pleids are agitated, the hydrogen peroxide bubbles from both openings, producing a froth around the bug. In a fish's mouth, the release can be much more explosive. We are uncertain whether the hydrogen peroxide serves as the primary repellent or merely acts to disperse another, as yet uncharacterized substance.

Creeping Water Bugs and Giant Water Bugs (Naucoridae and Belostomatidae)

In the naucorids, the metathoracic glands are slender, branched, tubular bodies, lacking muscles and nerves. They also are held in position by tracheae from the metathoracic spiracles. An exit duct empties into a median reservoir. A closing apparatus is located between the reservoir and the efferent openings (ostioles). The closing apparatus acts as a valve intermediate between the reservoir and vestibule (fig. 16.9) and has a single pair of opener muscles (Staddon and Thorne 1973).

The glands of *Ilycoris cimicoides* (Naucoridae) discharge a viscous, colorless, and odorless liquid. The substance is dominated by p-hydroxybenzaldehyde and menthyl-p-hydroxybenzoate (Staddon and Weatherston 1967), compounds found previously in dytiscids and notonectids. It has been found repellent to the cichlid *Tilapia* (Hepburn et al. 1973).

In the Belostomatidae, the genus *Lethocerus* is placed in the subfamily Lethocerinae, the genus *Horvathinia* in the subfamily Horvathiniinae, and several remaining genera (*Limnogeton, Hydrocyrius, Diplonychus, Abedus,* and *Belostoma*) in the subfamily Belostomatinae. Although it is well known that *Lethocerus* possesses metasternal scent glands (Staddon 1971), Brindley (1933) failed to find scent glands in *Belostoma.* Further investigations by Staddon (1971) confirmed the general lack of metathoracic glands in the subfamily Belostomatinae.

The metathoracic glands of *Lethocerus* are peculiar in many ways. The glands are dilated to store the secretion, as there is no separate reservoir (fig. 16.9). Moreover, the glands of the females are greatly reduced in size relative to those of males. Pattenden and Staddon (1970) estimated that males contain approximately 25 times the secretion of females, although both sexes contain compounds. The male glands secrete a clear and colorless but highly odoriferous liquid. Caillot and Boisson (1954) reported that males of *Lethocerus indicus* are strongly scented during the breeding season, prompting Butenandt (1955) to speculate that the secretions serve to excite the females.

Investigations by Butenandt (1955), Butenandt and Tam (1957), and Pattenden and Staddon (1970) have shown that 98% of the metathoracic secretion is *trans*-hex-2-enyl acetate. This compound is dominant in secretions of both males and females. Male secretions contain an additional eight compounds, one of which is *trans*-hex-2-enyl butyrate (Devakul and Maarse 1964).

Based on these observations, one is prompted to speculate that giant water bugs have increased to a size at which adults at least are no longer susceptible to the kinds of predators experienced by smaller water bugs. An alternative hypothesis would be that the sit-and-wait feeding mode adopted by giant water bugs acts, as in the case of Nepidae and Hydrometridae, to greatly lessen conspicuousness and hence reliance on defensive glands. In any case, the scent glands of *Lethocerus* seem no more than remnants fulfilling a secondary, sexual role.

DISCUSSION

Possession of prothoracic and pygdial glands in beetles and metathoracic glands in bugs is definitely linked with the production of noxious compounds. Many of these substances are highly repellent to potential predators, especially fish. Steroids, phenolics, and the compounds *p*-hydroxybenzaldehyde or p-hydroxybenzoic acid methyl ester in particular seem favored by highly aquatic groups. These two compounds have been isolated from 39 species in three families of two orders. The presence of these compounds in dytiscids, backswimmers, and creeping water bugs has been cited as strong evidence for evolutionary convergence (Staddon and Thorne 1974).

Some insight into the circumstances that favor compounds different from those found in terrestrial arthropods is gained by the preliminary findings of Hepburn et al. (1973). In their test of 25 arthropod defensive chemicals against the cichlid *Tilapia sparmanni*, the medium played a role in effectiveness. Given at a dosage of $6 \times 10-7$ moles on edible bits of food (bread pellets), responses allowed three generalizations: (1) chemicals produced by aquatic arthropods were more repellent to *Tilapia* than those of either cryptozoic or terrestrial arthropods; (2) compounds of only slight water solubility were more repellent than those of greater solubility; and (3) the particular phenolic compounds tested were more repellent than carbonyl or acidic compounds. The relatively poor performance of highly soluble compounds, especially acids, seemed due to rapid dispersion. Moderately soluble compounds seemed much more effective, less likely to be flushed away from membranous surfaces by buccal pumping.

The evolution of prothoracic glands in the Dytiscidae is the most striking modification against aquatic predators, especially fish. The use of steroids and the repellent properties of pygdial secretions allow these beetles to remain highly active despite their relatively large size and conspicuous behavior. In contrast, the loss of metathoracic glands in water measurers (Hydrometridae) and water scorpions (Nepidae) is associated with a morphology and behavior so cryptic that potential prey simply do not stimulate attacks from fish. The predators fail to recognize the bugs as edible items. The loss of those glands serves to underscore the fact that chemical defenses are only a part of a much broader repertoire of defensive tactics against predators. If behavior or morphology lowers the probability of recognition or encounter to nearly negligible levels, chemical defenses are of no practical use because they require contact to be effective as deterrents.

Although not considered in detail here, the general lack of chemical defenses among larval arthropods requires comment. It is curious that the most highly adapted aquatic life stages of beetles and bugs would be so poorly endowed with chemical defenses against fish. The palatability of larval beetles, for instance, is associated with their unexposed habits and general concealment. Yet among aquatic larval arthropods the reduction of phytophagy is noteworthy and may preclude the sequestering of secondary plant compounds as a possible option. At the very least, the sparse evidence for chemical defenses among aquatic larval arthropods contrasts with the strong development of glands and noxious substances in terrestrial counterparts and in more exposed, mobile adult stages.

ACKNOWLEDGMENTS

This research supported by NSF grant DEB 82–07007 to W.C.K.

REFERENCES

Balfour-Browne, F. 1940. *British water beetles*, vol. 1. London: Ray Society.

Benfield, E. F. 1972. A defensive secretion of *Dinuetes discolor*. *Ann. Entomol. Soc. Amer.* 65 : 1324–27.

Benke, A. C., and S. S. Benke. 1975. Comparative dynamics and life histories of coexisting dragonfly populations. *Ecology* 56 : 302–17.

Betten, H. 1943. Die Stinkdrüsen der Corixiden. *Zool. Jb.* 68 : 137–76.

Bettini, S. 1978. *Arthropod Venoms.* (Handbuch der Experimenteilen Pharmakologie, vol. 48.) New York: Springer-Verlag.

Blum, M. S. 1981. *Chemical defenses of arthropods.* New York: Academic Press.

Blunck, H. 1917. Die Schreckdrusen des Dytiscus und Ihr Secret. *Zeitschrift für Wissenschtliche Zoologie*. 117 : 205–56.

Borror, D. J., D. M. Delons, and C. A. Triplehorn. 1976. An introduction to the study of insects. 4th ed. New; York: Holt, Rinehart and Winston.

Brindley, M. D. H. 1930. On the metasternal scent-glands of certain Heteroptera. *Transactions of the Entomological Society of London* 78 : 199–207.

———. 1933. The development of thoracic stink-glands in Heteroptera. *Proceedings of the Royal Entomological Society of London* 8 : 1–2.

Brinkhurst, R. O. 1960. Studies on the functional morphology of *Gerris najus* (Gerridae). *Proceedings of the Zoological Society of London* 133 : 531–59.

Brönmark, C., B. Malmqvist, and C. Otto. 1984. Anti-predator adaptations in a nuestonic insect *Velia capria*. *Oecologia* 61 : 189–91.

Butenandt, A. 1955. Wirkstöffe des Insektenreiches. *Nova Acta Leopoldina* 17 : 445–71.

Butenandt, A., and N. Tam. 1957. Über einen geschlechtsspezifishen Dustoff Wasserwanze Belostoma indica Vitalis (*Lethocerus iodicus* Lep.), *Hoppe-Seylers Zeitschrift für Physiologische Chemie*, 308 : 277–83.

Caillot, Y., and C. Boisson. 1954. Développement larvaire du bélostome (*Lethocerus indicus* Lep.). Insecte Hémiptère, Hydrocoryse, Cryptocérate. *Annls Sci. Nat., Zool.* (11) 16 : 51–64.

Casper, A. 1913. Die Köeperdecke und die Drusen von *Dytiscus marginalis*. *Z. Wiss. Zool.* 107 : 387–508.

Chadha, M. S., N. K. Joshi, V. R. Mamdapur, and A. T. Sipahimalani. 1970. C-21 steroids in the defensive secretions of some Indian water beetles. *Tetrahedron* 26 : 2061–64.

Classen, R., and K. Dettner. 1983. Pygidial defensive titer and population structure of *Agabus bipustulatus* and *Agabus paludosus* F. (Coleoptera, Dytiscidae). *Journal of Chemical Ecology* 9 : 201–209.

Clayton, R. B. 1964. The utilization of sterols by insects. *J. Lipid Res.* 5 : 3–18.

———. 1970. The chemistry of non-hormonal interactions: Terpenoid compounds in ecology. In E. Sondheimer and J. B. Simeone (eds.), *Chemical ecology*, pp. 235–80. New York: Academic Press.

Dettner, K. 1979. Chemotaxonomy of water beetles based on their pygidial gland constituents. *Biochem. Syst. Ecol.* 7 : 129–40.

Dettner, K., and G. Schwinger. 1980. Defensive substances from pygidial glands of water beetles. *Biochem. Syst. Ecol.* 8 : 89–95.

Devakul, V., and H. Maarse. 1964. A second compound in the odorous gland liquid of the giant water bug *Lethocerus iodicus*. *Anal. Biochem.* 7 : 269.

Dierckx, F. 1899. Sur la structure des glandes angles des Dytiscides et le pretendu role defensif de ces glandes. *C. R. Acad. Sci., Paris* 127 : 1126–27.

Doyen, J. T., and G. Ulrich. 1978. Aquatic Coleoptera. In R. W. Merrit, and K. W. Cummins (eds.). *Aquatic insects of North America*, pp. 203–31. Dubuque, Iowa: Kendall/Hunt.

Eisner, T. 1970. Chemical defense against predation in arthropods. In E. Sondheimer and J. B. Simeone (eds.), *Chemical ecology*, pp. 235–80. New York: Academic Press.

Fescemyer, H. W., and R. O. Mumma. 1983. Regeneration and biosynthesis of dytiscid defensive agents (Coleoptera: Dyticidae). *J. Chem. Ecol.* 9 : 1449–63.

Forsyth, D. J. 1968. The structure of the defence glands in the Dytiscidae, Noteridae, Haliplidae and Gyrinidae (Coleoptera). *Transactions of the Royal Entomological Society of London* 120 : 159–82.

———. 1970. The structure of the defence glands of the Cirindelidae, Amphizoidae and Hygrobiidae (Coleoptera). *J. Zool.* 160 : 51–69.

Frost, W. E., and T. T. Macan. 1978. Corixidae (Hemiptera) as food of fish. *J. Anim. Ecol.* 17 : 174–79.

Hara, T. J. 1967. Electrophysiological studies of the olfactory system of the goldfish *Carassius auratus* L. III. Effects of sex hormones on olfactory activity. *Comparative Biochemistry and Physiology* 22 : 209–225.

Henrikson, L. and H. G. Oscarson. 1978. Fish predation limiting abundance and distribution of *Glaenocorisa p. propinqua*. *Oikos* 31 : 102–105.

Hepburn, H. R., N. J. Berman, H. J. Jacobson, and L. P. Fatti. 1973. Trends in arthropod secretions on aquatic predator assay. *Oecologia* 12 : 373–82.

Johnson, D. M., and P. H. Crowley. 1980. Odonate "hide and seek": habitat specific rules? In W. C. Kerfoot (ed.), *Evolution and ecology of zooplankton communities*, pp. 569–79. Hanover, N.H.: University Press of New England.

Kerfoot, W. C. 1982. A question of taste: Crypsis and warning coloration in freshwater zooplankton communities. *Ecology* 63 : 538–53.

Korschfelt, E. 1924. *Der Gelbrand Dytiscus marginalis L. 1*. Leipzig.

Macan, T. T. 1965. Predation as a factor in the ecology of waterbugs. *J. Anim. Ecol.* 34 : 691–98.

———. 1976. A twenty-one-year study of water bugs in a moorland fish pond. *J. Anim. Ecol.* 45 : 932–48.

———. 1977. The influence of predation on the composition of freshwater animal communities. *Biological Reviews of the Cambridge Philosophical Society* 52 : 45–70.

Maschiwitz, U. 1967. Eine neuartige Form der Abwer von Mikroorganismen bei Insekten. *Naturwissensschaften* 54 : 649.

———. 1971. Wassertoff als Antiseptikum bei einer Wasserwanze. *Naturwissenschaften* 58 : 572.

Meinwald, J., K. Opheim, and T. Eisner. 1972. Gyrinidal: A sesquiterpenoid aldehyde from the defense glands of gyrinid beetles. *Proc. Nat. Acad. Sci. U.S.A.* 69 : 1208–10.

Meinwald, J., K. Opheim, and T. Eisner. 1973. Chemical defense mechanisms of arthropods XXXVI: Stereospecific synthesis of gyrinidal, a nor-sesquiterpenoid aldehyde from gyrinid beetles. *Tetrahedron Letters* 4 : 281–84.

Merrit, R. W., and K. W. Cummins. 1978. An introduction to the aquatic insects of North America. Dubuque, Iowa: Kendall/Hunt. USA.

Miller, R. J., L. B. Hendry, and R. O. Mumma. 1975. Norsesquiterpenes as defensive toxins of whirligig beetles (Coleoptera: Gyrinidae). *J. Chem. Ecol.* 1 : 59–82.

Miller, R. J., and R. O. Mumma. 1973. Defensive agents of the American water beetles *Agabus seriatus* and *Grapboderus liberus* (Coleoptera: Dytiscidae.) *J. Insect Physiol.* 19 : 917–25.

———. 1974. Seasonal quantification of the defensive steroid titer in *Agabus seriatus* (Coleoptera: Dytiscidae). *Ann. Entomol. Soc. Amer.* 67 : 850–52.

———. 1976a. Physiological activity of water beetle defensive agents, I. Toxicity and anesthetic of steroids and norsesquiterpenes administered in solution to the minnow *Pimephales promales* Raf. *J. Chem. Ecol.* 2 : 115–130.

———. 1976b. Physiological activity of water beetle defensive agents. II. Absorption of selected anesthetic steroids and norsesquiterpenes across the gill membranes of the minnow *Pimephales promales* Raf. *J. Chem. Ecol.* 2 : 130–46.

Morin. D. J. 1984. The impact of fish exclusion on the abundance and species composition of larval odonates: results of short-term experiments in a North Carolina farm pond. *Ecology* 65 : 53–60.

Newhardt, A., and R. Mumma. 1979a. Defensive secretions of three species of *Acilus* (Coleoptera, Dytiscidae) and their seasonal variations as determined by high-pressure liquid chromatography. *J. Chem. Ecol.* 5 : 643–52.

———. 1979b. Seasonal quantification of the defensive secretions from Gyrinidae. *Ann. Entomol. Soc. Amer.* 72 : 427–29.

Oshima, K., and A. Gorbman. 1968. Modification by sex hormones of the spontaneous and evoked bulbar electrical activity in goldfish. *J. Endrocrinol.* 40 : 409–20.

Pattenden, G., and B. W. Staddon. 1968. Secretions of the metathoracic glands of the water bug *Notonecta glauca* (Heteroptera; Notonectidae). *Experientia* (Basel) 24 : 1209.

———. 1970. Observations of the metasternal scent glands of *Lethocerus spp.* (Hemiptera: Heteroptera: Belostomatidae). *Ann. Entomol. Soc. Amer.* 63 : 900–901.

Pennak, R. W. 1953. *Freshwater invertebrates of the United States.* New York: Ronald Press.

Pinder, A. R., and B. W. Staddon. 1965. Trans-4-oxohexen-2-al in the odoriferous secretion of *Sigara falleoi* (Hemiptera: Heteroptera). *Nature* 205 : 106.

Schildknecht, H. 1970. The defensive chemistry of land and water beetles. *Angewandte Chemie International Edition* 9 : 1–8.

———. 1971. Evolutionary peaks in the defensive chemistry of insects. *Endeavour* 30 : 136–41.

———. 1976. Chemical ecology—a chapter of modern products chemistry. *Angewandte Chemie International Edition* 15 : 214–22.

Schildknecht, H. and H. Birringer. 1969. Die Steroide des Schlammschwimmers *Ilybius fenestatus.* *Zeitschrift für Naturforschung,* Teil B. 24b : 1529–34.

Schildknecht, H., H. Birringer, and U. Maschwitz. 1967. Testosterone as a protective agent of the water beetle *Ilybius. Angewandte Chemie International Edition* 6 : 558–59.

Schildknecht, H., and D. Hotz. 1967. Identifizierung der Nebensteroide des Prothorakabwehrdrusen systems des Gelbrandkafers *Dytiscus marginalis. Angewandte Chemie International Edition* 6 : 881–82.

———. 1970. Das Prothorakalwehrsekret des Schwimmkäfers *Agabus bipustulatus. Chemiker-Ztg.* 94 : 130.

———. 1971. Naturally occurring steriod-isobutyrates, in James, V. and L. Martini (eds.), pp. 158–66, *Hormonal Steroids, Proceedings of the Third International Congress on Hormonal Steroids,* Hamburg, 1970. Excerpta Medica, Amsterdam.

Schildknecht, H., R. Siewerdt, and U. Maschwitz. 1966. A vertebrate hormone as defensive substance of the water beetle (*Dytiscus marginalis*). *Angewandte Chemie International Edition* 5(4) : 421–22.

Schildknecht, H., and H. Tacheci. 1971. Colymbetin, a new defensive substance of the water beetle *Colymetes fuscus* (Coleoptera: Dytiscidae) that lowers blood pressure. *J. Insect Physiol.* 17 : 1889–96.

Seyle, H., and R. D. H. Heard. 1943. The fish assay for the anesthetic effect of steroids. *Anesthesiology* 4 : 36–47.

Sipahimalani, A. T., V. R. Mamdapur, K. N. Joshi, and M. S. Chadha. 1970. Steroids in the defensive secretion of the water beetle *Cybister limbatus. Naturwissenschaften* 57 : 40.

Staddon, B. W. 1971. Metasternal scent glands in Belostomatidae (Heteroptera). *J. Entomol.* (A) 46 : 69–71.

———. 1973. A note on the composition of the scent from metathoracic scent glands of *Gelastocoris oculatus* (F.) (Heteroptera: Gelastocoridae). *Entomologist* 106(1326) : 253–55.

Staddon, B. W., and M. J. Thorne. 1973. The structure of the metathoracic scent gland system of the water bug *Ilyocoris cimicoides* (L.) (Heteroptera: Naucoridae) [Hem.]. *Transactions of the Royal Entomological Society of London* 124 : 343–63.

————. 1974. Observations on the metathoracic scent gland system of the back swimmer, *Notonecta glauca* L. (Heteroptera: Notonectidae). *J. Entomol. (A)* 48 : 223–27.

Staddon, B. W., and J. Weatherston. 1967. Constituents of the stink glands of *Ilyocoris cimicoides* (Heteroptera: Naucoridae). *Tetrahedron Letters* 46 : 4567–71.

Tucker, V. A. 1969. Wave making by whirligig beetles (Gyrinidae). *Science* 166(3907) : 897–99.

Weatherston, J., and J. E. Percy. 1978a. Venoms of Coleoptera. In S. Bettini (ed.), pp. 511–54. *Arthropod venoms*, New York: Springer-Verlag.

————. 1978b. Venoms of Rhyncota (Hemiptera). In S. Bettini (ed.), pp. 489–509, *Arthropod venoms*. New York: Springer-Verlag.

Wheeler, J. W., S. K. Oh, E. F. Benfield, and S. E. Neff. 1972. Cyclopentanoid norsequiterpenes from gyrinid beetles (Coleoptera). *Journal of the American Chemical Society* 94 : 7589–90.

Wilson, C. 1923. Water beetles in relation to pond-fish culture, with life histories of those in fish ponds at Fairport, Iowa, *U.S. Bureau Fish. Bull.* 38 : 231–345.

17. Predator-Induced Defenses: A Review

Prey species reduce their mortality to predation by timing their activities to avoid their predators, developing predator-resistant morphologies, producing chemical defenses, and adopting a variety of escape behaviors. Some prey species are capable of developmental polymorphisms in response to cues emitted by predators. These polymorphisms are expressed either in the individual receiving the cue or in their offspring. Initially described for a rotifer species, inducible defenses have now been reported for 5 protozoan species, 6 rotifers, 8 cladocera, 1 bryozoan, 1 barnacle, and 42 species of terrestrial plants. The extensive taxonomic and geographical distribution of prey employing predator-induced defenses suggests that these defenses may be an extremely common strategy for prey that encounter fluctuating predation intensity.

A number of studies have demonstrated the adaptive importance and effectiveness of the inducible features of animal prey species, but none has shown that there is any effect on predator fitness. In contrast, research on chemical defenses in terrestrial plants has shown important effects on herbivore fitness and counteradaptation by some specialist herbivores.

Prey species use a variety of tactics to avoid being eaten by their predators. Mobile animals are protected by armored and cryptic morphologies, noxious chemicals, and a variety of behaviors (Edmunds 1974; Pianka 1983). Sessile animals have similarly well developed suites of defenses (Barnes 1980). Plants are protected from grazers by their morphology, cell wall structure, mode of attachment, and chemical defenses (Gregory 1983; Futuyma 1983). Zooplankton antipredator tactics have been the focus of considerable research.

Zooplankton may avoid interactions with potential predators by their phenology (Dodson 1975), diurnal migration patterns (Ohman et al. 1983), and swimming speed and pattern, which influence mortality to both moving and stationary predators (Gerritsen and Strickler 1977; Li and Li 1979; Greene and Landry 1985). Prey mortality is influenced by body size (Lynch 1980; Scott and Murdoch 1983), body shape (Kerfoot et al. 1980; Havel and Dodson 1984), eye size (Zaret 1972), and body transparency (O'Brien et al. 1979). In addition, many water mites use distastefulness combined with warning coloration as a defense against fish predation (Kerfoot 1982), and some rotifers use mucous sheaths to protect against copepod predation (Stemberger and Gilbert, this volume).

Since the intensity of predation on many species varies both temporally and spatially (Pianka 1983), and defenses for at least some incur fitness costs (Cates 1975; Dodson 1984; Riessen 1984), prey genotypes that produce defenses only in the presence of predators should have a higher fitness than genotypes that maintain defenses continually. These preconditions favor the evolution of temporary defenses. Presence of defenses should be directly proportional to predation risk and indirectly proportional to cost (Rhoades 1979).

For temporary defenses to be useful, they must be accurately employed during periods of predator abundance and activity; i.e., prey must be able to predict the arrival of predators. If predators have seasonal patterns of foraging intensity, as fish and zooplankton do, then cues related to season can be used. Such may be the case for the helmeted *Daphnia* species that grow enlarged helmets in response to temperature and turbulence (Havel and Dodson

1985) and are partially protected against predation by copepods (Havel 1985a).

A more direct approach to predicting predator abundance is for prey species to respond directly to predator cues. In this chapter, I review the literature on these predator-induced polymorphisms, covering examples from freshwater zooplankton, marine nearshore, and terrestrial plant communities. Work on terrestrial plants is well developed and serves a useful contrast to studies of induced defenses in aquatic communities.

PREDATOR-INDUCED DEFENSES IN FRESHWATER ZOOPLANKTON

Protozoa

The freshwater ciliate *Euplotes octocarinatus* is polymorphic, with the larger, lateral-winged morph being difficult for ciliate predators to ingest (Kuhlmann and Heckmann 1985). *E. octocarinatus* and several congenerics change into the larger winged morph when water-soluble substances from *Lembadion lucens* (Ciliata) are present in the environment (Kuhlmann and Heckmann 1985). The authors reported that other ciliate predators of *Euplotes* were also capable of inducing this polymorphism (table 17.1) and that visible changes in *Euplotes* morphology occurred prior to cell division. Based on preliminary biochemical tests, the inducing substance (morphogen) is probably a heat-stable polypeptide (Kuhlmann and Heckmann 1985).

Rotifers

Predator-induced variation in an animal species was first reported by Beauchamp (1952a, b), who found that the predatory rotifer *Asplanchna brightwelli* released a morphogen into cultures of *Brachionus calyciflorus*, which caused offspring of *B. calyciflorus* to grow long spines not present in the parent. Two other species of *Asplanchna* have been shown to have a similar effect on *B. calyciflorus*, and many features of the interactions have been elucidated (Gilbert 1966, 1967, 1980a).

Asplanchna species release a water-soluble morphogen that stimulates uncleaved eggs carried by the *B. calyciflorus* mothers to develop later into individuals bearing long posteriolateral spines (Gilbert 1966). The *Asplanchna* factor does not cause spine production in *B. caly-*

ciflorus at any later developmental period. The chemical causing this induction is unidentified but is probably a heat-stable protein (Gilbert 1966). Different clones of *B. calyciflorus* vary in their sensitivity to spine induction by *A. brightwelli* (Halbach and Jacobs 1971). The long spined morphs are eaten less efficiently by both *Asplanchna* and copepods than are short-spined morphs, with the movable spines serving to interfere mechanically with feeding (Halbach 1971; Gilbert 1980a). Since long-spined morphs are not found in the absence of *Asplanchna*, one might expect the short-spined morph to have a higher fitness relative to the other. However, in controlled laboratory experiments with low and high food concentrations, both morphs had a similar intrinsic rate of growth (Gilbert 1980a), suggesting that fitness differences are either small or nonexistent in *B. calyciflorus*.

Further work by Gilbert and others has demonstrated that predator-induced polymorphisms are widespread among the rotifers (table 17.1; see also Stemberger and Gilbert, this volume). Several genera of rotifer prey respond to the *Asplanchna* factor, and more than one species of *Asplanchna* produces it. The recent report that two species of copepods, as well as *Asplanchna*, induce *Keratella cochlearis* spine elongation (Stemberger and Gilbert 1984), suggests that *Keratella* responds to a general type of cue that several of its predators produce.

Cladocera

Morphological variations in *Daphnia* species are induced by several predators (table 17.1). Two aquatic insects, *Anisops calcaratus* (hemiptera) and the larvae of *Chaoborus americanus* (diptera), each release water-soluble morphogens that affect *Daphnia* exoskeletal growth (Grant and Bayly 1981; Krueger and Dodson 1981; Hebert and Grewe 1985). *A. calcaratus* induced crest enlargement in two of three subspecies (or species; see Hebert 1977) in the *D. carinata* complex (Grant and Bayly 1981, P. D. N. Hebert, personal communication), and morphs with enlarged crests evaded *A. calcaratus* more efficiently than morphs with small crests (Grant and Bayly 1981). Australian *Daphnia* also respond to morphogens from North American *Notonecta* (P. D. N. Hebert, pers. comm.). *Chaoborus* induced enlargement of a toothed crest or

Table 17.1. *Reported cases of predator-induced polymorphisms in animals*

Community & prey taxon	Predator taxon	Feature induced	Reference
ZOOPLANKTON			
Protozoa			
Euplotes octocarinatus	*Lembadion lucens* (Protozoa, Ciliata)	Cell shape and size	Kuhlmann and Heckmann 1985
Euplotes octocarinatus	*Urostyla grandis*	Cell shape and size	Kuhlmann and Heckmann 1985
Euplotes octocarinatus	*Stylonychia mytilus*	Cell shape and size	Kuhlmann and Heckmann 1985
Euplotes octocarinatus	*Dileptus anser*	Cell shape and size	Kuhlmann and Heckmann 1985
Euplotes patella	*Lembadion lucens*	Cell shape and size	Kuhlmann and Heckmann 1985
Euplotes diadaleos	*Lembadion lucens*	Cell shape and size	Kuhlmann and Heckmann 1985
Euplotes plumipes	*Lembadion lucens*	Cell shape and size	Kuhlmann and Heckmann 1985
Euplotes aediculatus	*Lembadion lucens*	Cell shape and size	Kuhlmann and Heckmann 1985
Rotifera			
Brachionus bidentata	*Asplanchna brightwelli* (Rotifera)	Posterior spines	Pourriot 1964, 1974
Brachionus calyciflorus	*Asplanchna brightwelli*	Anterior and posterior spines	Beauchamp 1952a, 1952b; Halbach & Jacobs 1971
Brachionus calyciflorus	*Asplanchna girodi*	Anterior and posterior spines	Gilbert 1966, 1967, 1980a
Brachionus calyciflorus	*Asplanchna sieboldi*	Anterior and posterior spines	Gilbert 1966, 1967
Brachionus urceolaris sericus	*Asplanchna brightwelli*	Posterior spines	Pourriot 1964, 1974
Filinia mystacina	*Asplanchna brightwelli*	Tubular appendage	Pourriot 1964, 1974
Keratella cochlearis	*Asplanchna priodonta*	Posterior spine	Stemberger & Gilbert 1984
Keratella cochlearis	*Mesocyclops edax* (Crustacea, Copepoda)	Posterior spine	Stemberger & Gilbert 1984
Keratella cochlearis	*Tropocyclops prasinus*	Posterior spine	Stemberger & Gilbert 1984
Keratella slacki	*Asplanchna brightwelli*	Body length, anterior & posterior spines	Gilbert & Stemberger 1984
Keratella slacki	*Asplanchna girodi*	Body length, anterior & posterior spines	Gilbert & Stemberger 1984
Cladocera (Crustacea, Branchiopoda)			
Bosmina longirostris	*Epischura* spp. (Crustacea, Copepoda)	Mucro length	W. C. Kerfoot, in press
Daphnia ambigna	*Chaoborus americanus* (Insecta, Chaoboridae)	Helmet	Hebert & Grewe 1985
Daphnia cephalata[a]	*Anisops calcaratus* (Insecta, Notonectidae)	Crest, body length	Grant & Bayly 1981
Daphnia cephalata[a]	*Notonecta* spp. (Insecta, Notonectidae)	Crest	P. D. N. Hebert, pers. comm.
Daphnia hyalina	*Chaoborus americanus*	Crest	P. D. N. Hebert, pers. comm.
Daphnia longispina	*Chaoborus americanus*	Crest	P. D. N. Hebert, pers. comm.
Daphnia magniceps[a]	*Anisops calcaratus*	Crest, body length	Grant & Bayly 1981
Daphnia pulex	*Chaoborus americanus*	Crest	Krueger & Dodson 1981, Havel 1985b

Table 17.1 (continued)

Community & prey taxon	Predator taxon	Feature induced	Reference
Daphnia pulex	*Notonecta undulata*	Reduced body size	Dodson & Havel, unpublished
Daphnia rosea	*Chaoborus americanus*	Crest	P. D. N. Hebert, pers. comm.
MARINE NEARSHORE			
Bryozoa			
Membranipora membranacea	*Doridella steinbergae* (Mollusca, Nudibranchia)	Spines on zooecium, frontal membrane	Yoshioka 1982, Harvell 1984a
Membranipora membranacea	*Onchidoris muricata* (Mollusca, Nudibranchia)	Spines on zooecium, frontal ,e,brame	Harvell 1984a
Cirripedia (Crustacea)			
Chthamalmus anisopoma	*Acanthina angelica* (Mollusca, Pulmonata)	Bend in plates	Lively 1986

[a]Taxonomy of Hebert 1977.

helmet in several *Daphnia* species (Krueger and Dodson 1981, Hebert and Grewe 1985) but not others (Havel and Dodson 1985).

Chaoborus induction of crested *D. pulex* ("spined morphs") requires proper timing. *D. pulex* spined morphs are induced by a factor released from *C. americanus* during a short period of *Daphnia*'s embryonic development, the double red-eye stage (Krueger and Dodson 1981). In contrast, Gilbert (1966) found that the rotifer *B. calyciflorus* is responsive only prior to first cleavage of the egg, and Grant and Bayly (1981) found that induction in the *D. carinata* complex could occur during any stage of postnatal growth. Thus, *D. pulex* and *B. calyciflorus* are born with their induced features, whereas *D. carinata* develop theirs as free-swimming juveniles and adults. *D. pulex* spined morphs can appear in the population less than 2 days after *Chaoborus* reach the predaceous third and fourth instars and replace the typical morphs within the next week (Havel 1985b).

Induction of *D. pulex* spined morphs may be influenced by *Daphnia* genotype. In standard *Chaoborus*-induction laboratory experiments, clones founded from spring animals usually produced very few spined morph offspring, while those from summer animals produced high frequencies (Havel 1985b). These results suggest that a change in sensitivity of resident *Daphnia* occurred in these Wisconsin ponds during the year, perhaps through clonal suc-

cession. Induction of spined morphs in the laboratory also needs a high enough *Chaoborus* density (>0.5 *Chaoborus* per liter) and temperature ($\geq 10\,°C$), but patchy distributions of *Chaoborus* and *Daphnia* make extrapolation of these results to the field hard (Havel 1985b).

Although none of the morphogens released by *Lembadion*, *Asplanchna*, and *Chaoborus* have been completely identified, their preliminary characterizations suggest that at least the *Chaoborus* factor differs from the other two. Whereas the *Asplanchna* and *Lembadion* factors are probably polypeptides (Gilbert 1966; Kuhlmann and Heckmann 1985), the *Chaoborus* factor is probably a low-molecular-weight molecule (Hebert and Grewe 1985). A large difference in molecular weights is potentially important ecologically because low-molecular-weight compounds diffuse more rapidly than larger molecules (Lehninger 1970), resulting in differences in morphogen patchiness. All three factors are heat-stable and water-soluble.

Although the toothed crest of *D. pulex* provides only an 8% increase in head width, these individuals are eaten at lower rates (Krueger and Dodson 1981) and handled by *C. americanus* only half as efficiently as the typical morphs (Havel and Dodson 1984). Whether the crest or other associated features are responsible for this difference is unknown. The importance of this antipredator trait is reinforced by looking at size relationships. *D. pulex* retains the largest crest only during the first three instars (Krueger and Dodson 1981, personal observation), and these instars are within the preferred size range of *Chaoborus* prey (Pastorok 1981). The spined morph of *D. pulex* has not been found in ponds without *Chaoborus* (personal observation), and in laboratory tests at high food densities, it has an intrinsic rate of population increase 5% lower than the typical morph (Havel and Dodson, unpublished data).

Shifts in morphology of another cladoceran, *Bosmina longirostris*, also may be influenced by predator morphogens. At least one species of the copepod genus *Epischura* releases a factor that induces increased mucro length of *B. longirostris* (W. C. Kerfoot, in press). *Epischura* is known to selectively eat short- rather than long-featured morphs (Kerfoot 1975). Short-featured morphs in the field carry larger clutches of eggs than long-featured morphs, suggesting a difference in fitness (Kerfoot 1977).

Alternation between the opposing selective forces of protection against predation (favoring the spined and long-featured morphs) and differences in fitness (favoring the typical and short-featured morphs) may explain the evolution of these induced defenses in *Daphnia* and *Bosmina*.

D. pulex is also responsive to morphogens from predators other than *Chaoborus*. The minimum size at first reproduction for *D. pulex* decreased during the summer at the same time as a large population of predaceous *Notonecta undulata* (hemiptera: Notonectidae) developed, and in laboratory induction experiments, subcultures of several *D. pulex* clones exposed to *Notonecta* filtrate produced eggs, neonates, and youngest adults that were significantly smaller than those in subcultures grown without *Notonecta* filtrate (Dodson and Havel, unpublished data). Further work suggests that *D. galeata mendotae* also may reproduce at a smaller size when exposed to filtrates from *Lepomis macrochirus* (Centrachidae) (Dodson, unpublished data). These results suggest that some of the size frequency shifts in *Daphnia* populations that are commonly attributed to size-selective predation (cf. Threlkeld 1979) could be due instead to predator-induced reductions in body size. Since feeding rates of *Notonecta* and *Lepomis* increase with increased *Daphnia* body size (Scott and Murdoch 1983, Werner and Hall 1974), reduction of body size for the youngest adults may be adaptive in the presence of these predators. In the absence of such predators, optimal body size should be larger in order to maximize fecundity and minimize losses from smaller invertebrate predators (Lynch 1980). Predator-induced reductions in cladoceran body size should allow genotypes the flexibility to adopt the best life history strategy, depending on the presence or absence of large selective predators.

PREDATOR-INDUCED DEFENSES IN MARINE NEARSHORE COMMUNITIES
Bryozoa

Predator-induced defenses are also important in cheilostome bryozoa (Yoshioka 1982, Harvell 1984a). Colonies of *Membranipora membranacea* encrust blades of kelp along the northeast Pacific coast from Baja to Alaska (Seed 1976, Yoshioka 1982). Here they are exposed to heavy predation by nudibranchs, particularly *Doridella steinbergae* (McBeth 1968). Separate *Membranipora* colonies are single genotypes consisting of asexually derived zooids (Barnes 1980), and like many marine colonies, they are usually consumed only partially by their nudibranch predators (Harvell 1984b).

Several previously recognized species of *Membranipora*, *M. membranacea*, *M. serilamella*, and *M. villosa* (Kozloff 1974) may instead be inducible morphs of the same species (Yoshioka 1982). The major diagnostic characters separating these species are presence or absence of specific spines. These spines can be induced by two nudibranch genera, *Doridella* and *Onchidoris* (Harvell 1984a), and perhaps a third, *Corambe* (Yoshioka 1982). Spines appear within 36 hours of attack and require direct contact between *Membranipora* and the predator for induction to occur initially (Harvell 1984a). Spines are produced in zooids around the entire periphery of a colony in response to a single, localized attack. Unattacked colonies adjacent to attacked colonies also produce spines, suggesting the presence of diffusible morphogens (Harvell, unpublished data).

The induced spines protect zooids from nudibranch predation. Spined zooids are eaten at less than half the rate of unspined zooids, and mortality on partially spined colonies is restricted to unspined zooids from the interior of the colony (Harvell 1984a). Such responses are adaptive because nudibranchs feed slowly and intermittently and will move on to other colonies when food quality declines locally. The induced defense may bear a cost. Colonies with spined zooids in the field appear to grow at a slower rate than colonies with unspined zooids (Yoshioka 1982).

Barnacles

Another marine invertebrate also has predator-induced characters. The predatory snail *Acanthina angelica* induces the barnacle *Chthamalus anisopoma* to grow into a bent-over form rather than the typical upright form, and this bent-over form is more difficult for *Acanthina* to eat (Lively 1986).

HERBIVORE-INDUCED DEFENSES IN TERRESTRIAL PLANTS

Terrestrial plants are subject to attack by a variety of herbivores and pathogens, and their chemical defenses have received much attention (Levin 1976; Rhoades 1979; Harborne 1982; Futuyma 1983). In plants, defensive chemicals are commonly called secondary compounds and include a broad array of terpenes, waxes, quinones, phenolics, alkaloids, steroids, glucosinolates, and cyanogenic glycosides (Levin 1976). The combinations of these chemicals in plants appear to be species-specific and may be stored within secretory cells or in glandular trichomes (Levin 1973, 1976). In insects, the compounds act by inhibiting feeding or digestion, by poisoning, and by interfering with development or reproduction (Futuyma 1983). Some compounds, while inhibiting feeding in one insect, will stimulate feeding in another (Carroll and Hoffman 1980; Harborne 1982), contributing to the specificity of many insects to their host plants (Harborne 1982; Futuyma 1983).

The quality of plant tissue within a population can be highly variable. Significant variation in primary nutrients and secondary compounds can occur between individuals and between leaves on single individuals in trees (Schultz et al. 1982). The quality of leaves in many plant species also varies significantly over time; scales of hours, days, seasons, and years have been reported (Rhoades 1979). This spatial and temporal variation is probably an important feature of the defense of long-lived tree species, both imposing risk from other environmental factors to foraging insects and reducing the rate at which insects evolve detoxification mechanisms (Schultz 1983a, 1983b).

Inducible defenses against infection by bacteria, viruses, and fungi have been well documented by plant pathologists (Levin 1976; Deverall 1977). Infection of a broad variety of plants results in production of chemicals that allow local and sometimes distant tissues to resist later infection by the same or different pathogens (Levin 1976). Similar defenses against insect herbivores have recently been reported.

The methods employed to demonstrate herbivore-induced defenses involve first exposing one group of plants to herbivory or simulated damage, then assaying for a chemical defense, either by quantitative chemical analyses for suspected classes of compounds (Baldwin and Schultz 1983) or by some form of bioassay. Bioassays use an herbivore that naturally feeds on the host plant to test the quality of the host plant tissue by testing preference for feeding or oviposition (Carroll and Hoffmann 1980). Alternatively, some measure of herbivore fitness, such as growth, survival, or reproduction may be used (Karban and Carey 1984). The strongest evidence comes from studies where both chemical and bioassay methods were employed (Bryant 1981) because both antiherbivore activity and possible mechanism of action are revealed.

Inducible chemical defenses have been reported in studies from a broad array of plant taxa. Their distribution includes at least 42 species in more than a dozen families, from annuals to woody perennials (table 17.2). Many of the plants studied are important crop plants. The broad taxonomic distribution of plants employing herbivore-induced chemical defenses suggests that they may be universal in plants (Rhoades 1983b), although some studies suggest otherwise (Williams and Myers 1984).

Although in some plants the induced chemical remains localized (Bryant 1981), in most the chemical is systemic (Rhoades 1983b). Tomato and potato plants produced a proteinase inhibitor after feeding by Colorado potato beetles, and this substance accumulated within 24 hours in regions of the plant near and distant from the point of attack (Green and Ryan 1972; Ryan 1979). Proteinase inhibitors have been detected in a wide range of flowering

Table 17.2. *Reported cases of predator-induced polymorphisms in plants*

Growth form & prey taxon	Predator taxon or simulated damage	Feature increased	Detection method	Reference
ANNUALS				
Cucurbita moschata (squash)	Simulated damage	Chemical?	Feeding rate	Carroll & Hoffman 1980
Cucurbita pepo (zucchini)	Squash beetle (*Epilachna borealis*)	Cucubitacins	Chemistry, feeding, preference, fitness	Tallamy 1985
Glycine max (soybean)	Simulated damage	Polyribosomes	Radiolabel	Davies & Schuster 1981
Lycopersicon esculentum (tomato)	Colorado potato beetle or simulated damage	Proteinase inhibitor	Immunoassay	Green & Ryan 1972, Ryan 1979
Nicotiana tabacum (tobacco)	Simulated damage	Polyribosomes	Radiolabel	Davies & Schuster 1981
Pisum sativum (peas)	Simulated damage	Polyribosomes	Radiolabel	Davies & Schuster 1981
Zea mays (corn)	Simulated damage	Polyribosomes	Radiolabel	Davies & Schuster 1981
BIENNIALS				
Beta vulgaris (beet)	Beet fly (*Pegomya betae*)	Chemical?	Mortality	Röttger & Klingauf 1976
HERBACEOUS PERENNIALS				
Carex aquitilis	Simulated damage	Phenolics, proanthocyanidins	Chemistry	Rhoades 1979
Gossypium hirsutum (cotton)	Mites (*Tetranychus* spp.)	Chemical?	Mite pop. growth	Karban & Carey 1984
Helianthus maximiliani	Simulated damage	Terpenes	Chromatography	Gershenzon, unpublished
Medicago sativa (alfalfa)	Pea aphid, spotted aphid	Coumestrol	Chemistry, fecundity	Loper 1968
Medicago sativa	Voles (*Microtus* spp.)	Phenolics	Chemistry, weight gain	Lindroth & Batzli 1986
Rumex obtusifolius	Beetles (*Gastroidea viridula*)	Chemical?	Bioassay	Benz 1974
Senecio jacobaea	Simulated damage	Total alkaloids	Chemistry	Rhoades 1979
Solanum tuberosum (potato)	Colorado potato beetle or simulated damage	Proteinase inhibitor	Immunoassay	Green & Ryan 1972, Ryan 1979
GRASSES				
Agropyron intermedium	Black grass bug (*Labops hesperius*)	Fiber content	Gravimetric, feeding rate	Todd & Kamm 1974, Higgins et al. 1977
Andropogon greenwayi	Simulated damage	Silica content	Gravimetric	McNaughton & Tarrants 1983
Eustachys paspaioides	Simulated damage	Silica content	Gravimetric	McNaughton & Tarrants 1983
Panicum coloratum	Simulated damage	Silica content	Gravimetric	McNaughton & Tarrants 1983

Table 17.2 (continued)

Growth form & prey taxon	Predator taxon or simulated damage	Feature increased	Detection method	Reference
WOODY EVERGREENS				
Abies amabitis (true fir)	Balsam woody aphid (Adelges piceae)	Juvabione-related compounds	Chromatograph	Puritch & Nijholt 1974
Abies grandis (true fir)	Balsam woody aphid (Adelges piceae)	Juvabione-related compounds	Chromatograph	Puritch & Nijholt 1974
Picea sitchensis (spruce)	Sitka spruce weevil	Chemical?	Adult emergence	Overhulser et al. 1972
Pinus radiata (pine)	European wood wasp (Sirex noctilio)	Phenolics, resin	Chromatograph	Hillis and Inoue 1968
Pinus syvestris (pine)	European pine sawfly (Neodipron sertifer)	Phenolics	Chromatograph	Thielges 1968
WOODY DECIDUOUS				
Acer saccharum (sugar maple)	Simulated damage	Phenolics	Chemistry	Baldwin & Schultz 1983
Alnus crispa (green alder)	Snowshoe hare (Lepus americanus)	Terpenes, phenolics	Chemistry, feeding rate	Bryant 1981
Alnus rubra (red alder)	Tent caterpillar (Malacosoma californicum pluviale)	Proanthocyanidins	Larval growth rate, survivorship, fecundity	Rhoades 1983a
Betula allegheniensis (yellow birch)	Simulated damage	Phenolics	Chemistry	Baldwin & Schultz 1984
Betula papyrifera (paper birch)	Snowshoe hare (Lepus americanus)	Terpenes, phenolics	Chemistry, feeding rate	Bryant 1981

Plant species	Herbivore/treatment	Chemical	Response measured	Reference
Betula papyrifera	Spear-marked black moth (*Rheumaptera hastata*)	Chemical?	Survivorship	Wernes 1979
Betula pendula (European white birch)	Simulated damage	Tannins	Chemistry, feeding rate	Wratten et al. 1984
Betula populofolia (gray birch)	Simulated damage	Chemical?	Gypsy moth growth rate	Wallner & Walton 1979
Betula pubescens (birch)	Simulated damage	Tannins, phenolic trypsin inhibitors	Chemistry, feeding rate	Haukioja & Niemelä 1976, 1977; Wratten et al. 1984
Betula pubescens	Autumnal moth (*Oporinia autumnata*)	Chemical	Pupal weight	Haukioja 1980
Larix decidua (larch)	Larch budworm	Fiber, resins	Chemistry, mortality and fecundity	Benz 1974; Baltersweiler et al. 1977
Omphalea spp.	Moth (*Urania* spp.)	Chemical?	Larval growth rate, survivorship	Smith 1982
Populus balsamifera (balsam poplar)	Snowshoe hare (*Lepus americanus*)	Terpenes, phenolics	Chemistry, feeding rate	Klein 1977; Bryant 1981
Populus euroamericana (poplar)	Simulated damage	Phenolics	Chemistry	Baldwin & Schultz 1983
Populus tremuloides (quaking aspen)	Snowshoe hare (*Lepus americanus*)	Terpenes, phenolics	Chemistry, bioassay	Bryant 1981
Quercus rubrum (red oak)	Gypsy moth larvae (*Lymantria dispar*)	Tannins, phenolics, drymatter, toughness	Chemistry	Schultz & Baldwin 1982
Quercus velutina (black oak)	Simulated damage	Chemical?	Gypsy moth growth rate	Wallner & Walton 1979
Salix sitchensis (sitka willow)	Webworms (*Hyphantria cunea*)	Proanthocyanidins	Larval growth rate, survivorship, fecundity	Rhoades 1983a

plants, conifers, ferns, mosses, and fungi, where they can inhibit insect digestion and bacterial growth (Ryan 1979).

Most of the literature on herbivore-induced chemical defenses (table 17.2) has concentrated on the response within individual plants. There is some evidence that other individuals within a species may respond to cues emitted from the damaged plants; i.e., communication occurs between plants (Rhoades 1983a; Baldwin and Schultz 1983). Both red alder and sitka willow produced lower-quality leaves after attack by tent caterpillars; adjacent red alder, which were not attacked and had no root connections with attacked trees, also produced lower-quality leaves (Rhoades 1983a). Simulated damage to both poplar and sugar maple leaves increased concentrations of phenolics and tannins in other leaves of the same seedlings, and in both species, neighboring seedlings also responded (Baldwin and Schultz 1983). The authors suggest that an airborne cue may be released from damaged plants and received by their neighbors, although there are skeptics (Fowler and Lawton 1985).

Herbivore-induced chemical defenses reduce the fitness of many herbivore species. Reduced feeding rates, increased mortality, slower growth of body mass, reduced fecundity, and reduced rates of population increase have all been reported (table 17.2). Over evolutionary time, herbivores are capable of counteradaptation. Many insects have adapted to the chemical defenses of their host plants by biochemical detoxification pathways (Harborne 1982) and serve as the general model for the theory of coevolution (Ehrlich and Raven 1964; Futuyma 1983). Beetles that feed on squash are a good example of successful adaptation. Toxic chemicals called cucurbitacins are rapidly induced in squash by mechanical damage and have different effects on two different beetle species (Carroll and Hoffman 1980). Feeding of *Epilacha tredecimnotata* is inhibited, whereas that of *Acalymma vitta* is stimulated by the presence of these chemicals. *E. tredecimnotata* is still able to feed on damaged leaves, however, since it trenches the damaged spot before induced chemicals arrive. Tallamy (1985) found a similar behavioral adaptation in beetle-zucchini interactions.

The temporal changes in leaf quality for at least some tree species are probably due to induction by wounds produced by the herbivores (Haukioja and Niemelä 1977; Schultz and Baldwin 1982). Defoliation of a natural population of red oaks by gypsy moth larvae resulted in increased concentrations of tannins, phenolics, dry matter, and toughness in the remaining leaves, probably sufficient to significantly reduce gypsy moth fitness (Schultz and Baldwin 1982). Schultz (1983b) suggests that these induced responses, which operate in the scales of seasons and years in trees, are probably important in reducing the severity of future outbreaks of the insect.

Herbivore-induced defenses in plants may play a role in regulating the number of their herbivores, perhaps accounting for the cycling of many natural populations (Benz 1974; Haukioja and Hakala 1975; Bryant 1981; Rhoades 1983b; Lindroth and Batzli 1986). Such a mechanism may help explain the observation that herbivores frequently decline while their food remains abundant (Rhoades 1983b). Induced defenses may play a role in the cycling of snowshoe hares (Bryant 1981, 1982, cited in Rhoades 1983b). Snowshoe hare browsing of several Alaskan tree species resulted in the trees producing adventitious shoots containing higher concentrations of terpenes and phenolics, and these compounds caused severely reduced feeding by the hares. The 2- to 3-year time delay for production and maturation of juvenile stems is consistent with the 10-year period of the hare population cycle (Bryant 1982, cited in Rhoades 1983b).

CUES INITIATING PREY DEFENSES

Over long time scales, prey species adapt to predation through natural selection; genotypes with better defenses, on average, leave more offspring in future generations (Pianka 1983). Most descriptions of prey defenses (e.g., Edmunds 1974) implicitly assume that they are constitutive. In fact, many species use defenses that are induced by environmental cues, and, as shown in the present review, the activity of the predator is frequently the cue triggering the defense. These defenses develop rather rapidly, on the scale of a few hours in annual plants and protozoa (Davies and Schuster 1981; Kuhlmann and Heckmann 1985) to a couple of years in some trees (Overhulser et al. 1972). Once the cue is removed, the induced defense is gradually lost. In the absence of *Chaoborus*, spined morph *Daphnia* are replaced by typical morphs within a week (Havel 1985b). Induced chemical defenses of individual beet plants remain for 18 days (Röttger and Klingauf 1976); those of larch trees 5 years (Baltensweiler et al. 1977).

The origin of the cue (signal) may be either the predator or the prey. In plants, the cue originates from the attacked plant. Simulated damage of many plants can produce the induced secondary chemicals in the same plant (table 17.2) or in adjacent plants (Baldwin and Schultz 1983). In nature, herbivores would presumably cause similar damage. Chemical extracts of herbivores have not been reported to induce the production of defensive secondary compounds (J. C. Schultz, personal communication), although some work suggests that chemicals in saliva may stimulate vegetative growth in grasses (Dyer 1980).

Work on aquatic animal taxa suggests that the cue originates from the predator. Filtrates from media conditioned with predatory rotifers, insects, and protozoa can each be sufficient to induce morphological changes in zooplankton prey (Gilbert 1966; Krueger and Dodson 1981; Kuhlmann and Heckmann 1985). Physical damage alone has never been reported to produce such changes. Bryozoa require physical damage by their nudibranch predators to initiate spine development (Harvell 1984a), after which attacked colonies probably transmit a diffusible cue to other nearby colonies for developing defensive spines (Harvell, unpublished). The origin of the cue has interesting evolutionary implications. If the cue originates with the predator (or herbivore) and the predator's fitness is reduced as a result of subsequent prey defenses, then selection should favor predators who can either suppress recognition or overwhelm the defenses of the prey (Rhoades 1985). If the cue originates with conspecifics of the prey, then selection favoring the sender should depend on the genetic relatedness between sender and receiver (see next section).

The chemical composition of predator substances that cue prey defenses (morphogens) is still an open question, although preliminary work has been done on characterizing three of them (Gilbert 1966; Hebert and Grewe 1985; Kuhlmann and Heckmann 1985). Their identification should be a high priority because it would contribute insight into several inter-

esting developmental, ecological, and evolutionary questions. For instance: How are the morphogens bound to receptors on target cells and how are these translated into morphological changes? How specific is each morphogen to the particular predator releasing it? Is the morphogen a specialized compound or a necessary by-product of metabolism?

Predator morphogens are not the only interspecific signals operating in aquatic communities. Several instances of chemically mediated changes in behavior have been reported. Reduced glutathione, released from freshly killed zooplankton, stimulates the feeding response of *Hydra littoralis* (Lenhoff 1968). The lotic stonefly predators *Acroneuria lycorias*, *Megarcys signata*, and *Kogotus modestus* release chemicals that influence the swimming and drifting behavior of several downstream mayflies (Peckarsky 1980). *Epischura nevadensis* releases a substance that reduces the filtering rate of its competitor (and prey) *Diaptomus tyrrelli* (Folt and Goldman 1981).

DISTRIBUTION OF PREDATOR-INDUCED DEFENSES

The broad taxonomic distributions of prey with predator-induced defenses indicates that this phenomenon is widespread. Among the zooplankton, predator-induced defenses are common in protozoa, rotifers, and cladocera (table 17.1). Several studies have reported species in these taxa that were not responsive to induction by predators. Grant and Bayly (1981) showed that whereas several subspecies (species per Hebert 1977) of the *Daphnia carinata* complex grew enlarged crests after exposure to *Anisops* filtrate, several others did not. Helmet size of cyclomorphic *D. retrocurva* was not influenced by exposure to several predators: *Chaoborus*, *Hydra*, *Leptodora*, *Utricularia*, *Mesocyclops*, or *Acanthocyclops* (Havel and Dodson 1985); two of these predators have been shown to induce defenses in other species (Krueger and Dodson 1981; Stemberger and Gilbert 1984). Although several species of the ciliate *Euplotes* were induced to form predator-resistant morphs by *Lembadion* (table 17.1), *E. woodruffi* and *E. crenosus* did not (Kuhlmann and Heckmann 1985). These exceptions to the positive reports of predator induction suggest that such developmental responses are not universal.

Several theories have been proposed to suggest the predator–prey systems in which inducible defenses might evolve (Rhoades 1979; Gilbert 1980a; Harvell 1984a). Rhoades (1979) suggested that inducible defenses would be employed whenever the cost of maintaining permanent defenses is high. However, it is unclear to what level of fitness that cost corresponds. *Asplanchna*-induced defenses in *Brachionus* had no measurable fitness cost associated with them (Gilbert 1980a), whereas *Chaoborus*-induced defenses in *Daphnia* were associated with a 5% reduction in fitness (Havel and Dodson, unpublished data). Gilbert (1980a) proposed that predator-induced defenses should evolve when the defensive structure is effective against a single type of predator, and the prey organism bearing the structure has a short generation time relative to the speed with which the predator can reduce prey population levels. The *Chaoborus*/*D. pulex* system described earlier may fit Gilbert's (1980a) criteria. The *Chaoborus*-induced spines protect juvenile *D. pulex* against predation by *C. americanus*, but probably not *Notonecta*, which eats adult (and unspined) size classes at much higher rates than juvenile size classes (Scott and Murdoch 1983). Although natural densities of *Chaoborus* can reduce *Daphnia* populations by up to 30% per day (Dodson 1972), protected spined morphs are induced within 2 days.

Harvell (1984a) proposed that predator-induced defenses should evolve in clonal taxa because they often suffer intermittent nonfatal encounters with predators. Indeed, this is true with the bryozoa with which she works and with cladocerans, rotifers, and protozoa. If one considers clones as single organisms (Williams 1975), then mortality of some but not all individuals of a clone could be considered nonfatal to the genotype, and induced defenses in clone mates would improve the inclusive fitness of the signaler. On the other hand, induction of defenses in unrelated conspecifics could reduce the inclusive fitness of the signaler if the receivers of the signal are protected from predation and then leave behind more descendants than those of the signaler. Nonfatal encounters also occur in most terrestrial plants that produce chemical defenses following herbivory, and in both animals and plants that form antibodies and phytoalexins, respectively, after attack by bacteria, viruses, and fungi (Price 1980; Deverall 1977). Individuals are partially consumed, and the induced defense later protects the same individual.

Other asexual taxa are probably good candidates for future research on predator-induced defenses. Parthenogenesis and vegetative reproduction are found in taxa within most animal phyla; and in some taxa, such as rotifers, nematodes, acari, cladocera, ostracods, and aphids, these processes are very common (Bell 1982). Other polymorphic species within these taxa may develop induced defenses. Algae and macrophytes also would be worth examining from this perspective. Perhaps the spines, sheaths, and noxious qualities of some freshwater algae, known to reduce mortality to grazing zooplankton (Porter 1977), are induced by the presence of grazers. Such a process would augment the seasonal species succession of edible algae by inedible algae, and allow some species to persist longer. Aquatic macrophytes have physical and chemical characteristics that affect feeding by herbivorous molluscs, insects, fish, birds, and mammals (Sculthorpe 1967). Otto and Svenson (1981) reported that macrophytes are poorer-quality foods for leaf-eating caddisfly larvae than closely related terrestrial plants, and suggested that the presence of secondary compounds in the macrophytes was the factor influencing their preferences. Experimental assays of secondary compounds in macrophytes as a function of grazing may demonstrate herbivore-induced defenses analogous to those of terrestrial plants.

So far, the features of animals shown to be induced by predators have been morphological, whereas those of plants have been chemical (tables 17.1 and 17.2). This dichotomy may be artificial. The defensive chemistry of plants has received a lot of attention (Futuyma 1983), possibly at the expense of morphological features such as trichomes, which have known defensive functions (L. Gilbert 1971; Levin 1973). Morphological variation in zooplankton as an adaptation to predation is also a popular topic, so it is natural that the studies of predator induction should focus on easily visible morphology (e.g., Grant and Bayly 1981). Chemical defenses are known to be important in many terrestrial arthropods (Eisner 1970) and are now receiving more attention in studies of zooplankton (Kerfoot 1982). Perhaps a broader focus for future studies will reveal morphological features of plants and chemical features in animals that are induced by exposure to their predators.

THE PREDATOR'S PERSPECTIVE

Predator-induced defenses in prey reduce predator fitness in some species. This has been demonstrated in a variety of studies of induced defenses in plants where herbivores received that sole resource (table 17.2). Highly specialized herbivores are as vulnerable to the vagaries of their host plants as parasites are to their hosts (Price 1980). With the exception of the protozoan *Lembadion* (Kuhlmann and Heckmann 1985), fitness reduction has not yet been demonstrated for zooplankton predators. Perhaps having the variety of zooplankton food available to insects such as *Chaoborus* and *Anisops* allows diet switching and little loss; induction of prey defenses may be inconsequential to zooplankton carnivores.

Since predator-induced defenses can occur in prey species, then one would expect that prey-induced weapons might occur also in predators. Indeed, this phenomenon has been described for two predator–prey systems. Predaceous gilled fungi include two species that produce adhesive knobs on their conidia only when in the presence of their nematode prey (Thorn and Barron 1984). The adhesive discs allow the fungi to attach to moving nematodes and subsequently penetrate and digest them. The rotifer *Asplanchna sieboldi*, which is capable of inducing defensive morphological changes in its rotifer prey (Gilbert 1967), is itself polymorphic in response to food in its environment (Gilbert 1980b). Development of particular morphotypes is controlled by concentrations of vitamin E and prey type in the environment. When vitamin E is missing or at extremely low concentrations, only the small saccate morph develops; at higher concentrations, the cruciform morph develops, with size of lateral outgrowths positively related to concentration. Conspecific or crustacean prey are required to produce the large campanulate morph, which is an effective predator on these prey (Gilbert 1980b).

Predator-induced defenses probably can result in indirect interactions between species feeding on the same resource (Rhoades 1983b). Diffuse competition would occur if either species feeding resulted in the induction of a defensive response that inhibited the feeding of the second species. Facilitation would occur if the two species fed on different tissues of the same resource and if predator-induced defenses directed at one tissue were at the expense of defenses of the other tissues.

CONCLUSION

Predators clearly influence the species composition, size structure, and phenotypes of organisms living in many natural communities. They influence the development of prey defenses by both selecting for well-adapted genotypes and inducing changes in morphology or chemistry. Predator-induced defenses in both plants and animals are widespread, and future studies should reveal similar features in many more taxa.

ACKNOWLEDGMENTS

I thank Drs. C. D. Harvell, J. C. Schultz, and R. E. Smith for their helpful criticisms of the manuscript. Financial support was provided by the University of Wisconsin Graduate School.

REFERENCES

Baldwin, I. T., and J. C. Schultz. 1983. Rapid changes in tree leaf chemistry induced by damage: evidence for communication between plants. *Science* 221 : 277–79.

———. 1984. Damage- and communication-induced changes in Yellow Birch leaf phenolics. In R. Lenner (ed.), *Proceedings of eighth annual forest biology workshop*, pp. 25–33. Logan, Utah: Utah State University.

Baltensweiler, W., G. Benz, P. Bovey, and V. Delucchi. 1977. Dynamics of larch bud moth populations. *Annu. Rev. Entomol.* 22 : 79–100.

Barnes, R. D. 1980. *Invertebrate zoology.* Springfield, Ill.: W. B. Saunders.

Beauchamp, P. 1952a. Un facteur de la variabilité chez les Rotifères du genre *Brachionus. Compt. Rend.* 234 : 573–75.

———. 1952b. Variation chez les Rotifères du genre *Brachionus. Compt. Rend.* 235 : 1355–57.

Bell, G. 1982. *The masterpiece of nature: The evolution and genetics of sexuality.* University of California.

Benz, G. 1974. Negative Rückkoppelung durch Raum- und Nahrungskonkurrenz sowie Zyklische Veränderung der Nahrungsgrundlage als Regelprinzip in der Populationsdynamik des Gruen

Lärchenwicklers, *Zeiraphera diniana* (Guenée) (Lep., Tortricidae). *Z. Angew. Entomol.* 76 : 196–228.

Bryant, J. P. 1981. Phytochemical deterrence of snowshoe hare browsing by adventitious shoots of four Alaskan trees. *Science* 213 : 889–90.

———. 1982. The regulation snowshoe hare feeding behaviour during winter by plant antiherbivore chemistry. *Proc. Int. Lagomorph. Conf., 1st.* 1979.

Carroll, C. R., and C. A. Hoffman. 1980. Chemical feeding deterrent mobilized in response to insect herbivory and counteradaptation by *Epilachna tredecimnotata*. *Science* 209 : 414–16.

Cates, R. G. 1975. The interface between slugs and wild ginger: Some evolutionary aspects. *Ecology* 56 : 391–400.

Davies, E., and A. Schuster. 1981. Intercellular communication in plants: Evidence for a rapidly generated, bidirectionally transmitted wound signal. *Proc. Natl. Acad. Sci. (USA)* 78 : 2422–26.

Deverall, B. J. 1977. *Defence mechanisms of plants.* Cambridge: Cambridge University Press.

Dodson, S. I. 1972. Mortality in a population of *Daphnia rosea*. *Ecology* 53 : 1011–23.

———. 1975. Predation rates of zooplankton in arctic ponds. *Limnol. Oceanogr.* 20 : 426–33.

———. 1984. Predation of *Heterocope septentrionalis* on two species of *Daphnia*: Morphological defenses and their cost. *Ecology* 65 : 1249–57.

Dyer, M. I. 1980. Mammalian epidermal growth factor promotes plant growth. *Proc. Natl. Acad. Sci. USA* 77 : 4836–37.

Edmunds, M. 1974. *Defence in animals.* New York: Longman.

Ehrlich, P. R., and P. H. Raven. 1964. Butterflies and plants: A study in co-evolution. *Evolution* 18 : 586–608.

Eisner, T. 1970. Chemical defense against predation in arthropods. In E. Sondheimer and J. B. Simeone (eds.), *Chemical ecology*, pp. 157–218. New York: Academic Press.

Folt, C., and C. R. Goldman. 1981. Allelopathy between zooplankton: A mechanism for interference competition. *Science* 213 : 1133–35.

Fowler, S. V., and J. H. Lawton, 1985. Rapidly induced defenses and talking trees: The devil's advocate position. *Am. Nat.* 126 : 181–95.

Futuyma, D. J. 1983. Evolutionary interactions among herbivorous insects and plants. In D. J. Futuyma and M. Slatkin (eds.), *Coevolution*, pp. 207–31. Sinauer.

Gerritsen, J., and J. R. Strickler. 1977. Encounter probabilities and community structure in zooplankton: A mathematical model. *J. Fish. Res. Bd. Can.* 34 : 73–82.

Gilbert, J. J. 1966. Rotifer ecology and embryological induction. *Science* 151 : 1234–37.

———. 1967. *Asplanchna* and postero-lateral spine production in *Brachionus calyciflorus*. *Arch. Hydrobiol.* 64 : 1–62.

———. 1980a. Further observations on developmental polymorphism and its evolution in the rotifer *Brachionus calyciflorus*. *Freshwater Biol.* 10 : 281–94.

———. 1980b. Developmental polymorphism in the rotifer *Asplanchna sieboldi*. *American Scientist* 68 : 636–46.

Gilbert, J. J., and R. S. Stemberger. 1984. *Asplanchna*-induced polymorphism in the rotifer *Keratella slacki*. *Limnol. Oceanogr.* 29 : 1309–16.

Gilbert, L. E. 1971. Butterfly-plant coevolution: Has *Passiflora adenopoda* won the selectional race with Heliconiine butterflies? *Science* 172 : 585–86.

Grant, J. W. G., and I. A. E. Bayly. 1981. Predator induction of crests in morphs of the *Daphnia carinata* King complex. *Limnol. Oceanogr.* 26 : 201–18.

Green, T. R., and C. A. Ryan. 1972. Wound-induced proteinase inhibitor in plant leaves: A possible defense mechanism against insects. *Science* 175 : 776–77.

Greene, C. H., and M. R. Landry. 1985. Patterns of prey selection in the cruising calanoid predator, *Euchaeta elongata*. *Ecology* 66 : 1708–16.

Gregory, S. V. 1983. Plant-herbivore interactions in stream systems. In J. R. Barnes and G. W. Minshall (eds.), *Stream ecology*, pp, 157–89. New York: Plenum Press.

Halbach, U. 1971. Zum Adaptivwert der zyklomorphen Dornenbildung von *Brachionus calyciflorus* Pallas (Rotatoria). *Oecologia* 6 : 267–88.

Halbach, U., and J. Jacobs. 1971. Seasonal selection as a factor in rotifer cyclomorphosis. *Naturwissenschaften* 6 : 326–27.

Harborne, J. B. 1982. *Introduction to ecological biochemistry*. New York: Academic Press.

Harvell, C. D. 1984a. Predator-induced defense in a marine bryozoan. *Science* 224 : 1357–59.

———. 1984b. Why nudibranchs are partial predators: Intracolonial variation in bryozoan palatability. *Ecology* 65 : 716–24.

Haukioja, E. 1980. On the role of plant defenses in the fluctuation of herbivore populations. *Oikos* 35 : 202–13.

Haukioja, E., and T. Hakala. 1975. Herbivore cycles and periodic outbreaks: Formulation of a general hypothesis. *Rep. Kevo Subarct. Res. Stn.* 12 : 1–9.

Haukioja, E., and P. Neimelä. 1976. Does birch defend itself actively against herbivores? *Rep. Kevo Subarct. Res. Stn.* 13 : 44–47.

———. 1977. Retarded growth of a geometrid larva after mechanical damage to leaves of its host tree. *Ann. Zool. Fenn.* 14 : 48–52.

Havel, J. E. 1985a. Predation of common invertebrate predators on long- and short-featured *Daphnia retrocurva*. *Hydrobiologia* 124 : 141–49.

———. 1985b. Cyclomorphosis of *Daphnia pulex*

spined morphs. *Limnol. Oceanogr.* 30 : 853–61.

Havel, J. E., and S. I. Dodson. 1984. *Chaoborus* predation on typical and spined morphs of *Daphnia pulex*: Behavioral observations. *Limnol. Oceanogr.* 29 : 487–94.

———. 1985. Environmental cues of cyclomorphosis in *Daphnia retrocurva* Forbes. *Freshwater Biol.* 15 : 469–78.

Hebert, P. D. N. 1977. A revision of the taxonomy of the genus *Daphnia* (Crustacea: Daphnidae) in south-eastern Australia. *Aust. J. Zool.* 25 : 371–98.

Hebert, P. D. N., and P. M. Grewe. 1985. *Chaoborus* induced shifts in the morphology of *Daphnia ambigua*. *Limnol. Oceanogr.* 30 : 1291–97.

Higgins, K. M., J. E. Brown, and B. A. Haws. 1977. The black grass bug (*Labops hesperius* Uhler): Its effect on several native and introduced grasses. *J. Range Manage.* 30 : 380–84.

Hillis, W. E., and T. Inoue. 1968. The formation of polyphenols in trees. IV. The polyphenols formed in *Pinus radiata* after *Sirex* attack. *Phytochemistry* 7 : 13–22.

Karban, R., and J. R. Carey. 1984. Induced resistance of cotton seedlings to mites. *Science* 225 : 53–54.

Kerfoot, W. C. 1975. The divergence of adjacent populations. *Ecology* 56 : 1298–1313.

———. 1977. Competition in cladoceran communities: The cost of evolving defenses against copepod predation. *Ecology* 58 : 303–313.

———. 1982. A question of taste: Crypsis and warning coloration in freshwater zooplankton communities. *Ecology* 63 : 538–54.

Kerfoot, W. C. (in press). Translocation experiments: *Bosmina* responses to copepod predation. *Ecology.*

Kerfoot, W. C., D. L. Kellogg, and J. R. Strickler. 1980. Visual observations of live zooplankters: evasion, escape, and chemical defenses. *Am. Soc. Limnol. Oceanogr. Spec. Symp.* 3 : 10–27.

Klein, D. R. 1977. Winter food preferences of snowshoe hares (*Lepus americanus*) in Alaska. *Proc. Int. Congr. Game Biol.* 13 : 266–75.

Kozloff, E. N. 1974. *Keys to the marine invertebrates of Puget Sound, the San Juan Archipelago, and adjacent regions.* Seattle: University of Washington.

Krueger, D. A., and S. I. Dodson. 1981. Embryological induction and predation ecology in *Daphnia pulex*. *Limnol. Oceanogr.* 26 : 219–223.

Kuhlman, H. W., and K. Heckmann. 1985. Interspecific morphogens regulating prey-predator relationships in protozoa. *Science* 227 : 1347–49.

Lehninger, A. L. 1970. *Biochemistry.* Worth.

Lenhoff, H. M. 1968. Chemical perspectives on the feeding response, digestion, and nutrition of selected coelenterates. In M. Florkin and B. T. Scheer (eds.), *Chemical zoology*, pp. 157–221. New York: Academic Press.

Levin, D. A. 1973. The role of trichomes in plant defense. *Quart. Rev. Biol.* 48 : 3–15.

———. 1976. The chemical defenses of plants to pathogens and herbivores. *Annu. Rev. Ecol. Syst.* 7 : 121–59.

Li, J. L., and H. W. Li. 1979. Species specific factors affecting predator-prey interactions of the copepod *Acanthocyclops vernalis* with its natural prey. *Limnol. Oceanogr.* 24 : 613–26.

Lindroth, R. L., and G. O. Batzli. 1986. Inducible plant chemical defenses: A cause of vole population cycles? *J. Anim. Ecol.* 55 : 431–49.

Lively, C. M. 1986. Predator-induced shell dimorphism in the acorn barnacle *Chthamalus anisopoma*. *Evolution* 40 : 232–42.

Loper, G. M. 1968. Effect of aphid infestation on the coumestrol content of alfalfa varieties differing in aphid resistance. *Crop. Sci.* 8 : 104–106.

Lynch, M. 1980. The evolution of cladoceran life histories. *Quart. Rev. Biol.* 55 : 23–42.

McBeth, J. W. 1968. Feeding behavior of *Corambella steinbergae*. *Veliger* 11 : 145–46.

McNaughton, S. J., and J. L. Tarrants. 1983. Grass leaf silicification: Natural selection for an inducible defense against herbivores. *Proc. Nat. Acad. Sci. (USA)* 80 : 790–91.

O'Brien, W. J., D. Kettle, and H. P. Riessen. 1979. Helmets and invisible armor: Structures reducing predation from tactile and visual planktivores. *Ecology* 60 : 287–94.

Ohman, M. D., B. W. Frost, and E. B. Cohen. 1983. Reverse diel migration: an escape from invertebrate predators. *Science* 220 : 1404.

Otto, C., and B. S. Svensson. 1981. How do macrophytes growing in or close to water reduce their consumption by aquatic herbivores? *Hydrobiologia* 78 : 107–12.

Overhulser, D., R. I. Gara, and R. Johnsey. 1972. Emergence of *Pissodes strobi* (Coleoptera: Curculionidae) from previously attacked Sitka spruce. *Ann. Entomol. Soc. Amer.* 65 : 1423–24.

Pastorok, R. A. 1981. Prey vulnerability and size selection by *Chaoborus* larvae. *Ecology* 62 : 1311–24.

Peckarsky, B. L. 1980. Predator-prey interactions between stoneflies and mayflies: Behavioral observations. *Ecology* 61 : 932–43.

Pianka, E. R. 1983. *Evolutionary ecology.* New York: Harper and Row.

Porter, K. G. 1977. The plant-animal interface in freshwater ecosystems. *American Scientist* 65 : 159–70.

Pourriot, R. 1964. Étude expérimentale de variations morphologiques chez certaines espèces de rotifères. *Bull. Soc. Zool. Fr.* 89 : 555–61.

———. 1974. Relations prédateur-proie chez les rotifères: Influence du prédateur (*Asplanchna brightwelli*) sur la morphologie de la proie *Brachionus bidentata*). *Ann. Hydrobiol.* 5 : 43–55.

Price, P. W. 1980. Evolutionary biology of parasites. Princeton, N.J.: Princeton University Press.

Puritch, G. S., and W. W. Nijholt. 1974. Occurrence of juvabione-related compounds in grand fir and Pacific silver fir infested by balsam woolly aphid. Can. J. Bot. 52 : 585–87.

Rhoades, D. F. 1979. Evolution of plant chemical defense against herbivores. In G. A. Rosenthal and D. H. Jansen (eds.), Herbivores: Their interaction with secondary plant metabolites, pp. 3–54. New York: Academic Press.

———. 1983a. Responses of alder and willow to attack by tent caterpillars and webworms: Evidence for pheromonal sensitivity of willows. In P. A. Hedrin (ed.), Plant resistance to insects, pp. 55–68. Amer. Chem. Soc. Symp. Ser. 208.

———. 1093b. Herbivore population dynamics and plant chemistry. In R. F. Denno and M. S. McClure (eds.), Variable plants and herbivores in managed systems, pp. 155–220. New York: Academic Press.

———. 1985. Offensive-defensive interactions between herbivores and plants: Their relevance in herbivore population dynamics and ecological theory. Amer. Naturalist 125 : 205–38.

Riessen, H. P. 1984. The other side of cyclomorphosis: Why Daphnia lose their helmets. Limnol. Oceanogr. 29 : 1123–26.

Röttger, W., and F. Klingauf. 1976. Physiological changes in sugar beet leaves caused by the beet fly Pegomya betae Curt. (Muscidae: Anthomyidae). Z. Angew. Entomol. 82 : 220–27.

Ryan, C. A. 1979. Proteinase inhibitors. In G. A. Rosenthal and D. H. Jansen (eds.), Herbivores: Their interaction with secondary plant metabolites, pp. 599–618. New York: Academic Press.

Schultz, J. C. 1983a. Habitat selection and foraging tactics of caterpillars in heterogeneous trees. In R. F. Denno and M. S. McClure (eds.), Variable plants and herbivores in natural managed systems, pp. 61–90. New York: Academic Press.

———. 1983b. Impact of variable plant defensive chemistry on susceptibility of insects to natural enemies. In P. A. Hedrin (ed.), Plant resistance to insects, pp. 37–54. Amer. Chem. Soc. Symp. Ser. 208.

Schultz, J. C., and I. T. Baldwin. 1982. Oak leaf quality declines in response to defoliation by gypsy moth larvae. Science 217: 149–51.

Schultz, J. C., P. J. Nothnagle, and I. T. Baldwin. 1982. Seasonal and individual variation in leaf quality of two northern hardwoods tree species. Amer. J. Bot. 69 : 753–59.

Scott, M. A., and W. W. Murdoch. 1983. Selective predation by the backswimmer, Notonecta. Limnol. Oceanogr. 28 : 352–66.

Sculthorpe, C. D. 1967. The biology of aquatic vascular plants. New York: St. Martins Press.

Seed, R. 1976. Observations on the ecology of Membranipora (Bryozoa) and a major predator Doridella steinbergae (Nudibranchiata) along fronds of Laminaria saccharina at Friday Harbor, Washington. J. Exp. Mar. Biol. Ecol. 24 : 1–17.

Smith, N. G. 1982. Periodic migrations and population fluctuations by the neotropical day-flying moth Urania fulgens through the isthmus of Panama. In E. G. Leigh, Jr., A. S. Rand, and D. M. Windsor (eds.), The ecology of a tropical forest: Seasonal rhythms and long-term changes, pp. 331–34. Washington, D.C.: Smithsonian Institution.

Stemberger, R. S., and J. J. Gilbert. 1984. Spine development in the rotifer Keratella cochlearis: Induction by cyclopoid copepods and Asplanchna. Freshwater Biol. 14 : 639–47.

Tallamy, D. W. 1985. Squash beetle feeding behavior: An adaptation against induced cucurbit defenses. Ecology 66 : 1574–79.

Thielges, B. A. 1968. Altered polyphenol metabolism in the foliage of Pinus sylvestris associated with European pine sawfly attack. Can. J. Bot. 46 : 724–25.

Thorn, R. G., and G. L. Barron. 1984. Carnivorous mushrooms. Science 224 : 76–78.

Threlkeld, S. T. 1979. The midsummer dynamics of two Daphnia species in Wintergreen Lake, Michigan. Ecology 60 : 165–79.

Todd, J. G., and J. A. Kamm. 1974. Biology and impact of a grass bug Labops hesperius Uhler in Oregon rangeland. J. Range Manage. 27 : 453–58.

Wallner, W. E., and G. S. Walton. 1979. Host defoliation: A possible determinant of gypsy moth population quality. Ann. Entomol. Soc. Amer. 72 : 62–67.

Werner, E. E., and D. J. Hall. 1974. Optimal foraging and size selection of prey by the bluegill sunfish (Lepomis macrochirus). Ecology 55 : 1042–52.

Wernes, R. A. 1979. Influence of host foliage on development, survival, fecundity and oviposition of the spear-marked black moth Rheumaptera hastata (Lepidoptera: Geometridae). Can. Entomol. 111 : 317–32.

Williams, G. C. 1975. Sex and evolution. Princeton, N.J.: Princeton University Press.

Williams, K. S., and J. H. Myers. 1984. Previous herbivore attack of red alder may improve food quality for fall webworm larvae. Oecologia 63 : 166–70.

Wratten, S. D., P. J. Edwards, and I. Dunn. 1984. Wound-induced changes in the palatability of Betula pubescens and B. pendula. Oecologia 61 : 372–75.

Yoshioka, P. M. 1982. Predator-induced polymorphism in the bryozoan Membranipora membranacea (L.). J. Exp. Mar. Biol. Ecol. 61 : 233–42.

Zaret, T. M. 1972. Predators, invisible prey, and the nature of polymorphism in the cladocera (class Crustacea). Limnol. Oceanogr. 17 : 171–84.

B. Ecological Adjustments in Time and Space

18. Diapause as a Predator-Avoidance Adaptation

Nelson G. Hairston, Jr.

Predation avoidance is proposed as an explanation for the timing of diapause in two populations of the freshwater copepod *Diaptomus sanguineus*. Adaptation is demonstrated by showing a close fit between the observed dates that females switch from making subitaneous (immediately-hatching) eggs to diapausing eggs and the dates predicted by a theoretical model that incorporates seasonal catastrophic increases in predation. The sources of predation are identified from field investigations. A third population of D. *sanguineus* changed its timing of diapause after predator removal, indicating that predation had been an important selective force.

Diapause, dormancy, and cyst formation are traditionally viewed as adaptations to avoid periods in the environment that are physically harsh. Winter and dry seasons are the most commonly studied yearly catastrophes, and a broad variety of animals, plants, and other creatures are known to become inactive during these periods of low temperature and sunlight (e.g., Wareing and Phillips 1970; Beck 1980; Flint et al. 1981) or little water (e.g., Wiggins et al. 1980; Young 1982). Although my discussion here deals largely with diapause, I see no reason why the arguments should not apply to organisms that undergo other forms of suspended animation.

Formally, the relative merits of activity and diapause can be assessed in terms of immediate reproductive gain versus survival in the inactive state (Levins 1969). In general, diapause is favored when, over some period, the expectation of fitness accruing from active growth and reproduction is less than that from survival in diapause (Cohen 1970). Although Cohen and Levins couched their discussions in terms of insects avoiding winter, there is nothing in the theory that requires this interpretation. Any environmental catastrophe that fulfills Cohen's requirements may be expected to result in natural selection for a resistant, dormant stage. Specifically, the rate of change of abundance in a population of active individuals (r_A) is the difference between birth rate (b_A) and death rate (d_A) and can be positive or negative; but in a diapausing population, where birth rate (b_D) is zero, the rate of change (r_D) can only be zero or negative. Any event that either decreases b_A or increases d_A so that $r_A < r_D$ will select for diapause in that population (subject to variation between individuals in birth and death probabilities). Thus, the physical environment need not be the only source of selection for diapause. Slobodkin (1954) and Hutchinson (1967) have suggested that some Cladocera may make ephippia (diapausing eggs) in response to intense competition for algae. Likewise, absence of food stimulates diapause in some insects (Beck 1980), and a broad variety of algae are known to make resting cysts or spores in response to low nutrient availability (e.g., Sangren 1983; Hargraves and French 1983).

Southwood (1978), in his closing comments for a symposium on insect migration and diapause, remarked that the participants had not discussed the possibility that these traits might evolve as a means of avoiding natural enemies. To my knowledge, Strickler and Twombly (1975) were the first to suggest diapause as a predator-avoidance adaptation. They noted that immature *Cyclops scutifer* entered diapause in Bauline Long Pond, Newfoundland, when the predatory copepod *Epishura nordenskioldi* became abundant. Similarly, Nilssen (1977, 1980) found in his detailed study of *Cyclops*

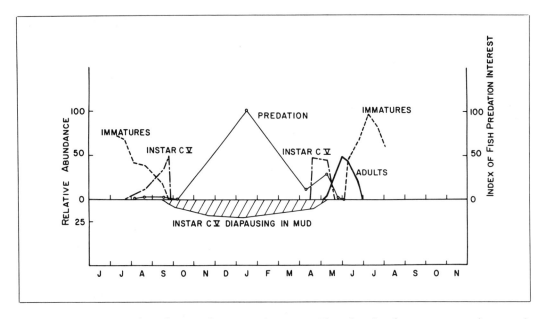

Figure 18.1. *Correspondence between the timing of diapause and predation in a population of* Cyclops abyssorum *in Lonavatn Lake, Norway. Life history of the copepods is redrawn from Nilssen and Elgmork (1977). The index of predation pressure on the copepods is derived from the stomach contents of Arctic char and is redrawn from Nilssen (1977).*

abyssorum in Lonavatn Lake, Norway, that fifth instar copepodids (the sixth instar is the adult stage) entered diapause and rested in the lake sediments during the period that predation by fish (arctic char) was focused on the zooplankton (fig. 18.1).

My colleagues and I have found comparable temporal correspondences between copepod diapause and predator activity in two small bodies of water in Rhode Island. In Bullhead Pond (a permanent lake), *Diaptomus sanguineus* adult females switch from making subitaneous (immediately-hatching) eggs to diapausing eggs at the time that sunfish activity begins to increase in early spring (fig. 18.2, top). The eggs hatch in autumn when fish activity declines. The same copepod species residing in a temporary pond (Pond C) switches to making diapausing eggs in late spring, more than 4 months before the pond dries up but very close to the time when the abundance of predatory midge larvae (*Chaoborus* and *Mochlonyx*) begins to increase (fig. 18.2, bottom).

ADAPTATION

Superficially then, these seasonal patterns appear to be instances where diapause serves as a means of remaining in the ecosystem while avoiding periods of particularly high predation. As they stand, however, the examples rely solely on the coincidence in time of two seasonal phenomena, the start of diapause and increased predation. Other seasonal events also occur in the habitats that could complicate or invalidate our interpretation. The arrival of winter at Lonavatn Lake, the arrival of summer at Bullhead Pond, and the decreasing water depth at Pond C each carries with it environmental correlates other than those affecting predators. The problem is that when studying adaptation by natural selection in field populations, it is usually easy to construct a plausible hypothesis assigning significance to whatever character one might choose, whereas it is much more difficult to establish the validity of one's explanation. Gould and Lewontin (1979) lament that investigators frequently make little effort to test their ideas. They object to the assumption, implicit in many studies, that every character in every environment has some perfectly adapted function, and they characterize this assumption as the "Panglossian paradigm," after Voltaire's Dr. Pangloss, whose philosophy held that "all is for the best in this best of all possible worlds." It is possible, for example, that *D. san-*

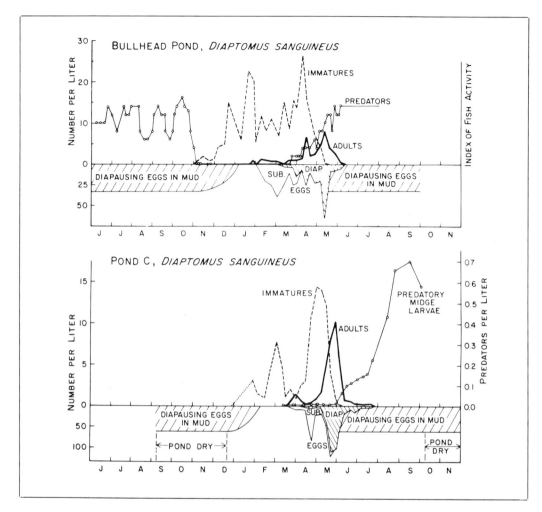

Figure 18.2. *Correspondence between the timing of diapause and predation in two populations of Diaptomus sanguineus in Rhode Island. For Bullhead Pond* **(top),** *life history data are from Hairston and Olds (1984) and Hairston and Munns (1984). The area labeled "eggs" represents either subitaneous (immediately-hatching) eggs or diapausing eggs carried by females in the plankton. Predator activity data are from Hairston et al. (1983). For Pond C* **(bottom),** *the copepod life history data are from Hairston et al. (1985) and from previously unpublished information.*

guineus makes diapausing eggs in the spring simply because some ancestral population made diapausing eggs at that time of year. Although the ecological or evolutionary forces molding the phenology of the ancestral population may no longer exist, present populations could retain the trait only because they are stuck with it. Gould and Lewontin's (1979) criticism is well founded, but demonstrating adaptation by natural selection in the field is a difficult proposition. Character changes may occur slowly, or from manifold causes. Indeed,

under stabilizing selection little change in a character would be expected. Yet it is difficult or impossible to infer causation from an observation of no change.

What kinds of data are needed, then, to raise an explanation from adaptive "storytelling" to believable hypothesis? Gould and Lewontin (1979) do not address this question, but several possibilities exist. The strongest evidence would come from a replicated, controlled, long-term-selection experiment in the field. For example, given an array of ponds, predator density might be manipulated in some and left unaltered in others, then response of copepod diapause timing monitored over several years in the ponds. Documentation would be clearest if the genetics of the diapause trait were known and if genotypic as well as phenotypic response to manipulation were observed. In short, it would be best to witness the process about which one is speculating. Such a study would require long-term access to the appropriate habitats and considerable luck in settling on a trait that could respond quickly to selection. Short of a scientific windfall of this kind, indirect evidence might be accumulated through field data showing that the presumed selective force does in fact affect survival or reproduction, laboratory data showing that the character under consideration is inherited, and laboratory experiments documenting that the trait responds to selection. Armed with unequivocal results (in the right direction) from these studies, it ought to be possible to build a believable case for the adaptive significance of the trait in question.

TELLING A BETTER STORY

Believability is ultimately at issue, and the "adaptive stories" criticized by Gould and Lewontin (1979) simply have not met the implicit standards set by these authors. If the story told is not good enough to be believed, a third approach is to tell a better story. From figure 18.2 we see that the start of diapausing egg production by D. sanguineus comes at *about* the right time for the copepods to avoid predation, but we can go beyond this and ask when, *exactly*, is the right time to begin making diapausing eggs. If it is possible to predict with reasonable precision when the switch to diapause should occur in order to avoid predation, then the closer the copepods come to matching this expectation, the less likely alternative explanations will be.

Hairston and Munns (1984) describe a simulation model of D. sanguineus life history that competes populations beginning production of diapausing eggs at different times before the season-ending catastrophe. Modeled patterns of reproduction that outcompete others and cannot be outcompeted themselves represent evolutionarily stable strategies (ESS) for diapause. The simulation incorporates details such as natural reproduction and survivorship

schedules, year-to-year variation in catastrophe date, and population growth in discrete cohorts. The important predictions are that if the catastrophe occurs on the same date every year (or the catastrophe date is perfectly predictable), the ESS for time to initiate diapause is one generation before that catastrophe takes place (here one generation is defined as the age of maturity plus the time to produce the first clutch of eggs). Taylor (1980) obtained the same result with an optimization model, and it is consistent with Cohen's (1970) conclusion because any female initiating diapause closer than one generation to the catastrophe will suffer a loss of reproduction through offspring mortality (high d_A). Any female making diapausing eggs earlier in time will face reduced reproduction (low b_A) because she could have made subitaneous eggs with time to hatch, mature, and make diapausing eggs of their own. Second, if there is annual variability in the time of the catastrophe that cannot be predicted by the copepods, the ESS for time to initiate diapause is greater than one generation before the mean catastrophe date. The greater the annual variation in the time of the catastrophe, the larger the time interval between the mean catastrophe date and the ESS time to begin diapause.

The Bullhead Pond Population

To make a prediction of the time D. sanguineus should start making diapausing eggs requires knowledge of the mean and variance of the onset of predation, and the copepod generation time. To test the model, the date that diapause begins in the field also must be known. At Bullhead Pond these data have been collected for between 5 and 7 years. Hairston et al. (1983) found that as water temperature and sunfish activity increase in spring, the copepods come under heavy predation pressure. Detailed field collections combined with experiments in the pond and laboratory showed that females carrying eggs are especially visible to fish and so suffer particularly high mortality. As a result, it is possible to establish the time at which predation begins to influence the copepod population dynamics by finding the date each spring when sex ratio begins to increase. Hairston and Munns (1984) point out that the slope of the natural logarithm of sex ratio as a function of time is equal to the difference between female and male mortality. They estimate the date of the onset of predation by using regression techniques to establish the point in time when the slope of the sex ratio-time curve becomes positive. They calculated catastrophe dates from 5 years of data (1978–1982) and found that fish predation began to affect copepod population dynamics on a mean date of 7 May ± 13 days (1 SD). Since their study, 2 more years of data have been collected, with catastrophe dates of 28 April 1983 and 25 May 1984. For all 7 years, the mean date of sex ratio increase is now 8 May ± 13 days (fig. 18.3, bottom). The ESS model predicts from these data that female *D. sanguineus* should switch from making subitaneous eggs to diapausing eggs between 1.15 and 1.45 copepod generations before 8 May. The estimate is variable because a random number generator was employed in the simulation to model the time of the catastrophes. The switching date that performs best in competition with others depends in part on the particular sequence of catastrophes generated (or presumably the particular sequence occuring in nature). Thus, a suite of ESS switching dates was obtained from the model when Hairston and Munns used 5 to 10 different random number sequences to delimit the range of competitively superior strategies. Within the range 1.15 to 1.45 generations, some switching dates were found to be better strategies than others because they won the majority of these computer runs.

Hairston and Munns (1984) estimated the generation time of *D. sanguineus* directly from six field cohort analyses of copepod abundance to be 33.8 days, using the method of Rigler and Cooley (1974). Hairston and Twombly (1985) have since criticized this method, but in the present case, the error is probably not greater than 10%. The bar at the center of figure 18.3 illustrates the period between 1.15 × 33.8 days and 1.45 × 33.8 days when diapause should begin. The two filled circles at the left of the bar show the switching dates within this range that, according to the ESS model, should win most often in competition. Finally, the timing of the initiation of diapause [determined according to the method described by Hairston and Olds (1984)] is plotted for 5 years of data at the top of figure 18.3. The mean date that diapause begins is calculated from the point in each year that 50% of the females have switched to making diapausing eggs. The

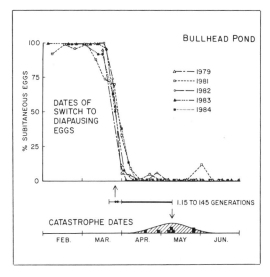

Figure 18.3. *The relationship between the observed and the predicted timing of the switch from subitaneous (immediately-hatching) eggs to diapausing eggs by D. sanguineus in Bullhead Pond. According to the model of Hairston and Munns (1984), the switch should occur 1.15 to 1.45 copepod generations before the onset of predation.* **Top:** *fraction of the population making subitaneous eggs as a function of time in 5 years; arrow pointing upward, mean date on which 50% of females were making subitaneous eggs.* **Bottom:** *squares give catastrophe dates when fish predation begins to affect copepod dynamics for 7 years of data; arrow pointing downward, mean catastrophe date.*

match between when the copepods should begin diapause, according to the ESS model, and when they actually do is extraordinarily good.

The Pond C Population

Fewer years of data have been collected at Pond C than at Bullhead Pond, but the picture is similar. The point in the season that predation becomes intense enough to constitute a catastrophe cannot be established using sex ratio changes because the midge larvae are not sex-specific in their feeding. Instead, death rate caused by predation from larval midges was calculated using consumption rates of *Chaoborus americanus* (the species in Pond C) determined by Fedorenko (1975). In her detailed study, midge larvae were fed two species of *Diaptomus* of different body sizes at a range of temperatures and prey densities. Knowing midge and copepod abundance in Pond C on

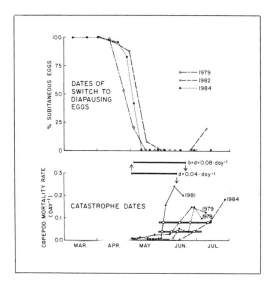

Figure 18.4. *The relationship between the observed and the predicted timing of the switch to diapause by* D. sanguineus *in Pond C. Depending on assumptions explained in the text, the model of Hairston and Munns (1984) predicts that the switch should occur during the first few days of May.* **Top:** *fraction of population making subitaneous eggs as a function of time in 3 years; arrow pointing upward, mean date (1 May) on which 50% of females were making subitaneous eggs.* **Bottom:** *increase in copepod mortality caused by predation from midge larvae in 4 years; arrow pointing downward, mean dates on which mortality reached either 0.04/day or 0.08/day. For reasons given in the text, death rates between these values are considered catastrophic to the population.*

different dates as well as pond temperatures, total copepod mortality was estimated for the 4 years for which I have data. The bottom panel in figure 18.4 shows calculated copepod mortality increasing during late spring in each year.

The problem then is to determine where along these mortality rate lines *Chaoborus* predation is sufficient to be called catastrophic. A logical place would seem to be the point at which mortality rate (d_A) exceeds reproduction (b_A) and population growth rate falls below zero. The potential birth rate of *D. sanguineus* in Pond C, calculated using field clutch sizes and development rates, and the Lotka–Euler equation (assuming no death), is estimated to be 0.08/day. The second open circle along each mortality rate line designates the point at which death rate exceeds 0.08/day, and thus the catastrophe date for that year. It is, of course, quite possible that there are sources of

mortality other than that caused by midge larvae, and so a lower level of predation may have been sufficient to drive $d_A > b_A$. The first open circles along the mortality rate lines show where catastrophe dates would be if predation of only 0.04/day were necessary to drive $d_A > b_A$. One difficulty with this approach is that *D. sanguineus* with discrete cohorts in Pond C violates the Lotka-Euler assumption of a stable age distribution. The birth rate obtained, however, is a logical first approximation of population reproduction.

If all *D. sanguineus* mortality in Pond C is caused by midge predation (d = 0.08/day), the mean catastrophe date is 20 June ± 17 days. If, on the other hand, midge predation makes up only half of the copepod mortality (d = 0.04/day), then the mean catastrophe date is 12 June ± 12 days (fig. 18.4, downward-pointing arrows).

The generation time of *D. sanguineus* in Pond C, calculated from 4 years of field data, is 35.2 days. With d = 0.08/day, the Hairston and Munns model predicts an ESS time for the switch to diapause of 3 May; and with d = 0.04/day, the model predicts diapause on 1 May (fig. 18.4). As in figure 18.3, ranges of ESS dates exist, but for clarity only the switching dates winning most often in competition are given here. The predicted time of diapause is robust for this example because increases of catastrophe variability in the ESS model lead to diapause dates farther from the mean catastrophe date. Even choosing predation rates as high as d = 0.15/day or as low as d = 0.025/day only give predicted diapause dates of 8 May and 24 April respectively. The date that diapause begins for *D. sanguineus* in Pond C, again taken as the mean day on which 50% of the females have switched to making diapausing eggs, is 1 May (fig. 18.4, top). As at Bullhead Pond, the match between the predictions of the model and the observed time of diapause is striking.

In both ponds, the start of diapausing egg production by *D. sanguineus* comes at nearly exactly the time predicted from theory if the copepods are to avoid seasonally high predation. Although this does not prove that the timing of diapause for these copepods is a predator-avoidance adaptation, it does show that, for the habitats in which the animals live, they could not possess a better-timed reproductive phenology. It would stretch credibility to

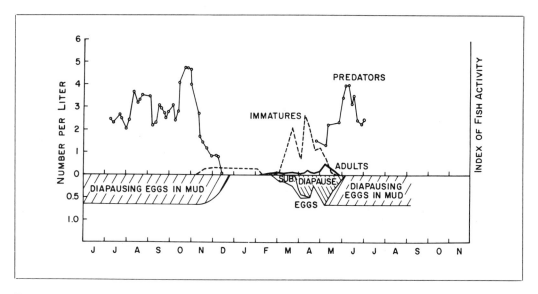

Figure 18.5. *Correspondence between the timing of diapause and predation in a population of* D. sanguineus *in Little Bullhead Pond with fish present. Data for* *the spring increase in fish activity are incomplete because the 1981 drought prematurely terminated the study. (From Hairston et al. 1983).*

suggest that the timing of diapause was set by random chance or even by some other unspecified selection force.

NATURAL SELECTION

As suggested earlier, the best evidence for adaptive evolution must come from witnessing the process as it occurs. Examples in nature are rare indeed, but I have been fortunate enough to record the response of the diapause trait in a population of *D. sanguineus* to a natural fish removal event. The details of this study are given elsewhere (Hairston and Walton 1986), and only the main points are described here.

The timing of diapause in a third population of *D. sanguineus* has been followed on and off since 1979. In Little Bullhead Pond, the copepods were initially exposed to predation pressure by redbreast sunfish very similar to that in Bullhead Pond (Hairston et al. 1983), and the onset of diapause presents a similar picture (fig. 18.5). In 1981, a very serious drought dried all the water in Little Bullhead Pond, killing all 14,000 resident sunfish (Hairston et al. 1983). In 1982 the pond was refilled, and no fish were reintroduced. Bullhead Pond is deeper, lies in a separate drainage basin, did not dry up, and no fish were killed. Thus, the copepods in Little Bullhead Pond were released from the selec-

tion pressure of fish predation, whereas those in Bullhead Pond were not. If the timing of diapause were heavily influenced by seasonal predation, as argued above, the copepods released from that selection pressure should be expected to begin production of diapausing eggs later in the summer. Without fish, any individual continuing to produce subitaneous eggs would have a large reproductive advantage over those diapausing. The results are dramatic. Within two copepod generations after the drought, the timing of diapause in the Little Bullhead population had shifted to 1 month later in the season (fig. 18.6). Over the same period, the timing of diapause in the Bullhead Pond population had not changed at all.

It is possible to show that natural selection could have produced the response seen in Little Bullhead Pond if we are willing to accept a second "Panglossian paradigm." Instead of suggesting that "all is for the best" for the organism we are studying, let us assume that this is the "best of all possible worlds" for population biologists. With genotypic variance and selection pressures constant over time, and assumptions of additive variance all correct, it is possible to show that quite reasonable combinations of heritability and selection would have resulted in the observed shifts in the timing of diapause (Hairston and Walton 1986). Because

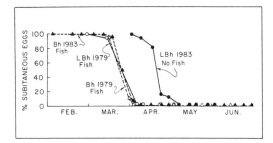

Figure 18.6. *The timing of the switch to diapause for* D. sanguineus *populations in Bullhead (Bh) and Little Bullhead (LBh) Ponds. In 1979, both ponds contained planktivorous sunfish. In 1983, Bullhead Pond retained its fish population, whereas Little Bullhead Pond had lost its fish in the 1981 drought.*

the *D. sanguineus* in Little Bullhead Pond responded to removal of fish rapidly and in the direction expected from the earlier modeling study, we may infer that seasonal predation by fish was an important selection force maintaining the timing of diapause in the predrought copepod population.

CONCLUSIONS

Although predation has been mentioned in some general discussions as potentially selecting for dormancy (Southwood 1978; Fryxell 1983), to date only copepod populations have been singled out as examples. One possible reason for this is that permanent aquatic environments are buffered against extreme physical stresses. Although terrestrial organisms may well experience seasonally intense predation (e.g., arrival of migratory birds or annual development of insect populations), the cold temperatures or drought conditions that induce seasonality in the predator may be severe enough to select for dormancy in the prey as well. In contrast, Bullhead Pond sunfish become quiescent during the winter, but water temperatures never get so low as to preclude copepod development. *D. sanguineus* in Bullhead Pond first attracted my attention precisely because it diapaused during physically benign summer conditions. Thus, other instances of dormancy as an adaptation to avoid intense community interactions will be most likely discovered in organisms that become inactive in environments that appear to be physically unstressful.

Another condition favoring the evolution of diapause for predator avoidance ought to be seasonal predictability of the catastrophe date. The more repeatable the arrival of predation, the simpler will be the adapted response. However, the studies of *D. sanguineus* just presented show that predation need not be perfectly predictable so long as it can be characterized by a stable mean and variance. The consistency of the switch date in Bullhead Pond and Pond C (figs. 18.3 and 18.4) suggests that it is physiologically cued by some highly repeated environmental phenomenon such as day length. Laboratory studies in controlled environments support this conclusion (Hairston and Olds 1986). The constancy of the distribution of dates when predation increases permits the copepods to use a correlated environmental cue as a reliable predictor of the future catastrophe. Annual seasonality is undoubtedly the most common cycle with which dormancy is associated, but it is possible to envision others that might confer predictability to catastrophes, such as the lunar cycle (e.g., Gliwicz 1986) or its associated tidal cycle.

It is not, however, a basic requirement that catastrophes be predictable for diapause to evolve. Strickler and Twombly (1975) suggest that the prey might be able to detect the presence of predators directly and respond by diapausing. They cite the induction of prey defences in rotifers by water-soluble substances released from predators (Gilbert 1966), and crustacean examples of this phenomenon have since been published (Grant and Bayly, 1981; Krueger and Dodson 1981), though not of diapause induction. Were the onset of predation unpredictable, initiating diapause when the predators are detected might still be a viable strategy for avoiding extinction, especially if predation pressure increases gradually, providing a lag period between the detection of predators and catastrophic mortality. It is, incidentally, difficult to envision such an explanation for the timing of diapause by *D. sanguineus* in either Bullhead Pond or Pond C since the switch from subitaneous eggs occurs more than a generation before predator activity begins to increase. Furthermore, Bullhead Pond copepods transferred to a pond containing no fish apparently could not detect the absence of these predators and continued to begin production of diapausing eggs, on schedule, in late March (Hairston and Olds 1984).

Setting aside predator detection, Hairston et al. (1985) have shown that at high levels of year-to-year variability in catastrophe date, the ESS for diapause is for females to lay both subitaneous and diapausing eggs regardless of the time of year. The example they use is a population of *D. sanguineus* living in a temporary pond in which desiccation determined by monthly rainfall is the catastrophe. The strategy should, however, work for any highly unpredictable source of mortality. Such a "bet-hedging" trait could evolve without either a correlated or a direct cue of predator activity. For *D. sanguineus*, there is an interesting developmental constraint on the expression of bet-hedging because females can make clutches of only one type of egg at a time (Hairston and Olds 1984). Thus, subitaneous and diapausing eggs must be produced sequentially rather than together. Which type of egg should be produced first appears, from the ESS model, to depend on the amount of variability in the catastrophe date.

One final point is that the best time to begin diapause may depend on the evolutionary perspective of the individual organism involved. The ESS for a female faced with a catastrophe, unpredictable in time but certain to occur at some point, may be to make both subitaneous and diapausing eggs because in this way she is represented regardless of what happens. However, for her individual offspring it would be better to begin diapause before the earliest possible catastrophe. For this reason there may be parent–offspring conflict (*sensu* Trivers 1974) for control of the diapause trait. In *D. sanguineus*, where diapausing eggs are provisioned and given a resistant shell by their mother, the parent is clearly in control. But in cyclopoid copepods, diapause occurs in late immature stages and is likely controlled by the individual rather than by its parent. For these animals a bet-hedging strategy would not be expected.

The range of ways that diapause can be imagined to evolve in response to periodic catastrophes is large. From the examples discussed here, it is clear that its expression in an animal's life history will depend in part on the nature and variability of the biotic and abiotic environment in which the animal resides and in part on the animal's inherent developmental and physiological properties. There are at present no general rules that would help decide, for any particular example, whether predation has been involved as a selective force. Without shortcuts, each discovery of a diapausing population will require extensive individual investigation in order to understand the origin and maintenance of this trait.

ACKNOWLEDGMENTS

E. J. Olds, W. R. Munns, Jr., W. E. Walton, and K. L. Van Alstyne helped with the research in many ways. W. C. Kerfoot and F. Taylor made helpful comments on the manuscript. G. H. Wheatley provided much appreciated cooperation at the W. Alton Jones Research Campus, and R. P. Clark and J. E. O'Brien allowed access to Bullhead and Little Bullhead Ponds. Fish were collected under Rhode Island Division of Fish and Wildlife permit 84-29. The research was funded by N.S.F. grants DEB-8010678 and BSR-8307350. I am particularly grateful to the U.R.I. Foundation for support during the critical summer of 1983.

REFERENCES

Beck, S. D. 1980. *Insect photoperiodism.* 2d. ed. New York: Academic Press.

Cohen, D. 1970. A theoretical model for the optimal timing of diapause. *Amer. Naturalist* 104 : 389–400.

Fedorenko, A. Y. 1975. Feeding characteristics and predation impact of *Chaoborus* (Diptera, Chaoboridae) larvae in a small lake. *Limnol. Oceanogr.* 20 : 250–58.

Flint, A. P. F., M. B. Renfree, and B. J. Weir. 1981. Embryonic diapause in mammals. *J. Reprod. Fertil.* (Suppl. 29).

Fryxell, G. A. 1983. Preface. In G. A. Fryxell (ed.), *Survival strategies of the algae*, pp. vii–x. Cambridge: Cambridge University Press.

Gilbert, J. J. 1966. Rotifer ecology and embryological induction. *Science* 151 : 1234–37.

Gliwicz, Z. M. 1986. A lunar cycle in zooplankton. *Ecology* 67 : 883–97.

Gould, S. J., and R. C. Lewontin. 1979. The spandrels of San Marco and the Panglossian paradigm: a critique of the adaptationist programme. *Proc. R. Soc. Lond.* [Biol.] 205 : 581–98.

Grant, J. W. G., and I. A. E. Bayly. 1981. Predator induction of crests in morphs of the *Daphnia carinata* King complex. *Limnol. Oceanogr.* 26 : 201–18.

Hairston, N. G., Jr., and W. R. Munns, Jr. 1984. The timing of copepod diapause as an evolutionarily stable strategy. *Amer. Naturalist* 123 : 733–51.

Hairston, N. G., Jr., and E. J. Olds. 1984. Population differences in the timing of diapause: Adaptation in a spatially heterogeneous environment. *Oecologia* 61 : 42–48.

Hairston, N. G., Jr., E. J. Olds, and W. R. Munns, Jr. 1985. Bet-hedging and environmentally cued diapause strategies of diaptomid copepods. *Int. Ver. Theor. Angew. Limnol. Verh.* 22 : 3170–77.

Hairston, N. G., Jr., and E. J. Olds. 1986. Partial photo-periodic control of diapause in three populations of the freshwater copepod *Diaptomus sanguineus. Biol. Bull.* 171 : 135–42.

Hairston, N. G., Jr., and S. Twombly. 1985. Obtaining life table data from cohort analyses: a critique of current methods. *Limnol. Oceanogr.* 30 : 886–93.

Hairston, N. G., Jr., W. E. Walton, and K. T. Li. 1983. The causes and consequences of sex-specific mortality in a freshwater copepod. *Limnol. Oceanogr.* 28 : 935–47.

Hairston, N. G., Jr., and W. E. Walton. 1986. Rapid evolution of a life history trait. *Proc. Nat. Acad. Sci. (U.S.A.)* 83 : 4831–33.

Hargraves, P. E., and F. W. French. 1983. Diatom resting spores: significance and strategies. In G. A. Fryxell (ed.), *Survival strategies of the algae,* pp. 49–68. Cambridge: Cambridge University Press.

Hutchinson, G. E. 1967. *A treatise on limnology,* vol. 2. New York: John Wiley.

Krueger, D. A., and S. I. Dodson. 1981. Embryological induction and predation ecology in *Daphnia pulex. Limnol. Oceanogr.* 26 : 219–23.

Levins, R. 1969. Dormancy as an adaptive strategy. *Symp. Soc. Exp. Biol.* 23 : 1–10.

Nilssen, J. P. 1977. Cryptic predation and the demographic strategy of two limnetic cyclopoid copepods. *Mem. Ist. Ital. Idrobiol.* 34 : 187–96.

———. 1980. When and how to reproduce: a dilemma for limnetic cyclopoid copepods. In W. C. Kerfoot (ed.), *Evolution and ecology of zooplankton communities,* pp. 418–26. Amer. Soc. Limnol. Oceanogr. Spec. Symp. Vol. 3. Hanover, N.H.: University Press of New England.

Nilssen, J. P., and K. Elgmork. 1977. *Cyclops abyssorum*—life cycle dynamics and habitat selection. *Mem. Ist. Ital. Idrobiol.* 34 : 197–238.

Rigler, F. H., and J. M. Cooley. 1974. The use of field data to derive population statistics of multivoltine copepods. *Limnol. Oceanogr.* 19 : 636–55.

Sangren, C. D. 1983. Survival strategies of chrysophycean flagellates: reproduction and the formation of resistant resting cysts. In G. A. Fryxell (ed.), *Survival strategies of the algae,* pp. 23–48. Cambridge: Cambridge University Press.

Slobodkin, L. B. 1954. Population dynamics in *Daphnia obtusa* Kurz. *Ecol. Monogr.* 24 : 69–88.

Southwood, T. R. E. 1978. Escape in space and time—concluding remarks. In H. Dingle (ed.), *Evolution of insect migration and diapause,* pp. 277–79. New York: Springer.

Strickler, J. R., and S. Twombly. 1975. Reynolds number, diapause and predatory copepods. *Int. Ver. Theor. Angew. Limnol. Verh.* 19 : 2943–50.

Taylor, F. 1980. Optimal switching to diapause in relation to the onset of winter. *Theor. Popul. Biol.* 18 : 125–33.

Trivers, R. L. 1974. Parent-offspring conflict. *Amer. Zool.* 14 : 249–64.

Wareing, P. F., and I. D. J. Phillips. 1970. The control of growth and differentiation in plants. New York: Pergamon.

Wiggins, G. B., R. J. Mackay, and I. M. Smith. 1980. Evolutionary and ecological strategies of animals in annual temporary pools. *Arch. Hydrobiol.* (Suppl.) 58 : 97–206.

Young, A. M. 1982. Population biology of tropical insects. New York: Plenum.

19. Vertical Migration of Freshwater Zooplankton: Indirect Effects of Vertebrate Predators on Algal Communities

Winfried Lampert

My collaborators and I tested the hypothesis that fish predation has indirect effects on algal communities by causing diel vertical migrations of grazing zooplankton. Evidence for fish predation as the driving force behind vertical migration in freshwater zooplankton is provided by comparison of two very similar *Daphnia* species in the field and in the laboratory. In Lake Constance *D. hyalina* performs large diel vertical migrations, whereas *D. galeata* stays near the surface. Despite the fact that *D. galeata* carries more eggs and has much higher birth rates, *D. hyalina* is more abundant in the lake. Laboratory experiments with simulated migration or nonmigration conditions showed that both species grew better and had higher reproductive rates when subjected to continuous surface conditions, compared to the food and temperature pattern experienced by migrating animals. These results, when combined with laboratory tests of the assumptions of the "metabolic advantage hypothesis" and stomach analyses of fish in the lake, suggest that vertical migration is a strategy to reduce mortality caused by fish predation.

In situ measurements of epilimnion grazing in a Holstein lake showed considerable fluctuation of values between day and night. These fluctuations were closely related to the changes in zooplankton biomass and thus were a consequence of the vertical migration. Since community grazing rates per unit biomass are not elevated during the night, vertical migration leads to significantly reduced grazing losses in the phytoplankton. However, a mathematical model shows that even if the zooplankton consume the same total amount of algae per day, regardless of whether they migrate or not, the nocturnal harvesting will result in enhanced algal biomass production.

Since grazer species that feed on different parts of the particle spectrum may exhibit different migration patterns, the effect of vertical migration has not only a quantitative but also a qualitative component and thus may influence the phytoplankton species composition.

Diel vertical migrations of zooplankton are a widespread phenomenon in both freshwater and marine habitats (Hutchinson 1967). Therefore, they have attracted the attention of researchers for many years. Besides the study of the proximate causes (e.g., control of migration by light; Siebeck 1960; Bainbridge 1961; Ringelberg 1964; McNaught and Hasler 1964; Enright and Hamner 1967), many investigators have focused on the ultimate causes of migration behavior. This is of special interest because the diel movement from the warm, food-rich surface layers to the cold nutritionally poor deep waters during daytime seems, at first sight, to be energetically disadvantageous for a variety of reasons: (1) the energy intake of the animals is reduced; (2) swimming up and down needs additional energy; and (3) the rate of development of the eggs carried by migrating animals is reduced because of low temperature. Nevertheless, there must be a selective advantage of the migration behavior because it has evolved in so many taxa.

Much less attention has been paid to a third aspect of the vertical migration: its impact on other components of the planktonic community. Diel vertical movements result in significant changes in zooplankton biomass and thus in changes of the effects related to zooplankton activity, such as grazing (Haney 1973; Crumpton and Wetzel 1982; Lampert and Taylor 1985), invertebrate predation (Ohman et al. 1983), and the excretion of nutrients (Eppley et al. 1967; Wright and Shapiro 1984). If migrating

291

herbivorous zooplankton harvest the epilimnetic algal crop only at night, phytoplankton can grow unimpeded by grazing during the light period of the day. So any selective force causing diel vertical migration indirectly also induces diel fluctuations of grazing pressure.

The hypotheses about the selective advantage of vertical migration for the zooplankton fall into two categories. They postulate either a metabolic or demographic advantage of migrating animals over nonmigrating ones (McLaren 1963, 1974; Enright 1977) or an advantage by avoidance of mortality through optically orientating predators (Zaret and Suffern 1976; Wright et al. 1980). Provided that the latter is really a driving force for vertical migration, the consequences of rhythmic changes of the grazing pressure for the algal community are an indirect effect of predation. In this contribution I will test the following hypotheses: (1) that vertical migration is a consequence of fish predation, (2) that the grazing intensity *in situ* is mainly a function of zooplankton biomass and thus related to the biomass changes, and (3) that rhythmic harvesting enhances the algal production, even if the migrating zooplankton consume the same total algal biomass per day as they would do under continuous feeding.

PREDATOR AVOIDANCE AS THE ULTIMATE REASON FOR VERTICAL MIGRATION

Although there is growing evidence that fish predation is the selective force for the evolution of diel vertical migration (Zaret and Suffern 1976; Wright et al. 1980; Ohman et al. 1983), this hypothesis is notoriously difficult to test directly because of the lack of controls.

Some assumptions of alternate hypotheses, however, can be tested in the laboratory. The model of Enright (1977), for example, predicts a metabolic advantage for animals that feed predominantly at the end of the day, when the algae contain large amounts of storage products. It is based on the observation that zooplankton after some time of starvation show increased feeding rates. We were able to demonstrate in the laboratory that hungry daphnids exposed to high food concentrations indeed exhibit extraordinarily high feeding rates. But contrary to the observations of Runge (1980) for marine copepods, this effect

cannot be seen below the incipient limiting level concentration in cladocerans, as would be required by Enright's model (Lampert et al., in preparation). Moreover, the starvation effect lasts only for a short time period, so that the starvation losses cannot be compensated for by prolonged "overshoot" feeding. Thus, the metabolic model of Enright (1977) cannot be valid for migrating daphnids.

Orcutt and Porter (1983) tested the effects of fluctuating temperature on life history parameters of *Daphnia parvula*. They found that spending part of the day in cool water resulted in lower realized rates of increase (r) for the daphnids, and so they rejected the hypothesis of McLaren (1974), which predicts a demographic advantage of vertical migration.

The predation hypothesis generates some predictions that can be tested (Wright et al. 1980), but the ideal test would require a study of reproductive rates and mortality losses in the same population under migratory and nonmigratory conditions. These exact specifications are probably impossible to attain. However, in his study of the vertical migration of zooplankton in Lake Constance, Stich (1985) was lucky to find an example that nearly fulfills the requirements. Among the various migration patterns of zooplankton, he observed two very similar *Daphnia* species that differed with respect to their diel vertical migration (Stich and Lampert 1981). The two species, *D. galeata* and *D. hyalina*, are the same size. They are also sometimes treated together as the "*longispina* group" because they are so similar. Moreover, morphologic intermediates frequently occur in Baltic lakes. Wolf and Mort (1986) have recently demonstrated by electrophoretic analysis that the morphologically intermediate specimens that live together with the "pure" types in some Holstein lakes are, in fact, hybrids.

D. galeata and *D. hyalina* coexist in Lake Constance, but *D. hyalina* performs pronounced diel vertical migrations with a maximum amplitude of up to 50 m, whereas *D. galeata* stays near the surface all day. Figure 19.1 clearly illustrates the different behavior of the two species. There is very little overlap between the two populations at noon but nearly complete overlap at midnight. However, the figure gives only one example (August) of a whole year cycle, each month with its own profile. Vertical migration of *D. hyalina* is not

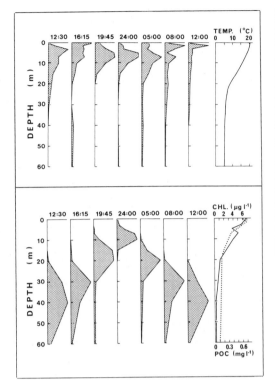

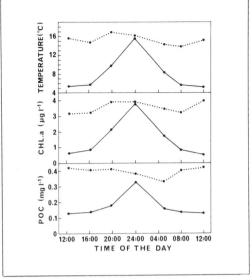

Figure 19.2. *Diel changes of the population-weighted means of environmental parameters for the example presented in fig. 19.1 Solid line, D. hyalina; dotted line, D. galeata.*

Figure 19.1. *Relative distributions of* Daphnia galeata **(upper panel)** *and* D. hyalina **(lower panel)** *in Lake Constance, August 1977. Time of the day indicated at the top of each profile. Vertical profiles of important environmental parameters are displayed at the right side. Concentrations of "edible" (< 30 μm) particles are given as chlorophyll a (solid line) and particulate carbon (dotted line). Data from Stich (1985).*

constant all year round. It starts in June, and its amplitude increases until October (Stich and Lampert 1981). The amplitude is reduced in November, and migration ceases during winter.

Figure 19.1 also presents the vertical profiles of two important environmental parameters: temperature and food. *D. galeata* stays in the warm epilimnion all day, whereas *D. hyalina* is in the cold water below the thermocline during the larger part of the day. *D. galeata* is, moreover, continuously exposed to good food conditions, but *D. hyalina* is exposed only a few hours per day. Not only is the concentration of edible organic particles (< 30 μm) very low in the deep layers, where *D. hyalina* lives during daytime, the quality of the food is also very poor because the particles are mainly detritus, as can

be seen from the low chlorophyll : carbon ratio (fig. 19.1). The diel variations of the weighted averages of these environmental parameters, as experienced by the populations, are summarized in figure 19.2. The differences between species caused by the differential migration behavior are striking. *D. hyalina* spends most of its time at very low temperatures, probably starving because of the low concentration and quality of the food particles.

This circumstance allows us to test whether the migrating species has any metabolic or demographic advantage over the nonmigrating one. The data indicate that a direct advantage has to be rejected. The nonmigrating *D. galeata* carries more eggs as a consequence of the better food availability (cf. Lampert 1978), and these eggs develop faster because of higher temperature. Our observations (Stich and Lampert 1981) were confirmed by Geller (1985), who calculated the total production of the two species. This means that the hypotheses based on metabolic or demographic advantages cannot be valid for *Daphnia* species.

Table 19.1 summarizes the consequences of vertical migration for the example presented in figures 19.1 and 19.2. Because of the higher egg numbers and shorter developmental times of

Table 19.1. *Environmental and population dynamic parameters for the two* Daphnia *populations in figure 19.1.*

	D. galeata	D. hyalina
Mean concentration of particles <30 μm (mg C/L)	0.34	0.16
Mean temperature (°C)	14.2	7.1
Number of eggs per adult	7.1	3.7
Egg development time (days)	8.8	14.5
Individuals per m²		
August	38,200	91,500
September	154,800	272,200
Birth rate (b) (day⁻¹)	0.147	0.055
Rate of increase (r) (day⁻¹)	0.047	0.036
Death rate (d) (day⁻¹)	0.100	0.019

the eggs, *D. galeata* has a much higher birth rate than *D. hyalina*. Without mortality this would lead to a numerical dominance of *D. galeata* in a short period of time. However, as long as the vertical migration takes place, *D. hyalina* is more abundant. From August to September, for example, both species increase in numbers. The realized rate of increase (*r*) is slightly higher in *D. galeata*, but the difference is not large enough for *D. galeata* to become the dominant species. The death rate (calculated as the difference of *b* and *d*) indicates that *D. galeata* suffers from a much higher mortality than *D. hyalina*, i.e., that the vertical migration reduces mortality. In fact, stomach analyses of pelagic whitefish and perch showed that the fish consumed 90% *D. galeata*, although *D. hyalina* was more abundant in the water column.

There remains, however, the problem that *D. hyalina* might be cold-stenothermic and thus could not exist continuously at the warm surface temperatures. Moreover, the question arises why *D. galeata* does not migrate. We addressed these questions in a series of laboratory experiments in which we simulated the conditions of migration and nonmigration (Stich and Lampert 1984). Both species were kept together in a flow-through system either at continuous high temperature (20°C) and high food concentration (1 mg C/L) (nonmigration) or at simultaneously fluctuating food and temperature [high food and high temperature at night, low food and low temperature during daytime] (migration).

Both species grew much faster and had more offspring under nonmigration conditions than under migration conditions. Thus, *D. hyalina* is not cold-stenothermic and would reach a higher birth rate in the lake if it were to stay near the surface. With respect to temperature, Orcutt and Porter (1983) found the same effect in *D. parvula*. Under strongly fluctuating conditions, both species showed retarded growth and reduced reproductive rates. The reduction, however, was more pronounced in *D. galeata*; *D. hyalina* could persist at low fluctuating food conditions, whereas the birth rate of *D. galeata* was too low to compensate for natural mortality. We therefore conclude that *D. galeata* is not able to adopt the migration strategy in the deep mesotrophic Lake Constance. It must stay near the surface where it suffers high mortality. When food is more abundant, both species are able to migrate vertically.

These field and laboratory results provide strong evidence that the avoidance of fish predation near the surface is the ultimate cause of diel vertical migration, at least in *Daphnia* species.

FLUCTUATIONS OF THE GRAZING PRESSURE AS A CONSEQUENCE OF VERTICAL MIGRATION

Diel changes in the herbivore biomass in the epilimnion must cause fluctuations of grazing pressure (Bowers 1979). Such day and night differences have been observed (Haney 1973; Haney and Hall 1975; Redfield 1980; Crumpton and Wetzel 1982). Increased night values of grazing, however, have been found not only as a consequence of the shift of grazer biomass but also of the increased filtering activity of the individuals (Haney and Hall 1975; Chisholm et al. 1975; Duval and Geen 1976). We studied the vertical distribution of herbivore biomass and community grazing at day and night over a whole year cycle in Schöhsee, a moderately eutrophic lake in Holstein, North Germany (Lampert and Taylor 1984, 1985). Thereby we measured the effect of the vertical migration on algal loss rates directly *in situ*.

Measurements were carried out using a modified method of Haney (1971), but instead of one radioactive tracer alga we used two different-size particles, the green alga *Scenedesmus acutus* (10 μm) labeled with ^{14}C and the small blue-green *Synechococcus elongatus* (1 μm) labeled with ^{32}P. Both particles were offered simultaneously (Lampert 1974) so that we were able to measure grazing rates as well as different intensities of grazing on particles of different sizes (for detailed methods see Lampert and Taylor 1985).

Typical day and night patterns are presented in figure 19.3. This series was made in late May when the grazing rate was maximal (clearwater phase) but vertical migration had just started. More examples from other seasons with and without vertical migration are given in Lampert and Taylor (1984, 1985). Figure 19.3 suggests a strong correlation between the local grazing rate and the zooplankton biomass. Grazing is very low at the surface during the day but very high at night. The depth profile closely resembled the biomass distribution, with the exception of the greatest depths during the day, where the grazing rate was lower than expected from the biomass. The lower grazing rates were probably due to the low temperature in the deep layers. Grazing rates for the very small *Synechococcus* were always lower than for the medium-size nanoplanktonic *Scenedesmus*.

We tried to correlate grazing rates measured during 12 day–night series in 1982–1983 with temperature, particulate carbon (<30 μm), and zooplankton biomass. The only significant correlation was found with the biomass. The effects of temperature and food concentration probably compensate each other, because they are highly correlated (high temperature and high particle concentration at the surface, low values of both in the depth). Increasing temperature, however, will stimulate the individual filtering rate, but increased food concentration will reduce it. The close correlation between zooplankton biomass and community grazing rate is clearly demonstrated in figure 19.4. This graph presents the average grazing rates and biomasses in the upper 5 m of Schöhsee during day and night for the 12 series at all seasons. Zooplankton biomass was the overwhelming factor that determined the community grazing rate.

We also compared the day and night values of community grazing per unit of zooplankton biomass and could not find a difference between the mass specific rates in the light and in the dark. This also is shown in figure 19.4; day

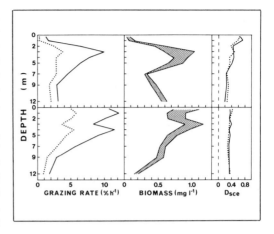

Figure 19.3. *Vertical profiles of grazing rate and zooplankton biomass in Schöhsee, May 20, 1983, during day* (**upper panel**) *and at night* (**lower panel**). *Grazing rates are shown for two different-size tracer algae, Scenedesmus (solid line) and Synechococcus (dotted line). Biomasses separated into Daphnia (white area) and Eudiaptomus (shaded area). Electivity index calculated for Scenedesmus* (D_{sce}) *from grazing rates (solid line) and predicted from the relative composition of the zooplankton biomass (dotted line) (see text).*

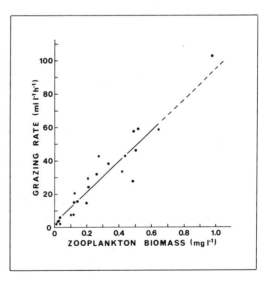

Figure 19.4. *Relationship between grazing rates (Scenedesmus) and zooplankton biomass in Schöhsee. Points represent average values for the uppermost 5 m of the lake on 12 dates during 1982 and 1983. Dots, night values; diamonds, day values. The regression (Y = 3.1 + 91.8 X; r = 0.926) does not include the single very high point.*

and night values are plotted with different symbols. Diel changes in the feeding activity of the individuals therefore are not important for the community grazing rates. Our data do not exclude the possibility that such nighttime increases of the feeding activity occur in certain components of the zooplankton community (e.g., in large daphnids; Haney 1985), but though this effect may be very important for the particular animal, its significance for algal loss rates is probably negligible in Schöhsee (Haney, personal communication).

Large differences between daytime and nighttime community grazing rates have been reported by Crumpton and Wetzel (1982). Although they found slight differences in the individual filtering rates of *D. pulex* and *D. retrocurva* between day and night, the magnitude of those changes can explain only a small portion of the diel variation in community grazing rates. Thus, zooplankton biomass changes, as a consequence of vertical migration, seem to be the overriding factor responsible for diel fluctuations of grazing losses experienced by phytoplankton.

In addition, figure 19.3 provides some information on the qualitative aspects of grazing. The selectivity index D (Jacobs 1974) indicates the relative selection of small and large particles. For *Scenedesmus* (D_{sce}), an index of $+1$ means that *Scenedesmus* is consumed exclusively; -1 would mean a complete preference for the small *Synechococcus*. Zero indicates that both particles are ingested in proportions corresponding to their relative abundance in the water. Figure 19.3 demonstrates not only that the total grazing rate varies between day and night but also that the relative grazing losses differ for small and large phytoplankton. For example, D_{sce} is homogeneous throughout the whole water column at night, indicating a slight preference for *Scenedesmus*. During the day, however, great differences exist, with high D_{sce} at the surface and low values at depth. This pattern was not consistent all year round. During July and August, when the daytime grazing rates at the surface were very low, slight negative values of D_{sce} were found in the uppermost meters, indicating a weak preference for the smaller particles.

These differences are caused by the relative contribution of different zooplankton species in the vertically migrating community. In Schöhsee, D_{sce} is mainly determined by the relative proportions of *Daphnia* and *Eudiaptomus*. Whereas the daphnids are generalists that filter a broad spectrum of particle sizes (Brendelberger 1985), the copepod is a much more selective feeder, and its feeding efficiency for very small particles is low (Muck and Lampert 1980; Vanderploeg et al. 1984). In the Schöhsee study (Lampert and Taylor 1985), we determined a D_{sce} of 0.18 for *Daphnia* and 0.88 for *Eudiaptomus* selected from the community samples. These values have been multiplied with the relative biomasses of *Daphnia* and *Eudiaptomus* in the example presented in figure 19.3 to predict the community D_{sce}. In fact, the predictions come close to the *in situ* measurements, indicating that the differences result from differences in the migration behavior of the species.

Diel vertical migrations of zooplankton, therefore, induce not only quantitative but also qualitative changes of the grazing pressure in the euphotic zone.

CONSEQUENCES OF DISCONTINUOUS GRAZING PATTERNS

Phytoplankton abundance is the result of growth and loss processes (for review see Reynolds 1984). Grazing contributes to the loss processes, but it has been suggested (Petipa and Makarova 1969; McAllister 1969, 1970) that the rhythm of particle elimination is important as well as the intensity. These authors constructed mathematical models to demonstrate that the production of phytoplankton is higher under a nocturnal feeding regime than under one of continuous grazing. McAllister (1970) also reported an experiment with different grazing patterns. We know that discontinuous nocturnal grazing is produced by the diel vertical migration of zooplankton, for during summer months approximately 80% of the epilimnetic grazing in Schöhsee takes place at night (Lampert and Taylor 1985).

The effect of vertical migration can be estimated if we compare algal production and loss rates for continuous and discontinuous grazing. There are three possible cases with respect to the feeding behavior of the zooplankton in the epilimnion:

1. Zooplankton do not migrate. They maintain a constant feeding rate per unit biomass all

day. Thus, grazing losses are uniform over the day.

2. Zooplankton maintain a constant grazing rate per unit biomass but perform diel vertical migrations. Grazing losses of phytoplankton and energy input of zooplankton are greatly reduced because grazing takes place at night only.

3. Zooplankton migrate but increase the feeding rate per unit biomass at night so that the total amount of algal mass consumed is the same as with continuous feeding.

The rate of algal production must be higher in cases 2 and 3 compared with case 1 because the algae can grow unimpeded during the light phase. The net gain of algal biomass will be highest in case 2 because the rate of production is higher and the loss rate is lower than in case 1. Case 2 reflects the maximum and case 3 the minimum effect of zooplankton vertical migration on phytoplankton.

The relative importance of different parameters for the resulting net algal gain can be illustrated by means of mathematical models. Figure 19.5 presents an example of case 3. The model (W. Gabriel, unpublished data) assumes that the algae grow at a constant growth rate during day hours and that grazing is the only loss factor. Zooplankton consume the same amount of algal biomass (as a fraction of the daily primary production) either distributed over the whole day or concentrated during the night hours. The figure displays the difference in algal biomass change for the two grazing patterns.

The outcome depends on the relative day length, the algal growth rate, and the rate of consumption. Parameters have been chosen to cover a realistic range for summer conditions in Schöhsee. The difference increases with increasing day length. Only one example (relative day length = 0.7, approximately 17 hours) is presented. Significant differences occur when the consumption rate equals or exceeds the rate of primary production (fig. 19.5 c,d). There are situations where the diel pattern of zooplankton grazing has a striking effect, although the total amount of algae harvested is the same. With the combination of an algal growth rate of 1.0/day and a relative consumption rate of 1.5, the algal biomass would decrease by approximately 50% under continuous grazing but would increase by 25% if algal losses were

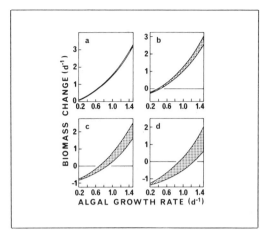

Figure 19.5. *Model calculations of the effect of different diel grazing patterns on phytoplankton net production. The model assumes that the total algal biomass consumed by the zooplankton per day is constant but that grazing takes place either continuously (no vertical migration) or only at night (with vertical migration). The shaded area indicates the difference in the relative change of algal biomass for the two grazing patterns. The relative day length is 0.7 in all examples. Examples a–d represent increasing amounts of phytoplankton consumed per day (as a fraction of unaffected primary production; a =0.2, b = 0.5, c = 1.0, d = 1.5).*

concentrated at night. Neither rate is unrealistic because algal growth rates of 1.0/day can be found in nature (Reynolds 1984), and average grazing rates of 170% per day have been measured in Schöhsee (Lampert and Taylor 1985).

As mentioned above, particles of different sizes may be affected differently by vertical migration. It is interesting to note in this respect that the algal growth rate influences the outcome of the model prediction. Small algae are usually considered to have higher growth rates than larger ones. They would, therefore, gain more profit from a discontinuous grazing scheme. As a result, the size structure of an algal community may shift to small particles. Algal communities can respond to herbivore grazing pressure by increasing the share of either fast-growing or inedible algae (Porter 1977). Vertical migration of the grazers will favor the small, fast-growing, edible forms, especially when the total grazing pressure is reduced as in case 2 above.

The model presented here calculates only the minimum advantage of the phytoplankton from vertical migration of grazers. This effect

may be balanced by reduced nutrient regeneration under strongly nutrient limited conditions
(Lampert, personal observations). In fact, our
Schöhsee study showed that community grazing rates per unit of biomass were not increased
at night. While this is not really a rigorous proof
because we were not able to compare the same
animals under surface conditions during day
and night (actually, we compared migrating
with nonmigrating animals), case 2 is more likely to be realistic in Schöhsee. Therefore, the
algal advantage is probably much higher than
the model indicates.

Fish predation can affect an algal community
in different ways. It may induce a diel vertical
migration or, if migration is not possible,
change the zooplankton community to smaller
or more evasive forms. Both effects go in the
same direction: enhancement of algal productivity and abundance. Which of these mechanisms has more influence on the algal community cannot be determined *a priori* because
feedbacks like nutrient regeneration, responses
of the algal species composition, and effects on
zooplankton fecundity have to be considered.
More complex models and experimental tests
are needed to understand fully the indirect effects of fish predation on phytoplankton communities.

ACKNOWLEDGMENTS

This paper summarizes the work of several
members of the former AG Planktonökologie.
I am especially grateful to H.-B. Stich who provided an excellent set of data on the vertical
migration in Lake Constance. B. E. Taylor
participated in the *in situ* grazing determina-
tions in Schöhsee. W. Gabriel kindly provided
some of his model results. I also thank W. C.
Kerfoot for his careful review and improvement of the English text.

REFERENCES

Bainbridge, R. 1961. Migrations. In T. H. Water-
man (ed.), *The physiology of Crustacea*, vol. 2, pp.
431–63. New York: Academic Press.

Bowers, J. A. 1979. Zooplankton grazing in simulation models: the role of vertical migration. In D.
Scavia and A. Robertson (eds.), *Perspectives in lake
ecosystem modelling*, pp. 53–73. Ann Arbor, MI:
Ann Arbor Science.

Chisholm, S. W., R. G. Stross, and P. A. Nobbs,
1975. Environmental and intrinsic control of filtering and feeding rates in arctic *Daphnia*. *J. Fish.
Res. Board Can.* 32 : 219–26.

Crumpton, W. G., and R. G. Wetzel. 1982. Effects
of differential growth and mortality in the seasonal succession of phytoplankton populations in
Lawrence Lake, Michigan. *Ecology* 63 : 1729–39.

Duval, W. S., and G. H. Geen. 1976. Diel feeding
and respiration rhythms in zooplankton. *Limnol.
Oceanogr.* 21 : 823–29.

Enright, J. T. 1977. Diurnal vertical migration:
Adaptive significance and timing. Part 1. Selective
advantage: A metabolic model. *Limnol. Oceanogr.*
22 : 856–72.

Enright, J. T., and W. M. Hamner. 1967. Vertical
diurnal migration and endogenous rhythmicity.
Science 157 : 937–41.

Eppley, R. W., E. H. Renger, E. L. Venrick, and
M. M. Muller. 1967. A study of plankton dynamics and nutrient cycling in the central gyre of
the North Pacific Ocean. *Limnol. Oceanogr.*
12 : 685–95.

Geller, W. 1985. Production, food utilization and
losses of two coexisting, ecologically different
Daphnia species. *Arch. Hydrobiol. Beih. Ergebn.
Limnol.* 21 : 67–79.

Haney, J. F. 1971. An in situ method for the measurement of zooplankton grazing rates. *Limnol.
Oceanogr.* 16 : 970–77.

———. 1973. An in situ examination of the grazing
activities of natural zooplankton communities.
Arch. Hydrobiol. 72 : 87–132.

———. 1985. Regulation of cladoceran filtering
rates in nature by body size, food concentration,
and diel feeding pattern. *Limnol. Oceanogr.*
30 : 379–411.

Haney, J. F., and D. J. Hall. 1975. Diel vertical
migration and filter-feeding activities of *Daphnia*.
Arch. Hydrobiol. 75 : 413–41.

Hutchinson, G. E. 1967. *A treatise on limnology. Vol.
2. Introduction to lake biology and the limnoplankton.*
New York: John Wiley and Sons.

Lampert, W. 1974. A method for determining food
selection by zooplankton. *Limnol. Oceanogr.*
19 : 995–98.

———. 1978. A field study on the dependence of
the fecundity of *Daphnia* spec. on food concentration. *Oecologia* 36 : 363–69.

Lampert, W., and B. E. Taylor. 1984. In situ grazing
rates and particle selection by zooplankton: Effects of vertical migration. *Int. Ver. Theor. Angew.
Limnol. Verh.* 22 : 943–46.

———. 1985. Zooplankton grazing in a eutrophic
lake: Implications of vertical migration. *Ecology*
66 : 68–82.

McAllister, C. D. 1969. Aspects of estimating zooplankton production from phytoplankton pro-

duction. *J. Fish. Res. Board. Can.* 26 : 199–220.

———. 1970. Zooplankton rations, phytoplankton mortality and the estimation of marine production. In J. H. Steele (ed.), *Marine food chains*, pp. 419–57. Berkeley, Calif.: University of California Press.

McLaren, I. A. 1963. Effects of temperature on growth of zooplankton and the adaptive value of vertical migration. *J. Fish. Res. Board Can.* 20 : 685–727.

———. 1974. Demographic strategy of vertical migration by a marine copepod. *Amer. Naturalist* 108 : 91–102.

McNaught, D. C., and A. D. Hasler. 1964. Rate of movement of populations of *Daphnia* in relation to changes in light intensity. *J. Fish. Res. Board Can.* 21 : 291–318.

Muck, P., and W. Lampert. 1980. Feeding of freshwater filter-feeders at very low food concentrations: Poor evidence for "threshold feeding" and "optimal foraging" in *Daphnia longispina* and *Eudiaptomus gracilis*. *J. Plankton Res.* 2 : 367–79.

Ohman, M. D., B. W. Frost, and E. H. Cohen. 1983. Reverse diel vertical migration—an escape from invertebrate predators. *Science* 220 : 1404–1407.

Orcutt, J. D. Jr., and K. G. Porter. 1983. Diel vertical migration by zooplankton: constant and fluctuating temperature effects on life history parameters of *Daphnia*. *Limnol. Oceanogr.* 28 : 720–30.

Petipa, T. S., and N. P. Makarova. 1969. Dependence of phytoplankton production on rhythm and rate of elimination. *Mar. Biol.* 3 : 191–95.

Porter, K. G. 1977. The plant-animal interface in freshwater ecosystems. *American Scientist* 65 : 159–70.

Redfield, G. W. 1980. The effect of zooplankton on phytoplankton productivity in the epilimnion of a subalpine lake. *Hydrobiologia* 70 : 217–24.

Reynolds, C. S. 1984. *The ecology of freshwater phytoplankton*. Cambridge: Cambridge University Press.

Ringelberg, J. 1964. The positively phototactic reaction of *Daphnia magna* Straus: A contribution to the understanding of diurnal vertical migration. *Neth. J. Sea Res.* 2 : 319–406.

Runge, J. A. 1980. Effects of hunger and season on the feeding behavior of *Calanus pacificus*. *Limnol. Oceanogr.* 25 : 134–45.

Siebeck, O. 1960. Untersuchungen über die Vertikalwanderungen planktischer Crustaceen unter besonderer Berücksichtigung der Strahlungsverhältnisse. *Int. Rev. Gesamten Hydrobiol.* 45 : 381–454.

Stich, H.-B. 1985. Untersuchungen zur tagesperiodischen Vertikalwanderung planktischer Crustaceen im Bodensee. Ph.D. thesis, University of Freiburg (Brsg.).

Stich, H.-B., and W. Lampert. 1981. Predator evasion as an explanation of diurnal vertical migration by zooplankton. *Nature* 293 : 396–98.

———. 1984. Growth and reproduction of migrating and non-migrating *Daphnia* species under simulated food and temperature conditions of diurnal vertical migration. *Oecologia* 61 : 192–96.

Vanderploeg, H. A., D. Scavia, and J. R. Liebig. 1984. Feeding rate of *Diaptomus sicilis* and its relation to selectivity and effective food concentration in algal mixtures and in Lake Michigan. *J. Plankton Res.* 6 : 943–52.

Wolf, H. G., and M. A. Mort. 1986. Inter-specific hybridization underlies phenotypic variability in *Daphnia* populations. *Oecologia* 68 : 507–11.

Wright, D., W. J. O'Brien, and G. L. Vinyard. 1980. Adaptive value of vertical migration: A simulation model argument for the predation hypothesis. In W. C. Kerfoot (ed.), *Evolution and ecology of zooplankton communities*, pp. 138–47. Hanover, N.H.: University Press of New England.

Wright, D. I., and J. Shapiro. 1984. Nutrient reduction by biomanipulation: An unexpected phenomenon and its possible cause. *Int. Ver. Theor. Angew. Limnol. Verh.* 22 : 518–524.

Zaret, T. M., and J. S. Suffern. 1976. Vertical migration in zooplankton as a predator avoidance mechanism. *Limnol. Oceanogr.* 21 : 804–13.

20. An Experimental Analysis of Costs and Benefits of Zooplankton Aggregation

Carol L. Folt

A major consequence of small-scale heterogeneity in the composition and abundance of zooplankton is that individuals experiencing different animal abundances are likely to experience different competitive pressures and predation risks. In this chapter, I present evidence for density-dependent reductions in the feeding rates and predation risks for two species of zooplankton (*Diaptomus tyrelli* and *Epischura nevadensis*). Density-dependent reductions in feeding (both herbivory and carnivory) by zooplankton are hypothesized here to be linked to a mechanism for avoiding predators. Moreover, both mixed- and single-species aggregations of zooplankton are hypothesized to reduce predation risk from some invertebrate predators.

The outcomes of interactions among co-occurring zooplankton in Lake Tahoe, California–Nevada, were measured under a range of competitor and predator densities. The "costs" of aggregating were measured as reductions in feeding rates and survival that occurred at representative field animal densities. Reductions were caused by both exploitative and interference mechanisms but were not due directly to predation. The "benefits" of aggregating were measured as density-dependent reductions in the risk of predation (by the invertebrate predator *Mysis relicta*) and were demonstrated for both preferred and nonpreferred species of prey.

Both the costs and the benefits associated with aggregation were greatest at the highest animal abundances. This result suggests that aggregation of zooplankton may be advantageous only in predator-prone environments.

Zooplankton distributions are extremely heterogeneous, both spatially and temporally. This patchiness has been well documented for zooplankton in the ocean and in many lakes (e.g., Cassie 1959; Healy 1967; Fasham 1978; Lewis 1978; Haury, et al. 1978; Brown et al. 1979). Large changes in the densities of plankton have been found over great distances (i.e., kilometers) (Cassie 1959; Cushing and Tungate 1963) and over smaller distances from several meters (Colebrook 1960; Wiebe 1970; Tessier 1983) to centimeters (Klemetsen 1970; Byron et al. 1983). Data on the size of patches and the processes responsible for forming them have been recently reviewed (Fasham 1978; Malone and McQueen 1983). It is surprising then, that the effects of spatial heterogeneity in animal abundance and species composition on the outcome of species interactions among zooplankton have rarely been measured.

A major consequence of patchiness is that individuals differ in the number and species of animals encountered and therefore are likely to experience different competitive pressures and predation risks. In this chapter I present evidence for density-dependent reductions in grazing rates and predation risks of two species of zooplankton. I hypothesize that both mixed- and single-species aggregations of zooplankton reduce predation risk from some invertebrate predators. I postulate a behavioral mechanism that links the density-dependent reductions in grazing with predator avoidance. These arguments suggest that aggregation of herbivorous zooplankton may be advantageous only in predator-prone environments.

Advantages and disadvantages of being in groups are known for many species of vertebrates, particularly mammals, birds, and fish. If predator density remains constant, predation risk in larger prey groups is reduced because of a dilution effect (Hamilton 1971; Foster and Treherne 1981) or to a reduction in the per capita efficiency of the predator when confronted with many prey (Neill and Cullen 1974; Milinski 1977a, 1977b; Kenward 1978; Buss 1981; Treherne and Foster 1981). Prey in mixed-species groups can face a lower predation risk than conspecifics in single-species groups of the same density if another species confounds the creation of a search image, confuses the predator (Milinski 1977a, 1979; Gillett et al. 1979; Burger 1981), or if the presence of a more preferred prey causes a relaxation in the risk on the other prey species (Murdoch 1969, Murdoch et al. 1975; Horsley et al. 1979; Viser 1982).

Being in an aggregation also can affect the foraging rates of group members. Feeding rates or efficiencies may decrease because of aggression or competition for resources (Pulliam 1976; Caraco et al. 1980; Burger 1981) but foraging efficiency increases in large groups under some conditions (Pulliam 1976; Robertson et al. 1976; Kenward 1978; Pitcher et al. 1982; Pitcher and Magurran 1983). The foraging rates of individuals in mixed-species groups can be greater than those in single-species groups at the same densities if resource partitioning between species reduces competition (Burger 1981). However, some interspecific interactions such as competition and predation can severely reduce feeding rates in mixed-species groups. It is clear then that composition and abundances of individuals in a group can influence both predation risks and foraging rates.

Although interactions within groups have been studied for many vertebrates, there are far fewer studies on aquatic invertebrates. Aggregation by the ocean skater (*Halobates robustus*) reduces its predation risk (Treherne and Foster 1980, 1981, 1982; Foster and Treherne 1981). As a predator approaches, *H. robustus* become agitated; this confuses the predator, reduces its probability of attack, and increases the probability of prey escape. For the heteropteran, *Velia caprai*, the risk of being eaten by brown trout declined as prey density increased (Brönmark et al. 1984). They suggested that this reduction was caused by active avoidance of distasteful prey by the trout, a behavior that requires learning and is reinforced at high prey densities. The same mechanism was proposed as an advantage of aggregation by another group of distasteful aquatic insects (whirligig beetles) (Heinrich and Vogt 1980) that are preyed on by fish.

Milinski (Milinski 1977a, 1977b; Heller and

Milinski 1979) examined the effects of aggregation by zooplankton prey (*Daphnia magna*) on their predation risk and on the foraging behavior of predatory three-spined sticklebacks. *D. magna* in a swarm had a reduced predation risk. The magnitude of the advantage depended on the location of the individual in the swarm and the hunger level of the fish. Similarly, a reduction in predation risk was suggested as a possible advantage to marine mysids that form mixed-species schools with postlarval grunts (McFarland and Kotchian 1982).

There are cases when the individual predation risk does not necessarily decrease as prey density increases. For example, salmon that were offered krill at several prey densities consumed the krill at a higher rate in the high-density treatments (Morgan and Ritz 1984), although predator efficiency was actually lower. In such cases, predation risks could increase or decrease with prey density depending on number of prey in the patch, length of time a predator feeds in a patch, and number of predators that locate each patch (e.g., Fenwick 1978; Hassell 1978).

The effects of zooplankton density on grazing rates have rarely been measured directly. Some studies suggest that zooplankton can deplete their resources. Tessier (1983) measured both the horizontal movements of a patch of cladocerans (*Holopedium gibberum*) and the phytoplankton concentration inside and outside the patch. The phytoplankton density within the patch was reduced to half of that outside the patch. Other studies (see Malone and McQueen 1983 for data and a review) have attempted to correlate chlorophyll densities with zooplankton densities spatially but have met with little success. The lack of success in the latter studies should not be taken as sufficient evidence that grazing within an aggregation cannot lead to local resource depletion and perhaps food limitation. This is an area of research that deserves more attention and experimental manipulation.

Whereas some investigators have measured zooplankton aggregations dominated by a single species (e.g., Klemetsen 1970; Byron et al. 1983; Tessier 1983), many others have documented mixed-species assemblages (Cushing and Tungate 1963; Wiebe 1970; Fenwick 1978; Hamner and Carleton 1979; Folt et al. 1982; Malone and McQueen 1983). Frequently, these aggregations contain potential competitors, preda-

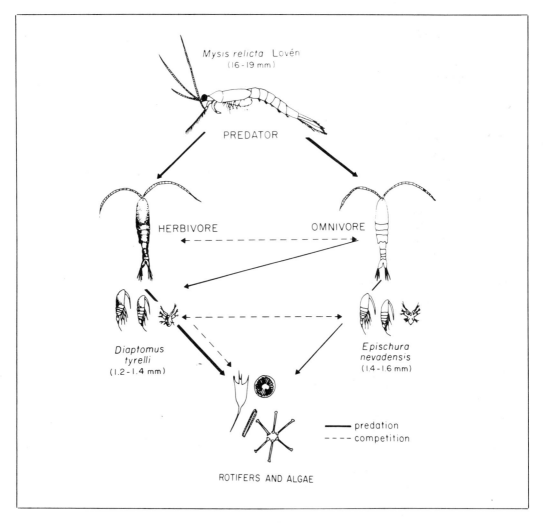

Figure 20.1. *Lake Tahoe zooplankton food web. Included are the dominant crustacean zooplankton and their feeding relationships.*

tors, and their prey so that interactions within the aggregation are likely to be very complex. For example, the mysids in the mixed-species schools discussed above (McFarland and Kotchian 1982) are prey of the fish with whom they school. Rotifers are often found in mixed assemblages with potential competitors (e.g., cladocerans) and predators (e.g., other rotifers and predatory copepods) (Malone and McQueen 1983). In many lakes during dawn and dusk vertical migrations through the thermocline, dense aggregations of zooplankton (e.g., *Daphnia*, *Chaoborus*, some copepods) form, composed of predators, prey, and potential competitors.

The identity of other individuals in close proximity within an aggregation is especially important for predator–prey and interference interactions. It has been shown for copepods that grazing algae creates a current (Strickler 1975, 1977) and possibly a chemical trail of excreted substances in a small volume surrounding an individual (Lehman and Scavia 1982). Zooplankton predators locate their prey largely by disturbances sensed mechanically at close range (Strickler 1975; Strickler and Twombly 1975), although there is evidence that they also can sense chemical signals (Poulet and Marsot 1978, 1980; Folt and Goldman 1981; Poulet and Ouellet 1982; Seitz 1984) and perhaps use them to locate food or mates (Hamner and Hamner 1977). Therefore, a prey species while foraging may produce the very signals used by its predators to locate it. In the presence of predators, the prey may either stop foraging and thereby reduce its risk, or it may continue to forage and increase its chance of being located. If food is limiting, the animal may be forced to continue feeding despite a possible increase in the risk of predation (Gillett et al. 1979; Sih 1980).

Here I examine interactions among adults of the three major species of crustaceans found in the pelagic waters of Lake Tahoe, California–Nevada. The adults of these species overlap in a narrow band near the thermocline (Rybock 1978; Folt et al. 1982). I present experimental evidence that, for these zooplankton, density and species composition in the region of overlap have strong effects on the feeding rates and risk of being killed by predators.

COMMUNITY DESCRIPTION

The three species of crustaceans in the pelagic waters of Lake Tahoe are two copepods, *Diaptomus tyrelli* (herbivore) and *Epischura nevadensis* (omnivore), and the mysid, *Mysis relicta* (predator). Relationships among these species are complex (fig. 20.1): both copepods graze a similar size range of algae, but adult *E. nevadensis* also prey primarily on the juvenile stages of *D. tyrelli* and conspecifics (Folt and Byron, unpublished MS). Adult *E. nevadensis* (1.5 mm) are slightly larger than adult *D. tyrelli* (1.3 mm). Filtering rates of *D. tyrelli* are generally two to four times greater than those for *E. nevadensis* (Folt and Goldman 1981). Since *E. nevadensis* is both a potential competitor with and predator on *D. tyrelli*, I will use the term *interference interaction* to describe their relationship. Adult mysids are 17 to 19 mm and prey on all size classes of both species of copepod (Rybock 1978; Cooper and Goldman 1980). This paper focuses on the interactions among the adults of all three species because they dominate in the aggregation near the thermocline. Smaller instars of the copepods are primarily found deeper in the water column (i.e., 30–110 m for the nauplii) (Folt et al. 1982).

The vertical and temporal distributions of these species strongly influence their potential interactions. *M. relicta* is well below the thermocline during the day; at this time there is no risk to the copepods of predation by mysids. At night, *M. relicta* migrates from below 100 m to forage in the waters near the thermocline ($\sim$ 30 m) where adult copepods are aggregated (Rybock 1978; Folt et al. 1982). Experiments presented below examine interactions among the copepods in the absence of *M. relicta* (analogous to the daytime distributions) and in the presence of *M. relicta* (analogous to the nighttime distributions).

METHODS

Several types of experiments were conducted to determine the effects of zooplankton density and species composition on individual feeding rates and predation risks. Experimental densities were within the range experienced by individuals in the lake, estimated from vertical tows (0 to 150 m) taken every 1 to 2 weeks since 1974 (Goldman 1981). Up to 90% of the large copepodids and adults in the water column are found in the upper 30 m for most of the year, and as a result the density experienced in this region is 5 to 10 times greater than the density estimated from vertical tows (Folt et al. 1982). Field densities range from 1 to 30 adults and copepodids per liter before correcting for stratification and from 1 to 230 after correction. Relative densities of *E. nevadensis* to *D. tyrelli* typically range from 0.1 to 6. The experimental densities were from 1 to 10 per 125 mls in the filtering rate experiments and from 2 to 20 per liter in the predation trials.

For all experiments, copepods were collected in vertical tows at a single location, with an 80 μm-mesh net towed from 150 m to the surface. Water (filtered to remove zooplankton) used in the experimental containers was taken from four depths where the zooplankton are found (8, 16, 32, and 64 m) and combined to form a "composite sample." Adults were separated by species and acclimated for 24 hours at 8°C in red light or in the dark (also the experimental conditions). This temperature and very low light or darkness represent the field conditions just below the thermocline.

Mysids were collected at the same station using Bongo nets towed from 150 m to the surface. Similarly sized mysids (16 to 18 mm) were placed individually into 1-gal glass jars with 3 L lake water (filtered to remove other zooplankton) and acclimated to the experimental conditions. They were starved prior to the trials for 12 to 24 hours.

Survival Rates of D. tyrelli *and* E. nevadensis

Survival and mortality rates of *D. tyrelli* and *E. nevadensis* in single- and two-species assemblages were measured in one experiment. Twenty adults of each species were placed together into each of six 1-gal jars containing 3 L of the composite lake water. Single-species treatments consisted of 2 replicates of 40 animals of each species alone. The living animals were counted every other day for 12 days, the water replaced to replenish the algal resources and the dead animals removed. Dead animals were not replaced.

Mortality rates were calculated as $q(x) = d(x)/l(x)$; where $d(x)$ = number dying over the time interval x to $x + 1$; $l(x)$ = number surviving to time x; and x = each sampling interval.

Filtering Rates

Filtering rates for *D. tyrelli* and *E. nevadensis* were measured using standard radiotracer techniques (Folt and Goldman 1981). A radioactively labeled food suspension was prepared by adding approximately 350 μCi of the radionuclide phosphorus 32 to 450 ml of the composite lake water sample, which was then incubated for 24 hours at 23°C under constant light. At the start of an experiment the experimental number of animals (1–10) was placed in 125 ml glass jars containing 120 ml fresh composite lake water. After 30 to 45 minutes 5 ml

of the labeled food suspension was pipetted into each jar and gently swirled. After feeding for 15 minutes the animals were heat-killed, rinsed, and soaked in a mild phosphate buffer for 1 to 2 hours. Radioactivity in the animals was counted in a Beckman LS-100 scintillation counter. Filtering rates (milliliters per animal per day) were calculated from the uptake of the radioactive label during the 15-minute trial and then converted to 24 hour (daily) rates.

Predation Trials

Individual mysids (in 3 L water) were presented with 2.7, 5.3, 13.3, or 20 prey per liter in one- or two-species combinations (table 20.1). The low-density treatments (2.7 per liter) were terminated after 3 or 4 hours because of high depletion rates. Trials at intermediate densities (5.3 per liter) lasted for 6, 12, or 14 hours. Trials at high densities (13.3 per liter) lasted 6 or 24 hours. The trials at the highest density (20 per liter) were run for 12 hours. All rates were converted to 6-hour rates.

M. relicta preferences were calculated using the following approximation, which is appropriate for cases where food does not remain constant during a trial (Manly et al. 1972; Manly 1974; Chesson 1978, 1983): $B(i) = \log[r(i)/A(i)]/\Sigma\log[r(s)/A(s)]$; where $B(i)$ = preference for prey type i; $r(i)$ = number of prey type (i) remaining at the end of the experiment; $A(i)$ = number of prey type (i) available initially; and s = total number of prey types available. This measure has a finite scale from 0 to 1, with a value of 0.5 indicating no preference for either species. It has a bivariate normal distribution and standard statistical tests can be used to test hypotheses about B's (Chesson 1983).

Table 20.1. *Experimental design for the 1- and 2-prey* M. relicta *predation trials*

Total prey density (no. per liter)	Density in 1-prey trials (no. D/no. E per 3 L)	Density in 2-prey trials (no. D/no. E per 3 L)
2.7	Ø/8; 8/Ø	2/6; 4/4; 6/2
5.3	Ø/16; 16/Ø	4/12; 8/8; 12/4
13.3	Ø/40; 40/Ø	5/35;10/30;20/20;30/10;35/5
20.0	Not run	10/50;20/40;30/30;40/20;50/10

Prey are adult *D. tyrelli* (D) and adult *E. nevadensis* (E) and the predators are adult M. *relicta*. All treatments were run in 3 L water (see Methods).

Table 20.2. *Effect of time (length of trial) and ratio (no. of E. nevadensis : no. of D. tyrelli) on M. relicta preferences in the 2-prey trials at 3 total prey densities; ANOVA (SAS, GLM)*

Source	df	SS	F	P	
At total prey density, 2.3 per liter; ratios, 0.33, 1.0, 3.0; times (h), 3,4					
Time	1	.020	1.11	0.31	NS
Ratio	2	.051	1.45	0.27	NS
Time*ratio	2	.077	2.18	0.15	NS
Error	14	.247			
At total prey density, 5.3 per liter; ratios, 0.33, 1.0, 3.0; times (h), 6,12,14					
Time	2	.098	0.87	0.43	NS
Ratio	2	.013	0.11	0.89	NS
Time*ratio	1	.001	0.02	0.90	NS
Error	20	1.12			
At total prey density, 13.3 per liter; ratios, 0.14, .33, 1.0, 7.0; times (h), 6,24					
Time	1	.001	0.05	0.82	NS
Ratio	3	.072	1.04	0.39	NS
Time*ratio	3	.101	1.46	0.24	NS
Error	28	.646			

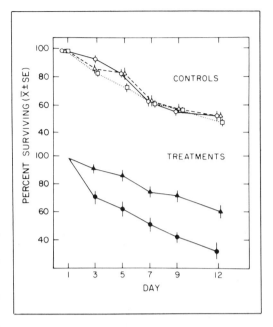

Figure 20.2. *Survival of D. tyrelli and E. nevadensis in single and two-species treatments. Depicted are the mean number of each species surviving (± SE) each time interval. Controls are single-species treatments run concurrently with the two-species treatments at the same total densities. D. tyrelli in the controls (◇); E. nevadensis in controls (△); sum of two species in the two-species treaments (⊟); D. tyrelli in the two-species treatments (◆); E. nevadensis in the two-species treatments (▲).*

Since several of the treatments were run for different lengths of time, the effect of time on preference was calculated with an analysis of variance (ANOVA). There were no significant differences in preference as a function of time (table 20.2), so the data for both time periods at a single density were pooled and the mean preferences at each density were calculated.

Predation risks in one- and two-prey species trials at each density (proportion of prey of each species killed during the trial) were analyzed. In this case the length of the feeding trial has a great effect on the estimate of risk. In all calculations of risk only the data collected in the trials that lasted 3 to 6 hours were included. Prey depletion was greater (50%) at the low density compared with the intermediate (45%) and high (40%) prey densities.

In most cases, at the completion of a feeding trial, dead or injured prey were removed, the original density reestablished, and the trial immediately repeated. In such cases, the predation rate or the preference for each mysid was calculated as the mean for that individual over all trials.

RESULTS

First, the results relating to the interaction between the copepods in the absence of the predator *M. relicta* are presented.

Survival in One- and Two-Species Treatments

In the two-species treatments the mean survival of *D. tyrelli* over 12 days was significantly lower than the survival of *E. nevadensis* ($p < 0.001$, Student's *t*-test; fig. 20.2). However, there were no significant differences in the survival rates between these species in the one-species treatments (fig. 20.2). Mortality rates of these species in the treatments were significantly different (table 20.3) on day 3 (*D. tyrelli* > *E. nevadensis*) and days 5 and 9 (*D. tyrelli* < *E. nevadensis*).

Table 20.3. *Mortality rates (±SE) for* D. tyrelli *and* E. nevadensis *in the 2-species survival experiments*

Species	Day of experiment					
	3	5	7	9	12	X
D. tyrelli	.27(.02)	.09(.03)	.13(.02)	.08(.02)	.10(.03)	.14(.02)
E. nevadensis	.18(.01)	.18(.01)	.15(.01)	.15(.01)	.13(.01)	.16(.01)
Significance	$p < .01$**	$p < .05$*	NS	$p < .05$*	NS	NS

Significance of difference between treatments measured with Student's *t*-test.

Table 20.4. *Linear regressions of filtering rates per individual* D. tyrelli *against treatment density in 1-species trials*

Date	Treatment density (no. *D. tyrelli*/ 125 mls)	No. replicates per treatment	N	r	p	Factor increase in filtering rates from mean highest to mean lowest density
12 June	2,3,4,6,10	16	66[a]	−.372	<.01**	2.8
24 July	2,4,6,8	6	23[b]	−.329	<.10*	1.9
4 Sept.	1,2,3,4,6,10	8	47[b]	−.202	>.10NS	1.6
4 Oct.	1,2,6,10	8	28[c]	−.531	<.01**	3.3
						x = 2.4

[a]4 replicates only at 2 per 125 mls.
[b]1 missing replicate.
[c]4 replicates only at 1 per 125 mls.

Filtering Rates in One- and Two-Species Treatments

There was a significant reduction (estimated by linear regression) in the filtering rates of individual *D. tyrelli*, with an increase in intraspecific animal density in three of four experiments (table 20.4). In the fourth experiment (4 September) the filtering rates at the intermediate and high densities were similar to each other and significantly ($p < 0.05$) lower than the rate measured at the lowest density (Folt 1982). On average, the filtering rates in the lowest-density treatments were twice as high as the filtering rates in the highest-density treatments. In all cases, the difference between the highest and lowest densities was significant ($p < 0.05$).

There was also a significant reduction ($p < 0.05$, Mann-Whitney U-test, table 20.5) in the filtering rates of *D. tyrelli* in the presence of *E. nevadensis* in four of six experiments conducted from June through October 1981. This effect was also documented in a separate set of experiments (Folt and Goldman 1981) that are not all discussed here. These experiments compared filtering rates of *D. tyrelli* in one-species treatments with two-species treatments at the same total animal density (*E. nevadensis:D. tyrelli* = 1). All data for the one-species trials were pooled and compared to the pooled two-species treatments. The filtering rates in the one-species treatments were significantly greater ($p < 0.04$, Mann-Whitney U-test) than in the two-species treatments. The average reduction in *D. tyrelli* filtering rates in the two-species trials was 64%.

Second, results relating to the interaction between the copepods in the presence of the predator *M. relicta* are presented.

Preferences of M. relicta in the Two-prey Trials

Preferences of *M. relicta* in two-prey trials at several absolute and relative densities were compared. Data from previously published work (Folt et al. 1982) have been combined with new data (5.3 per liter) and analyzed with a different and more appropriate measure of

Table 20.5. *Comparisons among filtering rates (FR = milliliters per animal per day) of* D. tyrelli *in 1- and 2-species treatments at the same total densities*

	1-species trials			2-species trials			
Date	FR (SE)	No. D/ 125 ml	No. repli- cates	FR (SE)	No. D, No. E/ 125 ml	No. repli- cates	Significance
12 June	8.28(1.34)	4	16	7.56(1.59)	2,2	16	$p < .15$ NS
24 June[a]	1.41(0.37)	4	4	0.94(0.22)	2,2	4	$p < .05*$
24 July	4.60(1.08)	4	8	1.28(0.22)	2,2	7	$p < .005**$
4 Sept.	0.91(0.13)	4	8	0.54(0.09)	2,2	8	$p < .05*$
4 Oct.	0.81(0.16)	2	8	0.43(0.20)	1,1	4	$p < .05*$
4 Oct.	0.56(0.05)	6	8	0.46(0.05)	3,3	4	$p < .15$ NS

The ratio of *D. tyrelli* (D) to *E. nevadensis* (E) in the 2-species trials was 1. Significance levels for the difference in filtering rates between the 1- and 2-species trials calculated with Mann-Whitney-U Test.
[a]In both treatments on this date only there was also 1 *M. relicta* in each experimental container.

preference. These data are presented as preferences by *M. relicta* for *E. nevadensis* [B(E)] or *D. tyrelli* [B(D)]. Preference for *D. tyrelli* is simply 1 minus B(E).

The effect of the relative abundances of the two prey types on the preference by *M. relicta* for *E. nevadensis* was examined by using analysis of variance (table 20.2). There were no significant effects of the relative abundance of prey on the preferences of *M. relicta*.

The preference as a function of total prey density was also compared with ANOVA by pooling preferences measured at all relative densities for each total prey density. There was a significant effect of total prey density on the preference of *M. relicta* for *E. nevadensis* ($F_{3,104}$; $p < 0.004$; fig. 20.3). Preference for *E. nevadensis* was significantly higher at the lowest total prey density (protected LSD, $p < 0.05$) (Jones 1984) than at other densities. Moreover, preferences for *E. nevadensis* were significantly greater than 0.5 at all densities.

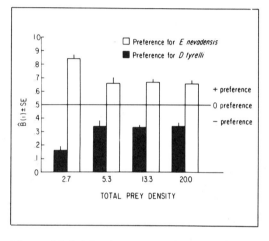

Figure 20.3. *M.* relicta *prey preferences in the two-prey trials. Prey are adult* E. nevadensis *and adult* D. tyrelli. *Depicted are the mean preferences (B(i)± SE) for each prey species at four different total prey densities (2.7 to 20 prey per liter). At each total prey density, treatments with several different proportions of either prey type were pooled (see text). A value of B(i) = 0.5 indicates no preference by the predator for either prey type.*

Predation Risks in One- and Two-prey Trials

The predation risks for *D. tyrelli* in the one-prey trials were not significantly affected by total prey density (fig. 20.4A). There was a significant reduction in the predation risk of *D. tyrelli* in the two- versus one-prey trials (ANOVA: $F_{1,73}$; $p < 0.04$; fig. 20.4A). The risks at each total density were then compared by multiple-comparison procedures (protected LSD; SAS/GLM), and only the risk at the lowest total

prey densities (2.7 per liter) was significantly lower in the presence of *E. nevadensis* ($p < 0.05$; protected LSD).

In the one-prey trials, predation risks for *E. nevadensis* were significantly reduced (ANOVA; $F_{2,50} = 17.7$; $p < 0.001$) with an increase in density (fig. 20.4B). The predation risks for individuals in the two-prey trials were significantly lower ($p < 0.05$, LSD) than in the one-prey trials

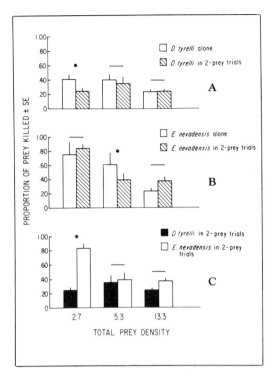

Figure 20.4. *Individual predation risks in single and two-prey species trials. Predators are M. relicta and prey are adult D. tyrelli and adult E. nevadensis. Depicted are the predation risks measured as the mean proportions (± SE) of each species killed relative to the number of that species available in single and 2-prey species trials.* **A:** *comparison of the predation risks to D. tyrelli in single and two-prey trials at three total prey densities.* **B:** *comparison of the predation risks to E. nevadensis in single and two-prey trials at the same total prey densities.* **C:** *comparison of the predation risks to D. tyrelli and E. nevadensis in the two-prey trials. Bars connect means that are not significantly different (protected LSD; p > .05; * connect means that are significantly different (protected LSD; p < .05).*

only at the intermediate density (5.3 per liter).

Predation risks for *E. nevadensis* and *D. tyrelli* in the two-prey trials at all three total prey densities were compared by ANOVA. There was a significant difference, with predation risks lower (*p* < 0.01; protected LSD) for *D. tyrelli* than *E. nevadensis* at the lowest prey densities (fig. 20.4C).

DISCUSSION

The results of this study clearly demonstrate that there are fitness-related costs and benefits associated with specific zooplankton density conditions. Density-dependent interactions between the two species of calanoid copepod and their mutual invertebrate predator affected the filtering or feeding rates of all three species and the mortality rates of the two prey species. *D. tyrelli* and *E. nevadensis* face different foraging constraints and predation risks depending on the density and the composition of the group in which they are found.

Filtering rates of *D. tyrelli* were significantly reduced at high conspecific densities (table 20.4). A similar reduction in both the filtering and predation rates of *E. nevadensis* with an increase in the density of conspecifics also was measured (Folt and Byron, unpublished MS) in experiments summarized in table 20.6. The effect of animal density on the grazing rates of zooplankton have rarely been measured at realistic densities. However, density-dependent reductions in grazing also were found by Helgren (personal communication) for *D. pulex* over a natural range of densities, by Hayward and Gallup (1976) for *D. schødleri*, and by Hargrave and Geen (1970) for the marine copepod *Acartia tonsa*.

Filtering rates of *D. tyrelli* also were significantly reduced at low densities by the presence of *E. nevadensis* (table 20.5). Folt and Goldman (1981) showed that a chemical released by *E. nevadensis* caused reduction in the filtering rates of *D. tyrelli*; we called this allelopathy. Allelopathy also has recently been measured by Seitz (1984) between two species of *Daphnia*. Skiver (1980) measured a significant reduction in the ingestion rate (of micronic spheres) of a marine calanoid copepod (*Acartia hudsonica*) in the presence of another copepod species (*Temora longicornis*). He suggested that physical interference mechanisms were responsible for the observed interactive effect. A third mechanism that could lead to a reduction in grazing or feeding rates is a depletion of resources (as measured by Tessier 1983) in a region of high grazer density.

The ultimate effect of a reduction in feeding rates on the fitness of individuals in a group depends on a number of factors, including the resource levels and renewal rates, the physiological status of individuals, and the length of time spent in the group. In Lake Tahoe, resource levels are extremely low (< 150 μg C per liter), and zooplankton are resource-limited throughout the year (Byron et al. 1984). The reduction in feeding for individuals in a mixed-species or dense single-species aggregation of 50% or greater as measured in these experiments is likely to have an impact on growth, reproduction, and survival of individuals. The abundance of *D. tyrelli* decreases rapidly in the early fall (Folt 1982) before the temperature and the algal concentrations drop appreciably. This mortality may be due in part to intra- and interspecific interactions in the aggregations. As shown (fig. 20.2), the survival of *D. tyrelli* was reduced in the presence of *E. nevadensis*, but examination of the carcasses proved that death was not the direct result of predation. At the densities of the survival experiments (1–2 per 125 ml) the filtering rates of *D. tyrelli* are predicted (from the filtering rate experiments) to be lower in the two-species treatments. Mortality of *D. tyrelli* in the two-species trials may have resulted from chemical and physical interference by *E. nevadensis* with feeding by *D. tyrelli*. Differences in age and in the past feeding histories of *D. tyrelli* could have caused differences in their abilities to withstand periods of reduced feeding, so that those most susceptible died in the first interval. This could explain the difference that was measured (table 20.3, day 3) in mortality rates of *E. nevadensis* and *D. tyrelli*. However on days 5 and 9, mortality rates of *E. nevadensis* were actually greater than those of *D. tyrelli*. Possibly, *E. nevadensis* were suffering from a lack of prey. Although algal foods were replenished, prey were not added. It is possible that algal resources provide insufficient nourishment for this omnivore over long periods of time. The difference in survival between the two species over the 12-day period resulted primarily from the large differences during the first-time interval.

The interference interaction of *E. nevadensis* could increase the mortality rates of *D. tyrelli* indirectly through exhaustion. *E. nevadensis* are tactile predators that locate prey largely by mechanical receptors (Strickler and Twombly 1975; Kerfoot 1977). The adults prey on juvenile *D. tyrelli* (Folt and Byron, ms). *D. tyrelli* may escape predation by *E. nevadensis* by stopping all movement (including filtering water) when *E. nevadensis* are sensed chemically or physically. In the laboratory whenever *E. nevadensis* get within 1 body length of *D. tyrelli*, the diaptomids take several long leaps or spirals and dive. This escape response may be costly in terms of energy expenditure (Koehl and

Table 20.6. *Filtering and predation rates of* E. nevadensis *adults in treatments at different densities of conspecifics*

Predation rates (# *d. tyrelli* nauplii killed • E. nevadensis^{-1} • day^{-1})

# E. nevadensis[a]	2	4	6	10
predation rate	1.12	0.40	0.61	0.25

Filtering rates (mls • E. nevadensis^{-1} • day^{-1})

# E. nevadensis[b]	1	2	4	8
filtering rate	0.123	0.005	0.028	0.011

Bars connect means that are not significantly different ($p < .05$, Student's *t*-test)
[a]No. in 500 mls water.
[b]No. in 125 mls water.
From Folt and Byron, unpublished MS.

Strickler 1981). After being chased by pipettes for several minutes *D. tyrelli* stay on the bottom of the container and no longer avoid the pipette, possibly indicating that they become exhausted after several escapes.

Despite the reduction in filtering rates, it may be necessary for *D. tyrelli* to respond as it does to *E. nevadensis*. If sensing *E. nevadensis* triggers an escape response by *D. tyrelli* that is correlated with a decrease in feeding, the gain from avoiding a potential predator may be more important over the short term than the cost of reduced feeding. This type of interaction is likely to be common among aquatic invertebrate predators and their potential prey. Prolonged exposure of predators and prey (i.e., during peak field densities or in dense mixed-species aggregations) should increase the negative impacts of the interaction.

For zooplankton, mixed species groups containing predators and prey are common (Cushing and Tungate 1963; Wiebe 1970; Fenwick 1978; Malone and McQueen 1983). Mechanisms such as diurnal vertical migration periodically aggregate species of predators, omnivores, and herbivores in dense layers. Although many zooplankton become largely invulnerable to invertebrate predators as adults (Hall et al. 1976), interference interactions (e.g., unsuccessful predation and predator avoidance) may frequently occur and influence fitness through reduced feeding or increased mortality. Reductions in foraging are therefore a likely consequence of grouping in zooplankton. Even among conspecifics, feeding rates can be reduced at high density. The obvious question is, are there advantages to single- and mixed-species aggregations for these animals? If so, do they outweigh the disadvantages?

During the night, M. relicta migrate into the region of the lake inhabited by the copepods (Rybock 1978). Other studies (Rybock 1978; Cooper and Goldman 1980) showed that M. relicta selected E. nevadensis over D. tyrelli. The working hypothesis was that the presence of the more preferred prey (E. nevadensis) in two-prey assemblages would result in a reduction in the predation risk for the less preferred prey (D. tyrelli). If so, the resulting increase in survival could offset the negative effects of reduced filtering rates that D. tyrelli experienced in the two-species assemblages.

Similarly, at high density the individual predation risks for the preferred prey (E. nevadensis) were hypothesized to be lower than at low density. Again, the reduction in predation risk due to aggregation could offset the reduction in feeding rates that E. nevadensis experienced at high densities of conspecifics. It was not obvious how the presence of a less preferred prey would affect the predation risks of E. nevadensis in two-species assemblages compared to one-species assemblages at the same total density.

One important consequence of a patchy prey distribution is that predators experience prey under different conditions. Prey that are taken under some conditions may be ignored at higher prey densities, when a more preferred prey is present, or when prey in a group interact in ways that affect their encounter rates with, or susceptibility to, the predator. The primary advantage of high density living for many species is a reduction in individual predation risk. This prediction is extended by my results to these species of zooplankton. However, if predators aggregate in areas where prey densities are greatest, the benefits are quite possibly reduced (Hassell and May 1974; Hassell 1978; Fenwick 1978). I have not addressed the question of predator clumping because M. relicta are not known to aggregate in Lake Tahoe. These data demonstrate that if predator density is constant across patches, individual prey (both preferred and nonpreferred species) face different risks as a function of group size and composition.

The abundance and composition of the zooplankton assemblage affected the feeding behavior of the predator (M. relicta) and the risk of predation to the prey. The predation rates and preferences of M. relicta were density-dependent. Predation rates on both prey increased with prey density, although they were significantly greater in single-prey E. nevadensis trials than in the two-prey trials at the lower densities (Folt 1984). At the lowest density of prey, M. relicta did not effectively prey on D. tyrelli (Folt et al. 1982; Folt 1985 and preferences for E. nevadensis over D. tyrelli were greatest. As the density of prey increased, the preferences of M. relicta for E. nevadensis decreased. This indicated that the two-prey species became more similar at higher densities in terms of their susceptibility to predation by M. relicta.

A decrease in selectivity with an increase in prey density is not predicted by simple foraging models (Werner and Hall 1974; Pyke et al. 1977; Pyke 1984). However, other models (e.g., Hughes 1979) that incorporate learning predict the inclusion of less preferred prey in the diet as encounter rates with those prey increase. This can cause a reduction in preference at high prey densities. Preference as normally measured (including the present study) does not necessarily indicate an active choice on the part of the predator but simply a greater abundance of one prey in the diet relative to the environment, compared with the other prey type. Indeed, whenever the ability of the predator to capture one prey type changes relative to the others available, a change in preference results (e.g., Attar and Maly 1980). The change observed in this study may have resulted from changes in the predation behavior of M. relicta (i.e., learning) or from changes in the behavior of the prey. For example, at higher densities D. tyrelli also have a greater encounter rate with E. nevadensis and conspecifics, both of which could affect their swimming and avoidance behavior and therefore their susceptibility to M. relicta.

The predation risk for both prey species varied with the density and the composition of the prey assemblage. For D. tyrelli (always the less preferred prey) the predation risks were lower in two-species assemblages at the lowest total prey densities. Predation risks to E. nevadensis (the preferred prey) were reduced at high density in one- and two-prey assemblages, and in the two-prey relative to the one-prey assemblages at intermediate densities. Predation risks to both species were never significantly greater in the two-prey versus the one-prey treatments at the same total density.

These results support the initial hypothesis that mixed-species aggregations of zooplankton are beneficial in terms of reduced predation risks. However, for both species of prey there was a cost measured as a significant reduction in filtering or feeding rates in an aggregation. The reductions in filtering rates by *D. tyrelli* are hypothesized to be linked to a mechanism for avoiding some predators, but no data testing this hypothesis are yet available. For *D. tyrelli*, aggregation is costly in the absence of the predator, *M. relicta*. Therefore, it is suggested that aggregation by herbivorous zooplankton in single- or mixed-species groups may be advantageous only in predator-prone environments or during reproduction.

In contrast, it may be advantageous for *E. nevadensis* to forage in mixed-species aggregations in the absence of *M. relicta* despite the reduction in filtering and feeding rates measured at high densities of conspecifics (table 20.6). *E. nevadensis* are omnivorous, and many omnivores require animal prey for growth and reproduction (Main 1961; Corner et al. 1976; Lonsdale et al. 1979). In addition, in Lake Tahoe, *E. nevadensis* select *D. tyrelli* nauplii over conspecifics (Folt and Byron, unpublished MS). Hence, foraging in a mixed-species and a mixed-age class assemblage may increase the chance of finding suitable animal prey.

In conclusion, patchiness, aggregation, and swarming by zooplankton are common. This study demonstrates costs and benefits associated with single- and mixed-species zooplankton assemblages at different densities. Measuring density-dependent interactions among zooplankton is necessary if we are to expand the study of zooplankton species interactions to include the striking spatial and temporal distribution patterns observed in nature.

SUMMARY

1. There was a cost measured as a reduction in filtering or feeding rates at high density for both the herbivore (*D. tyrelli*) and the omnivore (*E. nevadensis*).
2. There was a cost measured as a reduction in the filtering rates of the herbivore in the presence of the omnivore at low densities. This reduction can be induced by a chemi-

cal released by the omnivore (Folt and Goldman 1981).
3. There was a cost measured as a reduction in the feeding rates of the predator (*M. relicta*) in two-prey assemblages at low and intermediate densities (Folt 1985).
4. There was a benefit measured as a reduction in individual predation risks for the herbivore (the less preferred prey) at low prey densities in two-prey trials. The measured costs and benefits for the herbivore were both greatest at the lowest densities in the mixed-species assemblages. These conditions are most common in the field.
5. There was a benefit measured as a reduction in individual predation risks for the omnivore (the preferred prey) at intermediate densities in the two-prey trials and at high density in the one-prey trials.
6. The foraging behavior of the predator measured as preference varied with the density of the prey assemblage. Preference decreased as the total prey density increased.

ACKNOWLEDGMENTS

I am very grateful to E. Byron, W. Crumpton, and R. Richards for their assistance in the field; S. Boddy and C. Sims for their help with the *M. relicta* predation trials; S. Cooper for the use of his data on the predation rates of *M. relicta* in the highest density (20 per liter) trials. I have benefited tremendously from discussions with those mentioned above and with D. Goldberg, D. Hart, K. Hopper, C. Kerfoot, M. Morgan, D. Peart, and A. Sih. This manuscript was considerably improved by suggestions from C. Burns, W. Crumpton, M. Dionne, R. Hoenicke, C. Osenberg, D. Peart, and A. Sih; and was prepared by M. Audette. The research was generously supported by NSF BSR 8408173 to the author, NSF DEB 796221 to C. R. Goldman, and by The Kellogg Biological Station.

REFERENCES

Attar, E. N. and E. S. Maly. 1980. A laboratory study of preferential predation by the newt *Notophthalmus v. viridescens*. *Can. J. Zool.* 58 : 1712–17.

Brönmark, C., B. Malmqvist, and C. Otto. 1984. Anti-predator adaptations in a neustonic insect (*Velia caprai*). *Oecologia* 61 : 189–91.

Brown, R. G. B., S. P. Barker, and D. E. Gaskin. 1979. Daytime swarming by *Meganyctiphanes norvegica* (M. Sars) (Crustacea, Eupharisiacea) off Brier Island, Bay of Fundy. *Can. J. Zool.* 57 : 2285-91.

Burger, J. 1981. A model for the evolution of mixed-species colonies of ciconiiformes. *Q. Rev. Biol.* 56 : 143-67.

Buss, Leo W. 1981. Group living, competition, and the evolution of cooperation in a sessile invertebrate. *Science* 213 : 1012-14.

Byron, E. R., P. T. Whitman, and C. R. Goldman. 1983. Observations on copepod swarms in Lake Tahoe. *Limnol. Oceanogr.* 28(2) : 378-82.

Byron, E. R., C. L. Folt, and C. R. Goldman. 1984. Copepod and cladoceran success in an oligotrophic lake. *J. Plankton Res.* 6(1) : 45-65.

Caraco, T. S., S. Martindale, and H. R. Pulliam. 1980. Avian flocking in the presence of a predator. *Nature* 285 : 400-401.

Cassie, R. M. 1959. Microdistribution of plankton. *N. Z. J. Sci.* 2 : 398-409.

Chesson, J. 1978. Measuring preference in selective predation. *Ecology* 59 : 211-15.

————. 1983. The estimation and analysis of preference and its relationship to foraging models. *Ecology* 64(5) : 1297-1304.

Colebrook, J. M. 1960. Some observations of zooplankton swarms in Windermere. *J. Anim. Ecol.* 29 : 241-42.

Cooper, S. D. and C. R. Goldman. 1980. Opossum shrimp predation on zooplankton. *Can. J. Fisheries Aquat. Sci.* 37 : 909-19.

Corner, E. D. S., R. N. Head, C. C. Kilvington, and L. Pennycuick. 1976. On the nutrition and metabolism of zooplankton. X. Quantitative aspects of *Calanus helgolandicus* feeding as a carnivore. *Mar. Biol. Assoc. U. K.* 56 : 345-58.

Cushing, D. H., and D. S. Tungate. 1963. Studies on a *Calanus* patch. I. The identification of a *Calanus* patch. *J. Mar. Biol. Assn.* (UK) 43 : 327-37.

Fasham, M. J. R. 1978. The statistical and mathematical analysis of plankton patchiness. *Oceanogr. Mar. Biol. Ann. Rev.* 16 : 43-79.

Fenwick, G. D. 1978. Plankton swarms and their predators at the Snares Islands. *N.Z. J. Mar. Freshwater Res.* 12(2) : 223-24.

Folt, C. L. 1982. The effects of species interactions on the feeding and mortality of zooplankton. Ph.D. diss., University of California, Davis.

————. 1985. Predator efficiencies and prey risks at high and low prey densities. *Verh. int. Verein. Limnol.* 22 : 3210-14.

Folt, C. L., and C. R. Goldman. 1981. Allelopathy between zooplankton: a mechanism for interference competition. *Science* 213 : 1133-35.

Folt, C. L., J. T. Rybock, and C. R. Goldman. 1982. The effect of prey composition and abundance on the predation rate and selectivity of *Mysis relicta*. *Hydrobiologia* 93 : 133-44.

Foster, W. A., and J. E. Treherne. 1981. Evidence for the dilution effect in the selfish herd from fish predation on a marine insect. *Nature* 293 : 466-67.

Gillett, S. D., P. J. Hogarth, and F. E. Jane Noble. 1979. The response of predators to varying densities of *Gregaria* locust nymphs. *Anim. Behav.* 27 : 592-596.

Goldman, C. R. 1981. Lake Tahoe: Two decades of change in a nitrogen deficient oligotrophic lake. *Verh. int. Verein. Limnol.* 21 : 45-70.

Hall, D. J., S. T. Threlkeld, C. W. Burns, and P. H. Crowley. 1976. The size-efficiency hypothesis and the size structure of zooplankton communities. *Annu. Rev. Ecol. Syst.* 7 : 177-208.

Hamilton, W. D. 1971. Geometry for the selfish herd. *J. Theo. Biol.* 31 : 295-311.

Hamner, W. M., and J. H. Carleton. 1979. Copepod swarms: Attributes and role in coral reef ecosystems. *Limnol. Oceanogr.* 24 : 1-14.

Hamner, P., and W. M. Hamner. 1977. Chemosensory tracking of scent trails by the planktonic shrimp *Acetes sibogae australis*. *Science* 195 : 886-88.

Hargrave, B. T., and G. H. Geen. 1970. Effects of copepod grazing on two natural phytoplankton populations. *J. Fish. Res. Bd. Can.* 27 : 1395-1403.

Hassell, M. P. 1978. Arthropod predator-prey systems. *Monographs in Population Biology* 13. Princeton, N.J.: Princeton University Press.

Hassell, M. P., and R. M. May. 1974. Aggregation in predators and insect parasites and its effect on stability. *J. Anim. Ecol.* 43 : 567-94.

Haury, L. R., J. A. McGowan, and P. H. Wiebe. 1978. Patterns and processes in the time-space scales of plankton distributions. In J. H. Steele (ed.), *Spatial pattern in plankton communities*, pp. 227-327. NATO Conference Series IV. New York: Plenum Press.

Hayward, R. S., and D. N. Gallup. 1976. Feeding, filtering and assimilation in *D. schödleri* Sars as affected by environmental conditions. *Arch. Hydrobiol.* 2 : 139-63.

Healey, M. C. 1967. The seasonal and diel changes in distribution of *Diaptomus leptopus* in a small eutrophic lake. *Limnol. Oceanogr.* 12 : 34-39.

Heinrich, B., and D. Vogt. 1980. Aggregation and foraging behavior of Whirligig Beetles (Gyrinidae). *Behav. Ecol. Sociobiol.* 7 : 179-86.

Heller, R., and M. Milinski. 1979. Optimal foraging of sticklebacks on swarming prey. *Anim. Behav.* 27 : 1127-41.

Horsley, D. T., B. M. Lynch, J. D. Greenwood, B. Hardman, and S. Mosely. 1979. Frequency-dependent selection by birds when the density of prey is high. *J. Anim. Ecol.* 48 : 483-90.

Hughes, R. N. 1979. Optimal diets under the energy maximization premise: the effects of recognition time and learning. *Amer. Naturalist* 113–209, 221.

Jones, D. 1984. Use, misuse and role of multiple-comparison procedures in ecological and agricultural entomology. *Environ. Entomol.* 13(3): 635–49.

Kenward, R. E. 1978. Hawks and doves: attack success and selection in goshawk flights at woodpigeons. *J. Anim. Ecol.* 47: 449–60.

Kerfoot, W. C. 1977. Implications of copepod predation. *Limnol. Oceanogr.* 22: 316–25.

Klemetsen, A. 1970. Plankton swarms in Lake Gjokvatn, East Finland. Astarte, *J. Arctic Biol.* 3: 83–85.

Koehl, M., and J. R. Strickler. 1981. Copepod feeding currents: food capture at low reynolds numbers. *Limnol. Oceanogr.* 26: 1062–73.

Lehman, J. T., and D. Scavia. 1982. Microscale patchiness of nutrients in plankton communities. *Science* 216: 729–30.

Lewis, W. M. 1978. Comparison of spatial and temporal variation in the zooplankton of a lake by means of variance components. *Ecology* 59: 666–71.

Lonsdale, D. J., D. R. Heinle, and C. Seigfried. 1979. Carnivorous feeding behavior of the adult Calanoid copepod *Acartia tonsa* Dana. *J. Exp. Mar. Biol. Ecol.* 36: 235–48.

Main, R. A. 1961. The life history and food relations of *Epischura lacustris* Forbes. Ph.D. diss., Michigan State University.

Malone, B. J., and D. J. McQueen. 1983. Horizontal patchiness in zooplankton populations in two Ontario kettle lakes. *Hydrobiology* 99: 102–24.

Manly, B. F. J. 1974. A model for certain types of selection experiments. *Biometrics* 30: 281–94.

Manly, B. F., P. Miller, and L. M. Cook. 1972. Analysis of a selective predation experiment. *Amer. Naturalist* 106: 719–36.

McFarland, W. N., and N. M. Kotchian. 1982. Interaction between schools of fish and mysids. *Behav. Ecol. Sociobiol.* 11: 71–76.

Milinski, M. 1977a. Do all members of a swarm suffer the same predation? *Z. Tierpsychol.* 45: 373–88.

———. 1977b. Experiments on the selection by predators against spatial oddity of their prey. *Z. Tierpsychol.* 43: 311–25.

———. 1979. Can an experienced predator overcome the confusion of swarming prey more easily? *Anim. Behav.* 27: 1122–26.

Morgan, W. L., and D. A. Ritz. 1984. Effect of prey density and hunger state on capture of Krill, *Nyctiphanes austrialis* Sars, by Australian salmon, *Arripis trutta* (Bloch and Schneider). *Fish. Biol.* 24: 51–58.

Murdoch, W. W. 1969. Switching in general predators: experiments on prey specificity and stability

of prey populations. *Ecol. Monogr.* 39: 335–54.

Murdoch, W. W., S. Avery, and M. E. B. Symthe. 1975. Switching in predatory fish. *Ecology* 56: 1094–1105.

Neill, S. R. St. J., and J. M. Cullen. 1974. Experiments on whether schooling by their prey affects the hunting behavior of cephalopods and fish predators. *J. Zool.* 172: 549–69.

Pitcher, T. J., and A. E. Magurran. 1983. Schoal size, patch profitability and information exchange in foraging goldfish. *Anim. Behav.* 31: 546–55.

Pitcher, T. J., A. E. Magurran, and I. J. Winfield. 1982. Fish in larger shoals find food faster. *Behav. Ecol. Sociobiol.* 10: 149–51.

Poulet, S. A., and P. Marsot. 1978. Chemosensory grazing by marine Calanoid copepods (Arthropoda:Crustacea). *Science* 200: 1403–1405.

———. 1980. Chemosensory feeding and food-gathering by omnivorous marine copepods. In W. Charles Kerfoot (ed.), *The evolution and ecology of zooplankton communities*, pp. 198–218. Hanover, N.H.: University Press of New England.

Poulet, S. A., and G. Ouellet. 1982. The role of amino acids in the chemosensory swarming and feeding of marine copepods. *J. Plankton Res.* 4(2): 341–59.

Pulliam, H. R. 1976. The principle of optimal behavior and the theory of communities: In P. P. G. Bateson and P. H. Klopfer (ed.), *Perspectives in ethology*, Vol. 2. New York: Plenum Press.

Pyke, G. H. 1984. Optimal foraging theory: a critical review. *Annu. Rev. Ecol. Syst.* 15: 523–75.

Pyke, G. H., H. R. Pulliam, and E. L. Charnov. 1977. Optimal foraging: A selective review of theory and tests. *Annu. Rev. Biol.* 52: 137–54.

Robertson, D. R., H. P. Sweatman, E. A. Fletcher, and M. G. Cleland. 1976. Schooling as a mechanism for circumventing the territoriality of competitors. *Ecology* 57: 1208–20.

Rybock, J. T. 1978. *Mysis relicta* Loven in Lake Tahoe: Vertical distribution and nocturnal predation. PhD thesis, University of California, Davis.

Seitz, A. 1984. Are there allelopathic interactions in zooplankton? Laboratory Experiments with *Daphnia*. *Oecologia* 62: 94–96.

Sih, A. 1980. Optimal behavior: Can foragers balance two conflicting demands? *Science* 210: 1041–43.

Skiver, J. 1980. Seasonal resource partitioning patterns of marine calanoid copepod species interactions. *J. Exp. Mar. Biol. Ecol.* 44: 229–45.

Strickler, J. R. 1975. Intra- and interspecific information flow among planktonic copepods: receptors. *Verh. Int. Verein. Limnol.* 19: 2951–58.

———. 1977. Observation of swimming performances of planktonic copepods. *Limnol. Oceanogr.* 22: 165–69.

Strickler, J. R., and S. Twombly. 1975. Reynolds

number, diapause and predatory copepods. *Verh. Int. Verein. Limnol.* 19 : 2943–50.

Tessier, A. J. 1983. Coherence and horizontal movements of patches of *Holopedium gibberum* (Cladocera). *Oecologia* 60 : 71–75.

Treherne, J. E., and W. A. Foster. 1980. The effects of group size on predator avoidance in a marine insect. *Anim. Behav.* 28 : 1119–22.

———. 1981. Group transmission of predator avoidance behavior in a marine insect: The Tralfalgar Effect. *Anim. Behav.* 29 : 911–17.

———. 1982. Group size and anti-predator strategies in a marine insect. *Anim. Behav.* 32 : 536–42.

Viser, M. 1982. Prey selection by the three-spined Stickleback (*Gasterosteus aculeatus* L.). *Oecologia* 33 : 395–402.

Werner, E. E., and D. J. Hall. 1974. Optimal foraging and the size selection of prey by bluegill sunfish, *Lepomis macrochirus*. *Ecology* 55 : 1042–52.

Wiebe, P. H. 1970. Small-scale spatial distribution in oceanic zooplankton. *Limnol. Oceanogr.* 15 : 205–17.

21. Predation Risk: Indirect Effects on Fish Populations

Gary G. Mittelbach

Peter L. Chesson

Predators exert a strong influence on the diets and habitat use of their prey and consequently may indirectly affect resource competition within and between prey species. In this chapter, we explore several examples of how predator-mediated habitat use may affect species and size-class interactions in fish. At the intraspecific level, size-specific risk can lead to habitat segregation between size classes and a reduction in competition between large and small fish. This nonlethal effect of predators appears to play a major role in regulating population size structure in fishes such as the bluegill (Lepomis macrochirus) and perch (Perca spp.). At the interspecific level, predators often concentrate vulnerable size classes (usually small fish) from a number of species into a common protective habitat. As a result, competition may be intensified at certain stages in the life history of a species. We discuss a potential example of this effect in two sunfishes. We then develop a model, based on stock-recruitment relations, that illustrates how predator-induced competition among juveniles can lead to complex interactions between prey species. One particularly interesting feature of the model is the transmission of competitive effects between adults of two prey species, even when they use different resources.

Fish live highly flexible lives. Their diets, habitat use, and growth rates often change dramatically with ontogeny and also may vary between environments. For example, as lake trout (Salvelinus namaycush) in Lake Opeongo, Ontario, increase in size, they shift from feeding on zooplankton to insect larvae to fish. In nearby Cradle Lake, lake trout feed on zooplankton their entire life (Martin 1966, 1970). Such differences in diet may be due to size-specific foraging abilities and prey abundances and can have significant consequences for species interactions (Werner and Gilliam 1984). Recent work has shown that these same types of ontogenetic changes in diet or habitat also may be mediated by predation risk. In experimental studies, fish have been shown to respond to predators by moving to protective habitats (Cerri and Fraser 1983; Werner et al. 1983a; Power et al. 1985), reducing foraging distances (Dill and Fraser 1984), and/or limiting feeding time and intake (Milinski and Heller 1978; Power et al. 1985; Schmitt and Holbrook 1985). Because vulnerability varies with body size, these responses are often highly size-specific. Abundant field observations of habitat segregation by fish of different sizes also provide evidence of the importance of predation risk in determining size-specific resource use (e.g., Jackson 1961; Hobson and Chess 1976; Hall and Werner 1977; Helfman 1978; Hall et al. 1979; Laughlin and Werner 1980; Bray 1981; Mittelbach 1981; Haraldstad and Jonsson 1983; Power 1984; Jones 1984; Sandheinrich and Hubert 1984; Ebeling and Laur 1985). In each of these examples, fish of the most vulnerable sizes are found in the most protected habitats.

The studies cited above clearly document behavioral responses of fish to their predators, and much theoretical and empirical work is directed toward determining how well individuals can balance predation risks and foraging gains. However, little consideration has been given to how these predator-mediated changes in diet and habitat may affect competing species or size classes. For example, if two prey species move to separate habitats or refuges in the presence of the predator, competition between them is diminished; if they move to the same refuge, however, competition may be increased. We here term these effects on prey diets and habitat use as indirect or nonlethal effects of predation, in contrast to the direct effect of predators killing their prey. Note that this type of indirect effect is distinct from those considered by Levine (1976), Holt (1977, 1984) and others; in their examples, indirect effects involve prey deaths and numerical responses by predators and prey.

In this paper, we explore several examples of how predator-mediated habitat use may affect interactions between species and size classes of fish. We first deal with intraspecific consequences of size-class segregation. Bluegills (*Lepomis macrochirus*) and perch (*Perca* spp.) are considered as specific examples of fishes in which predation risk appears to play a major role in regulating population size structure. We then consider indirect effects of predation on interspecific competition. In this section, we discuss a specific example involving the interaction between two species of sunfish. Finally, we present a model illustrating some of the population consequences of predator-mediated habitat use in these and other fishes.

INTRASPECIFIC COMPETITION

Experimental studies by H. S. Swingle and his associates provide some of the clearest evidence for the importance of predation in mediating intraspecific competition in fish. In a series of studies begun in the late 1930s, Swingle experimentally stocked various combinations of predator and prey species in small Alabama ponds and lakes. The two species most frequently studied were the bluegill and its predator, the largemouth bass (*Micropterus salmoides*). Swingle and Smith (1940) found that when bluegills were stocked in the absence of predators, the stocked fish grew rapidly at first (gaining >10 times their weight in 3 months). However, after the bluegill population reproduced and large numbers of small fish were present, growth of the original bluegills ceased. The investigators repeated this experiment many times with the same result; bluegills stocked in the absence of predators inevitably developed populations dominated by small, slow-growing fish, commonly referred to as "stunted" in the fisheries' literature.

The causes for bluegill stunting in the absence of predators are now clear. First, bluegills

have relatively small mouths and are unable to shift to feeding on larger prey as they grow (Werner 1974). Thus, after bluegills reach a length of approximately 50 mm they are in direct competition with all larger conspecifics. Cannibalism is essentially absent because the bluegill is unable to feed effectively on larval fish (Werner 1977). Second, when prey resources are severely limited, small bluegills are at a competitive advantage over larger fish (Mittelbach 1981, 1983) because the higher metabolic costs of large bluegills significantly outweigh their small advantage in prey handling times and prey encounter rates (Mittelbach 1983). Thus, as bluegill numbers in a pond or lake increase, and prey become depressed, only small bluegills are able to maintain positive energy budgets. A related example of asymmetrical competition between size classes has recently been observed by Hamrin and Persson (personal communication) for Vendace (*Coregonus albula*). Finally, bluegills are able to reach sexual maturity at small sizes. Thus, in the absence of larger individuals, these small fish are capable of successfully spawning and maintaining the population.

When largemouth bass are stocked with bluegills, the pattern of stunting noted above may be altered. Studies by H. S. Swingle and others (review in Dillard and Novinger 1975) show that the presence of bass often results in more "balanced" bluegill populations, where a number of bluegill size classes coexist and large fish are able to maintain positive growth rates. Traditionally, the interaction between bluegills and largemouth bass has been viewed as a simple predator–prey system, with balanced populations occurring when bass predation rates effectively control bluegill numbers and reduce intraspecific competition. However, recent work has shown that bass have additional effects on bluegill populations beyond the simple removal of individuals and that these behavioral effects may be equally important in reducing intraspecific competition.

In the presence of largemouth bass, small bluegills show restricted habitat use, feeding only in or near protective vegetation (Mittelbach 1981; Werner et al. 1983a). When predators are removed, however, small bluegills shift to feeding in the open water on zooplankton (*Daphnia*) or from the bare sediments on infauna, if these habitats offer the highest foraging return (Werner et al. 1983b). Thus, small bluegills have the behavioral flexibility to use open habitats but do not do so in the presence of predators. Mittelbach (1981) estimated that small bluegills in a natural lake could increase their net energy gain up to 50% by feeding in the open water on zooplankton instead of feeding in the vegetation. Large bluegills, on the other hand, are relatively invulnerable to predators and feed in either vegetated or open habitats depending on relative foraging gain. Predation risk thus establishes habitat segregation between bluegill size classes, open habitats becoming exclusive resources for larger fish. Largemouth bass therefore have two major effects on bluegill populations: (1) the direct effect of predation on small fish and (2) the indirect effect of modifying bluegill behavior and causing large and small bluegills to partition resources. Both factors potentially reduce competition between size classes.

Recent studies on both European perch (*Perca fluviatilis*) and yellow perch (*P. flavescens*) provide a situation analogous to that in the bluegill. Perch, like bluegills, are prone to producing stunted populations in small lakes. Stunting is reported to be especially common in shallow lakes with relatively homogeneous habitat structure (Eschmeyer 1937; Alm 1946; Persson 1983). In lakes with little habitat diversity, diet overlap between large and small perch is high, and intraspecific competition is intense (Eschmeyer 1937; Persson 1983). However, in lakes containing both vegetated and open habitats, perch segregate by size classes, with small perch occupying the vegetation and adults feeding in the open water or in the more open bottom habitats (Keast 1977; Sandheinrich and Hubert 1984). In these lakes, adult perch grow well, and the populations are not stunted. Although there is no direct experimental evidence that size-class segregation in perch is due to predation risk, it is reasonable to assume that the smallest perch are most vulnerable and that vegetation provides them with significant protection from predators (Glass 1971; Savino and Stein 1982).

Keast (1977) suggested that the ability of perch to shift to feeding on continually larger prey as they grow is the major factor preventing the development of stunting. However, Sandheinrich and Hubert (1984) showed that even when large perch are feeding predominantly on zooplankton, they can grow at average or above average rates. In the population studied by Sandheinrich and Hubert, habitat segregation by different size classes of perch appeared to be the most important factor reducing intraspecific competition.

The studies with perch and bluegills illustrate the potential importance of predator-induced habitat partitioning in regulating competition between size classes. Additional examples of size-class segregation in fish populations abound (see citations in the opening section), and the restriction of vulnerable individuals to protective habitats appears to be a general phenomenon in fish and other size-structured taxa. Other environmental factors besides physical structure also act as prey refuges, e.g., differences in water depth, turbidity, and temperature. These factors also must be considered when one evaluates the importance of predators in promoting size-class segregation. It will be important in future research to disentangle the indirect effects of predators in maintaining resource partitioning from the direct effects of predators killing prey. Both kinds of effects tend to reduce intraspecific competition between size classes of prey.

INTERSPECIFIC COMPETITION

Although a behavioral response to predators may segregate size classes within a species, it is also likely to concentrate vulnerable size classes of different species into a common refuge. For example, as many as five to six species of sunfish co-occur in the vegetation of small lakes during their first 2 to 3 years of life (Werner et al. 1977; Keast 1978a; Laughlin 1979; Mittelbach 1984). While occupying the vegetation, these small fish feed on similar prey, and their diets overlap considerably (Keast 1978b; Laughlin 1979; Mittelbach 1984). Thus, a similar response to predation risk may increase interspecific competition early in a species' life history. McCabe et al. (1983) discuss an interesting case where juvenile chinook salmon (*Oncorhynchus kisutch*) pass through the Columbia River estuary before moving out to sea. While the salmon are in the estuary, their diets overlap broadly with the diets of other (nonsalmonid) species. McCabe et al. hypothesize that the estuary represents a refuge from many oceanic predators. If so, juvenile salmon may experience a period of interspecific competition in the estuary due to the antipredator responses of a number of fish spe-

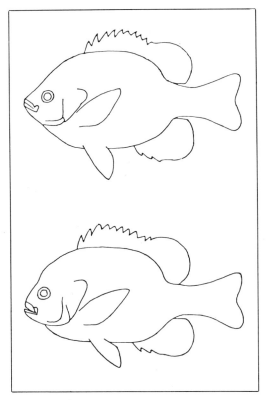

Figure 21.1. *Outline drawings of adult bluegill (top) amd pumpkinseed sunfish (bottom).*

cies. Many other co-occurring species of fish occupy a common protective habitat during vulnerable stages in their life history (Power 1984; Werner 1986; and others). If resources are limited in the refuge, interspecific competition may be intensified by the presence of predators. This effect runs counter to the traditional view that predation reduces interspecific competition only by removing individuals (but see also Holt 1984).

Below we present a specific example illustrating how predation risk may affect the interaction between two species of sunfish, the bluegill and its congener, the pumpkinseed (*L. gibbosus*). First we review the evidence for predator-mediated resource use in these species. Much of this work is drawn from Mittelbach (1984). We then discuss the evidence for resource limitation among juveniles during the time they are restricted to a refuge. Finally, we present a model illustrating how predator-mediated interactions among juveniles may affect species abundances and lead to the transmission of

competitive effects between adults, even though the adults of the two species feed on different prey.

The Bluegill–Pumpkinseed Interaction

Bluegill and pumpkinseed are native to the northeastern United States and southern Canada and commonly co-occur in a variety of small lakes and ponds. The species are quite similar in appearance (fig. 21.1) but differ significantly in their functional morphology and feeding efficiency (Keast 1978b; Mittelbach 1984). These differences in turn cause large bluegills and pumpkinseeds (>75 mm standard length) to feed on different prey types found in separate habitats. In a study of three Michigan lakes, large bluegills foraged primarily on open-water zooplankton (*Daphnia*), and large pumpkinseeds specialized on snails (table 21.1; see also Seaburg and Moyle 1964, Keast 1978b). This separation in diet is directly related to differences in the morphology and foraging ability of the two species. In laboratory experiments, pumpkinseeds feed on larger and stronger-shelled snails than do the bluegill, and take less time to handle the same-size snail. The bluegill, on the other hand, is more efficient at feeding on zooplankton (Mittelbach 1984).

In contrast to the differing diets and habitat use of large bluegills and pumpkinseeds, those of small fish (≤75 mm) are very similar (table 21.1). Small fish of both species feed predominantly in the vegetation, and 80 to 90% of their average seasonal diet is composed of vegetation-dwelling invertebrates (excluding snails). Calculated diet overlaps (Schoener 1970) average about 50% for small bluegills and pumpkinseeds but only 2 to 8% for large fish in lakes with true limnetic zones (Mittelbach 1984). Laughlin (1979) also measured diet overlaps between juvenile pumpkinseeds and juvenile northern longear sunfish (*L. megalotis peltastes*). He found that fish <75 mm total length shared about 50% of their prey in common, whereas diet overlaps among larger fish were <20%. Like young bluegills and pumpkinseeds, juvenile northern longear sunfish are concentrated in the vegetation refuge, whereas adults feed in areas of bare sediments.

The high overlap in diet among small bluegills and pumpkinseeds is due to two factors: (1) piscivorous fish restrict small sunfish to the

Table 21.1. *Average seasonal diets classified by habitat (prey) type for bluegills and pumpkinseeds in three Michigan lakes; sample sizes range from 21 to 44 fish per species per lake; average N = 32*

| | Average diet composition (% dry mass, $\bar{x}$ ± SE) | | | | | |
| | Lawrence Lake | | Three Lakes II | | Three Lakes III | |
	Bluegill	Pumpkinseed	Bluegill	Pumpkinseed	Bluegill	Pumpkinseed
SMALL FISH (≤75 mm SL)						
Vegetation-dwelling prey (nongastropods)	86 ± 4	79 ± 5	78 ± 5	76 ± 5	91 ± 4	84 ± 7
Gastropods	<1	19 ± 5	<1	22 ± 5	0	16 ± 7
Zooplankton	6 ± 3	0	3 ± 1	1 ± 1	5 ± 2	0
LARGE FISH (>75 mm SL)						
Vegetation-dwelling prey (nongastropods)	36 ± 6	25 ± 6	46 ± 9	36 ± 8	67 ± 7	33 ± 7
Gastropods	<1	73 ± 6	0	63 ± 8	<1	67 ± 6
Zooplankton	54 ± 7	0	46 ± 10	0	24 ± 6	<1

Some percents do not sum to 100 because not all prey could be classified among the three habitat types. Three Lakes III is very shallow and contains few large zooplankton.
Adapted from Mittelbach (1984).

vegetation, where they are less vulnerable, and (2) small pumpkinseeds are unable to crush effectively any but the tiniest snails and therefore cannot use the adult resource in the vegetation. Size-specific predation risks and foraging efficiencies therefore create what may be viewed as a two-stage life history for these species. Small fish are confined to a common habitat and share a common prey resource, whereas larger, less vulnerable bluegills and pumpkinseeds shift to feeding on different prey and have reduced diet overlaps. Whether the concentration of juveniles in response to predators influences interspecific competition will depend, in part, on whether juveniles are competing for limited resources in the refuge. For bluegills and pumpkinseeds, the available evidence indicates that resources are limiting in natural lakes.

Growth in fish is very sensitive to prey availability (Werner 1986). When we compared growth rates of young (age 1) bluegills and pumpkinseeds from a series of experimental ponds and natural lakes in southwest Michigan, we found that juvenile growth rates were three to eight times higher in the experimental ponds than in nearby natural lakes (Mittelbach 1986). In both environments, fish were feeding in the vegetation, and water temperatures were similar. However, no competitors were present in the ponds prior to the intro-

duction of bluegills or pumpkinseeds, and invertebrate prey were extremely abundant. Cage experiments conducted in the littoral zone of a natural lake also showed that growth rates of juvenile bluegills were density-dependent over the natural range of bluegill densities found in the lake (Mittelbach 1986). These studies therefore suggest that young bluegills and pumpkinseeds are competing for limited prey resources while they are restricted to the vegetation.

Interestingly, we have failed to detect any major difference in the juvenile competitive abilities of these species. In laboratory feeding experiments, small bluegills and pumpkinseeds harvested natural prey (amphipods) at identical rates from a habitat of *Chara* vegetation (Mittelbach 1984). Studies of growth rates in the field also show no difference in the foraging abilities of these small fish. While occupying the vegetation, both species grew at the same rate within a pond or lake (three natural lakes and two experimental ponds), although growth rates differed significantly between lakes or ponds (Werner and Hall 1979; Mittelbach 1984; 1986). Thus, juvenile bluegills and pumpkinseeds appear to have similar abilities to forage from the vegetation and grow through this stage of their life history at equal rates. The two species also show similar mortality rates when exposed to predation by large-mouth bass (Mittelbach, unpublished data). We would therefore expect juvenile competition between these species to be equal and symmetrical. This does not mean, however, that predator-induced juvenile interactions have no effect on the population dynamics or abundance of these species.

Survivorship and growth rate are positively correlated in many fish (for reviews see Backiel and LeCren 1978; Ware 1975; Werner 1986), and density-dependent growth has long been postulated as the principal factor regulating juvenile mortality in fishes (Ricker and Foerster 1948). Put simply, fish that grow quickly are vulnerable to predators for a shorter time and have higher survival. Gilliam (1982) has further shown mathematically that for stages of the life history where survivorship is already low (i.e., among juvenile fish), a small reduction in growth can cause a very large reduction in survivorship (see also Werner et al. 1983a). Thus, any density-dependent effects on growth that result from behavioral responses of juvenile fish to their predators are likely to have important consequences for survival, recruitment, and overall population density.

In the following section, we develop a model illustrating how predator-induced juvenile competition can lead to interesting and complex interactions between prey species. The model is essentially a two-species stock-recruitment model, and the single species components share much in common with Ware's (1980) bioenergetic approach to stock and recruitment. An important conclusion from the model is that strong competition in the juvenile stage can result in the transmission of negative effects between adults, even though the adults of two species use different resources.

A POPULATION MODEL

Above we have argued that the presence of predators may often cause juvenile fish to compete within a refuge, while having little impact on adult behaviors. To begin to explore the population consequences of this effect, we consider a simple model in which each species has two life stages, juveniles and adults. Juveniles are assumed to occupy the same habitat and compete for the same resources, whereas adults feed on different prey and do not compete directly. This is the essence of the bluegill–pumpkinseed interaction. The two main components of the model are density-dependent adult fecundity and density-dependent juvenile survival. By density-dependent, we mean that per capita fecundity declines with an increase in adult density, whereas the probability of surviving through the juvenile stage decreases as juvenile density increases. Thus, the per capita fecundity will be represented as a function $F(X)$ of the adult density X, and the juvenile survival will be a function $\ell(L)$ of juvenile density L. These functions are shown as monotonically decreasing in figure 21.2a.

The general phenomenon of density-dependent fecundity is well documented in fishes (for reviews, see Nikol'skii 1962; Schopka and Hempel 1973; McFadden 1977; Bagenal 1978; Ware 1980; for examples with bluegills and pumpkinseeds, Parker 1958; Cooper et al. 1971), although the exact shape of the fecundity curve is poorly known for any species. Survival of juveniles is by no means always density-dependent in fishes, but it is to be expected whenever competition for resources affects juvenile growth rates. Any factor that slows individual growth prolongs the time spent in vulnerable size classes and therefore increases the overall probability of death (Beverton 1962; Ricker 1979; Sheperd and Cushing 1980; Lasker 1981; Werner 1986). An added potential effect of reduced juvenile growth is an increase in time to maturity. For simplicity of presentation, the model given here does not allow for this effect. However, in Appendix C, we consider a more general model that includes the possibility that competition delays maturity. There it is shown that the results from the simple model given here carry over in essential details to the more general case. Yield data on density-dependent juvenile survival can be found in LeCren (1962, 1965), Egglishaw and Shackley (1977), Elliott (1984), and Beyerle and Williams (1972) for bluegills.

Per capita survival and fecundity are naturally expected to be monotonically decreasing functions, but total fecundity and survival, which are obtained by multiplying these per capita rates by density, need not be monotonic. There are two possibilities. These total rates may be monotonically increasing, or they may increase at first and then decrease to give a humped form (fig. 21.2b). The monotone case is simply the situation where increasing the number of individuals entering a life stage results in more output from that life stage. The outputs are reproduction or survival and maturation depending on whether the stage is adult or juvenile. The humped situation occurs when density dependence becomes so extreme at high densities that increased input eventually yields less output. Field evidence for the existence of monotonic and humped curves can be found in Burd and Parnell (1973), Lett et al. (1975), Lett (1980), and Ware (1980).

Putting together the juvenile survival and fecundity for a single species leads us to postulate the following dynamical equations:

$$L(t) = X(t)F(X(t)), \tag{21.1a}$$

$$X(t + 1) = L(t)\ell(L(t)), \tag{21.1b}$$

where $L(t)$ is the number of juveniles at time t, X equals the number of adults, F is the fecundity per individual and ℓ is the probability of surviving through the juvenile stage. These equations can be written as one difference equation describing the change in adult numbers from one time to the next:

$$X(t + 1) = X(t)F(X(t))\ell(X(t)F(X(t))). \tag{21.2}$$

The dynamics of the system can be understood by considering a graph of $X(t + 1)$ against $X(t)$, as in figure 21.3. There are three different possible forms of this relationship (A, B, C). The monotonic form A arises when the curves for total juvenile survival and total reproduction are both monotone. This form also can arise when larval survival is humped but peaks for an initial juvenile density higher than the possible total reproduction for the system. The form B, with just a single hump, occurs when either total reproduction or total survival is humped, and it also can occur in some cases where both are humped. The double-hump curve can occur only when both curves are humped but does not arise when the maximum reproduction occurs on an increasing part of the curve for total juvenile survival. Thus, if the humps of these component curves are broad and flat, a double hump is unlikely (see May and Oster 1976 for further discussion or iterated humped curves).

Equilibrium points are determined by the intersection of the curves in figure 21.3 with the 45-degree line, i.e., the line $X(t + 1) = X(t)$. There is at least one equilibrium point when-

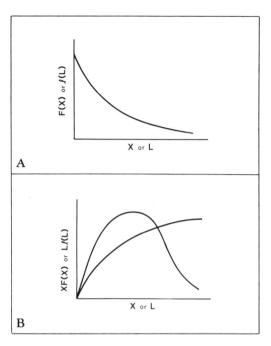

Figure 21.2. a: *Per individual fecundity, F(X), or juvenile survivorship ℓ(L), as a function of adult density, X or juvenile density, L, respectively.* **b:** *Two possible relationships between total fecundity, XF(X), and adult density, X, or between total juvenile survival [i.e., recruitment, Lℓ(L)] and juvenile density, L. Conditions leading to humped-shaped versus monotonic curves are described in the text.*

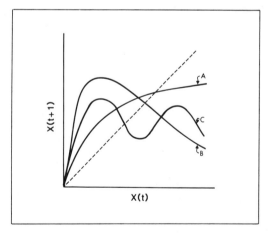

Figure 21.3. *Three possible forms (A,B,C) of the dynamical relationship between X(t) and X(t + 1). The intersection(s) of each curve with the dashed 45-degree line describes the condition where X(t + 1) = X(t) and the population is at equilibrium. The stability of the various equilibria are discussed in the text.*

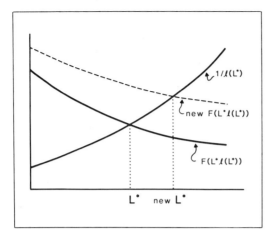

Figure 21.4. *An illustration of how the equilibrium number of juveniles of a single species changes with an increase in per capita fecundity. F(Lℓ(L)) describes per individual fecundity as a function of juvenile density, L, while 1/ℓ(L) describes the average number of juveniles needed to replace one adult. The intersection of the F(Lℓ(L)) and 1/ℓ(L) curves defines the point where per individual fecundity equals juvenile mortality and the population is at equilibrium. Increasing per individual fecundity (new F(Lℓ(L))) moves the intersection point to the right and increases the equilibrium juvenile density from L* to new L*. Since Lℓ(L) is assumed to be monotonic in L, an increase in juvenile density implies an increase in adult density at equilibrium.*

ever $\ell(0)F(0) > 1$, i.e., whenever the species can have positive growth at low density. If there is only one equilibrium and it occurs on a rising part of the curve, that equilibrium is locally stable. Intersections that occur on falling parts of the curve will be locally stable if the slope is no steeper than -1. Extreme density dependence at equilibrium can lead to a slope steeper than -1, and thus to instability. Cyclic or chaotic population dynamics are then likely (May and Oster 1976). Three equilibria will occur in case C if the 45-degree line cuts both humps. The middle equilibrium is necessarily unstable, but the other equilibria will be locally stable if they satisfy the criteria listed above for single equilibria. See May and Oster (1976), Fisher and Goh (1977), Rosenkranz (1983) for further information on stability of equations of this sort.

We now ask how the equilibrium density is changed when parameters of the model are changed. We assume that we are in a situation where stability occurs, which essentially

amounts to a constraint on the severity of density dependence. If X^* and L^* represent respectively the adult and juvenile densities at equilibrium, the following equation must hold:

$$\ell(L^*)F(X^*) = 1 \tag{21.3}$$

i.e., the product of per capita juvenile survival and per capita reproduction must equal 1. This means that a single adult gives rise on average to exactly one adult in the next generation. From equation (21.1b) we see that

$$X^* = L^*\ell(L^*), \tag{21.4}$$

and substituting in equation 21.3 we see that

$$F(L^*\ell(L^*))) = 1/\ell(L^*). \tag{21.5}$$

Figure 21.4 plots $F(L\ell(L))$ and $1/\ell(L)$ as functions of L, and by equation 21.4 their intersection determines the equilibrium juvenile density. The diagram depends on the assumption that $L\ell(L)$ is monotone in L, for otherwise $F(L\ell(L))$ would not be monotonic. This diagram can now be used to determine what would happen to the equilibrium if per capita reproduction, F(X), were increased. This is the sort of situation to expect if more resources were made available to adults. The dotted curve in figure 21.4 illustrates this case, and it is clear that the equilibrium number of juveniles must increase, as might naturally be expected. Also, the monotonicity of total juvenile survival implies that the equilibrium adult density obtained from equation 21.4 will also increase as resources for adults increase. These conclusions must be modified if the total juvenile survival function is not monotonic. Although the equilibrium juvenile densities can still be seen to increase if the reproduction curve is increased, it does not follow that the adult equilibria (there now can be more than one) will increase, because an adult equilibrium may correspond to a declining part of the total juvenile survival curve. In this case, an increase in adult resources will cause the adult equilibrium to decline.

Now consider a system of two interacting species, where juveniles compete interspecifically but adults do not. In this case, we leave the total reproduction functions the same as in the single-species model, except for the addition of a subscript to indicate the species and to allow the possibility that these functions differ between species. Thus we have:

$$L_i(t) = X_i(t)F_i(X_i(t)), \qquad (21.6a)$$

$i = 1, 2$. The juvenile survival equations become

$$X_i(t + 1) = L_i \ell(L(t)) \qquad (21.6b)$$

where now $L(t) = L_1(t) + L_2(t)$ is the total number of juveniles in the system. Here per capita survival of juveniles of species i declines as a function of total juvenile density, not just the density of juveniles of species i.

These equations demonstrate immediately how the adults of one species may affect adults of the other indirectly through juvenile competition. For example, an increase in the adult density of species 1 may increase juvenile output of that species and thereby increase competition among juveniles of both species. The outcome of this will be a reduction of juvenile survival of both species, but in particular the resulting adult density of species 2 will be less.

Simple analyses like the one above are useful in sorting out the effects of temporary perturbations to the system; however, they are of little use for looking at long-term effects, for instance, the effects of a permanent increase in the resources for the adults of one species. To begin to get an idea of the ramifications of permanent changes to a system, or alternatively, the effects of permanent differences between systems, we shall examine the behavior of the equilibrium densities of the two-species system. Throughout we continue to make the assumption that $L\ell(L)$ is monotonic.

Since the juvenile survival rate is the same for both species in this system, the equilibrium must satisfy

$$F_1(X_1^*) = F_2(X_2^*) = 1/\ell(L^*). \qquad (21.7)$$

Thus, the two species also must have equal per capita fecundities at equilibrium. Equilibrium per capita fecundities of the two species need not be equal when the species are present alone, i.e., in a single-species systems. Away from equilibrium the per capita fecundities will, of course, generally be different. In particular, the per capita fecundities at low adult densities, $F_1(0)$ and $F_2(0)$ may be different. The difference between $F_1(0)$ and $F_2(0)$ has impor-

tant consequences for the existence of the two-species equilibrium.

For the sake of argument, suppose $F_1(0) \le F_2(0)$, so that species 1 has a lower per capita fecundity than species 2 at low density. There will be some minimum value X_{2min} of X_2 so that $F_1(0) = F_2(X_{2min})$. The equilibrium value of X_2 in the two-species system must be greater than X_{2min}. The equilibrium total juvenile density must be greater than L_{min}, where $L_{min}\ell(L_{min}) = X_{2min}$; i.e., L_{min} is the number of juveniles of species 2 needed to produce X_{2min} adults of species 2. The sole condition for the existence of a two-species equilibrium then turns out to be

$$F_1(0)\ell(L_{min}) > 1, F_2(0)\ell(L_{min}) > 1, \qquad (21.8)$$

i.e., each species must be able to increase from low density when the number of juveniles in the system is at the infinum ("minimum") of possible equilibrium juvenile densities. If $F_2(0) < F_1(0)$, this condition continues to hold, but L_{min} is defined in terms of X_{1min}, where $F_1(X_{1min}) = F_2(0)$. This result can be understood from the discussion below on the behavior of the equilibrium. Naturally enough, condition 21.8 also can be shown to be equivalent to the requirement that each species can invade a system in which the other species is at its single-species equilibrium.

Local stability of the equilibrium is investigated in Appendix A, where it is shown that in the absence of severe density dependence the equilibrium is always locally stable. This result reflects the fact that intraspecific competition is always stronger than interspecific competition among adults, whereas intra- and interspecific competition are equal among juveniles. Local stability has been investigated in a number of other discrete-time competition models involving age structure (e.g., Hassell and Comins 1976; Travis et al. 1980). Fisher and Goh (1977) show how results on local stability for discrete time competition models may be extended to global stability.

To examine changes in the equilibrium that result from changes in the parameters of the model, we note that equation 21.7 establishes a relationship between X_1^* and X_2^*, and consequently the equilibrium adult density for species 2 can be written as a function of the equilibrium adult density for species 1. We can in fact go further than this: total adult equilibrium density can be written in terms of total ju-

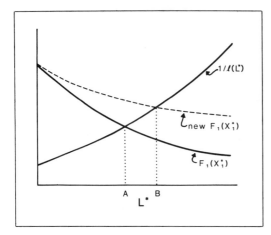

Figure 21.5. *A similar case to fig. 21.4 except that we now consider two interacting species with total juvenile density, L = L₁ + L₂. Illustrated here are the population consequences of increasing the per individual fecundity of species 2. By the arguments given in the text, increasing F₂ (X₂) results in an increase in F₁ (X₁*) as a function of L* and an increase in the total equilibrium juvenile density, L* from A to B. As a result of the increase in F₂(X₂) and L*, the equilibrium number of adults of species 1, X₁*, declines.*

venile equilibrium density according to the relationship

$$L^*\ell(L^*) = X_1^* + X_2^*. \tag{21.9}$$

This equation, together with the relationship between X_1^* and X_2^* allows us to write X_1^* as a function of $L^*\ell(L^*)$, i.e.,

$$X_1^* = f(L^*\ell(L^*)) \tag{21.10}$$

where f is some function. The precise details of how f is obtained are given in Appendix B. In figure 21.5 we use this equation (21.10) to plot $F_1(X_1^*)$ and $1/\ell(L^*)$ as functions of L^*. The intersection of these two curves then gives us the equilibrium total juvenile density, as in the single-species case.

We can now ask what happens to the equilibrium when the reproduction curve of one species is increased. This is not a simple procedure because the reproduction curve is used several times in the definition of the function f relating X_1^* and L^*. However, if F_2 is increased, the effect is to decrease the function $f(L^*\ell(L^*))$ for fixed L^* (see Appendix B). It follows that F_1 $(f(L^*\ell(L^*)))$ must be increased. Thus, the dotted line in figure 21.5 gives the curve $F_1(X_1^*)$

that applies with an increased value of $F_2(X_2)$. Figure 21.5 thus implies that increasing F_2 will shift L^* from the point A to the point B, i.e., the total equilibrium juvenile density increases from A to B. A number of other changes can be deduced:

a. Average juvenile survival of both species, $\ell(L^*)$ must decrease, because increased numbers of juveniles result in a lower per capita juvenile survival. But if equilibrium is to be maintained this means, by equation 21.7, that

b. Equilibirum per capita reproduction of species 1 and 2, $F_1(X_1^*)$ and $F_2(X_2^*)$, must increase. Because the reproduction curve of species 1 has not changed, an increase in per capita fecundity in species 1 can occur only if the equilibrium density of adults declines; therefore,

c. X_1^* must decrease. This is not true of species 2, however, because its reproduction curve has increased. Indeed, the fact that L^* increases means that $L^*\ell(L^*)$, which equals $X_1^* + X_2^*$, must increase. Together with (c) this means that

d. X_2^* must increase.

Thus, we have a complete solution to the changes in adult densities. Do the juvenile densities change in the same way? The answer for species 1 depends on whether the equilibrium adult density occurs on an increasing or decreasing part of the reproduction curve. If increasing, L_1^* will decrease, but if the reproduction curve is decreasing at equilibrium, L_1^* must increase. For species 2, however, these considerations do not apply. It is clear from (b) and (d) that both $F_2(X_2^*)$ and X_2^* must increase; therefore, $X_2^*F_2(X_2^*) = L_2^*$ must increase also.

In summary, a rise in the reproduction curve of species 2 increases the adult density of species 2 and its juvenile density, decreases the adult density of species 1, and decreases or increases the juvenile density of species 1. The equilibrium total juvenile density must, however, be increased.

Consequence (b) above is perhaps counterintuitive, for at first sight it does not seem reasonable that increasing the per capita reproduction of one species should increase the per capita reproduction of its juvenile-stage competitor. However, this is an important consequence of the indirect interaction that occurs between adult densities mediated by the juvenile stage. Total reproduction for species 2 is increased, which means that species 2 increases competition at the juvenile stage. This leads to a reduction in juvenile survival for both species, but in particular the number of adults of species 1 is reduced. Since resources for adults of species 1 remain the same, there is less competition at the adult stage and consequently a greater per capita reproductive output.

Our explanation of the behavior of equilibrium densities, though general in not specifying the particular functions involved, nevertheless deals with a qualitatively narrow range of situations. Fortunately, the conclusions are robust to changes in these specific features, as shown in Appendix C. It also would be useful to look at this sort of model in a stochastic setting because variation is a striking feature of reproduction and juvenile survival in many natural fish communities. However, experience with stochastic models suggests that the present analysis is adequate for small to moderate levels of stochastic variation.

DISCUSSION

The majority of taxa have populations containing numerous size classes (e.g., fish, amphibians, reptiles, most invertebrates, and plants). In many of these organisms, vulnerability to predators is strongly size-dependent (e.g., Dayton 1971; Ware 1975; Paine 1977; Peterson and Wroblewski 1984), the smallest individuals often being the most vulnerable (zooplankton show the reverse trend; see Brooks and Dodson 1965; Hall et al. 1976). Given that foraging in different habitats or on different prey types often carries with it differences in predation risk, we expect predators to routinely exert a strong size-specific influence on the diet and habitat use of a species. We have here considered some of the consequences of these predator-mediated behaviors to exploitative competition in fish. We have limited our discussion to fish in part because they provide some of the clearest examples available. However, recent studies on insects (Sih 1980, 1982), lizards (Stamps 1983), and snails (Schmitt 1982) indicate that many of the processes discussed here apply in broad fashion to other size-structured taxa as well. The chapters by Sih and Abrams in this volume also provide evidence for the importance of preda-

tors in modifying species interactions through changes in prey foraging behaviors and lifestyles.

We expect the behaviors of juveniles or larvae to be most commonly affected by predation risk, since prey vulnerability is inversely related to increasing body size in many species. When behavioral responses to predators lead to size-specific interactions, competition between species can become complex (Werner and Gilliam 1984). One interesting feature of predator-induced competition among juveniles is the potential transmission of competitive effects between adults, even when the adults of two species utilize separate resources. This result points out the danger in judging the strength of interspecific competition from studies in which only one stage in a species life history is examined. For example, many studies of fish food habits have examined only the diets of adult fish because large fish are often more easily caught or are available from commercial landings. These studies of adults then become the basis for inferences about the likelihood of interspecific competition. However, as the two-species stock-recruitment model demonstrates, adults may show strong negative effects even when there is no overlap in their resource use. Thus, studies of resource partitioning in which only a portion of a species life history is considered may be very misleading with regard to the potential for exploitative competition. This same caution of course applies to size-distributed taxa other than fish (e.g., lizards, snails, etc.).

In the specific case of the bluegill and pumpkinseed, it appears that juveniles of the two species have similar foraging abilities and growth rates while they feed in the vegetation refuge. It is interesting to consider the potential causes and consequences of this juvenile equivalence. Hubbell and Foster (1986) suggested that a common juvenile environment has selected for the extreme similarity observed among young tropical trees. By restricting species to a common protective habitat, predators also may indirectly lead to evolutionary convergence in phenotype and foraging behavior among their prey. This convergence may be a factor in the apparent equivalence of juvenile bluegills and pumpkinseeds. In addition, selection for divergence in resource use among juveniles may be constrained by factors that operate on adults. Without a radical metamorphosis, morphological features present in juveniles are carried directly into the adult stage (potentially modified by allometric growth). Thus, selection for differences in morphology and resource use at the juvenile stage may be opposed by selection for resource partitioning among adults.

When adults do exhibit pronounced resource partitioning (as in the bluegill and pumpkinseed), their differences in resource use actually may favor convergence at the juvenile stage. The reason is that compensation at the adult stage can at least partly make up for differences in feeding efficiencies of the two species as juveniles. This compensation lessens the disadvantage experienced by the less efficient species when the juveniles converge on the same resources. It must be kept in mind, however, that within any group of related species the earliest stages of ontogeny are most similar and development tends to proceed from the general to the special (Von Baer's law, Gould 1977). Thus, similarities among juveniles may represent developmental constraints as well as selection for convergence.

Whatever the ultimate causes, if species have similar competitive abilities as juveniles and use separate resources as adults, we might expect their local abundances to be determined largely by adult resource supply. In small lakes and ponds, the bluegill is generally dominant in numbers and biomass (Werner et al. 1977). Dominance by the bluegill may reflect the fact that it is an efficient planktivore and that the total production of zooplankton is generally much greater than the production of littoral prey (due to the relative volumes of the two habitats). Thus, the plankton resource may be able to support the greater biomass of adult fish. Numerous studies of salmonid species in Scandinavian and North American lakes also document a general pattern where the most planktivorous fish species is also the most abundant (Svardson 1976; Nilsson 1963; Andrusak and Northcote 1971; Nilsson and Northcote 1981; review in Werner 1986), although the exact nature of the competitive interactions between species is unclear. When species are not equal competitors as juveniles, bottlenecks to recruitment may result. Neill's (1975) laboratory study of competing cladocera demonstrates how strong, asymmetrical juvenile competition may lead to the elimination of a species from the community, even when adults of that species have abundant and ex-

clusive resources. Although no cases of asymmetrical juvenile competition and recruitment bottlenecks have been documented for fish, their existence seems likely.

Behavioral responses to predators affect not only the interactions within and between fish species but also the dynamics of the fishes' food resource. For example, by concentrating small fish in the vegetation, predation risk indirectly results in increased foraging intensity on littoral cladocera and decreased foraging on limnetic zooplankton. Kerfoot (1975) and Fulton (1985) showed that these differences in foraging intensity between habitats can strongly affect natural zooplankton communities. Werner et al. (1983a) manipulated predation risk on small bluegills and also found a significant indirect effect of a planktivore's predator on zooplankton populations. In their experiment, a small pond was divided in half, and bluegills and piscivorous largemouth bass were stocked on one side of the pond and only bluegills on the other. The bluegills quickly (< 10 days) eliminated Daphnia pulex from the side of the pond without bass. However, in the presence of the bass, the daphnids were able to coexist with bluegills for 20 days. These results are especially striking, considering the proximity of open water and vegetation habitats in this small pond and the fact that small bluegills could feed on Daphnia within a few meters from the vegetation. It is not inconceivable that in natural lakes, where habitats are more clearly separated, the impact of piscivores on the distribution of planktivores may be a major factor determining zooplankton species composition and abundance. In a series of recent studies, Power (1984 and this volume; Power and Matthews 1983; Power et al. 1985) has also shown that predator avoidance by grazing fishes can dramatically affect the distribution of algae in streams.

The only theoretical treatment of the effects of predator-mediated foraging behavior on food web dynamics is that of Abrams (1984). Abrams considered the situation in which a forager must expose itself to greater risk while foraging, and adjusts its foraging time adaptively to maximize fitness. He found that under this situation the adjustment of foraging time to predation risk can result in (1) interactions between the forager's predator and the forager's food, (2) predator self-limitation, and (3) potential interactions between food species.

From the above discussion it is clear that many of these same results can occur if foragers modify habitat use rather than foraging time in the presence of predators. Abrams was also able to show that these indirect effects can often be equal to or larger than the direct effects between trophic levels.

Although it can be shown that predators affect species' diets and habitat use, few general predictions can be made concerning what will happen to interactions if predators are removed. For example, in the absence of a size- or habitat-specific predator, prey may shift their resource use and no longer face the same suites of competitors. However, it is difficult in any reasonably complex community to predict the eventual competitive interactions that result because the response of each species will depend in part on the responses of other species and on changes in resource dynamics due to the shift in foraging pressure. In addition, the direct effect of the predator on prey mortality will be changed. Clearly, one way to begin to sort out these effects of predators on community structure is through field manipulations. Ecologists working in the intertidal zone have long employed predator removal-and-addition experiments, and their studies reveal a complex array of direct and indirect effects in many communities (e.g., Connell 1961; Paine 1966; Dayton 1971; Menge 1976; Lubchenco and Menge 1978; Garrity and Levings 1981). However, these studies have not generally measured the relative importance of these two factors. A major challenge in designing future experiments will be to separate out the direct and indirect effects of predators and to quantify their relative impact. If the impact of predators on prey foraging behaviors and habitat use turns out to be significant in regulating populations, ecologists will have to reevaluate many traditional views of how predation and competition act in structuring communities.

ACKNOWLEDGMENTS

We thank Peter Abrams, Jim Bence, Craig Osenberg, and Earl Werner for their helpful comments and discussion, Jim Gilliam for his contribution to the early formulation of the model, and Libby Marschall for drafting the figures. This work was supported in part by NSF grants DEB 81-04697 and BSR 81-04697, A01.

REFERENCES

Abrams, P. A. 1984. Foraging time optimization and interactions in food webs. *Amer. Naturalist* 124 : 80–96.

Alm, G. 1946. Reasons for the occurrence of stunted populations with special regard to the perch. *Inst. Freshwater Res. Drottningholm Rep.* 25 : 1–146.

Andrusak, H., and T. G. Northcote. 1971. Segregation between adult cutthroat trout (*Salmo clarki*) and Dolly Varden (*Salvelinus malma*) in small coastal British Columbia lakes. *J. Fish. Res. Board Can.* 28 : 1259–68.

Backiel, T., and E. D. LeCren. 1978. Some density relationships for fish population parameters. In S. D. Gerking (ed.), *Ecology of freshwater fish production*, pp. 279–302. New York: John Wiley.

Bagenal, T. B. 1978. Aspects of fish fecundity. In S. D. Gerking (ed.), *Ecology of freshwater fish production*, pp. 75–101. New York: John Wiley.

Beverton, R. J. H. 1962. Long-term dynamics of certain North Sea fish populations. In E. D. LeCren and H. W. Holdgate (eds.), *The exploitation of natural animal populations*, pp. 242–259. Oxford: Blackwell.

Beyerle, G. B., and J. Williams. 1972. *Survival, growth, and production by bluegills subjected to population reduction in ponds.* Michigan Department of Natural Resources Report No. 273.

Bray, R. N. 1981. The influence of water currents and zooplankton densities on daily foraging movements of blacksmith, *Chromis punctipinnis*, a planktivorous reef fish. *Fish. Bull.* 78 : 829–41.

Brooks, J. L., and S. J. Dodson. 1965. Predation, body size and composition of plankton. *Science* 150 : 28–35.

Burd, A. C., and W. G. Parnell. 1973. The relationship between larval abundance and stock in the North Sea herring. *Rapp. P.-V. Reun. Cons. Int. Explor. Mer.* 164 : 30–36.

Cerri, R. D., and D. F. Fraser. 1983. Predation and risk in foraging minnows: Balancing conflicting demands. *Amer. Naturalist* 121 : 552–61.

Connell, J. H. 1961. The influence of interspecific competition and other factors on the distribution of the barnacle *Chthamalus stellotus*. *Ecology* 42 : 710–23.

Cooper, E. L., C. C. Wagner, and G. E. Krantz. 1971. Bluegills dominate production in a mixed population of fishes. *Ecology* 52 : 280–90.

Dayton, P. K. 1971. Competition, disturbance and community organization: The provision and subsequent utilization of space in a rocky intertidal community. *Ecol. Monogr.* 41 : 351–89.

Dill, L. M., and A. H. G. Fraser. 1984. Risk of predation and the feeding behavior of juvenile coho salmon (*Oncorhynchus kisutch*). *Behav. Ecol. Sociobiol.* 16 : 65–71.

Dillard, J. G., and G. D. Novinger. 1975. Stocking largemouth bass in small impoundments. In H.

Clepper (ed.), *Black bass biology and management*, pp. 459–79. Washington, D.C.: Sport Fishing Institute.

Ebeling, A. W., and D. R. Laur. 1985. The influence of plant cover on surfperch abundance at an offshore temperate reef. *Environ. Biol. Fish* 12 : 169–79.

Egglishaw, H. J., and P. E. Shackley. 1977. Growth, survival and production of juvenile salmon and trout in a Scottish stream, 1966–1975. *J. Fish. Biol.* 11 : 647–72.

Elliott, J. M. 1984. Numerical changes and population regulation in young migratory trout *Salmo trutta* in a Lake District stream 1966–1983. *J. Anim. Ecol.* 53 : 327–50.

Eschmeyer, R. W. 1937. Some characteristics of a population of stunted perch. *Pap. Mich. Acad. Sci. Arts. Lett.* 22 : 613–28.

Fisher, M. E., and B. D. Goh. 1977. Stability in a class of discrete time models of interacting populations. *J. Math. Biol.* 4 : 265–74.

Fulton, R. S., III. 1985. Predator-prey relationships in an estuarine littoral copepod community. *Ecology* 66 : 21–29.

Garrity, S. D., and S. C. Levings. 1981. A predator-prey interation between two physically and biologically constrained tropical rocky shore gastropods: direct, indirect and community effects. *Ecol. Monogr.* 51 : 267–86.

Gilliam, J. F. 1982. Habitat use and competitive bottlenecks in size-structured fish populations. Ph.D. thesis, Michigan State University.

Glass, N. R. 1971. Computer analysis of predation energetics in the largemouth bass. In B. C. Patten (ed.), *Systems analysis and simulation in ecology*; Vol., 1, pp. 325–63. New York: Academic Press.

Gould, S. J. 1977. *Ontogeny and phylogeny*. Cambridge, Mass.: Harvard University Press.

Hall, D. J., and E. E. Werner. 1977. Seasonal distribution and abundance of fishes in the littoral zone of a Michigan lake. *Trans. Amer. Fish. Soc.* 106 : 545–55.

Hall, D. J., S. T. Threlkeld, C. W. Burns, and P. H. Crowley. 1976. The size-efficiency hypothesis and the size structure of zooplankton communities. *Annu. Rev. Ecol. Syst.* 7 : 177–208.

Hall, D. J., E. E. Werner, J. F. Gilliam, G. G. Mittelbach, D. Howard, C. G. Doner, J. A. Dickerman, and A. J. Stewart. 1979. Diel foraging behavior and prey selection in the golden shiner (*Notemigonous crysoleucas*). *J. Fish. Res. Board Can.* 36 : 1029–39.

Haraldstad, O., and B. Jonsson. 1983. Age and sex segregation in habitat utilization by brown trout in a Norwegian lake. *Trans. Amer. Fish. Soc.* 112 : 27–37.

Hassell, M .P., and H. N. Comins. 1976. Discrete time models for two-species competition. *Theor. Popul. Biol.* 9 : 202–21.

Helfman, G. S. 1978. Patterns of community structure in fishes: summary and overview. *Environ. Biol. Fish* 3 : 129–48.

Hobson, E. S., and J. R. Chess. 1976. Trophic interactions among fishes and zooplankters near shore at Santa Catalina Island, California. *Fish. Bull.* 74 : 567–98.

Holt, R. D. 1977. Predation, apparent competition and the structure of prey communities. *Theor. Popul. Biol.* 12 : 197–229.

———. 1984. Spatial heterogeneity, indirect interactions, and the coexistence of prey species. *Amer. Naturalist* 124 : 377–406.

Hubbell, S., and R. Foster. 1986. Biology, chance, and history and the structure of tropical rain forest tree communities. In J. Diamond and T. Case (eds.), *Ecological communities*, pp. 314–29. New York: Harper and Row.

Jackson, P. B. N. 1961. The impact of predation especially by the tiger fish (*Hydrocynus vittatus* Cast.) on African freshwater fishes. *Proc. Zool. Soc. Lond.* 136 : 603–22.

Jones, G. P. 1984. The influence of habitat and behavioral interactions on the local distribution of the wrasses, *Pseudolabrus celidotus*. *Environ. Biol. Fish* 10 : 43–58.

Keast, A. 1977. Diet overlaps and feeding relationships between the year classes in the yellow perch (*Perca flavescens*). *Environ. Biol. Fish* 2 : 53–70.

———. 1978a. Trophic and spatial interrelationships in the fish species of an Ontario temperate lake. *Environ. Biol. Fish* 3 : 7–31.

———. 1978b. Feeding interrelations between age groups of pumpkinseed sunfish (*Lepomis gibbosus*) and comparisons with the bluegill sunfish (*L. macrochirus*). *J. Fish. Res. Board. Can.* 35 : 12–27.

Kerfoot, W. C. 1975. The divergence of adjacent populations. *Ecology* 56 : 1298–1313.

Lasker, R. 1981. Marine fish larvae: Morphology, ecology and relation to fisheries. Seattle, Wash.: Washington Sea Grant Program.

Laughlin, D. R. 1979. Resource and habitat use patterns in two coexisting sunfish species (*Lepomis gibbosus* and *Lepomis megalotis peltastes*). Ph.D. thesis, Michigan State University.

Laughlin, D. R., and E. E. Werner. 1980. Resource partitioning in two coexisting sunfish: pumpkinseed (*Lepomis gibbosus*) and northern longear sunfish (*Lepomis megalotis peltastes*). *Can. J. Fisheries Aquat. Sci.* 37 : 1411–20.

LeCren, E. D. 1962. The efficiency of reproduction and recruitment in freshwater fish. In E. D. LeCren and H. W. Holdgate (eds.), *The exploitation of natural animal populations*, pp. 283–96. Oxford: Blackwell.

———. 1965. Some factors regulating the size of populations of freshwater fish. *Mitt. Internat. Verein. Limnol.* 13 : 88–105.

Lett, P. F. 1980. A comparative study of the recruitment mechanicms of cod and mackerel, their interaction, and its implications for dual stock assessment. *Can. Tech. Rep. Fish. Aquat. Sci.* No. 988.

Lett, P. F., A. C. Kohler, and D. N. Fitzgerald. 1975. The role of stock biomass and temperature in the recruitment of southern Gulf of St. Lawrence Atlantic cod *(Gadus morhua). J. Fish. Res. Board Can.* 32 : 1613–27.

Levine, S. H. 1976. Competition interactions in ecosystems. *Amer. Naturalist* 110 : 903–10.

Lubchenco, J., and B. A. Menge. 1978. Community development and persistence in a low rocky intertidal zone. *Ecol. Monogr.* 48 : 67–94.

Martin, N. V. 1966. The significance of food habits in the biology, exploitation, and management of Algonquin park, Ontario, lake trout. *Trans. Amer. Fish. Soc.* 95 : 415–22.

———. 1970. Long-term effects of diet on the biology of the lake trout and the fishery in Lake Opeonogo, Ontario. *J. Fish. Res. Board. Can.* 27 : 125–46.

May, R. M. 1974. Stability and complexity in model ecosystems. (2nd ed.). Princeton, N.J.: Princeton University Press.

May, R. M., and G. F. Oster. 1976. Bifurcations and dynamic complexity in simple ecological models. *Amer. Naturalist* 110 : 573–99.

McCabe, G. T. Jr., W. D. Muir, R. L. Emmett, and J. T. Durkin. 1983. Interrelationships between juvenile salmonids and nonsalmonid fish in the Columbia River estuary. *Fish. Bull.* 81 : 815–26.

McFadden, J. T. 1977. An argument supporting the reality of compensation in fish and a plea to let them exercise it. In W. VanWinkle (ed.), *Assessing the effects of power-plant-induced mortality in fish populations*, pp. 153–83. New York: Pergamon Press.

Menge, B. A. 1976. Organization of the New England rocky intertidal community: role of predation, competition, and environmental heterogeneity. *Ecol. Monogr.* 46 : 355–93.

Milinski, M., and R. Heller. 1978. Influence of a predator on the optimal foraging behavior of sticklebacks *(Gasterosteus aculeatus L.). Nature* 275 : 642–44.

Mittelbach, G. G. 1981. Foraging efficiency and body size: a study of optimal diet and habitat use by bluegills. *Ecology* 62 : 1370–86.

———. 1983. Optimal foraging and growth in bluegills. *Oecologia* 59 : 157–62.

———. 1984. Predation and resource partitioning in two sunfishes (Centrarchidae). *Ecology* 65 : 499–513.

———. 1986. Predator-mediated habitat use: some consequences for species interactions. *Envrion. Biol. Fish* 16 : 159–69.

Neill, W. E. 1975. Experimental studies of microcrustacean competition, community composition and efficiency of resource utilization. *Ecology*

56 : 809–26.

Nikol'skii, G. V. 1962. On some adaptations to the regulation of population density in fish species with different types of stock structure. In E. D. LeCren and H. W. Holdgate (eds.), *The exploitation of natural animal populations*, pp. 265–82. Oxford: Blackwell.

Nilsson, N.-A. 1963. Interaction between trout and char in Scandinavia. *Trans. Amer. Fish. Soc.* 92 : 276–85.

Nilsson, N.-A., and T. G. Northcote. 1981. Rainbow trout *(Salmo gairdneri)* and cutthroat trout *(S. clarki)* interactions in coastal British Columbia lakes. *Can. J. Fisheries Aquat. Sci.* 38 : 1228–46.

Paine, R. T. 1966. Food web complexity and species diversity. *Amer. Naturalist* 100 : 65–75.

———. 1977. Size-limited predation: an observational and experimental approach with the *Mytilus-Pisaster* interaction. *Ecology* 57 : 858–73.

Parker, R. A. 1958. Some effects of thinning on a population of fishes. *Ecology* 39 : 304–17.

Persson, L. 1983. Food consumption and competition between age classes in a perch *Perca fluviatilis* population in a shallow eutrophic lake. *Oikos* 40 : 197–207.

Peterson, I., and J. S. Wroblewski. 1984. Mortality rate of fishes in the pelagic ecosystem. *Can. J. Fisheries Aquat. Sci.* 41 : 1117–20.

Power, M. E. 1984. Depth distributions of armored catfish: predator-induced resource avoidance? *Ecology* 65 : 523–29.

Power, M. E., and W. J. Matthews. 1983. Algae-grazing minnows *(Campostoma anomalum)*, piscivorous bass *(Micropterus* spp.), and the distribution of attached algae in a small prairie-margin stream. *Oecologia* 60 : 328–32.

Power, M. E., W. J. Matthews, and A. J. Stewart. 1985. Grazing minnows, piscivorous bass and stream algae: dynamics of a strong interaction. *Ecology* 66 : 1448–56.

Ricker, W. E. 1979. Growth rates and models. In W. S. Hoar, D. J. Randall, and J. R. Brett (eds.), *Fish physiology*; vol. 8, pp. 677–743. New York: Academic Press.

Ricker, W. E., and R. E. Foerster. 1948. Computation of fish production. *Bull. Bingham Oceanog. Coll.* 11 : 173–211.

Rosenkranz, G. 1983. On global stability of discrete population models. *Math. Biosci.* 64 : 227–31.

Sandheinrich, M. B., and W. A. Hubert. 1984. Intraspecific resource partitioning by yellow perch *(Perca flavescens)* in a stratified lake. *Can. J. Fish. Res. Bd. Can.* 41 : 1745–52.

Savino, J. F., and R. A. Stein. 1982. Predator-prey interaction between largemouth bass and bluegills as influenced by simulated, submersed vegetation. *Trans. Amer. Fish. Soc.* 111 : 255–66.

Schmitt, R. J. 1982. Consequences of dissimilar defenses against predation in a subtidal marine community. *Ecology* 63 : 1588–1601.

Schmitt, R. J., and S. J. Holbrook. 1985. Patch selection by juvenile black surfperch (Embiotocidae) under variable risk: interactive influence of food quality and structural complexity. *J. Exp. Mar. Biol. Ecol.* 85 : 269–85.

Schoener, T. W. 1970. Nonsynchronous spatial overlap of lizards in patchy habitats. *Ecology* 51 : 408–18.

Schopka, S. A., and G. Hempel. 1973. The spawning potential of populations of herring (*Clupea harengus* L.) and cod (*Gadus morhua* L.) in relation to the rate of exploitation. *Rapp. Cons. Int. Expl. Mer.* 164 : 178–85.

Seaburg, K. G., and J. B. Moyle. 1964. Feeding habits, digestion rates, and growth of some Minnesota warm water fishes. *Trans. Amer. Fish. Soc.* 93 : 269–85.

Sheperd, J. G., and D. H. Cushing. 1980. A mechanism for density-dependent survival of larval fish as the basis of a stock-recruitment relationship. *J. Cons. Int. Explor. Mer.*

Sih, A. 1980. Optimal behavior: can foragers balance two conflicting demands? *Science* 210 : 1041–43.

———. 1982. Foraging strategies and the avoidance of predation by an aquatic insect, *Notonecta hoffmanni*. *Ecology* 63 : 786–96.

Stamps, J. A. 1983. The relationship between ontogenetic habitat shifts, competition and predator avoidance in a juvenile lizard (*Anolis aeneus*). *Behav. Biol. Sociobiol.* 12 : 19–33.

Svardson, G. 1976. Interspecific population dominance in fish communities of Scandinavian lakes. *Inst. Freshwater Res. Dorttningholm Rep.* 55 : 144–71.

Swingle, H. S., and E. V. Smith. 1940. Experiments on the stocking of fish ponds. *Transactions of the North Amer. Wildlife Conference* 5 : 267–76.

Travis, C. C., W. M. Post, D. L. DeAngelis, J. Perkowski. 1980. Analysis of compensatory Leslie matrix models for competing species. *Theor. Popul. Biol.* 18 : 16–30.

Ware, D. M. 1975. Relation between egg size, growth, and natural mortality of larval fish. *J. Fish. Res. Board. Can.* 32 : 2503–12.

———. 1980. Bioenergetics of stock and recruitment. *Can. J. Fisheries Aquat. Sci.* 37 : 1012–24.

Werner, E. E. 1974. The fish size, prey size, handling time relation in several sunfishes and some implications. *J. Fish. Res. Board Can.* 31 : 1531–86.

Werner, E. E. 1977. Species packing and niche complementarity in three sunfishes. *Amer. Naturalist* 111 : 553–78.

———. 1986. Species interactions in freshwater fish communities. In J. Diamond and T. Case (eds.), *Ecological communities*, pp. 344–58. New York: Harper and Row.

Werner, E. E., and J. F. Gilliam. 1984. The ontogenetic niche and species interactions in size-structured populations. *Annu. Rev. Ecol. Syst.* 15 : 393–426.

Werner, E. E., J. F. Gilliam, D. J. Hall, and G. G. Mittelbach. 1983a. An experimental test of the effects of predation risk on habitat use in fish. *Ecology* 64 : 1540–48.

Werner, E. E., and D. J. Hall. 1979. Foraging efficiency and habitat switching in competing sunfishes. *Ecology* 60 : 256–64.

Werner, E. E., D. J. Hall, D. R. Laughlin, D. J. Wagner, L. A. Wilsmann, and F. C. Funk. 1977. Habitat partitioning in a freshwater fish community. *J. Fish. Res. Board. Can.* 34 : 360–70.

Werner, E. E., G. G. Mittelbach, D. J. Hall, and J. F. Gilliam. 1983b. Experimental tests of optimal habitat use in fish: the role of relative habitat profitability. *Ecology* 64 : 1525–39.

APPENDIX A

The equations for the two-species model can be written as

$$\mathbf{X}(t + 1) = \mathbf{G}(\mathbf{H}(\mathbf{X}(t))), \qquad (21.A1)$$

where $\mathbf{X}(t) = (X_1(t), X_2(t))'$, and $\mathbf{G}$ and $\mathbf{H}$ are vector-valued functions with $G_i(\mathbf{L}) = L_i\ell(L)$, $H_i(\mathbf{X}) = X_iF_i(X_i)$. The local stability of the system is determined by the matrix $\mathbf{A}$ with $i - j$th element equal to $\partial X_i(t + 1)/\partial X_j(t)$,

evaluated at equilibrium. Application of the chain rule to (A1) shows that

$$\mathbf{A} = \mathbf{BC}, \qquad (21.A2)$$

where $\ell(L^*)\mathbf{B} = (\partial G_i/\partial L_j)$ and $F_i(X_i^*)\mathbf{C} = (\partial H_i/\partial X_j)$, evaluated at equilibrium. We make use here of the equilibrium equation $\ell(L^*)F_i(X_i^*) = 1$.

Defining $\beta = -d \log \ell(L)/dL|_{L=L}^*$ and

$\gamma_i = -d \log F_i(X_i)/dX_i|_{X_i=X_i^*}$, then β and γ_i represent the magnitude of density dependence in the juvenile and adult populations respectively. The matrices **B** and **C** can be now written as follows:

$$\mathbf{B} = \begin{bmatrix} 1 - L_1^*\beta & -L_1^*\beta \\ -L_2^*\beta & 1 - L_2^*\beta \end{bmatrix} \quad ; \quad (21.A3)$$

$$\mathbf{C} = \begin{bmatrix} 1 - X_1^*\gamma_1 & 0 \\ 0 & 1 - X_2^*\gamma_2 \end{bmatrix}.$$

The equilibrium will be locally stable if the eigenvalues of A are less than 1 in absolute value. Using the fact the $(L_1^* + L_2^*)\beta < 1$, because $L\ell(L)$ is monotone, the Schur-Cohn criterion (May 1974, p. 220) is easily applied to **A** to show that the equilibrium will be locally stable whenever $|1 - X_i^*\gamma_i| < 1$, $i = 1,2$. Since $\gamma_i > 0$, this can be rewritten as

$$X_i^*\gamma_i < 2, i = 1,2. \qquad (21.A4)$$

Condition 21.A4 is a sufficient condition for local stability, not a necessary condition. However, it becomes both necessary and sufficient whenever $X_1^*\gamma_1 = X_2^*\gamma_2$.

APPENDIX B

Equation 21.7 in the text allows us to write

$$X_2^* = g(X_1^*) \qquad (21.B1)$$

where $g(X_1^*) = F_2^{-1}(F_1(X_1^*))$. Because both F_1 and F_2 are decreasing functions, g is an increasing function. Also, if F_2 is increased as a function, then the function g must also increase. Combining equations 21.B1 and 21.9 in the text we get

$$L^*\ell(L^*) = X_1^* + g(X_1^*). \qquad (21.B2)$$

Since the RHS of (B2) is an increasing function, we can write X_1^* in terms of $L^*\ell(L^*)$,

$$X_1^* = f(L^*\ell(L^*)) \qquad (21.B3)$$

where f is the inverse function of $X + g(X)$. Note that f is necessarily an increasing function, and also that increasing the function g must necessarily decrease the function f, which means that the function $F_1(f(L^*\ell(L^*))$ must be increased. It follows that increasing the function F_2 necessarily increases $F_1(f(L^*\ell(L^*))$ as a function of L^*.

APPENDIX C

Generalizations of the Model

In the text, we presented a simplified model to illustrate the effects of juvenile competition and adult fecundity on species' abundances. Here we show that conclusions concerning changes in equilibrium abundances, derived from the simple model, also apply when we allow for overlapping generations and a delay in maturity due to juvenile competition.

We use the symbol $L_{ij}(t)$ to refer to the number of juveniles of species i and age j, $j = 0,1,2,\dots$ We assume that the total amount of competition that juveniles experience is determined by a weighted sum, L(t), of the juveniles densities:

$$L(t) = \Sigma_{ij} c_j(L_{1j}(t) + L_{2j}(t)), \qquad (21.C1)$$

where the c_j are constants allowing different age classes to contribute differently to competitive pressure.

The number of adults of species i is determined by the number of juveniles that mature, and the number of adults surviving from the previous time period:

$$X_i(t + 1) = \Sigma_j L_{ij}(t)a_j(L(t)) + (1 - d_i)X_i(t),$$
$$(21.C2)$$

where d_i is the adult death rate of species i and $a_j(L(t))$ is the density-dependent fraction of larvae of age j that mature before age $j + 1$.

Juvenile dynamics are described as follows:

$$L_{i0}(t) = X_i(t)F_i(X_i(t)), \qquad (21.C3)$$

$$L_{ij+1}(t + 1) = L_{ij}(t)p_j(L(t)), \tag{21.C4}$$

where $p_j(L(t))$ is the density-dependent fraction of juveniles of age class j surviving, but not maturing, from time t to time t + 1.

At equilibrium, equation 21.C4 implies

$$L_{ij}^* = L_{i0}^* \prod_{v=0}^{j-1} p_v(L^*) = L_{i0}^*\ell_j(L^*) \tag{21.C5}$$

Thus,

$$L^* = (L_{10}^* + L_{20}^*) \Sigma c_j\ell_j(L^*) \tag{21.C6}$$

$$= L_0^* s(L^*),$$

where L_0^* is the sum of age 0 juveniles and $s(L^*)$ is the weighted sum of the $\ell_j(L^*)$.

Expression 21.C2 implies

$$d_i X_i^* = \Sigma_j L_{ij}^* a_j(L^*) \tag{21.C7}$$

$$= L_{i0}^* \Sigma_j \ell_j(L^*)a_j(L^*),$$

$$= L_{i0}^*\theta(L^*), \text{ say,}$$

and therefore that

$$d_1 X_1^* + d_2 X_2^* = L_0^*\theta(L^*) \tag{21.C8}$$

$$= L^*\lambda(L^*),$$

where $\lambda(L^*) = \theta(L^*)/s(L^*)$. Equation 21.C8 is a generalization of equation 21.9 in the text.

To obtain a generalization of equation 21.7, we note from equation 21.C3 that

$$L_{i0}^* = X_i^* F_i(X_i^*). \tag{21.C9}$$

Combining this with equation 21.C7 we get

$$F_1(X_1^*)/d_1 = F_2(X_2^*)/d_2 = 1/\theta(L^*), \tag{21.C10}$$

which is the sought generalization of equation 21.7.

To obtain generalizations of the results in the text, we must assume that $L\lambda(L)$ is an increasing function of L and that $\theta(L)$ is a decreasing function of L. When these conditions hold, all of the conclusions in the text concerning changes in L^*, X_i^* and $F_i^*(X_i^*)$, $i = 1,2$, continue to apply. The conclusions in the text about the L_i^* become conclusions about the L_{i0}^* here.

To interpret the above assumptions, note that $\theta(L)$ will be a decreasing function of L if the total fraction of any cohort that survives to maturity decreases as a function of the competition it experiences, when the amount of competition is constant over time. This decrease in the fraction maturing is consistent with decreased age-specific survival rates, delayed maturity with no change in age-specific survival, or both of these in combination. The assumption that $L\lambda(L)$ increases with L merely requires that the competition be not too extreme.

We have not determined the conditions under which the equilibrium of this more complex model will be stable. It is likely, however, that density dependence must be milder for stability of this model as a result of the time lags introduced by allowing maturity over a number of ages.

22. Predator Avoidance by Grazing Fishes in Temperate and Tropical Streams: Importance of Stream Depth and Prey Size

Mary E. Power

Grazing fishes can strongly affect benthic algal distributions in streams. Distributions of grazing fishes along depth gradients and among pools in streams appear constrained by depth- and size-specific predators. Large (> 3 cm long) grazing fishes avoid shallow (< 20 cm) water, where wading and diving predators feed most commonly and effectively. As a result, bands of algae are maintained along shallow stream margins in a secondary rain forest stream in Panama and in a prairie-margin stream in south-central Oklahoma, despite intense grazing and scant algal standing crops in deeper areas. Small (< 3 cm long) grazing fishes, postulated to be less susceptible to wading and diving predators and more susceptible to swimming predators, were more abundant than large grazers in shallow water in Panamanian and Oklahoman streams. Shallow areas appear to be refuges from competitors and swimming predators for small grazers, which were nonetheless limited at densities below those necessary to deplete shallow algae. "Bigger–deeper" distributions of grazing catfishes in a Panamanian stream remained unchanged from the rainy to the dry season during 2 years, despite a two- to three-fold contraction of "critical habitat" (streambed under > 20 cm) and evidence from somatic growth rates of increasing food limitation.

On a larger scale, grazing fishes affect variation in algal standing crops among stream pools, as well as along depth gradients. In the Rio Frijoles of central Panama, algal standing crops were uniformly low in pools, despite large differences in primary productivities related to forest canopy cover. Large armored catfish in this stream appeared able to outgrow most swimming predators and could move among pools sufficiently to track and damp out incipient variation in algal standing crop. In contrast, pool-to-pool variation in algal standing crop was amplified by grazing minnows (*Campostoma anomalum*), which were excluded from some pools by piscivorous bass.

Grazing fishes in both Panama and Oklahoma avoid foraging in dangerous areas (shallow water, bass pools), even when their algal food is abundant there and scant elsewhere. Depth and size-specific predators, by restricting access to food for grazing stream fishes, can influence their demographic rates, carrying capacities of streams for their populations, and their effects on stream flora. Where grazing fishes are important, experimental studies of the effect of stream depth, an easily measured and manipulated variable, on foraging by grazing fishes and their predators should provide much insight into the complex ecological consequences of a single environmental variable.

Grazing fishes are abundant in many temperate and tropical streams, and are potentially voracious consumers of algae. Algae, in turn, can grow rapidly. High rates of algal growth and fish consumption often couple to produce patterns in streams that reveal the spatial distribution of herbivory. What behaviors and constraints underlie the distributions of grazing fishes in streams? Most streams meander and are therefore made up of sequences of pools separated by shallower riffles. Pools are habitats for larger species and size classes of stream fish. What factors control fish densities in particular pools relative to others in the stream? Can grazing fishes track pool-to-pool variation in resources such as food? To what extent are fish distributions constrained by factors like predators? One can ask similar questions about distributions of fishes within stream pools: along depth gradients or on different substrates. To answer these questions, one must learn about the risks and opportunities for fish in stream environments and also about fish capabilities. How mobile are grazing fishes and their predators; in particular, what are thresholds that determine whether they will cross barriers such as riffles? How good are their perceptions and memories of resource availability or predation risk? What compromises will they make when resource acquisition and risk avoidance require different behaviors? Do tradeoffs vary with hunger, age, body size, season, or among species?

Predators have been shown or postulated to restrict the foraging areas of their prey in marine and freshwater habitats (Stein and Magnuson 1976; Sih 1982; Seghers 1970; Randall 1965; Stein 1979; Cooper 1984; Garrity and Levings 1981; Mittelbach 1984; Werner et al. 1983; Power 1984a). By making portions of habitat too dangerous to use, predators may limit food available to prey and alter the effects of these prey on their communities. When fish that graze benthic algae avoid certain areas, the algae released from grazing often attain conspicuous standing crops (Randall 1965; Ogden and Lobel 1978; Power and Matthews 1983; Power et al., 1985). Clear patterns, such as grazed "halos" around rubble cover in coral reefs (Randall 1965) or bands of algae along stream margins (Power 1984a) may indicate the spatial distribution of predator-induced resource avoidance by grazing fishes (Ogden and Zieman 1977; Earle 1972; Hay 1984 and references therein). Significant indirect effects on other biota may result, because ungrazed algae may physically modify the habitat and provide food or cover for a wide range of organisms (Estes and Palmisano 1974; Duggins 1980; Paine 1980; Hynes 1970).

Here I will describe patterns of distribution of algae-grazing fishes, their algal food, and their predators in streams in Panama and Oklahoma. Armored catfish of the family Loricariidae graze algae in the Panamanian stream, whereas Oklahoma streams are grazed by schools of the minnow *Campostoma anomalum*. I will focus this review on distribution patterns of fish and algae that occur on two spatial scales: within pools along depth gradients, and among pools. Grazing minnows and armored catfish are constrained by predators with depth-specific foraging rates on some scales but not others. The presence or absence of such predators can account in large part for

the distribution and effects of these grazing fishes in stream communities.

DEPTH DISTRIBUTIONS OF GRAZING FISHES WITHIN STREAM POOLS

A very widespread pattern in fish assemblages is that small species and size classes occupy shallow habitats, and larger individuals occur at greater depths. This "bigger–deeper" distribution has been documented for fish in marine (Fishelson et al. 1971; Clarke 1977; Hobson 1968, 1974), estuarine (Hellier 1962), and freshwater (Hall 1972; Keast 1978; Werner et al. 1977; Bowen 1979; Jackson 1961) habitats (see Helfman 1978 for a review). The bigger–deeper distribution is well illustrated by four species of armored catfish (family Loricariidae) that graze algae in the Rio Frijoles of central Panama (9^0 9'N, 79^0 44'W). Large, noncryptic loricariids (*Ancistrus spinosus, Hypostomus plecostomus,* and *Chaetostoma fishcheri* > 5 cm SL) are rarely found in water < 20 cm deep and are most common at depths > 40 cm. Juveniles of these species and 2- to 7-cm long members of a thin cryptic species, *Rineloricaria uracantha,* occur frequently in water < 20 cm deep and are most common in depths < 40 cm (fig. 22.1; also see Power 1984a). Loricariid depth distributions were seasonally invariant despite a two- to three-fold contraction of stream habitat (area > 20 cm deep) from the rainy to the dry season (Power 1984a, and discussed below).

Similar patterns with depth occur among size classes of the grazing minnow *Campostoma anomalum* in prairie-margin and Ozark upland streams of Oklahoma. A school of adult *Campostoma* (> 4 cm SL) was observed on 20 occasions over a 6-day period in a stream pool where they were free of swimming predators. These *Campostoma* occurred in water > 20 cm deep on 20/20 scan samples, and in > 30 cm of water on 18/20 scans. In contrast, young-of-the-year *Campostoma* (2–3 cm SL) in the same pool were in water < 20 cm deep on 4/14 scans and were in < 30 cm of water on 6/14 scans (Power et al. 1985). Similar bigger–deeper patterns were seen in the distribution of size classes of *Campostoma* in a pool of Tyner Creek, an Ozark stream in northeast Oklahoma (Matthews et al., in press).

The bigger–deeper distributions of fish in

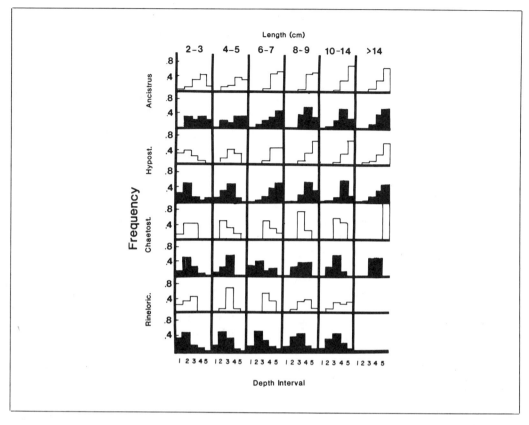

Figure 22.1. *Proportions of various species and size classes of loricariids that occurred within five depth intervals by day (open bars) and by night (solid bars). Loricariids were counted during two rainy-season, two dry-season, and two transitional-season census series of 150–168 L/m²*   *quadrats; proportions averaged from a total of 1,516 day quadrat counts and 1,412 night quadrat counts are plotted here. Depth intervals: 1 = <10 cm; 2 = 10–20 cm, 3 = 21–40 cm; 4 = 41–80 cm; 5 = >80 cm.*

streams and other aquatic habitats is consistent with the hypothesis that water depth and body size interact to influence vulnerability of fish to predators. Observations, and limited experimental evidence, suggest that fish in shallow water (< 20–30 cm) are more susceptible to wading or diving predators, whereas fish in deeper water are more susceptible to swimming predators. Moreover, wading and diving predators may take larger prey than "gape-limited" (Zaret 1980) swimming predators in streams. Where these conditions hold, large stream fish should avoid shallow areas, and small fish should avoid deep areas to reduce exposure to their most important predators.

There is much evidence that large, piscivo-rous fish exclude small fish from deep, open water (Jackson 1961; Goodyear and Ferguson 1969; Seghers 1970, 1974a; Goodyear 1973; Werner et al. 1977, 1983; Keast 1977; Mittelbach 1984; Power 1984a). Small fish emerge from littoral vegetation in tropical and temperate lakes when they have attained sizes at which they are less vulnerable to predatory fishes in deeper open habitats (Jackson 1961, Werner et al. 1983). Riffles in the Rio Frijoles of Central Panama are nursery areas for small (< 4 cm) armored catfish (family Loricariidae) which, when they are placed in deeper pools, are readily eaten by larger fish (Power 1984a). Shallow edges of pools in Brier Creek, Oklahoma, serve as refuges for larvae and young of the year of both minnows and centrarchids subject to predator by larger centrarchids in deeper areas (B. Harvey, unpublished data, personal observations).

Although there is ample evidence that swimming predators can exclude small fish from deeper habitats, evidence that wading and diving predators exclude larger fish from shallow habitat is, to date, largely circumstantial (e.g., Power 1984a). The potential threat to fish in shallow water from birds is apparent from severe depredations that occur when fish are concentrated artificially in shallow hatching pools (Kushlan 1978; Lagler 1939; Mott 1978; Naggiar 1974), or naturally in shrinking tropical stream pools during the dry season (Lowe-McConnell 1964, 1975; Williams 1971; Bonnetto 1975). Bird predation caused a 77% decrease in numbers of fish in a small Florida pond when water level dropped during the dry season (Kushlan 1976). Some experimental evidence also suggests a depth gradient in risk of predation for large fish. I tethered 20 large (7–16 cm SL) armored catfish (*Ancistrus spinosus*) in the Rio Frijoles, in water 11 to 25 cm deep, where herons and kingfishers had been seen fishing. After 24 hours, 0/6 catfish that had been tethered in water < 18 cm deep remained, whereas I recovered 10/14 catfish tethered in water ≥ 18 cm deep ($p = 0.005$; Fisher's Exact Test). In a subsequent experiment, size-matched groups of armored catfish were enclosed in open-topped pens that were similar in surface area and substrate but set at different depths in the stream. After 12 days, during which little blue herons (*Egretta caerulea*) were sighted foraging within a meter of one series of pens, loricariids were largely eliminated from those in 10 and 20 cm of water, but most remained in pens in 30 and 50 cm (Power, Dudley, and Cooper, unpublished data).

If diurnally foraging birds exclude fish from shallow water, fish depth distributions may change at night. Starrett (1950) observed minnows moving by night to forage in shallower parts of an Iowa stream. There was a slight (but insignificant) tendency for armored catfish in Panama to move into shallower water by night (fig. 22.1). Nocturnal movements of loricariids might have been inhibited by night-fishing tiger herons (*Tigrisoma rufescens*) and mammals around the Rio Frijoles (Power 1984a).

INFLUENCE OF SWIMMING PREDATORS ON DISTRIBUTIONS OF GRAZING FISHES AMONG STREAM POOLS

Bigger–deeper distributions of fish occur among pools in streams, as well as along depth gradients within them. Larger predatory fishes often occupy deeper stream pools, leaving shallower pools as potential refuges for smaller grazing species. Largemouth and spotted bass (*Micropterus salmoides* and *M. punctulatus* > 7 cm SL) are the major swimming predators of grazing minnows (*Campostoma anomalum*) in Brier Creek, Oklahoma. In a 1-km reach of Brier Creek, containing 14 pools, distributions of bass and *Campostoma* and maximum depths of stream pools sometimes changed, particularly after large floods (Power et al. 1985). During eight snorkelling censuses of the 14 pools conducted over a 2½-year period, large bass occurred in pools ≥ 50 cm deep on 72 pool-dates and were in shallower pools on only 3 occasions. The total occurrences of deep and shallow pools on the eight dates were 86 and 25, respectively, suggesting selectivity by bass for deeper pools (χ^2, 10.6; $p < 0.01$) *Campostoma* occurred in deep pools on 29 occasions and in shallow pools on 27 occasions. These minnow did not select shallow pools per se (χ^2, 0.42; $p > 0.50$), but their distributions among pools in Brier Creek were significantly complementary with those of large bass, which did select deeper pools, on seven of eight censuses (Power and Matthews 1983; Power et al. 1985, unpublished data). In one long pool where bass and *Campostoma* often co-occurred, they were spatially segregated with bass near the area of maximum depth, and *Campostoma* (and other minnows) in the long, shallow tails (Power and Matthews 1983; W. J. Matthews, B. Harvey, and M. E. Power, unpublished data). In pools that lacked bass, however, *Campostoma* spent most of their time in deepest areas.

These patterns suggest that large bass displace *Campostoma* from deeper pools and pool areas. During spring and autumn experiments, largemouth bass (18–28 cm SL) were introduced into a pool that formerly had contained *Campostoma* and no bass. Prior to bass addition, adult *Campostoma* had spent most of their time foraging over substrates in the

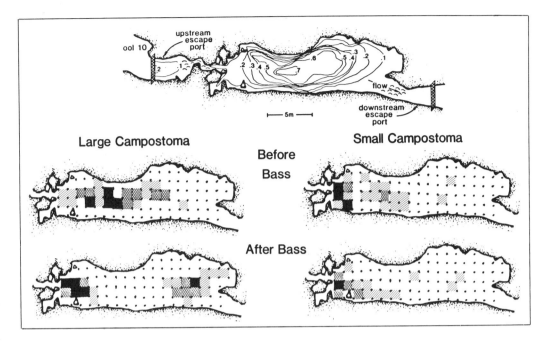

Figure 22.2. *Space use by large (4–8 cm SL) and young of the year (2–3 cm SL) Campostoma in a pool before and after the addition of two largemouth bass (Micropterus salmoides, 18 and 23 cm SL). The pool floor was marked with concrete nails into a grid of squares, 1 m on each side (indicated by dots). The num-bers of sightings of Campostoma within a particular square meter on scan samples carried out on 6 days before and 7 days after bass addition is coded as follows: solid, 5 or more sightings; cross-hatched, 3–4 sightings; single-hatched, 1–2 sightings.*

deepest parts of the pool (40–60 cm deep; fig. 22.2). After bass addition, these minnows were displaced within hours to shoals < 20 cm deep, while bass occupied the deepest area in about 60 cm of water (fig. 22.2, table 22.1; Power et al. 1985). During the spring experiment, *Campostoma* young of the year (2–3 cm TL) were present. Bass addition did not significantly change their depth distributions (fig. 22.2). Before bass addition, these small *Campostoma* had already occupied shallower habitats than did adults in the pool. At that small size, young of the year were subject to attack by sunfish (*Lepomis* spp.) in deeper areas (B. Harvey, unpublished data).

Predator avoidance by *Campostoma* and predation by bass both contributed to their complementary distributions among stream pools. Before experimental additions of bass to a pool in Brier Creek containing *Campostoma*, we blocked off adjacent portions of upstream and downstream pools, so that emigrating fish could be recovered. During fall experiments, 7

of 70 *Campostoma* were recovered from a blocked "escape port," which they had reached by crossing a narrow channel 2 m long and < 10 cm deep. This number may underrepresent the actual emigration, as the block net was displaced during a spate. Nets remained in place throughout the course of a spring experiment. Forty of 74 adult (4–8 cm SL) *Campostoma* originally in the pool were recovered in an "escape port" 6 hours after bass addition (Power et al. 1985). After 1 week, 19 adult *Campostoma* and 13 juveniles (2–3 cm SL) were unaccounted for and were presumed to have been eaten by the bass.

Campostoma are soft, thin fish that remained vulnerable to bass in Brier Creek throughout their lives. In contrast, large loricariids are well defended by spines and dermal armor, and they seemed able to outgrow most swimming predators in the Rio Frijoles. Four *Ancistrus*, 16 to 18 cm SL, developed eye lesions and became increasingly easy for me to capture. Yet I continued to sight these individuals for periods up

Table 22.1. *Remains of loricariids found along Rio Frijoles, presumed killed or mangled by predators*

Species, std. length	Wounds	Probable predator
Ancistrus, 15 cm	Triangular peck wounds, belly and head gone	Bird
Ancistrus, 10 cm	One triangular wound, 1.3 cm wide, through dorsal carapace	Bird
Ancistrus, 9 cm	Two round punctures, 1 mm diameter, spaced 2 cm apart	Mammal
Ancistrus, 14 cm[a]	Body gone except for head and left pectoral spine	Bird or mammal
Ancistrus, 13 cm[a]	Body gone except for chewed tail	Mammal
Ancistrus, 12 cm	Head bitten off	Bird or mammal
Rineloricaria, 12 cm	Two punctures, 1 mm wide, spaced 1 cm apart, above right pectoral fin	Fish or snake
Chaetostoma, 4 cm	Tail gnawed to stump, snout chewed	Mammal

[a]Length estimated from remaining body parts.
Maximum standard lengths (cm) of loricariids found in Rio Frijoles: *Ancistrus*, 20 cm; *Hypostomus*, 30 cm; *Chaetostoma*, 17 cm; *Rineloricaria*, 12 cm.

to 10 months after I first noticed their disease, a circumstance highly unlikely to occur had they been subject to predation. Large loricariids may have been nearly free of predators in deeper areas of the Rio Frijoles, but they could not outgrow vulnerability to predators that foraged in shallow areas. If captured, even the largest Rio Frijoles loricariids could be torn apart by birds or mammals, as evidenced by occasional fresh body fragments of the largest, most heavily armored size classes found along stream banks (table 22.2, fig. 22.3). Their freedom from swimming predators, however, enabled loricariids to move among pools in their stream and choose habitats on the basis of food availability, a prerogative with important consequences for algae in their stream.

EFFECTS OF GRAZING FISHES ON DISTRIBUTIONS OF STREAM ALGAE

Presence or absence of swimming predators that constrain their pool-to-pool distributions determines how grazing fishes will affect large-scale distributions of stream algae. Armored catfish, after they outgrow swimming predators, can move among pools and track variation in algal availability. Where forest canopy over pools is open and algal growth rates are high, grazing catfish are more dense than they are in dark, relatively unproductive pools. In fact, catfish are roughly six to seven times denser in sunny pools where algae grow about seven times faster than in dark pools (Power 1983, 1984b). Movement among pools is required to maintain this pattern, as algal productivity can change abruptly—for example, when a tree falls open-canopy over a pool or when pools are created or filled during floods. Because loricariids can find and exploit new feeding opportunities within weeks after they arise, these grazing fishes damp out incipient pool-to-pool variation in algal standing crop. As a result, standing crops of algae are uniformly scant among pools in the Rio Frijoles of Panama, despite large differences in primary productivity of algae in different pools (Power 1984b).

In marked contrast, striking variation in algal standing crop occurs among pools in Brier Creek, Oklahoma. Some pools are nearly barren, except for fringes of algae around their shallow margins. These barren pools contain schools of *Campostoma* and lack bass. Other, often adjacent pools that lack *Campostoma* and contain bass are at times nearly filled with filamentous green algae (*Rhizoclonium* sp. and *Spirogyra* sp.). Experimental transfers have shown that *Campostoma*, when introduced into pools from which bass have been removed, can denude large standing crops of algae in bass pools

Table 22.2 *Numbers, depth and activity of bass and minnows (Campostoma anomalum) before and after introduction of bass to a Campostoma pool*

	Numbers sighted	Campostoma				Bass	
		% in various depths activity				Depth	Activity
		< 20 cm	20–30 cm	> 30 cm	Activity		
9 Sept (a.m.)	70	0	0	100	Grazing		
15:00: Introduction of two bass, 18 and 23 cm SL							
15:30	70	0	0	100	Schooling	60	hiding
17:30	70	0	100	0	Grazing	60	hiding
10 Sept							
10:00	70	15	85	0	Milling	60	hovering
12:45	70	100	0	0	Milling	30–40	patrolling
13:19		100	0	0	Milling	60	hovering
16:35		100	0	0	Milling	60	hovering
13 Sept							
11:00	11	100	0	0	Milling	30–40	patrolling
12:00	11	100	0	0	Milling	30–40	patrolling
17 Sept: Bass escape during spate; replaced by two of same size							
18 Sept							
before bass added		0	0	100			
immediately after		100	0	0			
24 Sept							
15:20	12	100	0	0	Hovering	30–40	patrolling
7 *Campostoma* recovered from upstream escape port							
26 Sept	0					60	hovering
1 Oct	0					60	hovering
3 Oct	0					60	hovering
13 Oct Snorkeling census reveals 11 inactive *Campostoma* under cobbles, 10–20 cm deep							

within weeks. Similarly, when *Campostoma* grazing is inhibited by the introduction of bass into their pools, algal standing crops recover within weeks to levels typical of bass pools (Power et al. 1985). Consequences of the extreme pool-to-pool heterogeneity in algal standing crops that results from this predator–prey interaction for nutrient uptake and regeneration in Brier Creek are presently under investigation (A. J. Stewart, personal communication).

Whereas armored catfish in the Rio Frijoles and grazing minnows in Brier Creek produce opposite effects on pool-to-pool distributions of algae, they produce similar patterns within pools. In *Campostoma* pools of Brier Creek and throughout the Rio Frijoles, algal standing crops are often relatively high in water < 20 cm and decrease sharply at greater depths (fig. 22.4). This depth threshold corresponds to the

depths at which wading and diving birds forage most frequently and effectively (discussed below). Avoidance of these predators by both grazing minnows and catfishes is one explanation for higher standing crops of algae in shallow water. Light and nutrient availability is usually higher in shallow water, and differences in algal productivity also could produce depth gradients in standing crop. This explanation cannot, however, account for algal distributions along depth gradients in bass pools of Brier Creek. Algal standing crops in bass and *Campostoma* pools are similar in shallow water, but increase then asymptote with depth in bass pools, in contrast to standing crops in *Campostoma* pools that drop off abruptly in depths > 10 cm (Power and Matthews 1983: fig. 3). In bass pools that lack *Campostoma*, standing crops of algae are higher in water 11 to 60 cm deep than in wa-

Figure 22.3. *Remains of large loricariids found along banks of the Rio Frijoles. The stone on the leaf with the* *fresh, empty loricariid carapace is 6 cm wide.*

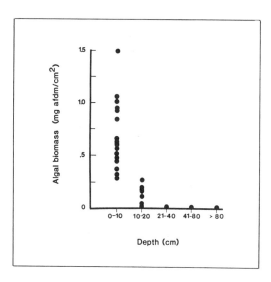

Figure 22.4. *Periphyton standing crops (ash-free dry mass) sampled from cobbles collected haphazardly from various depth intervals during the late dry season (March) in the Rio Frijoles, Panama. Sample sizes are N = 17 (<10 cm); N = 9 (10–20 cm); N = 10 (21–40 cm); N = 11 (41–80 cm); N = 10 (>80 cm).*

ter 0 to 10 cm deep ($p < 0.01$, Mann-Whitney U-test). The reverse is true in pools with *Campostoma* ($p < 0.03$, Mann-Whitney U-test. Brier Creek is frequented by raccoons (*Procyon lotor*), green herons (*Buturoides striatus*), belted kingfishers (*Ceryle alcyon*), and great blue herons (*Ardea herodias*; we commonly see tracks of these herons along shallow stream margins). Avoidance of depth-limited predators by *Campostoma* in Brier Creek and by loricariids in the Rio Frijoles may contribute to the maintenance of bands of algae along margins of pools in both streams.

COSTS OF PREDATOR AVOIDANCE FOR GRAZING STREAM FISHES

Predator-induced avoidance of dangerous feeding areas can impose energetic costs on food-limited prey (Sih 1982; Werner et al. 1983; Cooper 1984). For prey whose vulnerability changes with size, distributions of food and predation risk over the habitat can determine in large part the three fundamental demographic rates: size-dependent growth, size-dependent fecundity, and size-dependent mortality (Werner and Gilliam 1984). In size-structured populations, some size-classes may be more food-limited than others that have outgrown predators. Bluegills (*Lepomis macrochirus*) in experimental ponds in Michigan became less food-limited as they outgrew bass and moved from littoral vegetation into limnetic areas where they foraged more profitably (Werner et al. 1983; Werner and Gilliam 1984; Mittelbach 1984). In contrast, armored catfish became more food-limited as they grew and moved out of shallows of the Rio Frijoles, where algal standing crops were higher. To illustrate the size-dependency of their vulnerability and access to food, I will describe the life cycle of the most common loricariid in stream pools: *Ancistrus spinosus*.

Ancistrus begin their lives as eggs in their father's nest, typically a hollow log in a deep pool. He serves as an armored cork, blocking the nest entrance against potential egg and fry predators such as freshwater crabs (*Potamocarcinus richmondi*) and characin fishes. Hatchlings stay with their father until they are about 18 mm SL. Although a "fledging episode" of *Ancistrus* has not to my knowledge been observed in nature, it is likely that the young face a gauntlet of swimming predators when they leave the nest. Stream pools in the Rio Frijoles are densely populated by characin fishes, active by day and night, that devour *Ancistrus* 2 to 3 cm long when these are introduced into pools (Power 1984a). Young *Ancistrus* that survive enter shallow stream riffles, where the highest algal standing crops in the stream channel occur. The three noncryptic loricariids graze and grow in these shallow nursery areas until they are 3 to 4 cm long. (The cryptic *Rineloricaria* can be found in riffles until it is about 8 cm SL.) At this size, they begin to outgrow refuges available in the interstices of riffle cobbles and probably become more conspicuous to herons and kingfishers that commonly fish water <20 cm deep in the Rio Frijoles (Power 1984a). They also become less vulnerable to most swimming predators in the pools. At 4 to 5 cm SL, the catfish move into deeper habitats. After attaining lengths >5 cm, *Ancistrus* avoid water <20 cm deep by day and night, even during the dry season when algal food is abundant there and in short supply in deeper areas (Power 1984a).

Large loricariids (the three noncryptic species

>5 cm SL collectively made up 83% of the loricariid biomass in the Rio Frijoles) were food-limited for much of the year. In contrast to shallow algae, algae in deeper water had standing crops so scant that measurable amounts could not be scraped from any substrate in >20 cm of water within a 2.8-km channel at any season over a 2-year period (Power 1981). A total of 1,308 loricariids were individually marked and periodically recaptured to measure growth. Pre-reproductive *Ancistrus* (4–9 cm SL) showed their highest growth rates when early rainy season floods gave fish first access to algae in areas previously too shallow to graze. Growth declined but remained positive during the latter part of the rainy season (July–November) and stopped during the dry season (December–early April) (Power 1984a). [When deprived of food, loricariids lose fat but not live weight (Power 1984c). They may replace catabolized tissue with water, as do other fish.]

Loricariid catfish <4 cm SL were too delicate to mark with the technique I used, so growth data are not available for these small size classes. Higher standing crops of algae in their shallow habitats, however, suggests that small loricariids were not as food-limited as larger individuals in the more barren pools.

Because of their restricted depth distributions, it was possible to estimate seasonal food availability for particular species and size classes of loricariids. This food availability depended on the area and primary productivity of streambed within the depth interval grazed by given catfish. From a detailed bathymetric map (10-cm contour intervals) of the 2.8-km study reach and a hydrograph record for a 24-month period, I computed the area of streambed under various depths for periods of high (July–November), low (March–early April), and intermediate (January, June) flow. Taking into account the area of streambed under forest canopy that was <10% 10 to 25%, 26 to 50%, and >50% open; growth rates of stream algae under each canopy; light attenuation in the water column; and the proportions of grazeable substrate (excluding mud and sand), I estimated the daily renewal rate of algal organic matter available to loricariids within each depth interval (Power 1981: tables 2-4, 2-5, and 4-5). These estimates were compared with collective metabolic requirements of species and size classes of loricariids that grazed within each depth interval. I estimated numbers of loricariids with data from six series of 2 day and 2 night censuses over the 2.8-km reach. Census series were carried out during two dry, two rainy, and two transitional season periods over a 28-month period. Length–weight regressions, based on 159, 91, 50, and 41 measurements for *Ancistrus*, *Hypostomus*, *Chaetostoma*, and *Rineloricaria*, respectively, were used to estimate weights for loricariids in each length class. (Length : weight ratios did not vary seasonally.) For computations of caloric requirements of loricariids, I assumed that resting metabolic rates were similar to those measured for four species of tropical fishes of similar size at similar temperatures in the laboratory (Kaysner and Heusner 1964), that active metabolism was three times resting metabolism (Kramer 1983), and that loricariids assimilated 20% of the organic matter produced in a given depth interval (see Power 1981 for details).

Estimates of periphyton production and of metabolic needs of loricariids within various depth intervals (fig. 22.5) suggest that loricariids in water deeper than 20 cm were food-limited at all times of year except during the early part of the rainy season (late April–June). During the late rainy season (July–November), algal production just met loricariid needs, except in water > 40 cm deep. In the dry season, (December–early April) loricariid metabolic needs exceeded algal production except in very shallow water. These results are congruent with seasonal patterns of somatic growth rates in 4- to 9-cm long *Ancistrus*, which (as discussed above) peaked in the early rainy season, declined through the late rainy season, and stopped in the dry season.

In summary, three lines of evidence suggest that Rio Frijoles loricariids become increasingly food-limited with size and their shift to deeper water: steep depth gradients in algal standing crops; growth rates of loricariids; and comparisons of estimated depth-specific food renewal and collective metabolic needs of loricariids. Young fish in riffles are relatively safe from larger fish, their most important potential predators, but are limited, perhaps by bird predation, at densities below those necessary to deplete shallow algal standing crops. They may occupy shallows to avoid predatory fish, to exploit higher densities of food, or for both reasons. As catfish outgrow cover in shallow water and swimming predators in deeper water, they move down into pools where they are safe from wad-

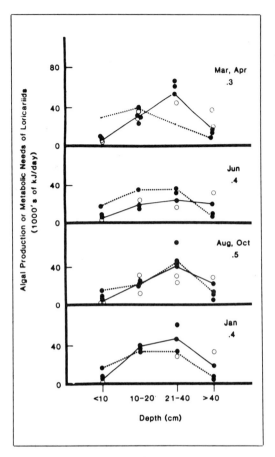

Figure 22.5. *Estimated harvestable periphyton production* (dotted lines) *and metabolic needs of loricariids grazing by day* (open circles) *and by night* (solid circles) *(solid lines drawn through mean 24-hour requirements) in four depth intervals during low (March, April), intermediate (January, June) and high (August, October) flow periods.*

ing and diving predators but where food is in short supply for much of the year. Large loricariids did not, over the range of food limitation I observed during a 28-month field study, compromise their safety by foraging in shallower water even when food was most limiting in the late dry season. As a result, mortality rates of large *Ancistrus* showed no seasonal variation, in contrast to the strong seasonal differences in rates of their somatic growth (Power 1984a).

Armored catfish that were experimentally starved for 8 and 18 days in their home pools were similarly conservative in their risk-taking behavior. After being released, starved fish showed no tendency to forage in shallower water than they had before starvation, or than was shown by fed control fish. Starved individuals did, however, spend more time out foraging on bedrock substrates in deep areas of pools that did fed controls (Power, Dudley, and Cooper, unpublished data). Some level of hunger might cause armored catfish to compromise their safety by foraging in shallow water. However, under the range of natural and experimental conditions that I have observed, predators increase food limitation for large armored catfish by excluding them from productive shallow areas of streams. Wading and diving predators therefore may intensify both intra- and interspecific competition for food, at least in the short term.

INTERSPECIFIC DIFFERENCES (COVER AND CRYPTICITY)

Differences in conspicuousness, and therefore perhaps in vulnerabilities to predators, occur among species as well as among size classes of grazing fishes. In the Rio Frijoles, *Ancistrus spinosus*, a stocky, spiny loricariid that outgrows most swimming predators, is common in deep (> 40 cm) pools with cover such as root tangles or undercut bedrock. In shallower pools that are relatively devoid of cover, *Ancistrus* is frequently outnumbered by a thin, cryptic species, *Rineloricaria uracantha*. When I tied a raft of small logs to the side of a pool 40 to 60 cm deep that formerly lacked cover, numbers of *Ancistrus* increased and those of *Rineloricaria* decreased over the following two weeks (fig. 22.6). Similar changes in *Rineloricaria* and *Ancistrus* numbers did not occur in a number of unmanipulated pools over the same time period. After cover addition and the influx of *Ancistrus*, *Rineloricaria* in the manipulated pool stopped using clay substrates in the deepest area, near the new shelter (fig. 22.7). Substrate use by *Ancistrus* was unchanged, except for their occurrence on the new wood of the raft. Replication, with observations of both species and of changes in algae, is needed to reveal whether interference or exploitative competition might change grazer guild composition after cover addition. Because behavioral interactions between *Ancistrus* and *Rineloricaria* were rarely observed even when the two species were in close proximity (Power 1984b), exploitative competition seems more likely than interference. These preliminary results are consistent with the hypothesis that

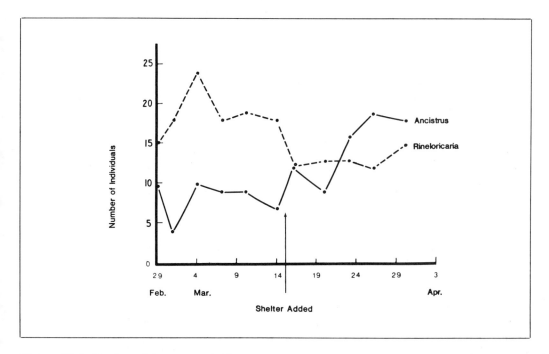

Figure 22.6. *Numbers of* Ancistrus (solid line) *and* Rineloricaria (dashed line) *counted in a sunny pool* *before and after the addition of a raft shelter.*

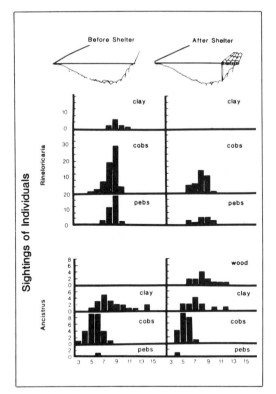

Figure 22.7. *Sightings of* Ancistrus *and* Rineloricaria *of various size classes on substrates in a sunny pool, before and after the addition of a raft shelter. Abscissa is standard length in centimeters.*

densities of *Ancistrus*, a potentially dominant competitor, can be limited by the availability of cover in sunny pools. When this limitation was relaxed, more *Ancistrus* colonized the pool, displacing *Rineloricaria*. If this interpretation is correct, it suggests a case in which predators create refuges from competition for subordinate competitors by making microhabitats too dangerous for dominant competitors to exploit. If predators promote coexistence of competing prey by altering their behavior rather than their relative numbers, the results may differ from patterns that arise from actual predation (Werner et al. 1983; Mittelbach, in press). Competition between prey species may be intensified or relaxed, depending on the degree to which one species will forage in areas avoided by the other.

DISCUSSION

Prey should avoid their most important predators or, if predictable, the areas where those predators forage most frequently and effectively. Predictability of the spatial distribution of risk depends on local abundances of various predators, on their behavioral flexibility, and on prey vulnerability. All three conditions change over long and short time scales. Prey may respond to changing risks in their habitats with learned or evolved changes in their behavior (Seghers 1974a, 1974b; Stein and Magnuson 1976; Stein 1979; Sih, this volume).

Spatial distribution of risk for stream fish is predictable when and where predators are jointly constrained by stream depth and prey size. Although common predators in streams such as bass (Savino and Stein 1982) and herons (Kushlan 1976, 1978; Meyerriecks 1962) are capable of a variety of hunting tactics, they are usually morphologically constrained to hunt most profitably while either wading, diving, or swimming. Each of these three hunting modes is effective only within certain parts of the stream channel (fig. 22.8).

Wading predators (e.g., raccoons, herons, and egrets) fish shallow areas along stream margins. Wading birds typically fish water no deeper than their leg lengths (Whitfield and Cyrus 1978; Kushlan 1978). Of 31 feeding behaviors used by 21 species of herons to catch aquatic prey, most (22 out of 31) are used in shallow water (Kushlan 1978). Wading depths for the little blue heron (*Egretta caerulea*) in a Panamanian stream ranged from 3 to 20 cm, with a mean of 11 (N = 15) (Power 1984a and unpublished data). Of nine species of wading birds in a South African lake, five fed in water < 10 cm deep, three fed in 20 cm or less, and one was seen feeding in water < 20 cm deep 85% of the time and in < 30 cm for the remainder (Whitfield and Cyrus 1978). In Ozark streams, green herons (*Buturoides striatus*) stood or walked along stream margins 90 to 95% of the time they were observed (Kaiser 1982). Mean foraging depths for adult herons were 30 cm (N = 36; range, 3–190 cm) and for juveniles, 8 cm (N = 8; range, 4–23 cm). The relatively high mean foraging depth for adults includes one observation of a heron jumping into water almost 2 m deep. Without this observation, mean foraging depth for adults would be 25 cm. Energetic feeding modes, such as plunging or diving into the water, are infrequently used by wading birds (Kushlan 1978). Although surprise may enhance the effectiveness of unusual hunting tactics, their infrequency and the awkwardness of a

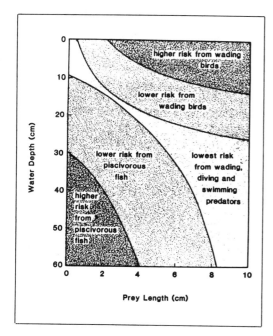

Figure 22.8. *Hypothesized distribution of depth- and size-specific risk for fish from swimming (e.g., fish) versus wading (e.g., herons) predators. Lines indicate "contours" along gradients of risk (adopted from Connell 1975, fig. 2-5). Quantitative limits for "Small and Large," "Deep and Shallow" are suggested for fishes and microhabitats in small (1st to 3rd order) streams.*

predator performing in defiance of its design constraints probably minimize their importance for potential prey. A laboratory study has shown that heron fishing effectiveness decreases with water depth. Capture success of a green heron fishing into a laboratory aquarium was 1.3 to 3.8 times higher when the bird fished within 10 cm of the water surface than when it fished at 10 to 20 cm or 20 to 30 cm, respectively (Kramer et al. 1983). These authors reason that near the surface birds can locate fish more precisely, there is less water resistance to slow the strike, and fish have less escape time after strike initiation. It therefore seems that although fish in deeper water are not completely free from attack by wading predators such as birds, their risk from such predators is much reduced with depth.

Diving predators such as kingfishers and skimming predators such as fishing bats fish near the water surface (Whitfield and Blaber 1978; Salyer and Lagler 1949). Kingfishers often initiate dives from perches but also will

dive after hovering (Skutch 1957; Betts and Betts 1977; Junor 1972; Whitfield and Blaber 1978; Boag 1982), thereby gaining access to surface waters away from shore. The European kingfisher (*Alcedo atthis*, whose diet is more than 95% fish) usually does not dive into water deeper than 20 to 25 cm (Boag 1982). In an experimental pool, European kingfishers readily dove into water 2 to 22 cm deep, using their wings and chests to brake when they took fish from very shallow water (Boag 1982). Green kingfishers (*Chloroceryle americana*) were observed on 10 dives into points where a Panamanian stream was 8, 8, 8, 10, 11, 12, 12, 17, 40, and 43 cm deep. On the 43-cm dive, the kingfisher skimmed the water surface; depth in the water column where the bird fished on the 40-cm dive is not known (Power 1984a and unpublished data). Fishing bats of the Neotropics (*Noctilio leporinus*) skim and dip into streams, taking fish that rest at night in a quiescent state just beneath the water surface (Bloedel 1955). Australian bats known to fish (*Myotes vivesci*, *M. adversus*) have similar foraging methods (E. Pierson, personal communication). In laboratory experiments, *Noctilio* could not detect food a few millimeters below the surface but readily dipped at water upwellings or minute wires extending just above the water surface (Suthers 1965). Bloedel (1955) believed that bats did not pursue identified fish, but rather skimmed and dipped over stream regions they had learned were profitable hunting grounds.

Therefore, a variety of predators fishing streams have depth-limited access to prey. How big are the prey they can take? Wading predators and some diving predators can capture and consume larger prey than can most swimming predators in streams. These land-based predators are often large relative to animals that live within stream channels. In addition, many wading and diving predators can manipulate their prey and, if it is too large to swallow whole, consume it piecemeal. Manipulative abilities of raccoons are legendary. All heron species can handle and subdue prey too large to swallow whole: "Large, hard, or dangerous prey may be battered, rubbed, shaken, dropped or stabbed, and may be picked apart and eaten in pieces" (Kushlan 1978, p. 254). Kaiser (1982) observed green herons dragging large prey away from the stream (making escape less probable), breaking off spines, and consuming fish in pieces. Bats also can consume prey piecemeal, and some carnivorous species (*Vampyrum spectrum*) have carried 150-g prey (almost as big as themselves) back to roosts (Vehrencamp et al. 1977). Kingfishers have less ability to manipulate prey, and small species, such as pigmy kingfishers (*Chloroceryle aenea*) may be limited in the size of prey they can take. At least three kingfisher species, however, can handle fish at least as long as half their own body lengths. European kingfishers 16.5 to 18.5 cm long swallowed fish up to 8 cm long (Boag 1982). Belted kingfishers [*Ceryle alcyon*, 28 to 37 cm long (Udvardy 1977)] took trout 5 to 13.5 cm long from a Michigan river (Alexander 1979) and trout up to 17.8 cm long from other freshwater habitats in Michigan (Salyer and Lagler 1949). A ringed kingfisher [*Ceryle torquata*, 40 cm long (MacArthur 1972)] captured a 25-cm-long cichlid (*Geophagus crassilabrus*) in a Panamanian stream and subdued it by beating it against a large branch on which it perched (Power 1981).

Larger fish may be more vulnerable to wading and diving predators than smaller fish. They are inherently more conspicuous and less able to hide within small cover such as interstices of cobbles in riffles. In addition, larger prey may be preferred by large predators such as ringed kingfishers (MacArthur 1972), gray herons (*Ardea cinerea*) and little egrets (*Egretta garzetta*) (Britton and Moser 1982), and wood storks (Kushlan 1978). Fish flee the site of bird strikes, and local behavioral depression (Charnov et al. 1976) may make it advantageous for fishing birds to select larger prey. Great blue herons fishing a Michigan river ate trout from 7.6 to 33 cm long but appeared to select those from 18 to 30.5 cm long (Alexander 1979). Stoneroller minnows (*Campostoma anomalum*) dominated fish remains in belted kingfisher nests, and dace (*Rhinichthys* spp.) were rare, despite the greater abundance of dace in a New York stream (Eipper 1956). The kingfishers may have selected the larger stonerollers over the small dace (Eipper 1956). In addition, stonerollers were breeding in shallow water while kingfishers were nesting, and may have been particularly available. [Although kingfishers increase the size of fish brought to nestlings as they grow (Skutch 1957; White 1938), wading birds do not select small prey for their young, but instead deliver well-digested food (Kushlan 1978).] Finally, large or wide fish may

be less likely to escape after capture by thin-billed herons, which tend to grasp small or thin fish but stab and seriously wound larger prey (Recher and Recher 1969b; Kushlan 1978).

For the reasons outlined above, it would be very difficult for prey in streams to "escape in size" from wading and some diving predators. Prey might, however, outgrow vulnerability to swimming predators, which must live or at least maneuver within stream channels and are therefore limited in body size. Moreover, piscivorous fish, the most common swimming predators in many streams, have limited ability to manipulate prey after capture and in most cases must swallow them whole (Popova 1978; Hyatt 1979). The length of prey many predatory fish can take varies linearly with their own body lengths. Maximum prey size found in predator guts is usually 30 to 50% as long as the predator; average prey size ranges from 10 to 20% of the predator's length, whereas minimum prey size does not vary systematically with predator length and is usually < 10% of it (table 22.3). Although larger predatory fishes can take larger size ranges of prey, they do not always eat the largest prey available, as handling time constraints may make smaller prey more profitable (Werner 1974, 1977; Hyatt 1979). Several small fish are digested more rapidly than one large fish, especially if larger prey have dense scaly armor or rigid skeletons (Popova 1978). Stream fish, particularly those like loricariid catfish with spines and armor that enhance their effective size, may outgrow vulnerability to "gape-limited" (Zaret 1980) swimming predators.

In contrast to wading and diving predators, many swimming predators [such as piscivorous fish and otters but not watersnakes (Mushinski et al. 1982)] fish primarily in deeper parts of streams. Piscivorous fish may hunt most commonly in deep water to avoid wading and diving predators themselves, or because they have more room to maneuver in deeper areas, or for clearer reception of pressure signals along their lateral lines (Clark Hubbs, personal communication). In six New Zealand rivers, adult brown trout (32–55 cm fork length) were observed feeding in depth from 14 to 122 cm, with frequency of sightings increasing sharply at depths ≥ 40 cm. The preferred feeding depth was estimated to be 65 cm (Shirvell and Dungey 1983). In a Michigan lake, 97% of bass counted

Table 22.3. *Linear regressions of prey length (Y) versus predator length (X)*

Predator	Maximum prey	Average prey	Minimum prey	Source
Esox lucius, pike, 4–75 cm	$Y = 0.51\,X + 0.13$ $r = .99\ N = 9$	$Y = 0.11\,X + 1.73$ $r = .93\ N = 9$	$Y = 0.005\,X + 2.37$ $r = .16\ N = 9$	(1)
Silurus glanis, sheatfish, 5–100 cm	$Y = 0.41\,X - 0.13$ $r = .98\ N = 9$	$Y = 0.09\,X + 1.96$ $r = .93\ N = 9$	$Y = 0.0002\,X + 2.10$ $r = .02\ N = 9$	(1)
Perca fluvi- atilis, perch, 3–41 cm	$Y = 0.41\,X + 5.47$ $r = .52\ N = 8$	$Y = 0.27\,X - 0.09$ $r = .97\ N = 8$	$Y = 0.07\,X + 0.42$ $r = .85\ N = 8$	(1)
Stizostedion lucioperca, zander, 4–70 cm	$Y = 0.35\,X + 1.00$ $r = .95\ N = 9$	$Y = 0.17\,X + 0.97$ $r = .98\ N = 9$	$Y = 0.04\,X + 0.70$ $r = .71\ N = 9$	(1)
Aspius aspius, asp, 6–57 cm	$Y = 0.34\,X + 0.56$ $r = .96\ N = 7$	$Y = 0.09\,X + 0.62$ $r = .95\ N = 7$	$Y = 0.01\,X + 0.34$ $r = .73\ N = 7$	(1)
Stizostedion vitreum, walleye, 11–23 cm	$Y = 0.53\,X - 1.90$ $r = .98\ N = 4$	$Y = 0.29\,X - 0.88$ $r = .63\ N = 44$	$Y = 0.15\,X - 0.29$ $r = .95\ N = 7$	(2)
Pomoxis annularis, white crappie, 17–32 cm	$Y = 0.44\,X - 11.51$ $r = 0.94\ N = 6$	$Y = 0.30\,X + 3.62$ $r = .81\ N = 7$	$Y = 0.13\,X + 12.75$ $r = .54\ N = 6$	(3)

Sources: (1) Popova 1978, Fig. 9.6; (2) Forsythe and Wrenn 1979; (3) Burris 1956, cited in Carlander 1977b.

in snorkeling censuses were in water > 30 cm deep, and 91% were in > 50 cm deep (Werner 1977). Depths in which predatory fish will forage may change with hunger, experience, light, season, cover, or other factors, but I know of no field data examining these relationships.

Stream fish may cross riffles shallower than areas where they would linger. Although movements of bass and trout through natural stream reaches have been studied (Gerking 1959; Larimore 1952; Fajen 1962), data on depths of intervening riffles at the time of pool-to-pool movements are not, to my knowledge, available. Such behavioral data would be useful in predicting flow conditions that would permit the coexistence of predatory fishes and their prey in streams with semi-isolated pools, like Brier Creek (Power et al. 1985).

The bigger–deeper distribution is one pattern predicted for prey that avoid predators subject to the size and depth constraints described above. Different predators could produce different patterns in prey assemblages. Watersnakes (*Nerodia* spp.) are "gape-limited" predators that forage in shallow as well as deep water (Mushinksi et al. 1982). Mergansers (*Mergus* spp.) pursue fish underwater in areas up to several meters deep (Lindroth and Bergstrom 1959). They are capable of swallowing large fish in a gradual, snakelike fashion. One merganser disgorged a trout 37 cm long (Salyer and Lagler 1949). White (1957) found suckers 28 cm long and an eel 41 cm long in merganser stomachs. It would be interesting to study the depth distributions of fish in habitats such as the Margaree River of Nova Scotia where these ducks are important predators (White 1936, 1939). Various crocodilians and otters are other swimming predators capable of taking large fish in streams. These predators may once have been important in many streams where they are now rare or absent because of human hunting and habitat destruction.

One of the few quantitative studies of the relative importance of various predators in streams was conducted in a managed trout stream in Michigan. Alexander (1979) estimated numbers and per capita feeding rates of four swimming predators [large brown trout (*Salmo trutta*), mergansers, mink, and otters]: one diving predator (belted kingfisher) and one wading predator (great blue heron) in a 32-km reach of the Au Sable River. Herons and large brown trout were the most important predators, taking, respectively, 19 to 20% and 12 to 24% of the annual trout production not caught by fishermen. American mergansers, belted kingfishers, mink, and otter each took

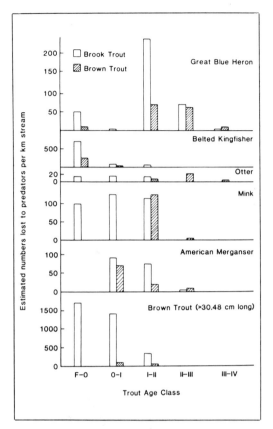

Figure 22.9. *Numbers of brook and brown trout of various age classes lost to predators. Data from tables 2–5 in Alexander 1979. Numbers of trout lost from two stream reaches subject to different angling regulations have been averaged. For given age classes, ranges of lengths (total or fork length in millimeters) reported by Carlander 1977a for North American brook trout are: F–0: < 15; 0–I: 15–236; I–II: 56–292; II–III: 66–343; and III–IV: 102–419. For brown trout, these lengths are: F–0: <25; 0–I: 25–198; I–II: 64–241; II–III: 132–366; III–IV: 157–495.*

< 10% of this production (computed from Alexander 1979: table 8). Herons, the most important wading predators, took larger size classes of trout, and brown trout, the most important swimming predator, took small brook trout (*Salvelinus fontinalis*) (fig. 22.9). In the Au Sable River in its contemporary managed condition, then, fish avoiding their two most important natural predators would maintain bigger–deeper distributions.

Stream fish distributions, their vulnerability to predators, and their access to food are affected by other physical factors that are not in-

dependent of depth in streams: in particular, current, cover, substrate, and surface rippling (e.g., Moyle et al. 1982; Binns and Eiserman 1979; Burns 1971; Jenkins 1969; Gibson 1966). Hypotheses I have presented are clearly oversimplifications of interactions of physical and biotic factors in stream communities. Nevertheless, quantitative and experimental studies of the relationships between depth, size, and predation risk for stream fish are likely to be of basic and applied interest. Depth is an easily measured and manipulated habitat variable that appears to affect space use by stream fish, and hence the carrying capacities of streams for their populations. Quantitative data on depth-specific foraging frequencies and size selectivities of predators on stream fish, and experimental tests of hypotheses regarding the extent and consequences of predator avoidance by particular classes of stream fish, should prove useful in explaining and predicting the dynamics and effects of fish populations in streams with different, and differing, flow regimes.

ACKNOWLEDGMENTS

I thank E. Pierson and W. Rainey for thoughtful comments and discussion that enlarged my perspective. W. Dietrich, A. Sih, E. Werner, W. Matthews, A. Stewart, and S. Cooper also made helpful comments on various drafts. This study was funded by NSF grant BSR-8307014.

REFERENCES

Alexander, G. R. 1979. Predators of fish in coldwater streams. In H. Clepper (ed.), Predator-prey systems in fisheries management, pp. 153–70. Washington, D.C.: Sport Fishing Institutes.

Betts, B. J., and D. L. Betts. 1977. The relation of hunting site changes to hunting success in green herons and green kingfishers. *Condor* 79 : 269–71.

Binns, A. N., and F. M. Eiserman. 1979. Quantification of fluvial trout habitat in Wyoming. *Trans. Amer. Fish. Soc.* 108 : 215–28.

Bloedel, P. 1955. Hunting methods of fish-eating bats, particularly *Noctilio leporinus*. *J. Mammalogy* 36 : 390–99.

Boag, D. 1982. The kingfisher. Dorset, England: Blandford Press.

Bonnetto, A. A. 1975. Hydraulic regime of the Parana river and its influence on ecosystems. *Ecological Studies* 10 : 175–97.

Bowen, S. H. 1979. A nutritional constraint on detritivory by fishes: the stunted population of *Saro-*

therodon mossambicus in Lake Sibaya, South Africa. *Ecol. Monogr.* 49 : 17–31.

Britton, R. H., and M. E. Moser. 1982. Size specific predation by herons and its effect on the sex-ratio of natural populations of the mosquito fish *Gambusia affinis* Baird and Girard. *Oecologia* 53 : 146–51.

Burns, J. W. 1971. The carrying capacity for juvenile salmonids in some northern California streams. *Calif. Fish and Game Bull.* 57 : 44–57.

Carlander, K. D. 1977. *Handbook of freshwater biology*, Vol. 2. Ames, Iowa: Iowa University Press.

Charnov, E. L., G. H. Orians, and K. Hyatt. 1976. Ecological implications of resource depression. *Amer. Naturalist* 110 : 247–59.

Clarke, R. D. 1977. Habitat distribution and species diversity of chaetodontid and pomacentrid fishes near Bimini, Bahamas. *Mar. Biol.* 40 : 277–89.

Connell, J. H. 1975. Some mechanisms producing structure in natural communities: a model and evidence from field experiments. In M. L. Cody and J. M. Diamond (eds.). *Ecology and Evolution of Communities*, pp. 460–490. Cambridge, Mass.: Harvard Univ. Press.

Cooper, S. D. 1984. The effects of trout on water striders in stream pools. *Oecologia* 63 : 376–79.

Duggins, D. O. 1980. Kelp beds and sea otters: An experimental approach. *Ecology* 61 : 447–53.

Earle, S. A. 1972. The influence of herbivores on the marine plants of Great Lameshur Bay. In B. B. Collette and S. A. Earle (eds.), *Results of the Tektite Program: Ecology of coral reef fishes*, pp. 17–44. L.A. County Natl. Hist. Mus. Sci. Bull. 14, Los Angeles.

Eipper, A. W. 1956. Differences in vulnerability of the prey of nesting kingfishers. *J. Wildlife Manage.* 20 : 177–83.

Estes, J. A., and J. F. Palmisano. 1974. Sea otters: Their role in structuring nearshore communities. *Science* 185 : 1058–60.

Fajen, O. F. 1962. The influence of stream stability on homing behavior of two smallmouth bass populations. *Trans. Amer. Fish. Soc.* 91 : 346–49.

Fishelson, L., D. Popper, and N. Gunderman. 1971. Diurnal cyclic behaviour of *Pempheris oualensis* Cuv. and Val. (Pemperidae, Teleostei.) *J. Natur. Hist.* 5 : 503–506.

Forsythe, T. D., and W. B. Wrenn. 1979. Predator-prey relationships among walleye and bluegill. In H. Clepper (ed.), *Predator-prey systems in fisheries management*, pp. 475–482. Washington, D.C.: Sport Fishing Institute.

Garrity, S. D., and S. C. Levings. 1981. A predator-prey interaction between two physically and biologically constrained tropical rocky shore gastropods: direct, indirect and community effects. *Ecol. Monogr.* 51 : 267–86.

Gerking, S. D. 1959. The restricted movement of fish populations. *Biol. Rev.* 34 : 221–42.

Gibson, R. J. 1966. Some factors influencing the distributions of brook trout and young Atlantic salmon. *J. Fish. Res. Bd. Can.* 23 : 1977–80.

Goodyear, C. P. 1973. Learned orientation in the predator avoidance behavior of mosquito fish, *Gambusia affinis*. *Behaviour* 45 : 191–220.

Goodyear, C. P., and D. E. Ferguson. 1969. Sun-compass orientation in the mosquito fish, *Gambusia affinis*. *Anim. Behav.* 17 : 636–40.

Hall, C. A. S. 1972. Migration and metabolism in a temperate stream ecosystem. *Ecology* 53 : 442–53.

Hay, M. E. 1984. Predictable spatial escapes from herbivory: How do these affect the evolution of herbivore resistance in tropical marine communities? *Oecologia* 64 : 396–407.

Helfman, G. S. 1978. Patterns of community structure in fishes: summary and overview. *Env. Biol. Fish.* 3 : 129–48.

Hellier, T. R. 1962. Fish production and biomass in relation to photosynthesis in the Laguna Madre of Texas. *Publ. Inst. Mar. Sci. Univ. Texas* 8 : 1–22.

Hobson, E. S. 1968. Predatory behavior of some shore fishes in the Gulf of California. *Bulletin of Sport Fish and Wildlife, Research Report* 73 : 1–92.

———. 1974. Feeding relationships of teleostean fishes on coral reefs in Kona, Hawaii. *Fish. Bull. U.S.* 72 : 915–1031.

Hyatt, K. D. 1979. Feeding strategy. In W. S. Hoar, D. J. Randall, and J. R. Brett (eds.), *Fish Physiology*; Vol. 8, pp. 71–119. *Bioenergetics and growth*. New York: Academic Press.

Hynes, H. B. N. 1970. *The ecology of running waters*. Toronto: University of Toronto Press.

Jackson, P. B. N. 1961. The impact of predation especially by the Tiger fish (*Hydrocyon vittatus* Cast.) on African freshwater fishes. *Proc. Zool. Soc. London* 136 : 603–22.

Jenkins, T. A. 1969. Social structure, position choice and microdistribution of two trout species (*Salmo trutta* and *Salmo gairdneri*) resident in mountain streams. *Anim. Behav. Monogr.* 2 : 56–123.

Junor, F. 1972. Estimates of daily food intake of piscivorous birds. *Ostrich* 43 : 193–205.

Kaiser, M. S. 1982. Foraging ecology of the green heron on Ozark streams. MS thesis, University of Missouri, Columbia.

Kaysner, C., and A. Heusner. 1964. Etude comparative du metabolism energetique dans la serie animale. *J. Physiol. (Paris)* 56 : 489–524.

Keast, A. 1977. Mechanisms expanding niche width and minimizing intraspecific competition in two centrarchid fishes. In M. K. Steere and B. Wallace (eds.), *Evolutionary biology 10*, pp. 333–95. New York: Plenum.

———. 1978. Trophic and spatial interrelationships in the fish species of an Ontario temperate lake. *Env. Biol. Fish.* 3 : 7–31.

Kramer, D. L. 1983. The evolutionary ecology of

respiratory mode in fishes: an analysis based on the cost of breathing. In T. M. Zaret (ed.), *Evolutionary ecology of neotropical freshwater fishes*, pp. 67–80. The Hague; W. Junk.

Kushlan, J. A. 1976. Wading bird predation in a seasonally fluctuating pond. *Auk* 93 : 464–76.

———. 1978. Feeding ecology of wading birds. In A. Sprunt IV, J. C. Ogden, and S. Winckler (eds.), *Wading birds*, pp. 249–97. Research Report 7. New York: National Audubon Society.

Lagler, K. F. 1939. The control of fish predators at hatcheries and rearing stations. *J. Wildlife Manage.* 3 : 169–79.

Larimore, R. W. 1952. Home pools and homing behavior of smallmouth black bass in Jordan Creek. *Ill. Nat. Hist. Surv. Biol. Notes* 28 : 1–12.

Lindroth, A., and E. Bergstrom. 1959. Notes on the feeding technique of the goosander in streams. *Rep. Inst. Freshw. Res. Drottningholm.* 40 : 165–75.

Lowe-McConnell, R. H. 1964. The fisehs of the Rupununi savanna district of British Guiana, South America. Part I. Ecological groupings of fish species and effects of the seasonal cycle on the fish. *J. Linn. Soc. (Zool.)* 45 : 103–44.

———. 1975. *Fish communities in tropical freshwaters.* London: Longman.

MacArthur, R. H. 1972. *Geographical ecology.* New York: Harper and Row.

Matthews, W. J., M. E. Power, and A. J. Stewart. 1986. Depth distribution of *Campostoma* grazing scars in an Ozark stream. *Environ. Biol. Fish.* (in press).

Meyerriecks, A. J. 1962. Diversity typifies heron feeding. *Natural History* 71 : 48–59.

Mittelbach, G. G. 1984. Predation and resource partitioning in two sunfishes (Centrarchidae). *Ecology* 65 : 499–513.

———. 1986. Predator-mediated habitat use: Some consequences for species interactions. *Env. Biol. Fish.* (in press).

Mott, D. F. 1978. Control of wading bird predation at fish-rearing facilities. In A. Sprunt IV, J. C. Ogden, and S. Winckler (eds.), *Wading birds.* pp. 131–32. Research Report No. 7. New York: National Audubon Society.

Moyle, P. B., J. J. Smith, R. A. Daniels, T. L. Taylor, D. G. Price, and D. M. Baltz. 1982. Distribution and ecology of stream fishes of the Sacramento–San Joaquin Drainage System, California. Berkeley, Calif.: University of Calif. Press.

Mushinski, H. R., J. J. Hebrard, and D. S. Vodopich. 1982. Ontogeny of water snake foraging ecology. *Ecology* 63 : 1624–29.

Naggiar, M. 1974. Man vs. birds. *Florida Wildlife* 27 : 2–5.

Ogden, J. C., and P. S. Lobel. 1978. The role of herbivorous fishes and urchins in coral reef communities. *Env. Biol. Fish.* 3 : 49–63.

Ogden, J. C., and J. C. Zieman. 1977. Ecological aspects of coral reef-seagrass bed contacts in the Caribbean. *Proc. Third. Internat. Coral Reef Symp., Univ. Miami I:* 377–382.

Paine, R. T. 1980. Food webs, linkage, interaction strength and community infrastructure. *J. Anim. Ecol.* 49 : 667–85.

Popova, O. A. 1978. The role of predaceous fish in ecosystems. In S. D. Gerking (ed.), *Ecology of freshwater fish production*, pp. 215–49. New York: Wiley.

Power, M. E. 1981. The grazing ecology of armored catfish in a Panamanian stream. Ph.D. diss., University of Washington, Seattle.

———. 1983. Grazing responses of tropical freshwater fishes to different scales of variation in their food. *Env. Biol. Fish.* 9 : 103–15.

———. 1984a. Depth distributions of armored catfish: Predator-induced resource avoidance? *Ecology* 65 : 523–28.

———. 1984b. Habitat quality and the distribution of algae-grazing catfish in a Panamanian stream. *J. Anim. Ecol.* 53 : 357–74.

———. 1984c. The importance of sediment in the feeding ecology and social interactions of an armored catfish, *Ancistrus spinsosus.* *Env. Biol. Fish.* 10 : 173–81.

Power, M. E., and W. J. Matthews. 1983. Algae-grazing minnows (*Campostoma anomalum*), piscivorous bass (*Micropterus* spp.) and the distribution of attached algae in a small prairie-margin stream. *Oecologia* 60 : 328–32.

Power, M. E., W. J. Matthews, and A. J. Stewart. 1985. Grazing minnows, piscivorous bass and stream algae: Dynamics of a strong interaction. *Ecology* 66 : 1448–56.

Randall, J. E. 1965. Grazing effects on sea grasses by herbivorous reef fishes in the West Indies. *Ecology* 46 : 255–60.

Recher, H. F., and J. A. Recher. 1969. Comments on the escape of prey from avian predators. *Ecology* 49 : 560–62.

Salyer, J. C., and K. F. Lagler. 1949. The eastern belted kingfisher, *Megaceryle alcyon alcyon* (Linnaeus) in relation to fish management. *Trans. Amer. Fish. Soc.* 76 : 97–117.

Savino, J. F., and R. A. Stein. 1982. Predator-prey interaction between largemouth bass and bluegills as influenced by simulated, submerged vegetation. *Trans. Amer. Fish Soc.* 111 : 255–66.

Seghers, B. H. 1970. Behavioral adaptations of natural populasions of the guppy, *Poecilia reticulata*, to predation. *Amer. Zool.* 10 : 89.

———. 1974a. Geographic variation in the responses of guppies (*Poecilia reticulata*) to aerial predators. *Oecologia* 14 : 93–98.

———. 1974b. Schooling behavior in the guppy (*Poecilia reticulata*): An evolutionary response to predation. *Evolution* 28 : 486–89.

Shirvell, C. S., and R. G. Dungey. 1983. Microhabitats chosen by brown trout for feeding and spawning in rivers. *Trans. Amer. Fish. Soc.* 112 : 355–67.

Sih, A. 1982. Foraging strategies and the avoidance of predation by an aquatic insect, *Notonecta hoffmani*. *Ecology* 63 : 786–96.

Skutch, A. F. 1957. Life history of the Amazon kingfisher. *Condor* 59 : 217–29.

Starret, W. C. 1950. Distribution of the fishes of Boone County, Iowa, with special reference to the minnows and darters. *Amer. Midland Naturalist* 43 : 112–27.

Stein, R. A. 1979. Behavioral response of prey to fish predators. In H. Clepper (ed.), *Predator-prey systems in fisheries management*, pp. 343–352. Washington, D.C.: Sport Fishing Institute.

Stein, R. A., and J. J. Magnuson. 1976. Behavioral response of crayfish to a fish predator. *Ecology* 57 : 751–61.

Suthers, R. A. 1965. Acoustic orientation by fish-catching bats. *J. Exp. Zool.* 158 : 319–48.

Udvardy, M. D. 1977. *The Audubon society field guide to North American birds.* New York: Knopf.

Vehrencamp, S. L., F. G. Stiles, and J. W. Bradburg. 1977. Observations on the foraging behavior and avian prey of the Neotropical bat, *Vampyrum spectrum*. *J. Mammal.* 58 : 469–78.

Werner, E. E. 1974. The fish size, prey size, handling time relation in several sunfishes and some implications. *J. Fish. Res. Bd. Can.* 31 : 1531–36.

———. 1977. Species packing and niche complementarity in three sunfishes. *Amer. Naturalist* 111 : 553–78.

Werner, E. E., and J. F. Gilliam. 1984. The ontoge-

netic niche and species interactions in size-structured populations. *Annu. Rev. Ecol. Syst.* 15 : 393–425.

Werner, E. E., J. F. Gilliam, D. Hall, and G. G. Mittelbach. 1983. An experimental test of the effects of predation risk on habitat use in fish. *Ecology* 64 : 1540–48.

Werner, E. E., D. J. Hall, D. R. Laughlin, D. J. Wagner, L. N. Wilsmann, and F. C. Funk. 1977. Habitat partitioning in a freshwater fish community. *J. Fish. Res. Bd. Can.* 34 : 360–370.

White, H. C. 1936. The food of kingfishers and mergansers on the Margaree River, Nova Scotia. *J. Biol. Bd. Can.* 2 : 299–309.

———. 1938. The feeding of kingfishers: Food of nestlings and effect of water height. *J. Fish. Res. Bd. Can.* 4 : 48–52.

———. 1939. *Bird control to increase the Margaree River salmon.* Bull. Fish. Res. Bd. Can. No. 58.

———. 1957. *Food and natural history of mergansers on salmon waters in the Maritime Provinces of Canada.* Bull. Fish. Res. Bd. Can. No. 111.

Whitfield, A. K., and S. J. M. Blaber. 1978. Feeding ecology of piscivorous birds at Lake St. Lucia, Part I: Diving birds. *Ostrich* 49 : 185–98.

Whitfield, A. K., and D. P. Cyrus. 1978. Feeding succession and zonation of aquatic birds at False Bay, Lake St. Lucia. *Ostrich* 49 : 8–15.

Williams, R. 1971. Fish ecology of the Kafue River and floodplain environment. *Fish. Res. Bull. Zambia* 5 : 305–30.

Zaret, T. M. 1980. *Predation in freshwater communities.* New Haven, Conn.: Yale University Press.

V. Local and Regional Extinctions: Historical Ramifications of Predation

23. The Paleoecological Significance of an Anachronistic Ophiuroid Community

Richard B. Aronson

Hans-Dieter Sues

Fossil evidence supports the hypothesis that a radical change in the composition of shallow-water marine communities occurred when a number of durophagous animal groups radiated in the Mesozoic. Predators and grazers, including teleostean fishes, largely eliminated dense populations of sessile, epifaunal suspension-feeders on soft substrates. Communities of stalked crinoids, for example, were driven from shallow-water habitats and today occur only in deep water. Similarly, although shallow-water, ophiuroid-dominated communities are known from the fossil record, dense extant populations occur only under rare circumstances. An examination of some of these dense populations reveals that they persist by virtue of low rates of predation. One example, Sweetings Pond, an isolated Bahamian saltwater lake, supports persistent, high-density populations of the epifaunal, suspension-feeding ophiuroid *Ophiothrix oerstedii* and the octopus *Octopus briareus*. Predatory reef fishes are virtually absent, and experiments and observations revealed that neither *Ophiothrix* nor *Octopus* was subject to more than negligible predation pressure. Sweetings Pond thus contains an anachronistic community of Paleozoic appearance, which persists because predatory fishes are not present.

From these ecological results, we make paleoecological predictions. First, in concurrence with the notion of a "Mesozoic marine revolution," dense ophiuroid communities should occur more commonly before the explosive Cretaceous radiation of teleosts. A review of the paleontological literature shows that dense, shallow-water assemblages of fossil ophiuroids are, with one exception of Late Pleistocene age, not known after the Late Jurassic; numerous examples occur in older deposits. Second, just as predation pressure on extant dense populations is low, the ancient populations should have experienced only limited predation pressure. Evidence from one Mesozoic and two Paleozoic occurrences suggests that the ophiuroids were subject to little, if any, predation. We postulate, therefore, that dense ophiuroid populations were drastically affected by biotic changes during the Mesozoic.

During the last few years, much attention has been paid to understanding large-scale shifts in community composition over geological time. Of particular interest have been recent efforts to uncover causal connections between physical events of global or even galactic magnitude and periodic, catastrophic extinctions (Alvarez et al. 1980; Silver and Schultz 1982; Raup and Sepkoski 1984). The human mind is perhaps so fascinated by the phenomenon of mass extinction because of the death-drama that it connotes. The intellectual attraction to extinction is amply reflected by numerous articles and two recent symposium volumes dealing solely with that subject (Silver and Schultz 1982; Nitecki 1984).

Equally interesting to both neontologists and paleontologists are the dynamics of biotic interactions that caused global-scale community reorganizations. Although predation may not occupy the elevated position of a cause for mass extinction (Stanley 1984), it is now apparent that predator activity was of great importance in at least one biotic upheaval, the "Mesozoic marine revolution" (Vermeij 1977). In this chapter, we hope to demonstrate how general ideas of the historical role of predation may be used to interpret the structure of an extant ophiuroid-dominated community in a Bahamian saltwater lake. Results from this living community are then used to formulate testable hypotheses concerning the structure of similar fossil assemblages. Williamson (1982) has virtually dismissed paleoecology as "a poor-man's applied ecology performed on inadequate data." Yet the fossil record, despite its inherent limitations, provides the only direct evidence for patterns of biotic change through time. We take the more constructive view of Peterson (1984) that paleoecology and ecology reciprocally illuminate each other and should be used in tandem to achieve a broad understanding of community evolution during the Phanerozoic.

Of particular interest to us are Paleozoic and Mesozoic communities that were dominated by epifaunal, suspension-feeding echinoderms. Great slabs covered with articulated stalked crinoids are the most spectacular documents of this ancient type of marine community, and assemblages of well-preserved fossil asterozoans capture the imagination as well. Whereas much has been written in recent years about the paleoecology of crinoids, ophiuroids have received scant attention.

A review of current thoughts on the temporal distribution of marine communities will set the stage for our discussion of ophiuroid paleoecology. The crinoids are important in this respect, and their paleoecology will be reviewed briefly. We shall also consider the cephalopods, which were common predators in some fossil communities and are the dominant carnivores in the Bahamian saltwater lake.

THE MESOZOIC MARINE REVOLUTION

Based on the factor analysis performed by Sepkoski, Phanerozoic diversity of marine organisms may be divided into three assemblages, each characterized by a distinct evolutionary fauna that dominated shelf communities on a global scale (Sepkoski 1981; Sepkoski and Sheehan 1983). The first of these, the "Cambrian" fauna, arose in the Vendian and Early Cambrian and was especially characterized by trilobites and inarticulate brachiopods. Suspension-feeders situated close to the sediment–water interface and surface deposit-feeders constituted Cambrian communities. The "Paleozoic" fauna replaced the "Cambrian" one, beginning with the great Ordovician radiations. This Paleozoic radiation produced diversified communities of epifaunal suspension-feeders, dominated by articulate brachiopods, stenolaemate bryozoans, anthozoans, asterozoans and crinoids (Sepkoski and Sheehan 1983). Dif-

ferent species fed on suspended matter at different heights above the substrate, and this epifaunal tiering has been interpreted as evidence for greater utilization and partitioning of available niche space by the "Paleozoic" evolutionary fauna than by its "Cambrian" predecessor (Ausich and Bottjer 1982; Bambach 1983). Dense assemblages of soft-substrate, epifaunal suspension-feeders, including crinoids and ophiuroids, are known from Paleozoic strata.

After the Ordovician diversification, the second major turning point in the history of marine biota was the mass extinction at the Permo-Triassic boundary (Schopf 1974). This event decimated the "Paleozoic" fauna and made way for the Mesozoic and Cenozoic radiations. Those radiations ultimately resulted in the "Modern" fauna, the characteristics of which pervade shelf communities to the present. Key groups within the "Modern" fauna include bony fishes, echinoids, malacostracans, bivalves, and gastropods (Sepkoski and Sheehan 1983). A number of profound biotic changes occurred during the Mesozoic, resulting in a drastic reorganization of marine communities.

Community structure in the Triassic was essentially an impoverished version of the "Paleozoic" type. The biotic changes that would eventually produce the "Modern" fauna commenced in the Jurassic, and its diagnostic community organization was established by Late Cretaceous times (Vermeij 1983). The evolution and diversification of durophagous crustaceans (decapods and stomatopods) and of teleostean fishes were correlated with, and probably caused, an increase in the frequency of nonlethal shell damage, a correlate of the efficacy of defensive architecture (Vermeij 1977, 1978, 1982, 1983; Vermeij et al. 1981). Predators and grazers, including acanthopterygian teleosts, had largely eliminated dense populations of epifaunal suspension-feeders on shallow-water soft substrates by the Late Cretaceous (Stanley 1977; Jablonski and Bottjer 1983). This sessile epifauna was increasingly replaced by infaunal and mobile epifaunal suspension-feeders (Meyer and Macurda 1977; Stanley 1977; Vermeij 1977). Increased bioturbation of the sediments contributed to the demise of sessile, soft-substrate suspension-feeders during the Mesozoic (Thayer 1979, 1983). The intensity of herbivory also increased during the Mesozoic, with important consequences for the evolution of encrusting coralline algae and perhaps erect, fleshy forms as well (Steneck 1983). The terminal Cretaceous mass extinctions did not alter the qualitative results of the Mesozoic marine revolution (Vermeij 1977).

Each successive fauna in this scheme made greater use of the total available resource space. As noted previously, the "Paleozoic" benthic biota produced their greatest innovations in the epifaunal realm. The three faunas originated in shallow-water habitats and then expanded offshore, each replacing the previous community type in progressively deeper shelf waters (Jablonski and Bottjer 1983; Jablonski et al. 1983). Of special importance in the present context is the fact that soft-substrate, suspension-feeding echinoderm communities of Paleozoic aspect were driven out of coastal waters. Although the onshore-modern/offshore-archaic pattern is readily apparent, explanations for both the nearshore origination and the subsequent offshore expansion of new community types are still unsatisfying, and a number of physical and biological hypotheses have been advanced (Jablonski and Bottjer 1983; Jablonski et al. 1983; Sepkoski and Sheehan 1983; Bretsky and Klofak 1985).

Signor and Brett (1984) have recently suggested that faunal changes similar to those in the Mesozoic occurred in the mid-Paleozoic. The Mid-Devonian radiation of durophagous crustaceans and primitive fishes apparently caused increases in the frequency of defensive adaptations in brachiopods, crinoids, bellerophont mollusks, and nautiloids. The authors hypothesize that "predation by shell-crushing predators has been an important control on the morphology and composition of the marine invertebrate fauna since at least the Middle Devonian."

CRINOIDS

Stalked crinoids occurred densely in a number of Paleozoic and Mesozoic shallow-water communities (Brett and Eckert 1982; Meyer and Ausich 1983). Their present-day restriction to deeper water (>100 m depth) has been explained in terms of the Cretaceous teleostean radiation (Meyer and Macurda 1977). Predatory fishes, it is suggested, eliminated stalked crinoids from shallow-water habitats, and the latter were replaced by unstalked forms, the comatulids. The mobile comatulids are able to take refuge from predators, and they possess structural features interpreted by Meyer and Macurda (1977) and Meyer (1985) as defensive adaptations. [Contrary to previously held beliefs, crinoids are subject to predation by various kinds of teleostean fishes in shallow water (Meyer 1985)]. Based on rather anecdotal evidence, Vasserot (1965) suggested a possible causal connection between the reduction in crinoid diversity and the diversification of palinurid lobsters during the Late Mesozoic. Dense populations of stalked crinoids do occur today in deep water (Lang 1974; Macurda and Meyer 1983), and this would seem to be a manifestation of the general onshore replacement–offshore restriction of the "Paleozoic" fauna. Comatulids occur in dense deep-water aggregations as well (Marr 1963; Fell 1966; Lang 1974), and one species, *Antedon bifida*, is very abundant in British coastal habitats (Reese 1966; Keegan and Könnecker 1980).

CEPHALOPODS

According to Packard (1972), the functional convergence found between coleoid cephalopods and fishes is, to a large extent, a product of selection pressure exerted by teleosts since the Mesozoic. The evolutionary loss of the external shell resulted from the needs to occupy deep ocean habitats and to swim more efficiently in the face of the diversification of marine vertebrate predators and competitors. The coleoids later reinvaded epipelagic waters and coastal habitats, where they currently compete with, prey on, and are preyed on by teleosts (Packard 1972). Packard's scenario appears questionable in light of the recent discovery of a Lower Devonian coleoid cephalopod (Stürmer 1985).

Cephalopods with external shells (ectocochliates) were common carnivores in Paleozoic and Mesozoic marine ecosystems. In fact, nautiloids and ammonoids may have contributed predation pressure to the mid-Paleozoic "prerevolution" (Signor and Brett 1984). Crop contents of Jurassic ammonites contained foraminiferans, ostracods, smaller ammonites and small comatulid crinoids but no remains of larger mobile animals (Lehmann and Weitschat 1973). Today, the deep-dwelling *Nautilus* is the only remaining ectocochliate; it is both a scavenger and a predator, feeding on crustaceans and fishes (Lehmann and Weitschat 1973). The increased development of coarse ornamentation on ammonoid shells, especially by Cretaceous times, is interpreted by Ward (1981) as the result of increased activity by durophagous predators, especially crustaceans. The ammonoids decreased in diversity for millions of years preceding their sudden extinction at the Cretaceous–Tertiary boundary. Both global-scale physical disruptions (Alvarez et al. 1984) and the activities of durophagous predators (including vertebrates; Ward 1983) may have contributed to this gradual decline.

ANACHRONISTIC MARINE COMMUNITIES AND PALEOECOLOGICAL HYPOTHESES

Under certain rare circumstances, we find modern analogues to ancient community types. Such anachronistic communities of organisms can be strongly reminiscent of fossil, "pre-Modern" assemblages in trophic patterns, relative species abundance, and even higher-level taxonomic composition. These rare circumstances are crucial to understanding why such assemblages persisted in the geological past. Stromatolites, which are structures composed of filamentous prokaryotes, trapped sediment, and associated biota, are a case in point. Common in Late Precambrian shallow marine environments, they are thought to have been excluded from most shallow-water habitats by burrowing organisms and grazing mollusks at the beginning of the Paleozoic (Garrett 1970; Knoll 1983). This inference is based in part on the fact that Recent stromatolite communities occur in a few supratidal, intertidal, and shallow subtidal settings where harsh physical conditions exclude bioturbators and grazers (Garrett 1970; Dravis 1983; Thayer 1983).

Sweetings Pond, an isolated saltwater lake on Eleuthera Island, Bahamas, contains another type of anachronistic community. This lake supports a persistent, high-density population of the epifaunal, suspension-feeding ophiuroid, *Ophiothrix oerstedii*. The ophiuroid density, which sometimes exceeds 400 individuals per square meter (figs. 23.1A and 23.2), is two orders of magnitude higher than that found in nearby coastal habitats and occurs because predatory fishes are virtually absent from the lake (Aronson and Harms 1985). When assem-

blages of ophiuroids comparable in density to those in Sweetings Pond were exposed in open arenas (from which they could not escape) at a coastal site off Eleuthera, the brittlestars were completely consumed within 48 hours. No significant ophiuroid mortality occurred in similar arenas in the lake. Gut content and fecal analyses of all possible Sweetings Pond predators of *Ophiothrix*, including the large majid crab, *Mithrax spinosissimus*, confirmed the virtual absence of predation. Through observation and experimentation, Aronson and Harms (1985) demonstrated that density variations within the lake are determined by variations in the degree of small-scale topographical heterogeneity, not by variations in predation pressure. In stark contrast to coastal conspecifics, Sweetings Pond brittlestars expose themselves day and night. This behavioral difference is causally related to the difference in predatory activity by fishes (Aronson 1985).

Another aspect of the lake's anachronistic character is its high density of *Octopus briareus*, the Caribbean reef octopus. The population density of this cephalopod is also orders of magnitude greater in Sweetings Pond than off the coast. Once again, the absence of predatory fishes appears to be responsible. The octopuses are limited by the availability of dens in Sweetings Pond (Aronson, in press).

It is perhaps more than coincidental that a slow-moving, epifaunal, suspension-feeding echinoderm, which carpets portions of the lake substrate and gives the benthos a distinctly Paleozoic appearance, is accompanied by a cephalopod (which does not feed upon ophiuroids; Aronson and Harms 1985). Ectocochliate cephalopod carnivores were common in Paleozoic and Mesozoic coastal benthic communities (see Discussion), and they may have exerted a relatively greater influence (in the absence of teleosts) than do the present-day coleoids. Therefore, the abundant octopuses in Sweetings Pond may be functional analogues of ammonoids and/or nautiloids in Paleozoic marine communities. The observations in Sweetings Pond support the hypothesis that increased fish predation in the Mesozoic contributed to the demise of dense ophiuroid as well as crinoid communities in coastal habitats. Where predation pressure from fishes (and crustaceans) is weak or absent, as in Sweetings Pond and some temperate and boreal coastal communities, exposed ophiuroids

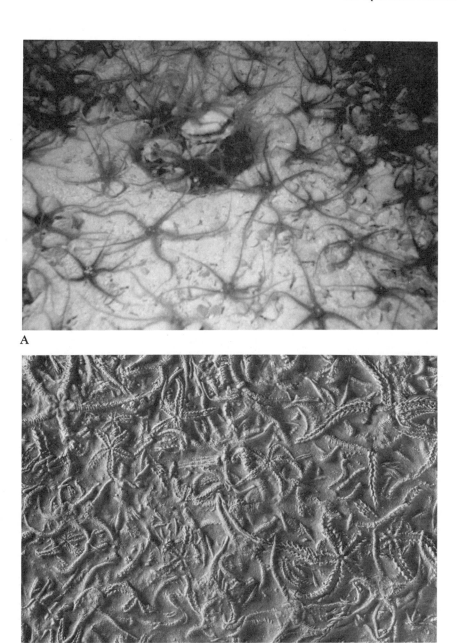

Figure 23.1. A: *Dense population of* Ophiothrix oerstedii *at 3.7 m depth in Sweetings Pond. From "Ophiuroids in a Bahamian saltwater lake: The ecology of a Paleozoic-like community" by R. B. Aronson and C. A. Harms,* Ecology *66 : 1472–83. Copyright © 1985 by the Ecological Society of America. Reprinted by permission.* **B:** Strataster ohioensis *from the early Mississippian. Photo courtesy of R. V. Kesling.*

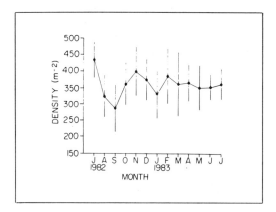

Figure 23.2. *Results of monthly censuses of Ophiothrix oerstedii at 3.7-m depth in Sweetings Pond. Based on eight random tosses of 1/9 m² quadrat. Modified by permission from Aronson and Harms (1985). Copyright © 1985 by the Ecological Society of America.*

still occur densely (Vevers 1952; Warner 1971; Wilson et al. 1977; Aronson and Harms 1985). In fact, fluctuations in the occurrence of dense beds of *Ophiothrix fragilis* in the English Channel over a period of several decades have been correlated with changes in predation pressure exerted by two species of the starfish *Luidia* (Holme 1984). In the next section, we shall review the fossil record of ophiuroids and attempt to address the following questions: (1) Was there a decline in the occurrence of dense ophiuroid assemblages in the fossil record after the Mesozoic marine revolution? (2) Is there evidence that predation pressure on dense populations of fossil ophiuroids was limited or absent?

OPHIUROID COMMUNITIES, EXTANT AND FOSSIL

Accumulations of articulated ophiuroids were occasionally preserved as fossil "starfish beds" (table 23.1). The essential condition for preservation of the delicate asterozoan skeletons in their entirety is rapid burial by sediment without subsequent hydraulic reworking or bioturbation (Goldring and Stephenson 1972). Because of this taphonomic requirement, dense assemblages of articulated ophiuroids are very rare in the fossil record. We have undertaken an extensive survey of the paleontological literature and have consulted with several students of fossil echinoderms in order to compile data concerning the distribution in time and space

of such fossil assemblages (table 23.1). "Dense assemblages" refers to aggregations of numerous articulated ophiuroids per unit area (usually square meter), rather than to localities characterized by the common occurrence of individual specimens. We are aware that there may be yet other examples, both published and unpublished, but an intriguing pattern of distribution has emerged.

A few dense ophiuroid assemblages (e.g., the Late Ordovician Girvan Starfish Bed; Goldring and Stephenson 1972) are apparently the result of turbulence in a shallow marine setting, where the benthic fauna and sediment were disturbed, shifted, and rapidly redeposited. However, as a consequence of the rapidity with which ophiuroids disarticulate after death (Schäfer 1962), the majority of starfish beds represent autochthonous death assemblages (thanatocoenoses) or at most the results of minor transport.

In the North Sea, burial and smothering is the leading cause of death for ophiuroids (Schäfer 1962). Schäfer noted that only 5 cm of sediment is required to trap and prevent the upward escape of these animals. Fine-grained sedimentary matter presumably obstructs the function of the water vascular system in ophiuroids (Rosenkranz 1971). Areas of rapid sedimentation are therefore avoided. Kesling (1969), Rosenkranz (1971), and Goldring and Stephenson (1972) have analyzed fossil examples of autochthonous thanatocoenoses of asterozoans due to rapid smothering by muddy sediment. Articulated ophiuroids are usually found in fine-grained Paleozoic and Mesozoic sediments, often with some clay content (see review in Rosenkranz 1971).

Ophiuroid skeletal structure regrettably permits few inferences regarding specific ecological roles. The mode of feeding can be variable even within a single species (Fell 1966). Most assemblages of fossil ophiuroids appear to consist mainly of epifaunal suspension-feeders that lived slightly covered by sediment in shallow water. Population densities in the autochthonous assemblages can be very high. Kesling and Le Vasseur (1971) estimated original population densities for the Early Mississippian *Strataster ohioensis* at 4,500 individuals per square meter at a water depth of approximately 30 m. Liddell (1975) reports densities for the Middle Ordovician *Stenaster salteri* greater than 440 individuals per square meter. Among

Table 23.1. *Occurrences of dense assemblages of articulated ophiuroids in the fossil record*

Locality	Age	Lithology	Associated fauna	Comments
Kirkfield, Ontario	Mid-Ordovician (Trentonian)	Shale	Abundant brachiopods, crinoids	Smothered, autochthonous assemblage; more than 440 ophiuroids per square meter
Girvan, Scotland[a]	Late Ordovician	Fine-grained sandstone	Other echinoderms	Considerable lateral displacement with rapid resedimentation
Leintwardine, Herefordshire[a]	Silurian (Ludlovian)	Calcareous siltstone	Inarticulate brachiopods and crinoids	Introduced into submarine channel-fill
Arkona, Ontario	Mid-Devonian	Mudstone	Platycerid gastropods	Smothered autochonous assemblage; up to 300 ophiuroids per square meter
Velbert, West Germany	Late Devonian (Famennian)	Siltstone	Little fauna in immediate association	
Angerbachtal, West Germany	Late Devonian (Famennian)	Siltstone	Virtually no associated fauna	
Royalton, Ohio	Mississippian	Shale	A few crinoids and bryozoans	Up to 4500 ophiuroids per square meter
Roitzka, Poland	Mid-Triassic (Muschelkalk)	Calcareous marl		80% of 200 specimens preserved on their backs
Weimar, East Germany	Mid-Triassic (Muschelkalk)	Limestone		More than 100 specimens on a slab area of 10 × 5 cm
Mergentheim, West Germany	Mid-Triassic (Muschelkalk)	Limestone		Ophiuroids occur "in flocks"
Dorset coast	Early Jurassic (Mid-Liassic)	Fine-grained sandstone		Smothered, autochthonous assemblage
Schofgraben, Switzerland	Late Jurassic (Oxfordian)	Silty Sandstone		Up to 1800 ophiuroids per square meter
La Voulte-sur-Rhône	Late Jurassic (Callovian)	Marls and claystones	Largely nectonic, coleoid cephalopods, fishes, crustaceans	Rapid burial; 500 ophiuroids on an area of 2.5 m^2
Solnhofen area, West Germany	Late Jurassic (Tithonian)	Limestone	Rich fauna with both nectonic and benthic components	*Geocoma* locally very abundant; lagoonal setting
Dunbar, East Lothian	Late Pleistocene	Clay	No associated fauna	

Data compiled from Kutscher (1940), Rosenkranz (1971), Kesling and Le Vasseur (1971), Dietl and Mundlos (1972), Haude and Thomas (1983), and Meyer (1984).

[a]Denotes assemblages with considerable lateral displacement of faunal components.

Recent epifaunal ophiuroids, densities in some populations are comparable. Warner (1971) studied an aggregation of Ophiothrix fragilis off the English coast with a mean density of 1,330 individuals per square meter (water depth: 14 m); Vevers (1952) counted about 340 individuals of this species per square meter at a depth of 48 m near Plymouth. Ophiocomina nigra occurs at up to 500 individuals per square meter in the Irish Sea at 10 to 30 m depth (Brun 1972). Even higher ophiuroid densities have been recorded off the British Isles by Keegan and Könnecker (1980). The highest mean density of Ophiothrix oerstedii recorded in Sweetings Pond was 434 individuals per square meter (Aronson and Harms 1985).

Aronson and Harms (1985) have suggested that the formation of dense assemblages of brittlestars (and other epifaunal suspension-feeders) may be related to low predation pressures. We found only one example of a post-Jurassic dense assemblage of articulated ophiuroids: Allman (1863) described an aggregation of Ophiolepis gracilis from the Brick Clay near Dunbar, East Lothian, apparently a late glacial deposit (Baird, personal communication).

From the Early Devonian Hunsrück shales in southwestern Germany, Lehmann studied more than 1,000 well-preserved ophiuroids, referable to 15 genera and 22 species. Only 23 specimens showed regeneration of arms or of arm tips in this large sample (Lehmann 1951). This is consistent with the hypothesis that predation pressures on Paleozoic ophiuroid populations were low. With the possible exception of a large placoderm fish, there are no potential predators in the well-preserved associated arthropod and fish faunas from the Hunsrück shales. The shales differ from other, more locally restricted occurrences of Paleozoic ophiuroids in that they were probably formed in the upper bathyal region (Seilacher and Hemleben 1966). The formation of this unusually rich and diversified assemblage of marine fossils was possibly the result of intermittent poisoning of the bottom water layers because at other times conditions for life (as indicated by trace fossils) were favorable.

Kesling and Le Vasseur (1971) found no evidence of predation, either in the form of regenerating arms or preserved predators, in their dense Strataster aggregation (fig. 23.1B). They estimate that this Strataster community persisted for at least 30 years.

Among the Mesozoic occurrences, that of the Late Jurassic Solnhofen lithographic limestones in Bavaria is of particular interest. These deposits formed in lagoonal basins between geologically older sponge reefs (Hemleben 1977). The small ophiuroids Geocoma carinata and Ophiurella speciosa are extremely abundant at Zandt (Kuhn 1963). The associated fauna contains a moderate radiation of small teleosts, along with numerous holostean fishes and numerous ammonoid and coleoid cephalopods. We know of no published details concerning predation on the ophiuroids.

Of 55 well-preserved specimens of Ophiomusium weymouthiense from the late Mid-Jurassic of Weymouth, Dorset, housed in the collections of the British Museum (Natural History), only one showed arm regeneration (Aronson, personal observation).

DISCUSSION

Where information is available, it is clear that extant ophiuroid communities have persisted for long periods of time, ecologically speaking. The Sweetings Pond Ophiothrix oerstedii community has been qualitatively the same for more than 10 years (Aronson and Harms 1985), and dense beds of O. fragilis have been known from the British Isles since before 1900. The same can be said for the fossil communities. Hence, these dense assemblages of ophiuroids are not examples of "explosive opportunism" (sensu Levinton 1970) but rather represent stable populations. Only one case conforms to Levinton's model. Kesling (1969) has postulated that the dense community of Eugasterella thorni from the Mid-Devonian Arkona Shale (table 23.1) was short-lived and fed on a previously destroyed crinoid "garden" along with other scavengers.

Even though direct evidence concerning the level of predation pressure on dense populations of fossil ophiuroids is scanty, the temporal distribution of those communities and the data on predation in some extant assemblages support our hypothesis that dense communities of epifaunal brittlestars were largely excluded from shallow water after the Mesozoic. In particular, we see a relation between the explosive Cretaceous radiation of teleostean fishes and the virtual disappearance of dense assemblages of fossil ophiuroids in the Cretaceous. Dense extant assemblages on soft substrates in shallow-water settings are frequently composed of infaunal species (see Thorson 1957). Exposed epifaunal brittlestars are more characteristic of deep than of coastal waters, and they can dominate the macrofaunas of deep-sea communities (Barham et al. 1967; Rowe 1971; Grassle et al. 1975; Tyler 1980).

A null hypothesis for the temporal distribution of brittlestar beds is that they occur at random in the fossil record. The more frequent (albeit still rare) occurrence of dense ophiuroid communities prior to the Mesozoic marine revolution could be simply due to the greater span of time elapsed from the Ordovician (time of origin for the "Paleozoic" fauna) to the end of the Jurassic (365 million years) than from the Cretaceous through the Recent (130 million years) (Odin et al. 1982). We found 12 autochthonous or presumed autochthonous "pre-revolution" assemblages in shallow water and one "post-revolution" one (table 23.1). Considering relative time spans only, the null probability of an ophiuroid community being a "pre-revolution" occurrence is $r = 0.737$ ($= 365/495$), and the probability of a $12:1$ distribution of communities is $p = 0.11$ (one-tailed binominal test) (Sokal and Rohlf 1969). Making some allowance for the relative amount of exposed sediments from various geological periods (Cretaceous and Cenozoic strata being more frequently exposed) and "pull of the Recent" (geologically younger faunas being more completely preserved) (Raup 1978) reduces p further. In fact, even a small reduction of r to 0.68 to take account of these two geological phenomena (Aronson, unpublished data) will yield $p < <0.05$, and we can reject the null hypothesis that ophiuroid communities are randomly distributed in the fossil record.

It is not unreasonable to imagine that cephalopods were common predators in some ancient ophiuroid-dominated communities, as they are in Sweetings Pond. Ectocochliates and coleoids are certainly common in the Late Jurassic deposits at Solnhofen (Kuhn 1963) and La Voulte-sur-Rhône (Dietl and Mundlos 1972). Based on data from Sweetings Pond, we suspect that many cephalopods in Paleozoic and Mesozoic communities did not consume brittlestars, even when the latter were extremely abundant. To our knowledge, the only living cephalopod that preys on ophiuroids is the deep-dwelling *Bathypolypus arcticus* (O'Dor and MacAlaster 1983). In the absence of fish, crustacean, and cephalopod durophagy, then, dense populations of ophiuroids could thrive in "Paleozoic" communities.

We have shown how extant anachronistic communities can be used to make paleoecological inferences. Saltwater lakes and other islands are promising places to look for such communities. The Bahama Islands contain a vast number of lakes of various degrees of salinity, and most of them remain unexplored. Great Oyster Pond, another lake on Eleuthera, is depauperate of fish, like Sweetings Pond. This body of water supports high densities of the epifaunal bivalve *Pinctada radiata*. In addition, a small apodid holothurian was found in great numbers on open substrate during the day (Aronson, personal observation). Release of benthic organisms due to the absence of predatory fishes has been noted anecdotally for the saltwater lakes of Palau, Micronesia (Hamner 1982; personal communication).

The study of ecological release in saltwater lakes has practical application. Information culled from Sweetings Pond and other lakes may enable us to predict the biotic consequences of large-scale overfishing in shallow-water marine habitats. The high abundance of *Octopus* in a back-reef community on the north coast of Jamaica may be related to the severe overexploitation of local fish resources (Aronson, in press). We recommend that overfished habitats be examined to ascertain whether the removal of predatory teleosts is increasing the abundance of epifaunal suspension-feeders and therefore is perhaps driving those communities toward anachronistic faunal compositions.

ACKNOWLEDGMENTS

We thank R. A. Allmon, W. D. Allmon, B. Baird, D. Fletcher, R. Goldring, S. J. Gould, L. S. Kaufman, J. A. Kitchell, A. H. Knoll, D. N. Lewis, W. D. Liddell, D. M. Raup, A. B. Smith, R. S. Steneck, C. W. Thayer, and P. G. Williamson for their advice and comments. R. B. A. acknowledges partial support by a National Science Foundation Graduate Fellowship.

REFERENCES

Allman. 1863. On a new fossil ophiuridian, from post-Pliocene strata of the valley of the Forth. *Proc. Roy. Soc. Edinburgh* 5 : 101–104.

Alvarez, L. W., W. Alvarez, F. Asaro, and H. V. Michel. 1980. Extraterrestrial causes for the Cretaceous-Tertiary extinction. *Science* 208 : 1095–1108.

Alvarez, W., E. G. Kauffman, F. Surlyk, L. W. Alvarez, F. Asaro, and H. V. Michel. 1984. Impact theory of mass extinctions and the invertebrate fossil record. *Science* 223 : 1135–41.

Aronson, R. B. 1985. Ecological release in a Bahamian salt water lake: *Octopus briareus* (Cephalopoda) and *Ophiothrix oerstedii* (Ophiuroidea). Ph.D. thesis, Harvard University.

———. 1986. The ecology of *Octopus briareus* in a Bahamian saltwater lake. *Stud. Trop. Oceanogr.* (in press).

Aronson, R. B., and C. A. Harms. 1985. Ophiuroids in a Bahamian saltwater lake: The ecology of a Paleozoic-like community. *Ecology* 66 : 1472–83.

Ausich, W. I., and D. J. Bottjer. 1982. Tiering and suspension-feeding communities on soft substrata throughout the Phanerozoic. *Science* 216 : 173–74.

Bambach, R. K. 1983. Ecospace utilization and guilds in marine communities through the Phanerozoic. In M. J. S. Tevesz and P. L. McCall (eds.), *Biotic interactions in Recent and fossil benthic communities*, pp. 719–46. New York: Plenum.

Barham, E. G., N. J. Ayer, Jr., and R. E. Boyce. 1967. Macrobenthos of the San Diego Trough: Photographic census and observations from bathyscape, *Trieste. Deep-Sea Res.* 14 : 773–84.

Bretsky, P. W., and S. M. Klofak. 1985. Margin to craton expansion of Late Ordovician benthic marine invertebrates. *Science* 227 : 1469–71.

Brett, C. E., and J. D. Eckert. 1982. Paleoecology of a well-preserved crinoid colony from the Silurian Rochester shale in Ontario. *Roy. Ont. Mus., Life Sci. Contrib.* 131 : 1–20.

Brun, E. 1972. Food and feeding habits of *Luidia ciliaris. J. Mar. Biol. Assn. U.K.* 55 : 191–97.

Dietl, G., and R. Mundlos. 1972. Ökologie und Biostratinomie von *Ophiopinna elegans* (Ophiuroidea) aus dem Untercallovium von La Voulte (Südfrankreich). *N. Jb. Geol. Paläont., Mh.* 1972 : 449–64.

Dravis, J. J. 1983. Hardened subtidal stromatolites, Bahamas. *Science* 219 : 385–86.

Fell, H. B. 1966. The ecology of ophiuroids. In R. A. Boolootian (ed.), *Physiology of Echinodermata*, pp. 129–43. New York: John Wiley.

Garrett, P. 1970. Phanerozoic stromatolites: Noncompetitive ecologic restriction by grazing and burrowing animals. *Science* 169 : 171–73.

Goldring, R., and D. G. Stephenson. 1972. The depositional environment of three starfish beds. *N. Jb. Geol. Paläont., Mh.* 1972 : 611–24.

Grassle, J. F., H. L. Sanders, R. R. Hessler, G. T. Rowe, and T. McLellan. 1975. Pattern and zonation: A study of bathyal megafauna using the research submersible *Alvin. Deep-Sea Res.* 22 : 457–81.

Hamner, W. M. 1982. Strange world of Palau's salt lakes. *National Geographic* 161 : 264–82.

Haude, R., and E. Thomas. 1983. Ophiuren (Echinodermata) des hohen Oberdevon im nördlichen Rheinischen Schiefergebirge. *Paläont. Z.* 57 : 121–42.

Hemleben, C. 1977. Autochthone und allochthone Sedimentanteile in den Solnhofener Plattenkalken. *N. Jb. Geol. Paläont., Mh.* 1977 : 257–71.

Holme, N. A. 1984. Fluctuations of *Ophiothrix fragilis* in the western English Channel. *J. Mar. Biol. Assn. U.K.* 64 : 351–78.

Jablonski, D., and D. J. Bottjer. 1983. Soft-bottom epifaunal suspension-feeding assemblages in the Late Cretaceous: implications for the evolution of benthic paleocommunities. In M. J. S. Tevesz and P. L. McCall (eds.), *Biotic interactions in Recent and fossil benthic communities*, pp. 747–812. New York: Plenum.

Jablonski, D., J. J. Sepkoski, Jr., D. J. Bottjer, and P. M. Sheehan. 1983. Onshore-offshore patterns in the evolution of Phanerozoic shelf communities. *Science* 222 : 1123–25.

Keegan, B. F., and G. Könnecker. 1980. Aggregation in echinoderms on the West coast of Ireland: An ecological perspective. In M. Jangoux (ed.), *Echinoderms: Present and past*, p. 199. Rotterdam: Balkema.

Kesling, R. V. 1969. A new brittle-star from the Middle Devonian Arkona shale of Ontario. *Contrib. Mus. Paleont. Univ. Mich.* 23 : 37–51.

Kesling, R. V., and D. Le Vasseur. 1971. *Strataster ohioensis*, a new early Mississippian brittle-star, and the paleoecology of its community. *Contrib. Mus. Paleont. Univ. Mich.* 23 : 305–41.

Knoll, A. H. 1983. Biological interactions and Precambrian eukaryotes. In M. J. S. Tevesz and P. L. McCall (eds.), *Biotic interactions in Recent and fossil benthic communities*, pp. 251–83. New York: Plenum.

Kuhn, O. 1963. *Die Tierwelt des Solnhofener Schiefers.* Wittenberg-Lutherstadt: Ziemsen.

Kutscher, F. 1940. Ophiuren-Vorkommen im Muschelkalk Deutschlands. *Z. dtsch. geol. Ges.* 92 : 1–18.

Lang, J. C. 1974. Biological zonation at the base of a reef. *American Scientist* 62 : 272–81.

Lehmann, U., and W. Weitschat. 1973. Zur Anatomie und Ökologie von Ammoniten: Funde von Kropf und Kiemen. *Paläont. Z.* 47 : 69–76.

Lehmann, W. M. 1951. Anomalien und Regenerationserscheinungen an paläozoischen Asterozoen. *N. Jb. Geol. Paläont., Abh.* 93 : 401–16.

Levinton, J. S. 1970. The paleoecological significance of opportunistic species. *Lethaia* 3 : 69–78.

Liddell, W. D. 1975. Ecology and biostratinomy of a

Middle Ordovician echinoderm assemblage from Kirkfield, Ontario. M.A. thesis, University of Michigan.

Macurda, D. B., Jr., and D. L. Meyer. 1983. Sea lilies and feather stars. *American Scientist* 71 : 354–65.

Marr, J. W. S. 1963. Unstalked crinoids of the Antarctic continental shelf. Notes on their natural history and distribution. *Phil. Trans. Roy. Soc. London (B)* 246 : 327–79.

Meyer, C. A. 1984. Palökologie und Sedimentologie der Echinodermenlayerstd'tte Schofgraben (mittleres Oxfordien, Weissenstein, Kt. Solothurn). *Ecl. geol. Helv.* 77 : 649–73.

Meyer, D. L. 1985. Evolutionary implications of predation on Recent comatulid crinoids from the Great Barrier Reef. *Paleobiology* 11 : 154–64.

Meyer, D. L., and W. I. Ausich. 1983. Biotic interactions among Recent and among fossil crinoids. In M. J. S. Tevesz and P. L. McCall (eds.), *Biotic interactions in Recent and fossil benthic communities*, pp. 377–427. New York: Plenum.

Meyer, D. L., and D. B. Macurda, Jr. 1977. Adaptive radiation of the comatulid crinoids. *Paleobiology* 3 : 74–82.

Nitecki, M. H. (ed.). 1984. *Extinctions.* Chicago: University of Chicago Press.

O'Dor, R. K., and E. G. MacAlaster. 1983. *Bathypolypus arcticus.* In P. R. Boyle (ed.), *Cephalopod life cycles*, Vol. I, pp. 401–12. London: Academic Press.

Odin, G. S., D. Curry, N. H. Gale, and W. J. Kennedy. 1982. The Phanerozoic time scale in 1981. In G. S. Odin (ed.), *Numerical dating in stratigraphy*, pp. 957–60. Chichester, U.K.: Wiley.

Packard, A. 1972. Cephalopods and fish: The limits of convergence. *Biol. Rev.* 47 : 241–307.

Peterson, C. H. 1984. The pervasive biological explanation [book review]. *Paleobiology* 9 : 429–36.

Raup, D. M. 1978. Cohort analysis of generic survivorship. *Paleobiology* 4 : 1–15.

Raup, D. M., and J. J. Sepkoski, Jr. 1984. Periodicity of extinctions in the geologic past. *Proc. Nat. Acad. Sci. U.S.A.* 81 : 801–805.

Reese, E. S. 1966. The complex behavior of echinoderms. In R. A. Boolootian (ed.), *Physiology of Echinodermata*, pp. 157–218. New York: John Wiley.

Rosenkranz, D. 1971. Zur Sedimentologie und Ökologie von Echinodermen-Lagerstätten. *N. Jb. Geol. Paläont., Abh.* 138 : 221–58.

Rowe, G. T. 1971. Observations on bottom currents and epibenthic populations in Hatteras Submarine Canyon. *Deep-Sea Res.* 18 : 569–81.

Schäfer, W. 1962. *Aktuopaläontologie nach Studien in der Nordsee.* Frankfurt: W. Kramer.

Schopf, T. J. M. 1974. Permo-Triassic extinctions: Relation to sea-floor spreading. *J. Geol.* 82 : 129–43.

Seilacher, A., and C. Hemleben. 1966. Beiträge zur Sedimentation und Fossilführung des Hunrückschiefers. 14. Spurenfauna und Bildungstiefe der Hunsrückschiefer (Unterdevon). *Notizbl. hess. Landesamt. Bodenf.* 94 : 40–53.

Sepkoski, J. J., Jr. 1981. A factor analytic description of the Phanerozoic marine fossil record. *Paleobiology* 7 : 36–53.

Sepkoski, J. J., Jr., and P. M. Sheehan. 1983. Diversification, faunal change, and community replacement during the Ordovician radiations. In M. J. S. Tevesz and P. L. McCall (eds.), *Biotic interactions in Recent and fossil benthic communities*, pp. 673–717. New York: Plenum.

Signor, P. W. III, and C. E. Brett. 1984. The mid-Paleozoic precursor to the Mesozoic marine revolution. *Paleobiology* 10 : 229–45.

Silver, L. T., and P. H. Schultz (eds.) 1982. Geological implications of impacts of large asteroids and comets on the Earth. *Geol. Soc. Amer., Spec. Pap.* 190.

Sokal, R. R., and F. J. Rohlf. 1969. *Biometry.* San Francisco: Freeman.

Stanley, S. M. 1977. Trends, rates and patterns of evolution in the Bivalvia. In A. Hallam (ed.), *Patterns of evolution as illustrated by the fossil record*, pp. 209–50. Amsterdam: Elsevier.

———. 1984. Marine mass extinctions: A dominant role for temperature. In M. H. Nitecki (ed.), *Extinctions*, pp. 69–117. Chicago: University of Chicago Press.

Steneck, R. S. 1983. Escalating herbivory and resulting adaptive trends in calcareous algal crusts. *Paleobiology* 9 : 44–61.

Stürmer, W. 1985. A small coleoid cephalopod with soft parts from the Lower Devonian discovered using radiography. *Nature* 318 : 53–55.

Thayer, C. W. 1979. Biological bulldozers and the evolution of marine benthic communities. *Science* 203 : 458–61.

———. 1983. Sediment-mediated biological disturbance and the evolution of marine benthos. In M. J. S. Tevesz and P. L. McCall (eds.), *Biotic interactions in Recent and fossil benthic communities*, pp. 479–625. New York: Plenum.

Thorson, G. 1957. Bottom communities. *Geol. Soc. Amer. Mem.* 67(1) : 461–534.

Tyler, P. A. 1980. Deep-sea ophiuroids. *Oceanogr. Mar. Biol. Ann. Rev.* 18 : 125–53.

Vasserot, J. 1965. Un prédateur d'echinodermes s'attaquant particulièrement aux ophiures: la langouste *Palinurus vulgaris. Bull. Soc. Zool. Fr.* 90 : 365–84.

Vermeij, G. J. 1977. The Mesozoic marine revolution: evidence from snails, predators and grazers. *Paleobiology* 3 : 245–58.

———. 1978. Biogeography and adaptation: patterns of marine life. Cambridge, Mass.: Harvard

University Press.

——. 1982. Unsuccessful predation and evolution. *Amer. Naturalist* 120 : 701–720.

——. 1983. Shell-breaking predation through time. In M. J. S. Tevesz and P. L. McCall (eds.), *Biotic interactions in Recent and fossil benthic communities*, pp. 649–69. New York: Plenum.

Vermeij, G. J., D. E. Schindel and E. Zipser. 1981. Predation through geological time: Evidence from gastropod shell repair. *Science* 214 : 1024–26.

Vevers, H. G. 1952. A photographic survey of certain areas of the sea floor near Plymouth. *J. Mar. Biol. Assn. U.K.* 31 : 215–21.

Ward, P. 1981. Shell sculpture as a defensive adaptation in ammonoids. *Paleobiology* 7 : 95–100.

——. 1983. The extinction of the ammonites. *Sci. Amer.* 249(4) : 136–47.

Warner, G. F. 1971. On the ecology of a dense bed of the brittle-star *Ophiothrix fragilis*. *J. Mar. Biol. Assn. U.K.* 51 : 267–82.

Williamson, P. G. 1982. Cinderella subject [book review]. *Nature* 296 : 99–100.

Wilson, J. B., N. A. Holme, and R. L. Barrett. 1977. Population dispersal in the brittle-star *Ophiocomina nigra* (Abildgaard) (Echinodermata: Ophiuroidea). *J. Mar. Biol. Assn. U.K.* 57 : 405–39.

24. Branchiopod Communities: Associations with Planktivorous Fish in Space and Time

W. Charles Kerfoot

Michael Lynch

Present and past distributions of the Branchiopoda and bony fish argue for a strong interaction between the two taxa. Initially, the Branchiopoda were globally distributed in marine, fresh, and estuarine environments. Concomitant with the supposed Mesozoic "revolution" in marine waters, there was a gradual restriction of conspicuous, large-bodied forms to fish-free environments. Many groups went extinct at this time. We suggest that these changes were a direct consequence of the evolutionary increase in the intensity of fish predation during this era. We also argue that the rise of the relatively small-bodied and inconspicuous Cladocera is connected to the increased dominance of fish predation in freshwater pelagic environments.

The argument that ecological interactions take place on a time scale much faster than evolutionary responses often is used to discredit the importance of evolutionary phenomena in determining patterns of species abundance. Yet one of the more frequent outcomes of predator–prey interactions in simple environments, prey extinction, can proceed rapidly at a local site and yet be prolonged on a regional scale because of such classical properties as predator dispersal, habitat heterogeneity, or marginal refugia (Gause 1934; Huffaker 1958). For this reason, predation is rarely cited as the cause of immediate mass extinctions (Stanley 1984), although it may be the driving force behind some major displacements in the marine environment (Vermeij 1977, 1978).

Currently, there is no known freshwater counterpart to the "mid-Mesozoic marine revolution" (Vermeij 1977). It is conceivable, however, that this is primarily a consequence of the modest attention given to the paleontology of freshwater relative to marine communities. The ability of predators to alter the structure of freshwater communities radically on a scale of months or years is well established (Zaret 1972, 1980, 1982; Werner et al. 1983). Here we take a broader view and review on an evolutionary time-scale the evidence that the temporal and geographic distribution of the Branchiopoda has been strongly constrained by the evolution of the bony fishes (Osteichthyes). The evidence we cite derives from three sources: (1) the current restriction of formerly cosmopolitan, larger-bodied taxa (Anostraca, Notostraca, Conchostraca) to ephemeral or fish-free habitats, (2) latitudinal gradients in smaller-bodied taxa (Cladocera) that suggest exclusion of more conspicuous species from regions of year-round fish activity (i.e., tropical lowlands), and (3) a temporal correspondence between the evolution of fish feeding modes, extinction patterns of large-bodied species, and the appearance of modern microcrustacean assemblages.

RESTRICTION OF ANCIENT COSMOPOLITAN GROUPS (ANOSTRACA, NOTOSTRACA, CONCHOSTRACA) TO EPHERMERAL AND FISH-FREE HABITATS

Living representatives of the crustacean class Branchiopoda fall into four orders: Anostraca (fairy shrimps), Notostraca (tadpole shrimps), Conchostraca (clam shrimps), and Cladocera (water fleas). All have in common the following features: flat, biramous, and setose trunk appendages; grinding mandibles associated with reduced first and second maxillae; last body segment with a pair of short to long cercopods; and legs modified for swimming and/ or food gathering. The first three orders are the most primitive from the standpoint of number of body appendages and are distinguishable by major morphological differences.

When compared with cladocerans (total length 0.2–18 mm; mean, <2 mm), the ancient orders are all relatively large-bodied and certainly conspicuous. Body lengths range from a few millimeters to centimeters: fairy shrimps (7–100 mm), tadpole shrimps (10–58 mm), and clam shrimps (2–16 mm). Body form is quite different among orders. Fairy shrimps lack a carapace, possess 20 trunk segments with 11 to 17 pairs of swimming legs, have elongated and sometimes elaborate second antennae (modified for mating), and have compound eyes placed on stalks. By contrast, tadpole shrimp are covered by a large, shieldlike carapace, possess numerous body segments (25–44) and 35 to 71 pairs of legs, have reduced or absent second antennae, and have compound eyes firmly embedded in the anterior shield region. Clam shrimps form a third distinctive group, with individuals tucked almost entirely within a hard, protective bivalve shell. There are numerous body segments with 10 to 32 pairs of legs, long second antenna used primarily for locomotion, and compound eyes situated in the frontal region close to the head (Pennak 1978; Dexter 1959). The initial impression of these three groups is that they are distant relatives, at best. Unfortunately, a molecular phylogeny is not yet available.

All three orders (Anostraca, Notostraca, Conchostraca) are widely distributed. Each has a limited number of species, some of which are very widespread and morphologically conservative (Tasch 1969; Longhurst 1954, 1955a, 1955b). The cosmopolitan distribution of anostracans, notostracans, and conchostracans, coupled with the widespread range of certain species, suggests either great antiquity or great dispersal powers. For example, *Triops longicaudatus* is the only species of the genus in the Americas (Linder 1952), yet this species extends from the West Indies through Central America and across the Pacific to Japan by way of the Galapagos, Oahu, and New Caledonia. Throughout this extensive distribution, there is little morphological change (Longhurst 1955a, 1955b).

Worldwide, species in the three groups are not generally found in ponds or lakes but are primarily restricted to ephemeral environments or harsh, marginal situations. Collections are reported from roadside ditches, grassy vernal ponds, cattail marshes, woodland pools, muddy ponds or puddles bordering cultivated fields, rice paddies, cattle troughs, flood-plain pools, rainwater depressions, peat bogs, moors, alkaline pools, springs, and temporary lakes (Tasch 1969, 1980; Pennak 1978).

Although it is rare to find two species from the same genus occupying a common site, species from two or more orders may occupy the same ephemeral pool (Tasch 1969, 1980). When they do, there is a suggestion of the niche differentiation noted by Longhurst (1954): "pelagic" Anostraca, "carnivorous" Notostraca, and "detritus-feeding" Concostraca. For example, in Ellis County, Kansas, the anostracan *Thamnocephalus*, the notostracan *Triops*, and the conchostracans *Leptestheria* and *Cyzicus* often occur in the same pool (Packard 1883). Over a broader area, frequent Kansas–Oklahoma associations also include the anostracan *Streptocephalus* with the notostracan *Triops* and the conchostracan *Cyzicus* (Tasch 1964). Tasch (1969) additionally observes that the notostracan *Triops* is found with the anostracan *Branchipus*, whereas the notostracan *Lepidurus* often co-occurs with the anostracan *Chirocephalus* and conchostracan young. In Algeria, it is not uncommon to note the co-occurrence of the notostracan *Triops* together with the anostracan *Streptocephalus* and the conchostracan *Leptestheria* (Gauthier, in Tasch 1969). In Australia, several conchostracan genera are recorded from the same pool (e.g., *Limnadopsis*, *Lynceus*, *Cyzicus*) (Tasch 1969).

While most common in shallow, ephemeral pools, members of all three orders can occupy shoreline or even open-water regions of large lakes, although they are rare in fish-occupied habitats. For example, Anderson's (1974) detailed survey of 340 ponds and lakes in the Canadian Rocky Mountain stretch near Banff National Park revealed species from each of the three orders in 1 to 18 bodies of water, illustrating their general scarcity. The most abundant species, the anostracan *Branchinecta paludosa*, was found primarily in shallow prairie ponds, yet it also occupied a few deep lakes. Although the anostracan *Branchinecta shant-*

zi is commonly reported from alpine lakes (Tasch 1969), lake occupations are not restricted exclusively to high montane lakes. For example, the anostracan *Artemia salina* thrives in the Great Salt Lake, Nevada, whereas *Branchinecta compestris* occupies saline ponds in Grant County, Washington, and was reported from Soap Lake during early surveys (Tasch 1969; Edmondson, cited in Palmer 1957).

Similarly, the vast majority of conchostracans sporadically occupy temporary, shallow, muddy depressions. In Tasch and Zimmerman's (1961) comprehensive survey of 550 Midwestern (Kansas, Oklahoma) ponds, only 6.4 to 16.4% contained conchostracans each year over a 3-year span of study, again illustrating infrequent occurrence on a regional basis. Most of the pools that contained conchostracans were shallow (< 12 inches), turbid (suspended clay), and small (< 500 sq. ft). Yet conchostracans also are reported from the margins of large African lakes (Tasch 1969), from the coastal salt flats of Brazil (Tasch 1969), and from the macrophyte-filled margins of Gatun Lake, Panama (Kerfoot, personal observation).

Notostracans also are found primarily in temporary pools over vast geographic regions (Longhurst 1955b). Yet of the two known genera (*Triops*, *Lepidurus*), *Triops* is frequently reported from large temporary lakes in arid regions, occupies persistent saline lakes in the Tibetan Plateau (Tasch 1969), and occasionally shows explosive development in large reservoirs following rotenone treatment (e.g., Flaming Gorge Reservoir, Wyoming) (Kerfoot, personal observation). Although the genus *Lepidurus* is also typical of arid pools, the species *L. arcticus* occurs only in the circumpolar arctic regions, where it also occupies an unusually wide range of habitats: temporary and permanent small pools, stream systems, and even large lakes (Longhurst 1954).

In summary, the biogeographic data clearly indicate that anostracans, notostracans, and conchostracans have the capacity to live in large, permanent lakes. Their general exclusion from such environments is almost certainly a function of fish predation. As Pennak (1953) remarks, "Phyllopods are almost defenseless, and it is also notable that they are not often abundant in ponds containing carnivorous insects, and are *never present along with carnivorous fishes*" (p. 337).

CONTEMPORARY LATITUDINAL GRADIENTS IN CLADOCERANS

Over the past decade, exploratory surveys of regional and worldwide zooplankton have revealed evidence for two striking patterns: (1) species richness and diversity of microcrustaceans increase toward the tropics in marine waters, whereas (2) species richness and diversity of freshwater assemblages actually decrease toward the tropics.

In the marine environment, both pelagic and benthic organisms show noticeable increases in species richness and diversity as one moves closer to the tropics. Oft-cited examples include foraminifera (Williams and Johnson 1975), coccolithophorids (Honjo and Okada 1974), bivalves (Sanders 1968), and copepods (Wimpenny 1966). Diversity and species richness increase markedly from the subarctic to central latitudes, then both indices are subject to major variation associated with circulation of water currents and basin morphometry.

In marked contrast, freshwater taxa show little evidence for a corresponding latitudinal increase toward tropical latitudes. Actual reduction of species richness in tropical equatorial regions is characteristic of the American continent (Deevey et al. 1980; Frey 1982; of Europe-Africa comparisons (DuMont 1980) and of Asia–Indonesia contrasts (Fernando 1980a, 1980b; Lewis 1979). The trends are evident in copepods and rotifers but are especially striking in the Cladocera.

Table 24.1 summarizes observed trends for pelagic Cladoceran species. The few large-bodied species found in the families Holopediidae, Leptodoridae, and Polyphemidae drop out toward the tropics. Large-bodied species in the genera *Daphnia* and *Simocephalus* are relatively rare in the tropics at low elevations and are usually restricted to temporary waters or occur in high elevational lakes (e.g., *Daphnia pulex*, 2–3 mm, in Lake Atitlan, Guatemala; Weiss 1971). The declines for *Daphnia* species richness in North American and southeastern Asian surveys are shown in figure 24.1. *Daphnia* are relatively species-poor in the subarctic, increase to a maximum species richness at mid-northern latitudes, then decline in the subtropical region. *Daphnia* are not entirely absent from many subtropical or tropical areas, only locally and regionally de-

Table 24.1. *Cladoceran species richness in selected tropical and temperate locations; only planktonic species listed under corresponding families*

Taxa	Temperate			Tropical			
	Holarctic	Ontario	Britain	Asia	Malaysia	India	Sri Lanka
Holopedidae	1	1	1	0	0	0	0
Polyphemidae	1	1	1	0	0	0	0
Leptodoridae	1	1	1	0	0	0	0
Daphniidae							
Daphnia	18	12	10	1	1	1	2
Ceriodaphnia	4	8	8	2	2	1	2
Sididae							
Diaphanosoma	2	2	2	5	4	4	5
Bosminidae	3	6	2	3	1	2	2
Moinidae	0	4	3	1	3	4	3
Total no. species	30	30	28	12	11	13	14

Data from Fernando 1980.

pressed. Species are rare in the lowland tropical region of Central America (Costa Rica–Panama–Colombia lowlands) (Frey 1982) and the evergreen rain forest zone and adjacent Guinean savanna of Africa (DuMont 1980). Yet large-bodied *Daphnia* are sporadically recorded from temporary tropical waters in the New World, (e.g. Florida; Crisman, personal communication; Frey 1982), in Africa (*D. similis* and *D. magna* in marginal pools; DuMont 1980), and in tropical Southeast Asia (*D. cephalata* and *D. similis*; Fernando 1980b).

There are several conceivable explanations for the geographical trends evident in freshwater Cladocera, but one is struck by the declines in large-bodied taxa toward the tropics. Certain taxa seem truly absent from subtropical or tropical environments (*Holopedium*, *Polyphemus*, and *Leptodora*), whereas others are restricted to temporary or fish-free waters (subgenus Ctenodaphnia) (Fernando 1980a, b).

The marine environment provides evidence that microcrustaceans can radiate at tropical latitudes, given adequate long-term stability of water masses. In contrast, the terrestrial environment and lakes are characterized by local and regional instability. On the geological time-scale, the vast fields of north temperate lakes are exceptional, recent products of Pleistocene glaciation. The rare and scattered lakes of the tropics are more typical of the long-term geographical norm. With the exception of the African Rift system, tropical regions are really river-dominated systems, characterized by oxbow

lakes and numerous temporary pools subject to wet-dry season fluctuations; (Fernando and Holcik 1982). The isolated nature of lake basins would promote speciation, yet their relatively short geological lifespan would favor life histories molded by immigration and regional patterns. Only in stable, long-lived lakes would speciation be evident. Environments such as Lake Biakal and several African lakes do show evidence for species differentiation and radiation, yet the striking cases includes only littoral or benthic taxa (e.g. the gammarids of Lake Biakal; the cichlids of Malawi, Tanganyika, and Victoria). Pelagic species do not show equivalent evidence for evolutionary radiation. The geographic pattern is thus one of loss toward the tropics, not balanced by high speciation rates in the tropics.

Here we ascribe the loss rates largely to increased incidence and activity of fishes. Because fry and adult fish are present and active throughout the year at tropical latitudes, open-water zooplankton communities are exposed to year-round planktivory in the tropics. The strong suppression of conspicuous zooplankton species by fish has been documented in several tropical lakes (Green 1967, 1971; Zaret 1972, 1980, 1982; Zaret and Kerfoot 1975; Moriarty et al. 1973; Gliwicz 1986). No other hypothesis offers such a simple, straightforward explanation for the latitudinal geographic trends.

The geographic distribution of Cladocera has suggested a Pangea presence and subsequent Laurasia–Gondwanaland radiation. For

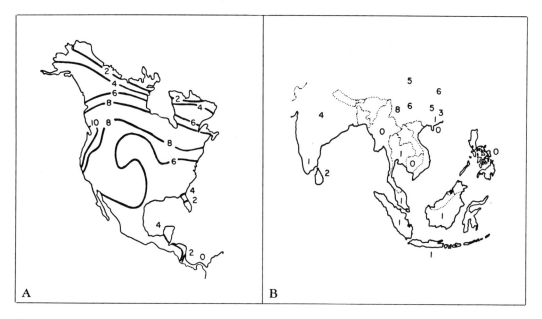

Figure 24.1. *Species richness for the genus* Daphnia *on two continents: (a) North and Central America (data from Brooks 1957) and (b) southeastern Asia (data from Fernando 1980a; except for China, T. Crisman, personal communication). Note the apparent decline in number of* Daphnia *species as one approaches lowland tropical regions. The mid-latitude high in western China coincides with approach to the Himalaya Mountains.*

example, most of the species in the subgenus *Ctenodaphnia* seem restricted to Gondwanaland (Sergeev and Williams 1983). Likewise, the distribution of subgenera in the family Bosminidae also suggests widespread distribution followed by radiation during the Laurasia-Gondwanaland split (Lieder 1982, 1983). The fossil record, however, reveals more direct information about ancestry.

THE FOSSIL RECORD

Marine and freshwater deposits suggest a diverse and thriving plankton community during the Late Precambrian and Early Paleozoic (fig. 24.2). The great morphological diversity of arthropods in the Early Cambrian, along with the various chemical compositions of their exoskeletons, suggest that the phylum arose in the Late Precambrian (Bergström 1980). Ancestral crustaceans were established by that time (McKenzie 1983). The exceptionally well preserved record of the Burgess Shale (Middle Cambrian, 550×10^6 years BP) even suggests relatives of the modern class Branchiopoda (Simonetta and Della Cave 1975; Briggs 1976, 1977, 1978).

The Branchiopoda were well established by the Late Devonian and flourished during the Late Paleozoic. The class included seven orders (Tasch 1980; Schram 1982) in three subclasses (four living, three or four extinct): subclass Sarsostraca (order Anostraca, living; order Lipostraca, Devonian), Calmanostraca (order Notostraca, living; order Kazacharthra, Jurassic; order Enantipoda, Carboniferous; and order Acercostraca, Devonian), and Diplostraca (order Conchostraca).

The earliest diplostracan or calmanostracan-like fossils come from marine sediments: *Protocaris marshi* from a Lower Cambrian slate deposit in Vermont, *Branchiocaris pretiosa* from the Middle Cambrian Burgess Shale (Briggs 1976). The earliest undisputed conchostracan fossils, some assigned to living genera, come from marine, estuarine, and freshwater sediments of Lower Devonian strata (Tasch 1969). The subclass Calmanostraca (Notostraca, Kazacharthra, Acercostraca) likewise is represented in the lower Devonian by the genus *Vachonisia* (order Acercostraca, family Vachonisiidae), as are the Anostraca (Tasch 1969), both from marine beds. At present, however, there are no marine notostracans, anostracans, or conchostracans,

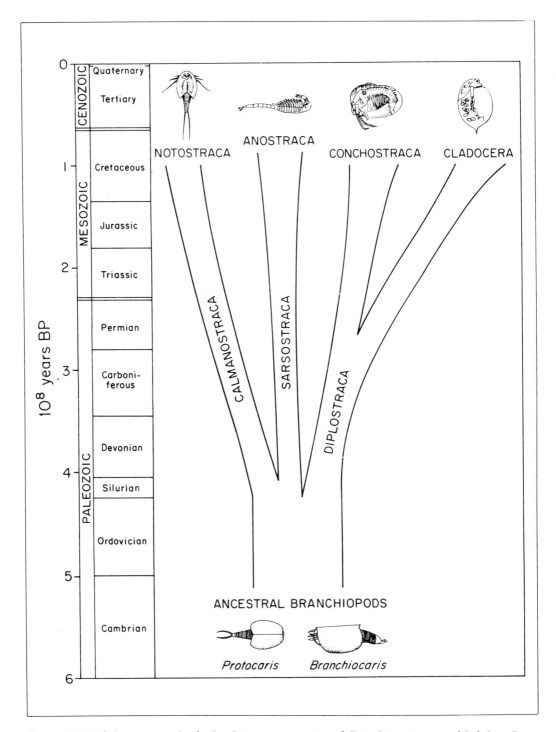

Figure 24.2. *Phylogenetic tree for the four living orders of the Class Branchiopoda and some extinct branches. Approximate times of divergence are based on Tasch (1963, 1969) and data discussed in the text. The three subclasses follow Schram (1982). Illustrations for Pro-* *tocaris and Branchiocaris are modified from Briggs (1976), for* Panacanthocaris *(Kazacharthra) and Va-* *chonisia (Acercostraca) modified from Tasch (1969). Illustrations of living orders are modified from Brooks (1959) and Tasch (1969).*

and pitifully few marine cladocerans, although all four groups can occupy saline lakes and seas.

While modern branchiopods are distinguishable by what appear to be major morphological differences, the early history of the group appears to reflect various attempts at developing hard, protective shields during the interval between the Cambrian and Devonian. It is interesting to note that the dominant predators prior to the Devonian were eurypterids (subclass Eurypterida, class Merostomata). These predaceous crustaceans were originally widely distributed, occupying marine, brackish, and fresh waters during the Devonian (Størmer 1958). Many species were benthic, but some were streamlined genera (e.g., *Hughmilleria*, *Pterygotus*) undoubtedly nektonic. Curtailment of the eurypterids during the Devonian has been attributed to the rise of early jawed fishes (Romer 1966). By Permo-carboniferous time, most eurypterids were restricted to fresh waters.

Although the crustacean fossil record is heavily biased toward heavily chitinized prey species, the predator–prey record is detailed enough to allow distinction of three major periods: (1) dominance of large-bodied invertebrate predators (Cambrian–Devonian), (2) replacement of large invertebrate predators by early jawed fishes (Devonian–Triassic), and (3) development of suction feeding within Holosteans and Teleosts, an event that coincided with the appearance of the modern microcrustacean assemblages (Jurassic–Cretaceous). These three intervals saw dramatic shifts in the diversity and distribution of branchiopods: (1) geographic radiation and evolutionary diversification during the Devonian–Jurassic interval (Fig. 24.2); (2) widespread mid- to late-Paleozoic occurrence in shallow marine, brackish, and freshwater habitats, followed by mid-Mesozoic restriction of many groups to fresh waters and extinction of others; and (3) initial co-occurrence of notostracans and conchostracans with fish (Carboniferous–Triassic) followed by later disjunct distributions (Cenozoic–Recent).

At the end of the Paleozoic, conchostracans were particularly diverse (nine families in the Carboniferous) and widespread in both freshwater and marine deposits. For example, the conchostracan *Quadriasmussia* (lower Carboniferous, Germany) occurs in beds with marine trilobites and brachiopods, whereas a species of *Cyzicus* (lower Devonian, Germany) is fossilized with marine ostracods (Tasch 1963). In Tasch and Zimmerman's (1961) careful study of the Wellington Formation (Permian; Kansas and Oklahoma), fossil clam shrimps were preserved predominately in sediments derived from small, temporary ponds. However, remains were also found in argillaceous limestone deposits, indicating lacustrine environments, as well as in cross-bedded silt- and sandstones, indicating moving water, and evaporite facies, indicating alkaline pools. Regional reconstructions suggested a coastal swamp area where the sea occasionally invaded brackish to freshwater habitats.

The subclass Calmanostraca is a little more difficult to interpret but certainly includes "living fossils." The Acercostraca are very primitive, and their assignment to the subclass Calmanostraca is disputed because fossils lack the postabdominal telson and furca characteristic of the group (Tasch 1969). The order is based on a series of fossils found in the Hunsrück Shale (lower Devonian, Germany). These shales are presumably of marine origin. In contrast, the Order Kazacharthra is clearly closely associated with the Order Notostraca. The order is based on an excellent series of fossils from the Ketmen Mountains, southeastern Kazakhstan, Russia (Tasch 1969). The lower Jurassic beds revealed at least seven genera, some (e.g., *Almatium*) preserved with coleopteran elytra, cockroach remains, and conchostracan valves—just the sort of assemblage that would indicate the temporary pond habitats of today. The most complete fossil record of the group, however, belongs to the Notostraca. Remains of notostracans are recorded from a wide range of deposits, reaching from the Carboniferous to the Recent. This record is unusual in that many of the remains are assignable to species in the two extant genera, *Triops* and *Lepidurus*, making these genera "living fossils." For example, *Lepidurus* remains from the upper Triassic of South Africa, the Triassic of South America, and the lower Cretaceous of Turkestan, as well as *Triops* from the upper Carboniferous of Germany, the Permian of Oklahoma, and the Triassic of south Africa, are all referable to modern species (Longhurst 1955a, 1955b; Tasch 1969).

Carboniferous to Triassic deposits document the co-occurrence of early bony fish (coelocanths, chondrosteans) with conchostracans and notostracans (table 24.2). Not only are fishes preserved alongside the valves and shields of branchiopods, but coprolites of coelo-canths contain the remains of conchostracans (Tasch 1969). By the early Jurassic, however, all but three families within the Conchostraca had gone extinct, and the remaining taxa were re-stricted to fresh waters (Tasch 1963). This re-striction coincided with the first appearance of modern Cladocera.

The Conchostraca and the Cladocera ap-pear to be much more closely related than the various other orders of branchiopods (except perhaps the Notostraca and Kazacharthra) and consequently are grouped together in the subclass Diplostraca. Closeness of the latter two groups is suggested by (1) basic structural similarity of the body plan, (2) examples of tru-ly transitional species, and (3) incidence of life history traits, particularly parthenogenesis.

The existence of an exposed head and an ab-breviated trunk with a reduced number of limbs in larval conchostracans results in a rather strik-ing structural similarity with the cladoceran body plan and led Brooks (1959) to suggest that the Cladocera are neotenic derivatives of an early conchostracan. Subsequently, Tasch (1963) pointed out that the living conchostra-can *Cyclestheria hislopi* (Sars 1887) is similar to an expected transitional species. Indeed, it is strikingly so. Unlike other conchostracans, *C. hislopi* has a single cyclopean eye, and there are no naupliar stages. The offspring are born with a complete set of appendages (16 pairs) and a well-developed shell. The most remarkable fea-ture of *C. hislopi*, however, is its mode of prop-agation. It is a cyclical parthenogen with a life cycle virtually identical to that of cladocerans. During the greater part of the growing season, reproduction is via unfertilized female-bearing eggs which develop directly in a dorsal brood chamber. As conditions deteriorate, males are produced parthenogenetically and diapausing "winter eggs" are generated sexually. Partheno-

Table 24.2. *Reported co-occurrences of branchiopods and fishes in the fossil record*

Source	Age	Association	Formation
Gall 1971	Triassic, lower	Notostracans: *Triops cancriformis* Conchostracans: *Palaeolimnadia alsatica* Fishes: chondrosteans *Praesemionotus, Dipteronotus. Pericentrophorus.* undescribed coelacanth	Bunter Sandstone, Vogesian Mts., Germany
Tasch 1964	Permian, lower	Notostracans: *Triops*	Wellington Formation, Oklahoma and Kansas, USA
Tasch and Zimmerman 1961		Anostracans: *Streptocephalus* Conchostracans: *Cyzicus* Other: xiphosurans, eurypterids, ostracods Fishes: fragmental, teeth attached to jaws, scales	
Tasch 1969	Permian	Conchostracans: "Estheriid valves have been found in coelocanth coprolites or in beds containing coelocanth remains"	Not specified
Olson et al. 1978	Triassic, upper	Concostracans: *Cyzicus (princetonensis?), Palacolimnadia?* Fishes: unidentified holostean (pholidophorid?), holostean *Scmionotus brauni*, subholostean *Synorichthys*; paleoniscid *Turseodus*; Coelacanth *Diplurus newarki*	Newark Series, Eastern USA

genesis is known to occur in a few other conchostracans, but in these cases it is obligatory and generates only resting eggs, and juvenile development involves naupliar stages.

The existence of *Cyclestheria hislopi* tempts the suggestion that the Cladocera are derived from the family Cyclestheriidae. However, circumstantial evidence suggests otherwise. The restriction of cyclestherids to Australia, Africa, India, and Ceylon implies that the family may have arisen some time in the Triassic after Pangaea began to break up. Tasch (1969), a conchostracan paleontologist, suggests a Tertiary origin. As we shall see below, this appears to significantly postdate the probable time of origin of the Cladocera. Thus, the existence of cyclestherids merely provides suggestive evidence that alterations to the basic developmental pathway of conchostracans can lead to a very cladoceran-like animal.

Until very recently, the fossil record for cladocerans was very meager. *Lynceites ornatus* Golddenberg from the Carboniferous of Transbaykal was possibly a cladoceran, although this is disputable. Cladoceran ephippia are certainly known from Tertiary deposits: *Daphnia* from the Oligocene of West Germany (*Daphnia fossilis* Heyden; Braunkohle), the Oligocene and Miocene of both Colorado and Nevada (*Daphnia florissantensis* Cockerell; Florissant and Humbolt Formations), and *Moina* from the Miocene (Frey 1964, 1967; Goulden 1968; Tasch 1969). Based on this scattered and fragmentary evidence, Frey (1967) suggested that Cladocera might extend back at least to the Mesozoic.

Recently described remains by Smirnov (1970, 1971) now confirm Frey's conjecture. Smirnov has reported some remarkable fossils from two Russian strata. Both deposits are from freshwater lakes, one from the late Permian (eastern Kazakhstan) and the other from the Jurassic (Transbaykal). In the Permian material, Smirnov assigned remains of presumably three species to the Daphniidae (new genus *Archedaphnia*) and one possibly to the Chydoridae (new genus *Propleuroxus*). However, the supposed Permian daphniids are very unusual. The head shield is not separate from the valve, the bodies seem encased in an envelope, and there are no accompanying impressions of mandibles. Both of the Permian genera are larger than modern taxa (*Archedaphnia*, 5–8.5

mm total length; *Propleuroxus*, 3.6 mm). These remains occur with abundant insect (orders Blattoidea, Plecoptera, Paraplecoptera, Orthoptera, Homoptera, Psocoptera, Coleoptera, Neuroptera, Mecoptera, and Trichoptera), bivalve (*Microdontella*), and conchostracan (Limnadiidae) fossils. Although the associated taxa suggest that the habitat was suitable for cladocerans, more work is required to clarify the anatomical details of the two primitive "cladoceran" genera.

On the other hand, the Jurassic fossils (also assigned to the genus *Archedaphnia*) are undeniably Modern in appearance. Remains contain excellent impressions of mandibles, eggs, antennae, carapace, and head shield. Individuals were relatively large (4 mm), helmeted, and abundant. No details on co-occurring taxa were included in the original description of material, although many hundreds of eggs and mandibles were recorded at the village of Novospasskoye (Smirnov 1971).

Given the antiquity of both the class and some orders, the relatively poor fossil record for some representatives, and the inclusion of aberrant genera (e.g., *Leptodora* in the Cladocera), detailed reconstructions of phylogeny are tentative, at best. There is a great likelihood that many higher taxa will be found to be polyphyletic on closer examination.

We suggest that the increased trophic specialization of fishes throughout the late Paleozoic and early Mesozoic set the stage for intensified conflict during the Mesozoic. The evolution of suction feeding in particular must have exerted strong selective pressures on most aquatic invertebrates. To quote Liem (1984):

With the emergence of the Halecostomes during the Mesozoic, prey and food capture was first accomplished by a highly effective suction mechanism. Any kind of prey or food floating or swimming in or on the water column, an enormous diversity of organisms, which are loosely attached to the substratum, and even prey items within cracks, cavities, and the substratum, can be effectively collected by the basic teleostean suction mechanism. (p. 273)

Body size is not necessarily indicative of planktivory. Although the Carboniferous palaeoniscids were certainly small-sized, their cranial characteristics were decidely primitive. However, Jurassic holosteans such as *Hetero-*

lepidotus show cranial, maxilla, and lower jaw features that approach the teleost condition (Romer 1966). One of the surviving holosteans, the living genus *Amia*, is capable of suction feeding, although the upper jaw does not show protrusion (Lauder and Liem 1983, Liem 1984). The rise of the teleosts in the Upper Jurassic, which included herring-like representatives (e.g., *Leptolepis*), marked the beginning of suction-feeding dominance (Romer 1966).

SUMMARY

Judging from the fossil record, ancient branchiopod groups (fairy, tadpole, and clam shrimps) once occupied a much broader range of habitats than they do today. These groups ranged from marine to freshwater habitats, from small to large lakes, and coexisted with fishes in several cases. In the Mesozoic, however, conchostracans and notostracans became limited largely to temporary pools. Given that the ancient taxa once ranged from marine to freshwater environments, it is not surprising that modern species of fairy, tadpole, and clam shrimps can tolerate highly saline environments. The progressive restriction of large-bodied branchiopods to temporary or highly saline waters is interpreted as forced by the evolution of more versatile feeding within bony fishes. The origin and development of suction feeding (especially in advanced teleosts) coincided with the development of the modern microcrustacean community.

In a broader perspective, the branchiopod history could be interpreted as merely one component in a major co-evolutionary interaction between fishes and invertebrates. That is, early jawed fishes radiated to exploit the abundance of then-available invertebrates. Progressive diversity and specialization on invertebrates brought about density and species shifts in the forage base, until the evolution of suction feeding allowed even the efficient harvesting of small pelagic species. The latter stages of this scenario coincided in time with the so-called Mesozoic Revolution in marine environments. Increased efficiency of fish predation brought about extinction of some groups, yet many groups persisted in fish-free refugia. This displacement opened pelagic environments for the evolution of lineages with inconspicuous body form, early reproduction, and small body size. Many of these lineages are included in the taxonomic category Cladocera.

ACKNOWLEDGMENTS

We thank D. G. Frey and R. B. Aronson for helpful comments on the original manuscript draft.

REFERENCES

Anderson, R. S. 1974. Crustacean plankton communities of 340 lakes and ponds in and near the National Parks of the Canadian Rocky Mountains. *J. Fish. Res. Bd. Canada* 31(5) : 855–69.

Bergstrøm, J. 1980. Morphology and systematics of early arthropods. Abh. naturwiss. *Ver. Hamburg (NF)* 23 : 7–42.

Briggs, D. E. G. 1976. The arthropod *Branchiocaris* N. Gen., Middle Cambrian, Burgess Shale, British Columbia. *Geol. Surv. Can. Bull.* 264 : 1–29.

———. 1977. Bivalved arthropods from the Cambrian Burgess Shale of British Columbia. *Palaeontology* 20 : 595–621.

———. 1978. The morphology, mode of life, and affinities of *Canadaspis perfecta* (Crustacea; Phyllocarida), Middle Cambrian, Burgess Shale, British Columbia. *Phil. Trans. Roy. Soc. London B* 281 : 439–87.

Brooks, J. L. 1957. *The systematics of North American Daphnia.* Mem. Conn. Acad. Arts Sci., Vol. 13. New Haven, Conn.: Yale University Press.

———. 1959. Cladocera. pp. 587–656 In W. T. Edmondson (ed.), *Fresh-water biology.* New York: John Wiley and Sons.

Deevey, E. S., Jr., G. B. Deevey, and M. Brenner. 1980. Structure of zooplankton communities in the Peten Lake District, Guatemala. In W. C. Kerfoot (ed.), *Evolution and ecology of zooplankton communities,* pp. 669–78. Hanover, N.H.: University Press of New England.

Dexter, R. W. 1959. Anostraca. In W. T. Edmondson (ed.), *Fresh-water biology,* pp. 558–571. New York: John Wiley and Sons.

DuMont, H. J. 1980. Zooplankton and the science of biogeography: The example of Africa. In W. C. Kerfoot (ed.), *Evolution and ecology of zooplankton communities,* pp. 685–96. Hanover, N.H.: University Press of New England.

Fernando, C. H. 1980a. The freshwater zooplankton of Sri Lanka, with a discussion of tropical freshwater zooplankton composition. *Int. Rev. Gesamten Hydrobiol.* 65 : 85–125.

———. 1980b. The species and size composition of tropical freshwater zooplankton with special reference to the Oriental region (South East Asia). *Int. Rev. Gesamten Hydrobiol.* 65 : 411–25.

Fernando, C. H., and J. Holcik. 1982. The nature of fish communities: A factor influencing the fishery potential and yields of tropical lakes and reservoirs. *Int. Rev. Gesamten Hydrobiol.* 67 : 127–40.

Frey, D. G. 1964. Remains of animals in Quaternary lake and bog sediments and their interpretation. *Arch. Hydrobiol., Beih., Ergebn. Limnol.* 2 : 1–114.

———. 1967. Cladocera in space and time. In Symposium on Crustacea, Part 1 : 1–9. Marine Fisheries P.O., Mandapam Camp: Marine Biol. Assoc. of India.

———. 1982. Cladocera. In S. H. Hulbert and A. Villalobos-Figueroa (eds.), Vol. 3: *Aquatic biota of Mexico, Central America and the West Indies.* Iztapalapa, Mexico: Universidad Autonoma Metropolitana.

Gall, J. 1971. *Faunes et Paysages du grès à Voltzia du Nord des Vosges. Essai Paléoécologique sur le Buntsandstein Supérieur.* Mem. Serv. Carte geol. Als. Lorr. 34.

Gause, G. F. 1934. *The struggle for existence.* Baltimore: Williams and Wilkins.

Gliwicz, Z. M. 1986. A lunar cycle in zooplankton. *Ecology* 64 : 883–97.

Goulden, C. E. 1968. The systematics and evolution of the Moinidae. *Amer. Phil. Soc. Trans. (NS)* 38 : 1–101.

Green, J. 1967. The distribution and variation of *Daphnia lumholtzi* (Crustacea: Cladocera) in relation to fish predation in Lake Albert, East Africa. *Zool. Lond.* 151 : 181–197.

———. 1971. Association of Cladocera in the zooplankton of the lake sources of the White Nile. *J. Zool.* 165 : 373–414.

Honjo, S., and H. Okada. 1974. Community structure of coccolithophorids in the photic layer of the mid-Pacific Ocean. *Micropaleontology* 20 : 209–30.

Huffaker, C. B. 1958. Experimental studies on predation: dispersion factors and predator-prey oscillations. *Hilgardia* 27 : 343–83.

Lauder, G. V., and K. F. Liem. 1983. The evolution and interrelationships of the actinopterygian fishes. *Bull. Mus. Comp. Zool.* 150 : 95–197.

Lewis, W. M. 1979. Zooplankton community analysis. New York: Springer Verlag.

Lieder, U. 1982. Aspekte der Zoogeographie der *Bosmina*-Untergattungen (Crustacea, Cladocera). *Zool. Jb. Syst.* 109 : 511–19.

———. 1983. Revision of the genus *Bosmina* Baird, 1845 (Crustacea, Cladocera). *Int. Rev. Gesamten Hydrobiol.* 68 : 121–39.

Liem, K. F. 1984. Functional versatility, speciation, and niche overlap: Are fishes different? In D. G. Meyers and J. R. Strickler (eds.), *Trophic interactions within aquatic ecosystems,* pp. 269–305. (AAAS Selected Symposium 85). Boulder, CO.: Westview Press.

Linder, F. 1952. Contributions to the morphology

and taxonomy of the Notostraca, with special reference to the North American species. *Proc. U.S. Nat. Mus.* 102 : 1–69.

Longhurst, A. R. 1954. Reproduction in the Notostraca. *Nature* 173 : 781.

———. 1955a. Evolution in the Notostraca. *Evolution* 9 : 84–86.

———. 1955b. A review of the Notostraca. *British Museum (Natural History) Bulletin (Zoology)* 3(1) : 1–54.

McKenzie, K. G. 1983. On the origin of Crustacea. In F. R. Schram (ed.) *Crustacean phylogeny,* pp. 21–43. Rotterdam: Balkema.

Moriarty, D. J. W., J. P. E. C. Darlington, I. G. Dunn, C. M. Moriarty, and M. P. Tevlin. 1973. Feeding and grazing in Lake George, Uganda. *Proc. R. Soc. Lond. B* 184 : 299–319.

Olsen, P. E., C. L. Remington, B. Cornet, and K. S. Thomson. 1978. Cyclic change in Late Triassic lacustrine communities. *Science* 201 : 729–33.

Packard, A. S. 1883. A monograph of the phyllopod crustacea of North America with remarks on the order Phyllocardia. In *U.S. Geol. Survey, Terr.* 12 Ann. Rept., pp. 295–590.

Palmer, A. R. 1957. Miocene arthropods from the Mojave Desert, California. *U.S. Geol. Survey Prof. paper* 294-G, pp. 237–275.

Pennak, R. W. 1953. *Freshwater invertebrates of the United States.* New York: Ronald Press.

———. 1978. *Freshwater invertebrates of the United States* (2nd ed.). New York: John Wiley and Sons.

Romer, A. S. 1966. *Vertebrate paleontology.* (3rd ed.) Chicago: University of Chicago Press.

Sanders, H. L. 1968. Marine benthic diversity: A comparative study. *Amer. Naturalist* 102 : 243–82.

Schram, F. R. 1982. The fossil record and evolution of Crustacea. In L. G. Abele (ed.), *The biology of Crustacea.* Vol. 1: *Systematics, the fossil record, and biogeography,* pp. 93–147. New York: Academic Press.

Sergeev, V., and W. D. Williams. 1983. *Daphniopsis pusilla* Serventy (Cladocera: Daphniidae), an important element in the fauna of Australian salt lakes. *Hydrobiology* 100 : 293–300.

Simonetta, A. M., and L. Delle Cave. 1975. The Cambrian non-trilobite arthropods from the Burgess Shale of British Columbia. A study of their comparative morphology, taxonomy, and evolutionary significance. *Palaeontol. Ital.* 69 : 1–37.

Smirnov, N. N. 1970. Cladocera (Crustacea) from the Permian of Eastern Kazakhstan. *Paleont. J.* 3 : 376–82. (In Russian)

———. 1971. A new species of *Archedaphnia* (Cladocera, Crustacea) from Jurassic deposits of Transbaykal. *Paleont. J.* 5 : 391–92. (In Russian)

Stanley, S. M. 1984. Marine mass extinctions: A dominant role for temperature. In M. H. Nitecki (ed.), *Extinctions,* pp. 69–117. Chicago: University of Chicago Press.

Størmer, L. 1958. Chelicerata, Merostomata. *Treatise on Invertebrate Paleontology, (P) Arthropoda 2* : Pl, P4. Geol. Soc. Amer., Boulder, Colo.

Tasch, P. 1963. Evolution of the Branchiopoda. In H. B. Whittington and W. D. I. Rolfe (eds.), *Phylogeny and evolution of Crustacea*, pp. 145–57. Mus. Comp. Zool., Spec. Publication, Harvard University.

———. 1964. Periodicity in the Wellington Formation of Kansas and Oklahoma. *Symposium on Cyclic Sedimentation, Bulletin* 169(2) : 481–96.

———. 1969. Branchiopoda. In R. C. Moore (ed.), *Treatise on invertebrate paleontology*, pp. R128–R191. Part R. Arthropoda 4. Geol. Soc. Amer., Boulder, CO.

———. 1980. *Paleobiology of the invertebrates*. New York: John Wiley and Sons.

Tasch, P., and J. R. Zimmerman. 1961. Fossil and living conchostracan distribution in Kansas-Oklahoma across a 200-million-year time gap. *Science* 133 : 584–85.

Vermeij, G. T. 1977. The Mesozoic marine revolution: Evidence from snails, predators and grazers. *Paleobiology* 3 : 245–58.

———. 1978. *Biogeography and adaptation: Patterns of marine life*. Cambridge, Mass.: Harvard University Press.

Weiss, C. M. 1971. Water quality investigations, Guatemala, Lake Atitlan 1968–1970. Environmental Sciences and Engineering Publication No. 274, pp. 1–175. Chapel Hill, University of North Carolina.

Werner, E. E., J. F. Gilliam, D. J. Hall, and G. G. Mittelbach. 1983. An experimental test of the effects of predation risk on habitat use in fish. *Ecology* 64 : 1540–48.

Williams, D. F., and W. C. Johnson II. 1975. Diversity of Recent planktonic Foraminifera in the Southern Ocean and Late Pleistocene paleotemperatures. *Quaternary Research* 5 : 237–50.

Wimpenny, R. S. 1966. *The plankton of the sea*. New York: American Elsevier.

Zaret, T. M. 1972. Predator-prey interactions in a tropical lacustrine ecosystem. *Ecology* 53 : 248–57.

———. 1980. *Predation and freshwater communities*. New Haven, CT: Yale University Press.

———. 1982. The stability/diversity controversy: A test of hypotheses. *Ecology* 63 : 721–31.

Zaret, T. M., and W. C. Kerfoot. 1975. Fish predation on *Bosmina longirostris*: Body-size selection versus visibility selection. *Ecology* 56 : 232–37.

Contributors

Peter Abrams
Department of Ecology and
  Behavioral Biology
University of Minnesota
318 Church Street S.E.
Minneapolis, MN 55455

S. M. Adams
Environmental Sciences Division
Oak Ridge National Laboratory
Oak Ridge, TN 37831

Richard B. Aronson
The Biological Laboratories
Harvard University
Cambridge, MA 02138

Current address:
  Department of Pure and Applied Zoology
  University of Reading
  Whiteknights
  P.O. Box 228
  Reading RG6 2AJ, England

James Bence
Department of Biological Sciences
University of California
Santa Barbara, CA 93106

Stephen R. Carpenter
Department of Biology
University of Notre Dame
Notre Dame, IN 46556

Peter L. Chesson
Department of Zoology
Ohio State University
Columbus, OH 43210

D. L. DeAngelis
Environmental Sciences Division
Oak Ridge National Laboratory
Oak Ridge, TN 37831

Michael L. Dungan
Department of Biological Sciences
University of California
Santa Barbara, CA 93106

Carol L. Folt
Division of Environmental Studies
University of California
Davis, CA 95616

Current address:
  Department of Biological Sciences
  Dartmouth College
  Hanover, NH 03755

John L. Forney
Department of Natural Resources
Cornell University
Ithaca, NY 14853

John J. Gilbert
Department of Biological Sciences
Dartmouth College
Hanover, NH 03755

Nelson G. Hairston, Jr.
Department of Zoology
University of Rhode Island
Kingston, RI 02881

Current address:
  Ecology and Systematics
  Cornell University
  Corson Hall
  Ithaca, NY 14853-2701

John E. Havel
Department of Zoology
University of Wisconsin
Madison, WI 53706

Current address:
  Department of Biology
  Central Michigan University
  Mt. Pleasant, MI 48859

W. Charles Kerfoot
Great Lakes Research Division
Institute of Science and Technology and
   Department of Biology
University of Michigan
Ann Arbor, MI 48109

James F. Kitchell
Center for Limnology
University of Wisconsin–Madison
Madison, WI 53706

Winfried Lampert
Max Planck Institute for Limnology
Department of Physiological Ecology
Postfach 165, 2320 Plon
West Germany

Charles Levitan
Division of Environmental Studies
University of California
Davis, CA 95616
   and
Science Department
Sierra Nevada College
P.O. Box 4269
Incline Village, NV 89450

Michael Lynch
Department of Ecology, Ethnology,
   and Evolution
University of Illinois
Champaign, IL 61820

Thomas E. Miller
Kellogg Biological Station
Michigan State University
Hickory Corners, MI 49060

Current address:
   School of Biological Science
   University of East Anglia
   Norwich, England NR4 7TJ

Edward L. Mills
Department of Natural Resources
Cornell University
Ithaca, NY 14853

Gary G. Mittelbach
Department of Zoology and Ohio
   Cooperative Fishery Research Unit
Ohio State University
Columbus, OH 43210

Peter Jay Morin
Department of Biological Sciences
P.O. Box 1059
Rutgers University
Piscataway, NJ 08854

William W. Murdoch
Department of Biological Sciences
University of California
Santa Barbara, CA 93106

W. John O'Brien
Department of Systematics and Ecology
University of Kansas
Lawrence, KS 66045

Mary E. Power
University of Oklahoma Biological
   Station
Kingston, OK 73439
   and   ·
Marine Science Institute
University of California
Santa Barbara, CA 93106

Current address:
   Division of Entomology and Parasitology
   University of California
   Berkeley, CA 94720

Steve Scrimshaw
Department of Biological Sciences
Dartmouth College
Hanover, NH 03755

Andrew Sih
Behavioral and Evolutionary Ecology
   Research Group
T. H. Morgan School of Biological
   Sciences
University of Kentucky
Lexington, KY 40506

Richard S. Stemberger
Department of Biological Sciences
Dartmouth College
Hanover, NH 03755

Hans-Dieter Sues
Museum of Comparative Zoology
Harvard University
Cambridge, MA 02138
Current address:
  Department of Paleobiology
  National Museum of Natural History
  Room NHB E-207 MRC NHB 121
  Smithsonian Institution
  Washington, D.C. 20560

Steven T. Threlkeld
Biological Station and Department of
  Zoology
University of Oklahoma
Kingston, OK 73439

Michael J. Vanni
Department of Fisheries & Oceans
Freshwater Institute
501 University Crescent
Winnipeg, Manitoba, Canada
  R3T 2N6

Kenneth J. Wagner
Department of Natural Resources
Cornell University
Ithaca, NY 14853

Index